Proceedings of the Conference on Complex Analysis

This Conference was supported by the United States Air Force Office of Scientific Research

Proceedings of the Conference on Complex Analysis

Minneapolis 1964

Edited by

A. Aeppli · E. Calabi · H. Röhrl

Springer-Verlag
Berlin · Heidelberg · New York
1965

Professor Dr. ALFRED AEPPLI
School of Mathematics, Institute of Technology, University of Minnesota, Minneapolis
Professor Dr. EUGENIO CALABI
Department of Mathematics, University of Pennsylvania, Philadelphia
Professor Dr. HELMUT RÖHRL
Department of Mathematics, University of California at San Diego, La Jolla

ISBN 978-3-642-48018-8 ISBN 978-3-642-48016-4 (eBook)
DOI 10.1007/978-3-642-48016-4

Softcover repring of the hardcover 1st edition 1965

Library of Congress Catalog Card Number 65-14592

Title-No. 1257

Preface

This volume contains the articles contributed to the Minnesota Conference on Complex Analysis (COCA). The Conference was held March 16–21, 1964, at the University of Minnesota, under the sponsorship of the U. S. Air Force Office of Scientific Research with thirty-one invited participants attending. Of these, nineteen presented their papers in person in the form of one-hour lectures. In addition, this volume contains papers contributed by other attending participants as well as by participants who, after having planned to attend, were unable to do so.

The list of participants, as well as the contributions to these Proceedings, clearly do not represent a complete coverage of the activities in all fields of complex analysis. It is hoped, however, that these limitations stemming from the partly deliberate selections will allow a fairly comprehensive account of the current research in some of those areas of complex analysis that, in the editors' belief, have rapidly developed during the past decade and may remain as active in the foreseeable future as they are at the present time.

In conclusion, the editors wish to thank, first of all, the participants and contributors to these Proceedings for their enthusiastic cooperation and encouragement. Our thanks are due also to the University of Minnesota, for offering the physical facilities for the Conference, and to Springer-Verlag for publishing these proceedings.

Finally, a word of gratitude is due to the U.S. Air Force Office of Scientific Research, and in particular to Dr. R. G. Pohrer, for the financial help and the efficient administrative procedure which enabled us to go all the way from the original proposal of the conference, in November 1963 to the appearance of these Proceedings in such a short time.

A. Aeppli
E. Calabi
H. Röhrl

Minneapolis, in December 1964

Contents

List of Participants

Abhyankar, S.
Aeppli, A.
Ahlfors, L. V.
Bergman, S.
Bers, L.
Bishop, E. A.
Bungart, L.
Calabi, E.
Griffiths, P. A.
Guggenheimer, H. W.
Gunning, R. C.
Hironaka, H.
Holmann, H.
Ise, M.
Kallin, E. M.
Kohn, J. J.
Kuhlmann, N.
Kuranishi, M.
MacLane, S.
Maskit, B.
Matsumura, H.
Murakami, H.
Pohl, W. F.
Pumplün, D.
Röhrl, H.
Rossi, H.
Royden, H. L.
Satake, I.
Spallek, K. H.
Stein, K.
Stoll, W.

On Factorization of Holomorphic Mappings*

By

K. Stein

Introduction

Let X, Y be reduced complex spaces, $\tau : X \to Y$ a holomorphic mapping, denote by R the equivalence relation in X defined by the level sets (i.e. the connected components of the fibres) of τ. If the level sets are compact then by a theorem of H. Cartan [*1*] the quotient space X/R carries naturally the structure of a complex space and the natural projection $\varepsilon : X \to X/R$ is a proper holomorphic mapping; thus τ admits a factorization $\tau = \tau^* \circ \varepsilon$ where $\tau^* : X/R \to Y$ is a nowhere degenerate holomorphic mapping.

It is known that things become more complicated if the assumption on the compactness of the level sets of τ is dropped. In this case the quotient space X/R need not be Hausdorff, not even if the local rank of τ is constant. Consider the following example: Put $X := \mathbf{C}^2(z_1, z_2) - \{(z_1, z_2) : |z_1| \leqq 1, |z_2| = 1\}$, $Y := \mathbf{C}^1(z_1)$, let $\tau : X \to Y$ be the holomorphic mapping defined by $X \ni (z_1, z_2) \to z_1 \in Y$. The fibres $\tau^{-1}(z_1)$ consist of two connected components if $|z_1| \leqq 1$ and otherwise they are connected. Obviously two points of the quotient space X/R which correspond to the two connected components of a fibre $\tau^{-1}(z_1)$ with $|z_1| = 1$ do not satisfy the Hausdorff separation axiom. But X/R is a non Hausdorff manifold, moreover a complex structure can be introduced on X/R in an evident manner such that X/R becomes a non Hausdorff Riemann surface and that the mappings $\varepsilon : X \to X/R$ and $\tau^* : X/R \to Y$ become holomorphic mappings. Thus there is a factorization $\tau = \tau^* \circ \varepsilon$ as in the case of compact level sets but with the restriction that X/R is not a complex space in the usual sense.

Now this statement is a special case of a more general proposition. In this paper we consider holomorphic mappings $\tau : X \to Y$ of constant local rank where X is a complex manifold. We introduce the notion of *quasi complex space* (§ 1); it turns out that the space X/R carries the structure of a quasi complex space (§ 2). We then get a factorization theorem (§ 3) which corresponds to an earlier statement on factorizations of proper

* Received June 8, 1964.

holomorphic mappings ([5]). The proofs of the statements given below are sketched.

Notations. By a complex space we always mean a reduced complex space i.e. a complex space in the sense of J. P. SERRE ([2], [7]). For basic notions connected with complex spaces and more generally with ringed spaces compare [1], [3]; for equivalence relations in ringed spaces see [1]. — The local rank of an holomorphic mapping $\tau: X \to Y$ in a point $x \in X$ is denoted by $rk_x \tau$, the global rank of τ is $rk\,\tau := \sup_{x \in X} rk_x \tau$ [5].

1. Quasi complex spaces

Let $X_1 = (X_1, \mathfrak{O}_1)$, $X_2 = (X_2, \mathfrak{O}_2)$ be puredimensional complex spaces of equal dimension, $\sigma: X_1 \to X_2$ a proper surjective nowhere degenerate holomorphic mapping (thus (X_1, σ, X_2) is an analytic covering). Consider an equivalence relation R_1 in X_1 such that

(a) R_1 is open (i.e. the saturation of any open set is open)

(b) σ is constant on each R_1-equivalence class.

Let $X_1/R_1 = Z$ be the quotient space of X_1 with respect to R_1 and put $\mathfrak{Q} = \mathfrak{O}_1/R_1$ (compare [1]). Then the ringed space $(Z, \mathfrak{Q})$ is called a *quasi analytic covering space.*

We say that the ringed space $(X, \mathfrak{O}) = X$ is a *quasi complex space* if X is locally isomorphic to quasi analytic covering spaces, i.e. if each point $x \in X$ has an open neighbourhood U such that $(U, \mathfrak{O} | U)$ can be bimorphically mapped onto a quasi analytic covering space $(Z, \mathfrak{Q})$. Clearly a quasi complex space is a locally quasi compact T_1-space.[1]

With respect to quasi complex spaces one has the usual notions of complex analysis like *holomorphic* and *meromorphic function, holomorphic mapping, nowhere degenerate holomorphic mapping, local analytic* and *analytic subset, normality* etc. If A is a local analytic subset of the quasi complex space X then the ringed structure of X induces a ringed structure on A, and it is easily seen that A supplied with this ringed structure is a quasi complex space again, therefore A can be called a quasi complex subspace of X.

Every quasi complex space X contains an open dense subset M such that M has locally the structure of a complex manifold. Furthermore X can be reduced in each point $x_0 \in X$ to a germ of a complex space in the following sense: There is an open neighbourhood U of x_0 and a nowhere degenerate holomorphic mapping $\tau: U \to Y$ onto a *complex* space $Y = (Y, \widetilde{\mathfrak{O}})$ such that the induced homomorphism ${}^*\tau_{x_0}: \widetilde{\mathfrak{O}}_{\tau(x_0)} \to \mathfrak{O}_{x_0}$

[1] In [4] H. HOLMANN introduced the notion of pseudo complex spaces; these spaces are locally complex spaces and in particular locally Hausdorff. Any pseudo complex space is quasi complex, but a quasi complex space need not to be locally Hausdorff.

is an isomorphism. This statement is a conclusion of the following Proposition 1.

We define: Let X be a quasi complex space, X' a complex space. A holomorphic mapping $\varrho: X \to X'$ is called a *c-reduction of* X if

(i) ϱ is nowhere degenerate and surjective,

(ii) ϱ majorizes each mapping $\varphi: X \to Y$ of X into a complex space Y (i.e. there is a holomorphic mapping $\mu: X' \to Y$ such that $\varphi = \mu \circ \varrho$).

It is immediate that two c-reductions $\varrho: X \to X'$ and $\varrho_1: X \to X_1$ are holomorphically equivalent: There exists a biholomorphic mapping $\alpha: X' \to X_1'$ such that $\varrho_1 = \alpha \circ \varrho$.

Now one has the

Proposition 1. *Let X be a quasi complex space. Assume that a nowhere degenerate holomorphic mapping $\psi: X \to Y$ of X into a complex space Y exists. Then each of the following conditions is sufficient for the existence of a c-reduction of X:*

(1) *X is normal,*

(2) *ψ is proper.*

In case (1) *the reduction space X' is normal.*

To prove this one shows first that under both conditions ψ can be assumed as a surjective mapping. Then one considers the class $\mathfrak{C}$ of all nowhere degenerate holomorphic mappings of X onto complex spaces; $\mathfrak{C}$ is ordered in a natural manner. By means of Zorn's lemma one sees that a maximal element exists in $\mathfrak{C}$ and this element gives a c-reduction $\varrho: X \to X'$ of X. If X is normal then X' must be normal too, otherwise ϱ could be lifted to a holomorphic mapping into the normalization of X' and this would contradict (ii).

Remark. There are quasi complex spaces which have no c-reduction. Example: Take the unit disk $U := \{z : |z| < 1\}$ in the complex plane, choose a real irrational number ξ and a real number $\varepsilon > 0$ such that the domains $D_1 = \{z: 1/2 < |z| < 1, -\varepsilon < \varphi < \varepsilon\}$ $(z = |z| \cdot e^{2\pi i \varphi})$ and $D_2 = \{z: 1/2 < |z| < 1, -\varepsilon + \xi < \varphi < \varepsilon + \xi\}$ are disjoint. Consider the mapping $\beta: U \to U$ defined by

$$\begin{aligned} z &\to z && \text{if } z \in U - (D_1 \cup D_2), \\ z &\to z \cdot e^{2\pi i \xi} && \text{if } z \in D_1, \\ z &\to z \cdot e^{-2\pi i \xi} && \text{if } z \in D_2, \end{aligned}$$

and identify points of U which are related by β. This gives a quasi complex space X (which is a non Hausdorff Riemann surface) and one has a natural holomorphic mapping $\mu: U \to X$. Now if f is a meromorphic function in X then $\tilde{f} := f \circ \mu$ is a meromorphic function in U, and $\tilde{f}$ satisfies in U the relation $\tilde{f}(z) = \tilde{f}(z \cdot e^{2\pi i \xi})$, therefore $\tilde{f}$ and consequently f is constant. It follows that every holomorphic mapping of X into a Riemann surface is constant, therefore X cannot have a c-reduction.

2. A theorem on equivalence relations in complex manifolds

Theorem 1. *Let $X = (X, \mathfrak{O})$ be an irreducible complex manifold, Y a complex space, $\tau: X \to Y$ a holomorphic mapping such that $rk_x \tau = rk\,\tau$ for all $x \in X$. Let R be the equivalence relation in X defined by the level sets of τ. Then the ringed space $(X/R, \mathfrak{O}/R) = X^*$ is a normal quasi complex space, furthermore the natural mapping $\varepsilon: X \to X^*$ and the mapping $\tau^*: X^* \to Y$ such that $\tau = \tau^* \circ \varepsilon$ are holomorphic.*

We indicate the proof: Every fibre $F_x = \tau^{-1}(\tau(x))$, $x \in X$, is a pure-dimensional analytic set in X of dimension $n - r$ if $n = \dim X$ and $r = rk\,\tau$. Choose an irreducible local analytic set M of dimension r in X such that $F \cap M = \{x\}$. There is an open irreducible subset N of M with the following properties: (1) $x \in N$, (2) the holomorphic mapping $\tau | N: N \to Y$ is nowhere degenerate, (3) $\tau(N) = N'$ is a local analytic set in Y and the restriction mapping $\tau': N \to N'$ is proper. Then (N, τ', N') is an analytic covering. The equivalence relation R induces an equivalence relation R_N on N and one has an injective mapping $\mu: N/R_N \to X/R$. Now one uses the

Lemma. *The equivalence relation R is open.*

By means of the lemma one shows

(a) The ringed space $(N/R_N, \mathfrak{O}_N/R_N) = N/R_N$ ($\mathfrak{O}_N$ the structure sheaf of N) is a quasi analytic covering space,

(b) $\mu: N/R_N \to X/R$ is locally biholomorphic.

Therefore $(X/R, \mathfrak{O}/R)$ is a quasi complex space. Then X^* must be normal because X is a manifold.

The Lemma follows immediately from the

Proposition 2. *Let X be an irreducible complex manifold, Y a complex space, $\tau: X \to Y$ a holomorphic mapping such that $rk_x \tau = rk\,\tau$ for all $x \in X$, furthermore let $L, L_\nu (\nu = 1, 2, \ldots)$ be level sets of τ. Assume that there is a point $x_0 \in L$ such that every neighbourhood of x_0 is met by almost all L_ν. Then every point of L has the same property.*

The steps to prove this proposition are:

(1) Let $L = \bigcup_i L^{(i)}$, $i \in I$, be the decomposition of L into irreducible components, denote by S the set of all points $x \in L$ which have the same property as x_0. Then $x_1 \in L^{(i_1)} \cap S$ and $x_1 \notin \bigcup_{i \neq i_1} L^{(i)}$ imply $L^{(i_1)} \subset S$ (this can be shown by applying a theorem on essential singularities of analytic sets, cf. [6]). Therefore there is a subset J of I such that $S = \bigcup_{j \in J} L^{(j)}$.

(2) Assume $J \neq I$. Take a point $z^{(0)} \in \bigcup_{j \in J} L^{(j)} \cap \bigcup_{i \in I-J} L^{(i)}$ and choose an open relatively compact coordinate neighbourhood U of $z^{(0)}$ with local coordinates $z_1, \ldots, z_l$. In order to study the local behaviour of τ in $z^{(0)}$ one can suppose that $\tau | U$ is a mapping of U into some number space C^m and that $\tau | U$ is given by $(z_1, \ldots, z_l) \to (w_1, \ldots, w_m) = (f_1(z_1, \ldots, z_l), \ldots, f_m(z_1, \ldots, z_l))$ where $f_\mu(z_1, \ldots, z_l)$ $(\mu = 1, \ldots, m)$ are holomorphic functions in U. Let M be the set of points of U where the rank of the matrix $\left(\frac{\partial f_\mu}{\partial z_\lambda}\right)$ is less than $r = rk\,\tau$, put $N = (\tau | U)^{-1}(\tau(M))$ and denote by $\bar{N}$ the closure of N in U. Then $U - \bar{N}$ is a dense open connected subset of U. Every fibre of $\tau | U - \bar{N}$ is free of singularities. Take now irreducible components $L^{(j_0)}, L^{(i_0)}$ $(j_0 \in J, i_0 \in I - J)$ of L passing through $z^{(0)}$, choose points $z^{(j_0)} \in L^{(j_0)} \cap U$, $z^{(i_0)} \in L^{(i_0)} \cap U$ which are ordinary points of L. Furthermore take irreducible local analytic sets $A^{(j_0)}, A^{(i_0)}$ of dimension r without singularities passing through $z^{(j_0)}$ resp. $z^{(i_0)}$ such that $A^{(j_0)} \cap L^{(j_0)} = \{z^{(j_0)}\}$, $A^{(i_0)} \cap L^{(i_0)} = \{z^{(i_0)}\}$ and $rk_z \tau | A^{(j_0)} = rk\,\tau$, $rk_{z'} \tau | A^{(i_0)} = rk\,\tau$ for all $z \in A^{(j_0)}$ resp. $z' \in A^{(j_0)}$. Then $(U - \bar{N}) \cap A^{(j_0)}$ resp. $(U - \bar{N}) \cap A^{(i_0)}$ is dense, open and connected in $A^{(j_0)}$ resp. $A^{(i_0)}$. If $V^{(j_0)}$ is any neighbourhood of $z^{(j_0)}$ it is easily seen that $V^{(j_0)} \cap A^{(j_0)}$ is intersected by almost all L_ν.

(3) Choose points $'z^{(j_0)} \in (U - \bar{N}) \cap A^{(j_0)}$, $'z^{(i_0)} \in (U - \bar{N}) \cap A^{(i_0)}$ and a curve $\mathfrak{C} \subset U - \bar{N}$ connecting $'z^{(j_0)}$ and $'z^{(i_0)}$. One shows: If $'z^{(j_0)}$, $'z^{(i_0)}$, $\mathfrak{C}$ are contained in a suitably small neighbourhood of $(L^{(j_0)} \cup L^{(i_0)}) \cap U$ then for each point $z \in \mathfrak{C}$ the level set of $\tau | U$ passing through z intersects $A^{(j_0)}$ and $A^{(i_0)}$. In particular: If a level set of τ intersects $(U - \bar{N}) \cap A^{(j_0)}$ in points sufficiently near $z^{(j_0)}$ then it intersects also $A^{(i_0)}$. Now the same must be true for all level sets of τ intersecting $A^{(j_0)}$ in points near $z^{(j_0)}$ since every convergent sequence of level sets certainly converges towards a connected part of a level set. It follows that almost all sets L_ν intersect $A^{(i_0)}$. Hence $z^{(i_0)} \in S$ and therefore $L^{(i_0)} \subset S$ because of (1); thus there is a contradiction.

Remark. One can prove another statement on convergent sequences of analytic sets:

Let L, L_ν $(\nu = 1, 2, \ldots)$ be different puredimensional connected analytic sets of the same dimension in a complex manifold X such that any two different irreducible components of L are in relative general position[2]. *Assume that the following conditions are fulfilled*

(a) $(\bigcup_\nu L_\nu) \cap (X - L)$ *is an analytic set in* $X - L$

---

[2] Two puredimensional analytic sets S, S' in a complex manifold X are in relative general position in a point $x \in S \cap S'$ if $\dim_x(S \cap S') = \dim_x S + \dim_x S' - \dim_x X$. S and S' are in relative general position if they are in relative general position in every point of $S \cap S'$.

(b) *there is a point* $x_0 \in L$ *such that every neighbourhood of* x_0 *is met by almost all* L_ν.

Then every neighbourhood of each point of L *is met by almost all* L_ν.

If the assumption about the relative general position of the irreducible components of L is omitted then the statement above becomes false as can be shown by simple counterexamples. But without that assumption the statement remains true by Proposition 2, provided L, L_ν are level sets of a holomorphic mapping.

3. A factorization theorem

By means of the statements of § 1 and § 2 one gets

Theorem 2. *Let* X *be an irreducible complex manifold,* Y *a complex space,* $\tau: X \to Y$ *a holomorphic mapping of constant local rank. Then* τ *admits a factorization* $\tau = \gamma \circ \beta \circ \varrho \circ \varepsilon$ *such that*

(1) $\varepsilon: X \to X^*$ *is a simple holomorphic mapping of* X *onto a normal quasi complex space,*

(2) $\varrho: X^* \to X'$ *is a c-reduction of* X^* *where* X' *is a normal complex space,*

(3) $\alpha = \varrho \circ \varepsilon: X \to X'$ *is a maximal holomorphic mapping strictly related to* τ[3],

(4) $\beta: X' \to Y'$ *is a nowhere degenerate holomorphic mapping of* X' *onto a normal complex space* Y',

(5) $\gamma: Y' \to Y$ *is a weakly quasi injective holomorphic mapping.*

A factorization of τ *with these properties is uniquely determined up to holomorphic equivalence.*

Here a holomorphic mapping $\gamma: Y' \to Y$ is called *weakly quasi injective* if the following two conditions are satisfied:

(a) γ is locally quasi injective (cf. [5]),

(b) if y_1', y_2' are different points in Y' then there exists a neighbourhood $U(y_1')$ such that y_2' is not contained in the interior of $\gamma^{-1}(\gamma(U(y_1')))$.

For the proof of Theorem 2 take X^* and $\varepsilon: X \to X^*$ as in Theorem 1, then one has $\tau = \tau^* \circ \varepsilon$ where $\tau^*: X^* \to Y$ is a nowhere degenerate holomorphic mapping of the normal quasi complex space X^* into the complex space Y. By Proposition 1 there is a c-reduction $\varrho: X^* \to X'$ where X' is normal. The surjective holomorphic mapping $\alpha = \varrho \circ \varepsilon: X \to X'$ is maximal because it clearly majorizes every holomorphic mapping $X \to Z$ related to τ into any complex space Z. The mapping $\tau^*: X^* \to Y$ can be lifted to a holomorphic mapping $\hat{\tau}^*: X^* \to \mathfrak{P}(Y)$ into the normal complex space $\mathfrak{P}(Y)$ of the irreducible germs

---

[3] Cf. [8].

of local analytic sets in Y; the image $\hat{\tau}^*(X^*) = Y'$ is an open subspace of $\mathfrak{P}(Y)$. Now let $\beta: X^* \to Y'$ be the restriction of τ^* and $\gamma: Y' \to Y$ the restriction of the natural projection of $\mathfrak{P}(Y)$ into Y, then β and γ have the properties stated in (4) and (5). The uniqueness of the factorization up to holomorphic equivalence is rather obvious.

It seems that Theorem 2 (as well as Proposition 2 and Theorem 1) remains true if X is assumed to be a normal complex space.

References

[1] Cartan, H.: Quotients of complex analytic spaces. International Colloquium on Function Theory, Tata Institute Bombay, Jan. 1960, pp. 1—15.
[2] Grauert, H.: Ein Theorem der analytischen Garbentheorie und die Modulräume komplexer Strukturen. Publications Mathématiques de l'Institut des Hautes Études Scientifiques, No. 5 (1960), pp. 233—292.
[3] —, and R. Remmert: Komplexe Räume. Math. Ann. **136**, 245—318 (1958).
[4] Holmann, H.: Quotientenräume komplexer Mannigfaltigkeiten nach komplexen Lieschen Automorphismengruppen. Math. Ann. **139**, 383—402 (1960).
[5] Remmert, R., and K. Stein: Eigentliche holomorphe Abbildungen. Math. Zschr. **73**, 159—189 (1960).
[6] — Über die wesentlichen Singularitäten analytischer Mengen. Math. Ann. **126**, 263—306 (1953).
[7] Serre, J. P.: Géométrie algébrique et géométrie analytique. Ann. l'Institut Fourier, G (1955/56), 1—42.
[8] Stein, K.: Maximale holomorphe und meromorphe Abbildungen, I. Amer. J. Math. **85**, 298—315 (1963).

Department of Mathematics
University of Chicago
Chicago, Illinois

München, Ulmenstraße 14

Cauchy Integral Formulas and Boundary Kernel Functions in Several Complex Variables*

By

L. Bungart

1. Introduction

In the present paper we attempt to coordinate some of the recent works concerned with the construction of domain-dependent Cauchy integral formulas in several complex variables, and then to point out an intimate relationship to boundary kernel functions. In particular, we are looking for formulas that are holomorphic in the parameters involved and not only real analytic as for instance the Bochner-Martinelli formula which has been established in [2] and [13]. The circle of ideas got probably

* Received May 25, 1964.

started with the establishment of the so-called Cauchy-Weil formula in [15]. A somewhat related formula was later discussed by S. BERGMAN in [1] where also its relationship to boundary kernel functions is discussed. However, these formulas are valid only for domains with very special boundary properties. This prompted the search for a formula for any type of domain. The existence of such a formula has recently been established by A. GLEASON [9] and the author [3], and to a certain extent also by O. FORSTER [8].

Before we go into technical details, let us explain our objective using Cauchy's integral formula for the unit disc

$$D = \{z \in \boldsymbol{C} : |z| < 1\}.$$

We denote by A the algebra of continuous functions on the closure $\bar{D}$ of D that are holomorphic in D. For $f \in A$ we have

$$f(z) = \frac{1}{2\pi i} \cdot \int_B f(w) \frac{1}{w - z} dw, \quad z \in B,$$

where B denotes the boundary of D. If we let $d\Theta$ be the Lebesgue measure on B, then we can write this formula in a somewhat more convenient way,

$$f(z) = \frac{1}{2\pi} \int_B f(w) \frac{w}{w - z} d\Theta(w), \quad z \in D,$$

or, by observing that $|w| = 1$ on B,

$$f(z) = \frac{1}{2\pi} \int_B f(w) \frac{1}{1 - z\bar{w}} d\Theta(w), \quad z \in D. \tag{1.1}$$

The first step toward the solution of our general problem consists of reading formula (1.1) in the proper way. If, for fixed $z \in D$, we let

$$d\mu_z = \frac{1}{2\pi} \cdot \frac{1}{1 - z\bar{w}} d\Theta$$

then μ_z is a Radon measure on B, and we have

Theorem A. *There is a holomorphic mapping*

$$z \to \mu_z$$

from D into the Banach space of (complex) Radon measures on B such that for any function $f \in A$

$$f(z) = \int_B f(w) \, d\mu_z(w), \quad z \in D.$$

Or, if we consider $\frac{1}{2\pi} \cdot \frac{1}{1 - z\bar{w}}$ as a function on $D \times B$, we have

Theorem B^p. *There is a positive Radon measure σ on B and a measurable function k on $D \times B$ such that*

(a) *$k(\cdot, w)$ is holomorphic for $w \in B$,*

(b) *$k(z, \cdot) \in L^p(d\sigma)$ for $z \in D$,*

(c) $\int_B f(w)\, k(z,w)\, d\sigma(w)$ *is a holomorphic function in* z *for each* $f \in L^q(d\sigma)$,

(d) $f(z) = \int_B f(w)\, k(z,w)\, d\sigma(w)$, $z \in D$, *for every function* $f \in A$.

In this theorem q is the number determined by the equation

$$\frac{1}{p} + \frac{1}{q} = 1,$$

and a function on $D \times B$ is measurable if it is measurable with respect to the σ-field generated by

$$\{F \times G : F \subset D \text{ open and } G \subset B \text{ measurable for } \sigma\}.$$

Finally, for $p = 2$ the last theorem can be amended by

Theorem C^2. *If* σ *is as in* Theorem B^2 *then* k *can be chosen such that* $\overline{k(z,\cdot)}$ *belongs to the closure* $H^2(d\sigma)$ *of* A *in* $L^2(d\sigma)$ *for each* $z \in D$. k *has then a holomorphic "extension" to* $D \times D^*$, *where* D^* *denotes the domain* D *with the conjugate holomorphic structure. This extension is called the boundary kernel function.*

In the present situation, k is actually continuous on $D \times B$ and the "extension" into $D \times D^*$ is continuous on $D \times (D^* \cup B)$. In general we cannot expect to get so strong a result. We will explain later how the "extension" in Theorem C^2 is to be understood in general.

Our aim, now, is to establish Theorems A, B^p and C^2 for domains in several complex variables where the integration is taken over any fixed closed subset B of the topological boundary that contains the Silov boundary. Then we will study the question of how to decide for a given measure σ whether it is a candidate for Theorem C^2. This will apply to some special domains to which a natural measure is attached, like for instance to domains with differentiable boundary in $\mathbf{C}^n$ where we have the Lebesgue measure on the boundary.

2. Cauchy integral formulas

In this section we will discuss the solutions to Theorems A and B^1 mentioned in the introduction. The proof of Theorem A is given in the author's paper [*3*] as theorem 19.1, and in a weaker form also in Forster [*8*]. A. Gleason proves Theorem B^1 in [*9*] as his main theorem which of course includes Theorem A. We will present here the idea of this very elegant proof.

First let us state the hypotheses. X is a complex analytic space (or a complex analytic manifold if the reader so wishes) which we assume to be separable. D is a relatively compact open subset of X. A denotes any algebra (with identity) of continuous functions on the closure $\bar{D}$ of D which are holomorphic on D, and we assume that A is closed in the uniform norm on $\bar{D}$. Then there is a smallest closed subset $S \subset \bar{D}$, called the

Silov boundary of $\bar{D}$ (whith respect to A), on which all the functions in A assume the maximum modulus. The maximum modulus principle implies that S must be a subset of the topological boundary ∂D of D. We let B be any closed subset of $\bar{D}$ satisfying

$$S \subset B \subset \partial D .$$

Let $\mathcal{M}$ be the Banach space of (complex) Radon measures on B. $\mathcal{M}$ can be viewed as the dual Banach space of the Banach space $\mathcal{C}(B)$ of all continuous complex valued functions on B. We use the notation

$$\langle f, \mu \rangle, \quad f \in \mathcal{C}(B), \quad \mu \in \mathcal{M}$$

for the natural pairing of $\mathcal{C}(B)$ and $\mathcal{M}$.

Theorem A requires us to construct a certain holomorphic $\mathcal{M}$-valued function

$$\mu : z \to \mu_z$$

on D. That μ be holomorphic means that in a local coordinate system, μ can be expressed by a convergent power series with coefficients in $\mathcal{M}$. By a local coordinate system at a point x we mean any set $(p_1, \ldots, p_n)$ of holomorphic functions in a neighborhood of x that vanish at x and embed some neighborhood of x as a closed subvariety of the unit polycylinder in $\boldsymbol{C}^n$. The power series expansion of a holomorphic function in terms of $(p_1, \ldots, p_n)$ is of course not unique unless X is a manifold in a neighborhood of x and n is the dimension of X at x.

If μ is a holomorphic $\mathcal{M}$-valued function on D, then for each $f \in \mathcal{C}(X)$,

$$z \to \langle f, \mu_z \rangle$$

is a (complex valued) holomorphic function on D as is easily seen by looking at the power series expansions of μ. This means that μ is scalar holomorphic. The converse is also true (and holds for any function with values in the dual space of any Banach space). This is by now a well known theorem if D is a manifold, and the proof is found for instance in A. Gleason's paper [*9*, Theorem 2.8]. For analytic spaces it is proved in the author's paper [*3*] (a combination of Theorems 16.1 and 16.2 will do it).

Let A^* be the dual Banach space of the Banach algebra A. If $A^\perp$ denotes the subspace of $\mathcal{M}$ of measures that annihilate the functions in A, then A^* can be identified with the quotient space $\mathcal{M}/A^\perp$. Thus we have a natural continuous linear quotient mapping

$$\eta : \mathcal{M} \to A^* .$$

If our aim was to construct a holomorphic A^*-valued function ν on D with

$$f(z) = \langle f, \nu_z \rangle \quad \text{for} \quad f \in A ,$$

the task would be quite easy. For each $z \in D$, the evaluation δ_z at z,

$$\langle f, \delta_z \rangle = f(z) \quad \text{for} \quad f \in A\,, \tag{2.1}$$

is a continuous linear functional on A and thus belongs to A^*. If we let

$$\nu_z = \delta_z\,,$$

then ν is an A^*-valued function, and equation (2.1) tells us that ν is holomorphic and satisfies

$$f(z) = \langle f, \nu_z \rangle\,, \quad f \in A\,.$$

The problem to construct μ reduces thus to a lifting problem:

$$\begin{array}{ccc} & & \mathscr{M} \\ & \overset{\mu}{\nearrow} & \downarrow \eta \\ D & \xrightarrow[\nu]{} & A^* \end{array}$$

This problem is easily solved locally. One expands ν into a power series

$$\nu_z = \sum a_{k_1 \ldots k_n} p_1(z)^{k_1} \ldots p_n(z)^{k_n}$$

with coefficients bounded by 1 in A^*. By the open mapping theorem there is a constant K such that the $a_{k_1 \ldots k_n}$ are the images under η of elements $b_{k_1 \ldots k_n}$ in $\mathscr{M}$ with norm less than K.

$$\mu_z = \sum b_{k_1 \ldots k_n} p_1(z)^{k_1} \ldots p_n(z)^{k_n}$$

defines now locally an $\mathscr{M}$-valued holomorphic function with $\eta \circ \mu = \nu$. One could now try to use cohomological arguments to construct a global function μ. This is essentially what is done in [*3*]. However, A. GLEASON had a better idea; it works like this.

First we find a continuous $\mathscr{M}$-valued function μ^0 on D with $\eta \circ \mu^0 = \nu$. This can be done by lifting ν locally and patching up the result by a partition of unity. Now we put a hermitian metric on D which gives us a volume element dv. (For a discussion of hermitian metrics on analytic spaces see [*5*].) Let h be any positive continuous weight function such that $\|\mu^0\|$ is square integrable with respect to $h\,dv$. Notice that then

$$\langle g, \mu^0 \rangle \in L^2(h\,dv) \quad \text{for} \quad g \in \mathscr{C}(B)\,.$$

We denote by H^2 the space of holomorphic functions on D which are in $L^2(h\,dv)$. H^2 is a Hilbert space and possesses a Bergman kernel function $K(z, w)$,

$$g(z) = \int_D g(w)\, K(z, w)\, dv(w)\,, \quad z \in D\,,$$

for $g \in H^2$. (If D is an analytic space the existence of K can be proved as in [*6*].)

$$\mu_z = \int_D \mu_w^0\, K(z, w)\, h(w)\, dv(w)\,, \quad z \in D\,,$$

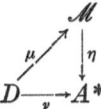

defines an $\mathcal{M}$-valued function. μ is scalar holomorphic since for $g \in \mathcal{C}(X)$,

$$\langle g, \mu_z \rangle = \int_D \langle g, \mu_w^0 \rangle K(z, w) h(w) dv(w)$$

is holomorphic in z. Thus μ is a holomorphic function and we have for $f \in A$,

$$\begin{aligned}\langle f, \mu_z \rangle &= \int_D \langle f, \mu_w^0 \rangle K(z, w) h(w) dv(w) \\ &= \int f(w) K(z, w) h(w) dv(w) \\ &= f(z),\end{aligned}$$

which proves Theorem A.

Theorem B^1 is now established very easily. The method is again due to A. Gleason. One notes that, since D is separable, the range of the function μ in Theorem A is separable. Let $\mu_1, \ldots$ be a dense sequence in the range of μ and define

$$\sigma = \sum_{n=1}^{\infty} 2^{-n} \frac{|\mu_n|}{\|\mu_n\|},$$

where $|\mu_n|$ denotes the total variation of μ_n. By the Radon-Nikodym theorem, $\mu_n \in L^1(d\sigma)$ for all n. Thus μ has values in the closed subspace $L^1(d\sigma)$ of $\mathcal{M}$. If $L^1(d\sigma)$ were a space of functions we could interpret now μ as a function on $D \times B$. But $L^1(d\sigma)$ consists of classes of functions, so we have to make a choice $k(z, \cdot) \in \mu_z$ for each z. This can be done in the following way. We expand μ locally into a power series

$$\sum f_{k_1 \ldots k_n} p_1(z)^{k_1} \ldots p_n(z)^{k_n}$$

where $f_{k_1 \ldots k_n} \in L^1(d\sigma)$. Since this power series converges absolutely in $L^1(d\sigma)$, it converges a. e. for any choice of the $f_{k_1 \ldots k_n}$ by Beppo Levi's theorem. Since we can cover D by a countable number of coordinate patches, it is legitimate to define locally

$$k(z, w) = \sum f_{k_1 \ldots k_n}(w) p_1(z)^{k_1} \ldots p_n(z)^{k_n}.$$

3. Boundary kernel functions

A modification of the trick that proved Theorem B^1 will actually prove Theorem B$^\infty$, and thus Theorem B^p for any p.

We cover D by a countable number of coordinate neighborhoods U_m with coordinates $p_1^m, \ldots, p_n^m$ $(n = n(m))$ of modulus less than one such that the function μ of Theorem A has a power series expansion

$$\sum \mu_{k_1 \ldots k_n}^m [p_1^m(z)]^{k_1} \ldots [p_n^m(z)]^{k_n}$$

in U_m with the property that

$$K_m = \sum \|\mu_{k_1 \ldots k_n}^m\|$$

converges. Let

$$\sigma = \sum_m \sum 2^{-m} K_m^{-1} \,|\, \mu_{k_1\dots k_n}^m \,|.$$

Then

$$\left|\mu_{k_1\dots k_n}^m\right| \leqq 2^m K_m \sigma ,$$

so that by the Radon-Nikodym theorem

$$d\mu_{k_1\dots k_n}^m = f_{k_1\dots k_n}^m \, d\sigma ,$$

with $f_{k_1\dots k_n}^m \in L^\infty(d\sigma)$ and

$$\| f_{k_1\dots k_n}^m \|_\infty \leqq 2^m K_m .$$

Since this bound is independent of $k_1, \dots, k_n$,

$$\sum f_{k_1\dots k_n}^m [p_1^m(z)]^{k_1} \dots [p_n^m(z)]^{k_n}$$

converges absolutely in $L_\infty(d\sigma)$ for $z \in U_m$ and

$$k(z, w) = \sum f_{k_1\dots k_n}^m (w) \, [p_1^m(z)]^{k_1} \dots [p_n^m(z)]^{k_n}$$

is the kernel required in Theorem B^∞.

Theorem B^p implies immediately the following

3.1 Corollary. *Suppose σ is a positive Radon measure on B for which* Theorem B^p *holds. Then for each compact set $K \subset D$ there is a constant c such that*

$$|f(z)| \leqq c \| f \|_p \quad \text{for} \quad f \in A, \quad z \in K. \tag{3.2}$$

Now let $H^p(d\sigma)$ be the closure of A in $L^p(d\sigma)$ and $H(D)$ the space of holomorphic functions on D with the topology of uniform convergence on compact sets. The statement in Corollary 3.1 is equivalent to

3.3 Corollary. *The natural injection*

$$A \to H(D)$$

extends to a continuous linear map

$$i : H^p(d\sigma) \to H(D) .$$

One is tempted to call a function $f \in H^p(d\sigma)$ a boundary value for the holomorphic function $i(f) \in H(D)$, and $i(f)$ a holomorphic extension of f into D. Notice, however, that it is not clear that i is injective, and thus a function f in the range of i might have several "boundary values".

3.4 Problem. *Find conditions on σ which imply that i is an injection.*

Only in a very special case discussed in the next section, we are able to prove that i is injective.

There is a converse to Corollary 3.1.

3.5 Theorem. *Suppose σ is a positive Radon measure on B such that equation* (3.2) *holds for a suitable constant c depending only on the compact set K. If $p < \infty$, then* Theorem B^p *holds for σ.*

Proof: Equation (3.2) implies that A can be considered as a subspace of $L^p(d\sigma)$; zero is the only function in A that vanishes a. e. Thus the natural injection $A \to H(D)$ extends to a continuous linear map

$$i: H^p(d\sigma) \to H(D)$$

as in Corollary 3.3. Let E be the dual space of $H^p(d\sigma)$; E is a quotient space of $L^q(d\sigma)$. If δ_z denotes the evaluation at $z \in D$ on $H(D)$ then

$$g(z) = \delta_z \circ i$$

defines a map $D \to E$ such that

$$\langle f, g(z) \rangle = i(f)(z), \quad z \in D,$$

for $f \in H^p(d\sigma)$ i. e., g is scalar holomorphic. g is therefore a holomorphic E-valued function on D. Solving the lifting problem, we find a holomorphic $L^q(d\sigma)$-valued function k on D such that for $f \in A$,

$$\langle f, k_z \rangle = f(z), \quad z \in D.$$

As in the proof of Theorem B^1 we can interpret k as a measurable function on $D \times B$.

In case $p = 2$ we get a much sharper result.

3.6 Theorem. *Let σ be a positive Radon measure such that the estimate (3.2) is valid for $p = 2$. Then* Theorem B^2 *holds for σ with a function k on $D \times B$ that satisfies*

$$(b') \qquad k(z, \cdot) \in \bar{H}^2(d\sigma) \quad \text{for} \quad z \in D,$$

where $\bar{H}^2(d\sigma) \subset L^2(d\sigma)$ are the conjugates of the functions in $H^2(d\sigma)$.

Proof: This assertion follows immediately from the proof of the last theorem since the dual E of the Hilbert space $H^2(d\sigma)$ can be identified with $\bar{H}^2(d\sigma)$ in a canonical way:

$$\langle f, \bar{g} \rangle = \int_B f \bar{g} \, d\sigma, \quad f, g \in H^2(d\sigma).$$

This proves part of Theorem C^2. If

$$i: H^2(d\sigma) \to H(D)$$

is the continuous linear map that "extends" the functions in $H^2(d\sigma)$ holomorphically into D then the function k_0 defined by

$$\overline{k_0(z, w)} = i(\bar{k}(z, \cdot))(w)$$

is a holomorphic function on $D \times D^*$ since

$$z \to k_0(z, \cdot)$$

is a holomorphic $\overline{H(D)}$-valued function. This completes the proof of Theorem C^2.

The following proposition is easily verified.

3.7 Proposition. *Let σ be as in* Theorem C² *and suppose $\{h_k\}$ is an orthonormal base for $H^2(d\sigma)$.*

Then

$$k(z, w) = \sum i(h_k)(z) \cdot \overline{h_k(w)},$$
$$k_0(z, w) = \sum i(h_k)(z) \cdot \overline{i(h_k)(w)}.$$

There are, of course, a lot of questions that could be asked and to which I do not have an answer. Here is a list of some of them.

3.8 Problems. (1) *Under which conditions does a* Theorem C^p, $1 \leq p < \infty$, *hold with a given measure σ?*

(2) *Under which conditions is $\bar{H}^p(d\sigma)$ the dual space of $H^p(d\sigma)$, $1 < p < \infty$?*

(3) *When is $H^p(d\sigma)$ a direct factor of $L^p(d\sigma)$ $1 < p < \infty$?*

Before turning to the discussion of some special cases, we should perhaps point out the generality of the method described in this exposition. A could be any uniformly closed (real or complex) linear space of continuous functions on a compact closure $\bar{D}$ of a locally compact σ-compact Hausdorff space such that there is a closed maximum modulus set $B \subset \bar{D} - D$. $H(D)$ would then be the closure of A in the space $\mathscr{C}(D)$ of continuous functions on D equipped with the topology of uniform convergence on compact sets. If E^* is the dual of a Banach space E, one would call an E^*-valued function μ on D an $H(D)$-function if

$$\langle f, \mu \rangle \in H(D)$$

for $f \in E$. If the lifting problem can be solved for $H(D)$-functions then the method described above for holomorphic functions yields a Theorem A. This is the case for instance when D is a domain in $\boldsymbol{R}^n$ (or $\boldsymbol{C}^n$) and the elements in A are harmonic (pluriharmonic) on D; and since the vector valued $H(D)$-functions are actually analytic in this case, one can prove Theorems B^p and C² as before. Actually, the proper assumption on A that yields Theorems A, B^p and C², is that $H(D)$ be a nuclear space.

4. Some special cases

We want to close our exposition by mentioning a few special, but very important, cases that have been treated in the literature before, and which fit into the general framework of the last section. We do, however, not claim that this discussion will be in any way complete.

In all examples, we will be concerned with the case $p = 2$ and the establishment of Theorem C² for a certain positive Radon measure σ on B. Thus for each compact set $K \subset D$, the basic estimate

$$|f(z)| \leq c \|f\|_2, \quad z \in K, \quad f \in A \tag{4.1}$$

has to be established. It is also clear that if we have already an integral formula at our disposal that satisfies

(b) $k(z, \cdot) \in L^2(d\sigma)$ depends weakly continuous on $z \in D$,

(d) $f(z) = \int_B f(w)\, k(z, w)\, d\sigma(w)$, $z \in D$, for all $f \in A$,

then an estimate (4.1) will automatically hold.

First there are the Cauchy-Weil integral formula [*15*] and a related formula by S. BERGMAN [*1*]. In these cases D is a bounded domain in $\boldsymbol{C}^2$ (or more generally in $\boldsymbol{C}^n$) which is bounded by a finite number of analytic hypersurfaces, and $B = S$ is the Silov boundary. There are a lot of further hypotheses that can be found in [*1*] and [*15*] respectively. These hypotheses imply the regularity of the Silov boundary; σ is then the Lebesgue measure on $B = S$. In BERGMAN's paper [*1*], the estimate (4.1) is proved in Lemma 2.2, while for the situation in WEIL's paper, (4.1) follows from the existence of the Cauchy-Weil formula (which is a meromorphic formula). S. BERGMAN discusses also the boundary kernel function k_0 for his domains.

If D is a bounded symmetric homogeneous (Cartan) domain embedded into $\boldsymbol{C}^n$ in such a way that $0 \in D$ and the stable subgroup at 0 of the group G of automorphisms of D is a subgroup of the linear group (see HARISH-CHANDRA [*10*]), then the Silov boundary S of D is an orbit of G in ∂D (Theorem 3.6 in [*12*]). Let σ be the G-invariant measure on S (S is actually a compact submanifold of $\boldsymbol{C}^n$ and σ is the Lebesgue measure on S). By Theorem 4 in [*14*], FURSTENBERG's construction [*7*] yields a Poisson kernel on $D \times S$ for the continuous functions on $\bar{D}$ that are harmonic in the BERGMAN metric on D. We have therefore an estimate (4.1). It is perhaps interesting to note that if M is any other orbit of G on ∂D and σ_M the G-invariant measure on M, then an estimate (4.1) does hold with $\sigma = \sigma_M$ and $B = \bar{M}$ (which contains S by Proposition 3.2 in [*12*]), and hence a kernel function can be constructed. C. MOORE has established this fact; he is presently considering these kernel functions in connection with automorphic functions on D.

Finally let D be a bounded domain in $\boldsymbol{C}^n$ with piecewise differentiable boundary. We let $B = \partial D$ and σ the Lebesgue measure on B. All differentiable forms on B are easily seen to be absolutely continuous with respect to $d\sigma$. Therefore we can use the Bochner-Martinelli formula (see BOCHNER [*2*] or MARTINELLI [*13*]) to obtain an estimate (4.1). This method can be used to establish (4.1) also when we take D to be a relatively compact domain with piecewise differentiable boundary $\partial D = B$ on a complex manifold (or an analytic space) X on which the global holomorphic functions separate points. σ is then the volume on ∂D associated with any Rimannian metric on X. The proofs are in [*4*].

A special case are the Reinhardt circular domains D with piecewise differentiable boundary in $\boldsymbol{C}^n$; these are the domains that satisfy

$$(z_1 t_1, \ldots, z_n t_n) \in D \quad \text{if} \quad (z_1, \ldots, z_n) \in D \quad \text{and}$$
$$|t_1|, \ldots, |t_n| \leqq 1 .$$

In this particular case the normalized monomials

$$z_1^{\alpha_1} \ldots z_n^{\alpha_n}$$

form an orthonormal base for $H^2(d\sigma)$. The expansion of a function $f \in H^2(d\sigma)$ in terms of these monomials gives then the power series expansion of $i(f)$ in D, i being the canonical map from $H^2(d\sigma)$ into $H(D)$. Hence the uniqueness of the power series expansion implies that i is injective, thus solving Problem 3.4 in this special case. The kernel function k_0 has also a nice property not shared by kernel functions of other domains. For each $z \in D$, $\overline{k_0(z, \cdot)}$ has a holomorphic extension into a neighborhood of $\bar{D}$ and

$$k_0(z, w) = k(z, w) \quad \text{for} \quad (z, w) \in D \times \partial D .$$

Thus a Theorem C^p, $1 \leqq p \leqq \infty$, is actually valid in this situation. The existence of the kernel function for Reinhardt circular domains has been shown in F. KAMBARTEL [11] for $n = 2$ and in the author's paper [4] for general n. In these papers one also finds the proof that the kernel function for the unit ball in $\boldsymbol{C}^n$ is

$$k_0(z, w) = \frac{2^{n-1}}{(2\pi)^n} \frac{(n-1)!}{(1 - z\bar{w})^n}$$

(where $z\bar{w} = \sum z_j \bar{w}_j$) which has a resemblance to the Cauchy kernel (1.1).

References

[1] BERGMAN, S.: Bounds for analytic functions in domains with a distinguished boundary surface. Math. Zschr. **63**, 173—194 (1955).

[2] BOCHNER, S.: Analytic and meromorphic continuation by means of Green's formula. Ann. Math. (2) **44**, 652—673 (1943).

[3] BUNGART, L.: Holomorphic functions with values in locally convex spaces and applications to integral formulas. Trans. Am. Math. Soc. (2) **111**, 317—344 (1964).

[4] — Boundary kernel functions for domains on complex manifolds. To appear in Pacific J. Math.

[5] — Integration on real analytic varieties I. To appear in J. Math. Mech.

[6] —, and H. ROSSI: On the closure of certain spaces of holomorphic functions. Math. Ann. **155**, 173—183 (1964).

[7] FURSTENBERG, H.: A Poisson formula for semi-simple Lie groups. Ann. Math. **77**, 335—386 (1963).

[8] FORSTER, O.: Funktionswerte als Randintegrale in komplexen Räumen. Math. Ann. **150**, 317—324 (1963).

[9] GLEASON, A.: The abstract theorem of Cauchy-Weil. Pacific J. Math. **12**, 511—525 (1962).
[10] HARISH-CHANDRA: Representations of semi-simple Lie groups VI. Amer. J. Math. **78**, 564—628 (1956).
[11] KAMBARTEL, F.: Orthogonale Systeme und Randintegralformeln in der Funktionentheorie mehrerer Veränderlicher. Schr. Math. Inst. Univ. Münster. **18** (1960).
[12] KORANYI, A., and J. WOLF: Realization of hermitian symmetric spaces as tube domains. To appear in Ann. Math.
[13] MARTINELLI, E.: Alcuni teoremi integrali per le funzioni analitiche di più variabili complesse. Rend. Accad. Italia **9**, 269—283 (1938).
[14] MOORE, C.: Compactification of symmetric spaces II: The Cartan domains. (forthcoming).
[15] WEIL, A.: L'intégrale de Cauchy et les fonctions de plusieurs variables. Math. Ann. **111**, 178—182 (1935).

Department of Mathematics
University of California
Berkeley, California

Extrinsic Complex Projective Geometry*

By

WILLIAM F. POHL

With one figure

§1. Let P_N denote N-complex-dimensional complex projective space, and M a complex manifold. Consider complex analytic mappings $f: M \to P_N$. A general problem of extrinsic geometry is to relate the intrinsic invariants of M with the extrinsic invariants of such maps[1].

An example of such a relation is provided by one of the Plücker formulas: Let M be a compact Riemann surface, $N = 2$, f not constant (i.e., consider a plane curve); then

$$2g - 2 = \gamma_1 - 2\xi + \varkappa, \tag{1}$$

where g is the genus of M, ξ is the order (i.e. the number of points of intersection of the curve with a generic line), γ_1 is the class (i.e., the number of tangents to the curve passing through a generic point of P_2), and $\varkappa$ is the number of cusps (counted with multiplicities). Note that in practice this formula might be used either to compute the intrinsic invariant g from the extrinsic invariants, or else to relate the extrinsic invariants if g were known.

* Received May 5, 1964.

[1] For a general discussion and survey of this problem, as well as further information, cf. papers of CHERN [3, 4].

A similar formula holds for complex surfaces, though the situation is more complicated, for numerical invariants no longer suffice. Instead, we must consider invariants lying in the Chow ring, or the homology ring, of M^2.

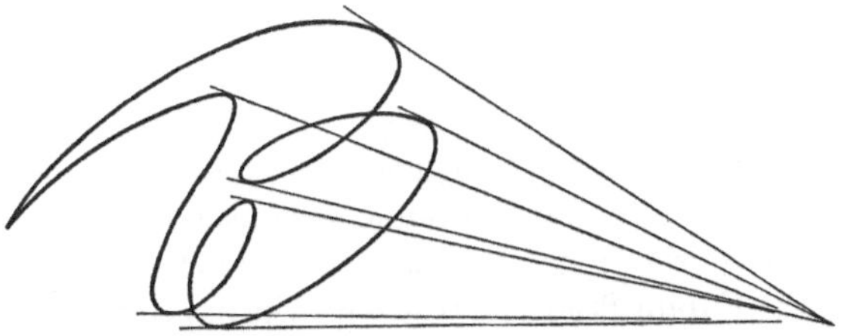

Let $f: M \to P_N$ be a surface in P_3, and assume that M is a non-singular algebraic variety, though f may have arbitrary singularities. Let $x \in P_3$ be a generic point, and let γ_1 denote the homology class of the closure of the locus of points $y \in M$ with the properties that f is non-singular at y, and the tangent plane at y passes through x. Let ξ denote the homology class of a generic hyperplane section, let $\varkappa_1, \ldots, \varkappa_r$ denote the homology classes of the cuspidal edges, and let $i_1, \ldots, i_r$ denote the multiplicities of the cuspidal edges (these will be made explicit later); then the homology class

$$c_1 = -\gamma_1 + 3\xi - (i_1 \varkappa_1 + \cdots + i_r \varkappa_r) \tag{2}$$

[2] We shall encounter shortly the vexatious question whether a closed algebraic variety V of (complex) dimension n lying in a Kähler manifold M carries a non-zero homology cycle of real dimension $2n$. Thanks to Hironaka's work [6] this can now be treated easily. Blow up V to a non-singular complex manifold $\tilde{V}$, and let $\pi: \tilde{V} \to V$ be the projection map. For each differential form Φ_{2n} of degree $2n$ consider

$$V(\Phi_{2n}) = \int_V \Phi_{2n},$$

by which we understand the integral over the non-singular part of V. If $\Phi_{2n} = d\psi_{2n-1}$, then

$$\int_V \Phi_{2n} = \int_{\tilde{V}} \pi^* \Phi_{2n} = \int_{\tilde{V}} \pi^* d\psi_{2n-1} = \int_{\tilde{V}} d\pi^* \psi_{2n-1} = 0,$$

by Stokes' theorem, since $\tilde{V}$ is compact. Thus $V(\Phi_{2n})$ vanishes on exact forms, and represents a linear functional on the vector space of $2n$-forms which can be identified with a $2n$-dimensional homology class. This class is independent of the way in which V is blown up, for blowing up affects only a subset of V of measure zero. This homology class is non-zero; for if Ω is the associated two-form of the Kähler metric on M, the restriction of Ω to the non-singular part of V is the associated form of the induced metric. Now $d\Omega = 0$, hence $d\Omega^n = 0$. But $1/n!\,\Omega^n$ is the volume element on V, so that the integral $V(\Omega^n) \neq 0$.

is an intrinsic invariant of M, known variously as the canonical class, or first Chern class, of M.

The main purpose of this lecture is to generalize formulas (1) and (2) to arbitrary dimension.

§2. The formulas of the preceding paragraph admit an interpretation in integral geometry. Taking dual cohomology classes, this is clear *a priori*, for any cohomology class can be realized as integration of a differential form. But the cohomology classes considered here can be represented by forms having local geometric significance.

Let $Z = (Z_0, \ldots, Z_N)$ be the homogeneous coordinates of P_N, and let $(,)$ denote hermitian inner product in C^{N+1}. P_N has a natural hermetian metric, the Fubini-Study metric, with the expression

$$ds^2 = \frac{(Z,Z)(dZ,dZ) - (Z,dZ)(dZ,Z)}{(Z,Z)^2}.$$

If $ds^2 = \sum g_{ij} dz_i d\bar{z}_j$ is an hermitian metric on a complex manifold N of complex dimension n, the associated two-form is defined to be $\Omega = \frac{i}{2} \sum g_{ij} dz_i \wedge d\bar{z}_j$. The volume element is $1/n!\,\Omega^n$.

If $M \subset N$ is an analytic subvariety of dimension m, the associated two-form of the induced metric on M is clearly just the restriction of Ω. Hence the volume of M is given by

$$\frac{1}{m!} \int_M \Omega^m .$$

Now suppose that the hermitian metric on N is Kähler, i.e., $d\Omega = 0$. Then $d\Omega^m = 0$. Hence if $M, M' \subset N$ are homologous closed analytic subvarieties, then

$$\int_M \Omega^m = \int_{M'} \Omega^m ,$$

so that M and M' have the same volume.

Returning to P_N now, the Fubini-Study metric is Kähler, and the volume of an n-plane is known to be $(2\pi)^n$. $\frac{1}{2\pi}\Omega$ is dual to a hyperplane. Thus the volume of an algebraic variety of dimension n is $(2\pi)^n/n!$ times its order, which is a theorem of WIRTINGER [*14*].

We can now reinterpret formula (1) of the previous section. $2\pi\xi$ is the arc length of the curve, and $2\pi^2\gamma_1$ is the area swept out by the tangent lines. The interpretation of (2) is somewhat more difficult. We must make further geometric constructions.

§3. Let $G_{n,N-n}$ denote the Grassmann manifold of n-dimensional linear subspaces of P_N. Using the Cayley-Grassmann coordinates, $G_{n,N-n}$ can be embedded in some projective space, from which embedding it derives a complex structure. $G_{n,N-n}$ also has a natural Kähler metric [*2*, *13*] and the volume in this metric has an interesting relationship with the volume in P_N. Let $f': M \to G_{n,N-n}$ be an m-complex-parameter family of n-planes. Then the volume of the $m+n$-dimensional variety in P_N consisting of the union of these planes is the volume of $f'M$ times the volume of P_n. (This is shown, in a special case, in [*13*].)

The homology of $G_{n,N-n}$ is well known [*5*]. It is determined as follows: Choose a flag in P_N, i.e., a nested sequence of linear subspaces

$$P_0 \subset P_1 \subset \cdots \subset P_N .$$

For each sequence of integers $a_0, \ldots, a_n$, $a_0 \leqq a_1 \leqq \cdots \leqq a_n \leqq N-n$, consider the family of n-planes $Q \subset P_N$ satisfying the conditions $\dim (Q \cap \cap P_{a_j+j}) \geqq j$; denote it by $(a_0 \ldots a_n) \subset G_{n,N-n}$. These subvarieties are called Schubert varieties, and they form an additive basis for the homology of $G_{n,N-n}$. Their intersection products are given by universal formulas (the Schubert calculus). In particular, $(a_0 \ldots a_n)$ and $(N-n-a_n \ldots N-n-a_0)$ are of complementary dimensions and their intersection number is one. All other intersection numbers are zero. The Schubert varieties $(00 \ldots 011 \ldots 1)$ generate the homology *ring* of $G_{n,N-n}$.

Now P_N can be thought of as the family of complex lines through the origin of C^{N+1}, the $N+1$-dimensional complex number space. Thus any $P_n \subset P_N$ can be regarded as an $n+1$-plane through the origin of C^{N+1}. There is a canonical complex $n+1$-plane bundle $E_{n+1} \to G_{n,N-n}$, the fibre of which over a point $x \in G_{n,N-n}$ is just the $n+1$-plane corresponding to x in C^{N+1}. Let us agree to adopt Hirzebruch's definition [*7*] of the Chern classes. Then the i-th Chern class of E_{n+1} is $(-1)^i$ times the cohomology class which takes the value one on $(0 \ldots 01 \ldots 1)$ ($n-i+1$ 0's followed by i 1's), and zero on every other cycle [*1*]. These classes generate the cohomology ring, by what we stated above.

There is only one Schubert cycle of (complex) dimension one. Its dual is $-c_1(E_{n+1})$, which is represented by $1/2\pi\Omega$, where Ω is the associated two-form of the Kähler metric.

Consider now a plane curve. The Schubert variety of lines (01) consists of all lines through a fixed point. The class, γ_1, is just the number of tangent lines belonging to (01).

Let $f: M \to P_2$ be the curve. By the *associated map* $f': M \to G_{1,1}$, we mean the map which assigns to each point $x \in M$ the tangent line to f at x. Note that the singularities of f' are removable, for at a cusp of f there is always a well-defined limiting position of the tangent line. We can thus interpret the class as the evaluation of the inverse image of $-c_1(E_2)$ under the associated map on the fundamental cycle of M. And so the class is also the arc length of f' times 2π.

$G_{1,1}$ is itself a projective plane, the so-called dual projective plane. f' is called the dual curve.

It is worthwhile to summarize the various interpretations we have given of the class, γ_1, of a plane curve:

$\gamma_1 =$ 1) the number of tangent lines passing through a fixed point,
2) $1/2\pi^2$ times the area swept out by the tangent lines,
3) the intersection number of the image of the fundamental cycle of M under the associated map, with the Schubert cycle (01),
4) the evaluation of the inverse image under the associated map f' of $-c_1(E_2)$ on the fundamental cycle of M,
5) $1/2\pi$ times the arc length of the curve f',
6) the order of the dual curve,
7) $1/2\pi$ times the arc length of the dual curve.

The main purpose of this lecture is to generalize the Plücker formula (1) to higher dimensions. We shall introduce a generalization of the class, γ_1. The interpretations 1) to 5) above will generalize, and 6) and 7) as well, for hypersurfaces.

§ **4.** We now discuss the higher dimensional case. We shall begin by considering only complex-analytic mappings $f: M \to P_N$ which are immersions, i.e., the Jacobian matrix is assumed to have rank $n = \dim M$ everywhere. Self intersections are allowed, however.

Choose a flag $P_0 \subset P_1 \subset \ldots \subset P_{N-1} \subset P_N$, and let f be as above. For each Schubert condition $(a_0 \ldots a_n)$ we can consider the locus of points $x \in M$ with the property that the tangent projective plane in P_N to f at x satisfies the condition $(a_0 \ldots a_n)$ with respect to the given flag. If another flag is chosen, the resulting locus will be rationally equivalent to the first, hence homologous. Thus any Schubert condition gives rise to a homology class of M, and this does not depend on a choice of flag.

Let $f': M \to G_{n,N-n}$ denote the associated map to f, i.e., the map which assigns to each point x of M the tangent plane to f at x, and let $f'^! E_{n+1}$ denote the inverse image of the canonical $n+1$-plane bundle on $G_{n,N-n}$. The homology class of M determined by the Schubert condition $(N-n-1 \ldots N-n-1\; N-n \ldots N-n)$ (i $(N-n-1)$'s followed by $n-i+1$ $(N-n)$'s) is the Poincaré dual of $(-1)^i c_i(f'^! E_{n+1})$ in M, as can be seen from the considerations of the last paragraph. Call it γ_i.

Let ξ denote the homology class in M of a generic hyperplane section. If $E_1 \to P_N$ is the standard line bundle over P_N (the fibre of which over a point $x \in P_N$ consists of the corresponding line through the origin in C^{N+2}), then ξ is the Poincaré dual of $-c_1(f^! E_1)$.

If M has dimension one, then the class $2\xi - \gamma_1$ is an intrinsic invariant of M, the Euler class, by formula (1). In the late 1930's EGER, TODD, and others showed that in the higher dimensional case, as well, certain combinations of the extrinsic invariants ξ^j, γ_i, are intrinsic invariants of M, namely

$$\Gamma_r = \sum_j \binom{n+1-r+j}{j} \gamma_{r-j} \xi^j .$$

In the late 1940's and early 1950's S.S. CHERN, using the topological and differential geometric properties of bundles, discovered cohomology invariants of complex vector bundles. Since, for the tangent bundle of an algebraic variety, these agreed with the Poincaré duals of Todd's classes in a number of cases, it was conjectured that they agreed in general. After attempts by various people, this was completely proved by KODAIRA and SERRE, and by S. NAKANO [9].

NAKANO directed his attention to the bundle $f'^! E_{n+1}$ and showed that it is topologically (but not analytically) isomorphic to $f^! E_1 \oplus (f^! E_1 \otimes T)$, where T denotes the tangent bundle of M, and $\oplus$, $\otimes$ are Whitney sum, tensor product, respectively. Thus T plus a product line bundle is isomorphic to $(f^! E_1)^{-1} \otimes (f'^! E_{n+1})$, and the Chern classes can be computed by standard formulas [7], which yield

$$c_r(T) = \mathscr{D} \sum_j \binom{n+1-r+j}{j} \gamma_{r-j} \xi^j , \tag{3}$$

where $\mathscr{D}$ denotes Poincaré dual. Thus $c_r(T) = \mathscr{D}\Gamma_r$. We should point out that this formula gives a concrete geometric interpretation of the Chern classes of an algebraic variety.

Formula (3) contains formula (1) as a special case for curves without cusps. It is thus natural to ask whether formula (3) can be generalized to contain (1) as a special case for general curves. Work in this direction has been done by S. S. CHERN, using integral geometry, and by H. I. LEVINE, using topological methods. (This work is unpublished.) There is work of I. R. PORTEOUS [11, 12] closely related to this problem.

LEVINE and PORTEOUS make certain genericity, or transversality, assumptions on the singularities. Thus, in the case of curves, PORTEOUS' conditions would exclude cusps altogether, and LEVINE's would exclude cusps of multiplicity higher than one (e.g., the plane curve $y^3 = x^5$). Now in the study of C^∞ mappings of real manifolds it is reasonable to make transversality assumptions, since non-transversal singularities may be eliminated by approximation. The approximation techniques break down, however, in the complex-analytic case, so that a theory of analytic maps with transversal singularities seems somewhat unrealistic, though such theories certainly point the way to a general theory.

We shall give a generalization of formula (3) to maps with arbitrary singularities, for $r = 1$. We should point out that some care must be taken in defining the classes γ_i for singular mappings, since there is in general no well-defined tangent projective n-plane to f at a point where the rank of the Jacobian falls.

§ 5. In [*10*] we studied the relations between the extrinsic and intrinsic invariants of maps $f: M \to P_N$ in the singular case. Our beginning point was the study of the bundle $f'^{!}E_{n+1}$ of the previous paragraph. We showed that this bundle depended only on the line bundle $f^{!}E_1$, and not on the map f.

Our fundamental construction was called the "first derivative". Given a line bundle $V \to M$, a new bundle $\Delta_1 V$ was defined which contains V as a vector subbundle, and which is topologically, but not analytically, isomorphic to $V \oplus (T \otimes V)$, where T denotes the tangent bundle of M. The crucial geometric property of this construction is the following: let $\varphi: V \to C^{N+1}$ be a complex analytic map which is a vector space homomorphism on each fibre; such a map is called a *realization*, and may be singular; then there is an induced realization $\Delta_1 V \to C^{N+1}$ which agrees with the original one on $V \subset \Delta_1 V$, and which maps the fibre of $\Delta_1 V$ above a point $x \in M$ onto the linear span of the tangent spaces to φ at all points in the fibre of V above x.

If $E_1 \to P_N$ is the standard bundle, the fibre over a point $x \in P_N$ is just x, itself, regarded as a line through the origin of C^{N+1}. Thus E_1 has a canonical realization, and so $f^{!}E_1$ has a canonical realization, for any $f: M \to P_N$. It is not hard to see that the induced realization $\Delta_1 f^{!}E_1 \to$ $\to C^{N+1}$ maps the fibre of $\Delta_1 f^{!}E_1$ over a point $x \in M$ onto the plane corresponding to the tangent projective space to f at x. This induces a bundle map $\Delta_1 f^{!}E_1 \to f'^{!}E_{n+1}$ over those points of x where f' is defined. Thus $\Delta_1 f^{!}E_1$ plays the role of a bundle of universal tangent spaces. If f is singular, a comparison of $\Delta_1 f^{!}E_1$ with the bundle of concrete tangent spaces yields the desired generalization of the Plücker formula (1).

In [*10*] we carried this out, under the assumption that the associated map f' could be defined over the singularities of f. In the meantime, the famous work of HIRONAKA [*6*] has become available, and makes it possible to extend our result without any assumptions on the singularities.

Proposition 1. *Let $V \to M$, $W \to M$ be complex-analytic vector bundles of fibre dimensions n and m, respectively, and let $\varphi: V \to W$ be an analytic homomorphism which has rank n, except over a subvariety of M of lower dimension. If there is a vector subbundle $W' \subset W$ of fibre dimension n such that $\varphi(V) \subset W'$, then the locus of points of M over which φ is singular has codimension one, and W' is uniquely determined.*

On the other hand, if the singular locus has codimension one, and is locally irreducible (i.e., if φ is singular over $x \in M$, then the singular locus is given by $f = 0$, for some holomorphic function f, in a neighborhood of x; and if $f = gh$, for g, h holomorphic functions, either $g \neq 0$ or $h \neq 0$ at x, *for suitably chosen f), then there is a vector sub-bundle W' such that $\varphi(V) \subset W'$, and W' is uniquely determined.*

Proof. Suppose such a sub-bundle $W' \subset W$ exists. Since it is exactly $\varphi(V)$ over the non-singular points, it is unique by continuity. If $\varphi' : V \to W$ is φ with the range restricted, the singular locus is given locally by the vanishing of the determinant of φ', hence has codimension one.

On the other hand, suppose the singular locus has codimension one, and is given locally by the irreducible condition $f = 0$. Take n local sections of V, $S_1, \ldots, S_n$, which span the fibre at each point, and similarly m sections of W, $T_1, \ldots, T_m$. Then

$$\varphi S_1 \wedge \ldots \wedge \varphi S_n = \sum_{i_1 < \cdots < i_n} f_{i_1 \ldots i_n} T_{i_1} \wedge \ldots \wedge T_{i_n},$$

where $\wedge$ denotes the exterior product. Now all $f_{i_1 \ldots i_n}$ vanish on the singular locus, so each is essentially a power of f. Let f^q be the least such power. Then $f^{-q} \varphi S_1 \wedge \ldots \wedge \varphi S_n$ is non zero, and defines a sub-bundle $W' \subset W$ of fibre dimension n. It is unique by continuity, *q.e.d.*

Corollary. *Let $f : M \to P_N$ be a complex-analytic map of a complex manifold of which the Jacobian has rank n (= dim. M) except on a subvariety of M of lower dimension. If the associated map $f' : M \to G_{n, N-n}$ can be extended over this subvariety, then it has codimension one. On the other hand, if this subvariety is locally irreducible of codimension one, then the associated map is everywhere defined.*

Proof. Apply the proposition to the bundle map $\Delta_1 f^! E_1 \to C^{N+1}$. The condition that $\Delta_1 f^! E_1$ map into a sub-bundle of fibre dimension $n+1$ is just the condition that there is a unique tangent projective n-plane in limiting position at each singular point.

We observe that the singularities of a curve are always points, and are thus locally irreducible of codimension one, so that the associated map always exists.

We shall need the following

Proposition 2 *(The Riemann-Hurwitz Formula). Let $\varphi : V \to W$ be an analytic homomorphism of complex vector bundles of the same fibre dimension n which has rank n except over an analytic subvariety of lower dimension of the base. Then this subvariety has codimension one, since it is determined by the condition that the determinant of φ is zero, and*

$$c_1(W) - c_1(V) = c_1(LD\varphi),$$

where $LD\varphi$ denotes the line bundle associated to the divisor determined locally by $\det \varphi = 0$.

Corollary. *If $f : M \to N$ is an analytic map of complex manifolds of the same dimension which has rank n except on a subvariety of M of lower dimension, then this subvariety has codimension one and*

$$f^* c_1(N) - c_1(M) = c_1(LDJ_f),$$

where LDJ_f is the line bundle associated to the divisor, determined locally by the condition Jacobian determinant equals zero.

The proof is given in [*10*]. The classical Riemann-Hurwitz formula is just the corollary applied to maps of compact Riemann surfaces. It can be shown that $c_1(LD\varphi)$ is the cohomology class dual to the divisor itself.

We can now state and prove the main result.

Theorem 3. *Let M be a compact algebraic complex manifold (i. e., a Kähler variety of restricted type, or a Hodge manifold) of complex dimension n, and let $f: M \to P_N$ be a complex-analytic map such that the Jacobian matrix has rank n except for a subvariety of M of lower dimension. Let $Q \subset P_N$ be a generically situated linear subspace of dimension $N-n-1$. Consider the locus of points $x \in M$ at which the Jacobian of f has rank n and the tangent projective n-plane to f at x meets Q. Let γ_1 denote the homology class of its closure. Let ξ denote the homology class in M of a generic hyperplane section, c_1 the Poincaré dual of the first Chern class, and $\sigma_1, \ldots, \sigma_r$ the homology classes of the irreducible components of codimension one of the singular locus of f. (By the singular locus we mean the locus of points where the rank of the Jacobian is less than n.) Then*

$$\gamma_1 = (n+1)\xi - c_1(M) - (i_1\sigma_1 + \cdots + i_r\sigma_r), \tag{4}$$

where the i's are positive integers determined as follows. Let $x \in M$ be a point on the j-th component of codimension one of the singular locus, and let $z_1, \ldots, z_n$ be local coordinates about x. Take inhomogeneous coordinates in P_N about $f(x)$. Then f can be represented as a vector function $\mathbf{X}(z_1, \ldots, z_n)$ in a neighborhood of x. Consider

$$\frac{\partial \mathbf{X}}{\partial z_1} \wedge \cdots \wedge \frac{\partial \mathbf{X}}{\partial z_n} = \sum f_{k_1 \ldots k_n} C_{k_1} \wedge \cdots \wedge C_{k_n},$$

where $C_1, \ldots, C_N$ are vectors forming a frame for the given inhomogeneous coordinate system. Each of the $f_{k_1 \ldots k_n}$ vanishes on the singular locus, so each is essentially a power of the irreducible equation defining the locus. Let i_j be the least such power occurring.

Proof. M is algebraic; hence $f: M \to P_N$ is a rational map, by Chow's theorem. Consider the associated map $f': M \to G_{n,N-n}$; that is the map which assigns to a point of M the tangent projective n-plane to f at that point. This map is a rational map, and will be undefined on the part of the singular locus which has codimension greater than one, and may be undefined at the points where the irreducible components of codimension one of the singular locus cross, by Proposition 1. According to a theorem of HIRONAKA [*6*, p. 140] one can perform a sequence of monoidal transformations with non-singular centres on M and obtain a new manifold $\tilde{M}$, with projection $\pi: \tilde{M} \to M$, such that $f'\pi$ is everywhere defined on $\tilde{M}$. Thus $f'\pi = (f\pi)'$, by continuity.

As we have already observed there is a bundle map $\varphi: \Delta_1(f\pi)^! E_1 \to \to (f\pi)'^! E_{n+1}$ defined over the points at which $f\pi$ is non-singular. $(f\pi)'^! E_{n+1}$ may be regarded as a sub-bundle of a trivial bundle C^{N+1}, and there is a map $\Delta_1(f\pi)^! E_1 \to C^{N+1}$ defined everywhere over $\tilde{M}$, and agreeing with the first bundle map at non-singular points. Hence by continuity $\Delta_1(f\pi)^! E_1$ maps into $(f\pi)'^! E_{n+1}$ everywhere, so φ is everywhere defined.

These bundles have the same fibre dimension, and the singular locus of φ is precisely the singular locus of $f\pi$.

By Proposition 2,

$$c_1((f\pi)'^! E_{n+1}) - c_1(\Delta_1(f\pi)^! E_1) = c_1(LD\varphi),$$

where c_1 denotes the first Chern cohomology class. Since $\Delta_1(f\pi)^! E_1$ is topologically isomorphic to $(f\pi)^! E_1 \oplus (T \otimes (f\pi)^! E_1)$, where T is the tangent bundle of M, $c_1(\Delta_1(f\pi)^! E_1) = (n+1)c_1((f\pi)^! E_1) + c_1(\tilde{M})$. By the corollary of Proposition 2, $c_1(\tilde{M}) = \pi^* c_1(M) - c_1(LDJ_\pi)$. Hence $c_1((f\pi)'^! E_{n+1}) = (n+1)c_1((f\pi)^! E_1) +$

$$+ \pi^* c_1(M) - c_1(LDJ_\pi) + c_1(LD\varphi). \tag{5}$$

We must interpret each of these terms. Taking Poincaré duals and blowing the resulting homology classes down to M will yield formula (4).

Let $\mathscr{D}$ denote Poincaré dual, and let π_* denote the induced map from the homology of $\tilde{M}$ to the homology of M. Then $\pi_* \mathscr{D} \pi^* \mathscr{D}$ is the identity on the homology of M, since π is of degree one. Thus $\pi_* \mathscr{D} c_1((f\pi)^! E_1) = \mathscr{D} c_1(f^! E_1) = -\xi$, since $\mathscr{D} c_1(E_1)$ is the negative of a hyperplane. And $\pi_* \mathscr{D} \pi^* c_1(M) = \mathscr{D} c_1(M)$.

Now the singularities of π are precisely the inverse images of the centers of dilatation; hence under π_* their homology classes go to zero. But $\mathscr{D} c_1$ (LDJ_π) consists precisely of these homology classes. Hence $\pi_* \mathscr{D}(LDJ_\pi) = 0$. The singularities of φ consist precisely of the singularities of π and the inverse images of the singularities of codimension one of f, since singularities of f of higher codimension need to be blown up, by Proposition 1. Therefore $\pi_* \mathscr{D} c_1(LD\varphi) = i_1\sigma_1 + \cdots + i_r\sigma_r$, where the σ's are the irreducible components of codimension one of the singular locus of f, and the i_j's are multiplicities which can easily be shown to be exactly those given in the statement of the theorem.

We assume the $N - n - 1$-plane Q so chosen that not all the limiting tangent planes along any component of the singular locus of $f\pi$ meet Q. The homology class $\mathscr{D} c_1((f\pi)'^! E_{n+1})$ is precisely the homology class of the locus of points x of M such that the tangent plane to $f\pi$ at x satisfies the Schubert condition $(N - n - 1 \;\; N - n \;\; N - n \ldots N - n)$, and this is just the condition that the tangent plane meet Q. This locus has codimension one, and contains no components of the singular locus of $f\pi$. Hence $f\pi$ is non-singular at almost all points of it, and hence

$$\pi_* \mathscr{D} c_1((f\pi)'^! E_{n+1}) = -\gamma_1.$$

So we now apply $\pi_* \mathscr{D}$ to formula (5) and obtain formula (4), *q. e. d.*

PH. GRIFFITHS has asked about the relationship of this formula with the adjunction formula, a formula which concerns the Chern classes of the normal bundle of a submanifold of a complex manifold. This can be used to obtain formula (1), since the class can be interpreted as the order of the dual curve. In higher dimension and codimension the adjunction formula seems to be unrelated to ours, since the normal bundle admits no analytic geometric realization.

Theorem 3 is noteworthy in that it gives an "enumerative" formula which holds in arbitrary dimension and codimension, and without any restrictions on the singularities. Let us now interpret it in terms of volumes.

Proposition 4. *Let $f: M \to P_N$ be as in Theorem 3, and suppose that the variety $f'(M) \subset G_{n,N-n}$ has dimension n, or, equivalently, that the variety in P_N consisting of all tangent projective n-planes to f at all points of M has dimension $2n$. Then the volume swept out by f in $G_{n,N-n}$ is just the n-fold self-intersection number γ_1^n times $(2\pi)^n/n!$*

Proof: We follow the proof of Theorem 3. It suffices to determine the volume of $(f\pi)'$, since blowing up affects only a subset of measure zero. We have seen that $\frac{1}{2\pi}\Omega$ represents $-c_1(E_{n+1})$, where Ω is the associated two-form of the Kähler metric in $G_{n,N-n}$. The volume element for n-dimensional submanifolds is $1/n!\Omega^n$, and this represents $(-2\pi)^n c_1(E_{n+1})^n$. Observing that π_* preserves intersection numbers proves the proposition.

Theorem 5. *Let $f: M \to P_N$ be a complex-analytic map of an algebraic complex manifold of complex dimension n. Suppose that the variety of tangent projective n-planes to f at all points of M has dimension $2n$. Then the volume swept out by the tangent projective n-planes in P_N is $(2\pi)^{2n}/n!$ times the n-fold self-intersection number γ_1^n of the class γ_1 of Theorem 3.*

Proof: The Theorem follows from Proposition 4 and the observation, already made, that the volume of an n-parameter family of n-planes in P_N is just the volume of a single n-plane times the volume swept out by the associated map in $G_{n,N-n}$, *q. e. d.*

Some explanation is needed of the hypothesis that the tangent planes constitute a $2n$-dimensional variety. In the case of a surface this is equivalent to the condition that neither is the surface developable, nor does it lie in a three dimensional projective subspace [*8*]. It seems that similar considerations must hold in higher dimensions. For surfaces in P_3, the volume swept out by the tangent planes is not a meaningful notion.

The appropriate bundle constructions for a theory of the extrinsic invariants of developable surfaces were made in [*10*], and they lead to a generalization of Theorem 5.

Many questions remain unanswered. For instance, can analogues of Theorem 3 be found for the γ_i, $i > 1$, generalizing formula (3)? Can these homology formulas be generalized to integral formulas valid for maps of open manifolds? Can generalizations be found of the Plücker formula

$$\gamma_1 = \xi(\xi - 1) - 2\delta - 3\varkappa,$$

where δ is the number of double points of the curve? Such a generalization would involve the difficult study of the secants.

References

[*1*] Borel, A., and F. Hirzebruch: Characteristic classes and homogeneous spaces, II. Am. J. Math. **81**, 363—371 (1959).

[*2*] Chern, S. S.: Complex Manifolds. (Notes of E. H. Spanier.) Textos de Matemática, No. 5. Instituto de Física e Matemática, Universidade do Recife, 1959.

[*3*] — Geometry of submanifolds in a complex projective space. Symposium Internacional de Topologia Algebraica, Mexico City, 1958, 87—96.

[*4*] — Holomorphic mappings of complex manifolds. L'Enseign. Math. **7**, 179—187 (1961).

[*5*] — On the characteristic classes of complex sphere bundles and algebraic varieties. Amer. J. Math. **75**, 565—597 (1953).

[*6*] Hironaka, H.: Resolution of singularities of an algebraic variety over a field of characteristic zero. Ann. Math. **79**, 109—326 (1963).

[*7*] Hirzebruch, F.: Neue topologische Methoden in der algebraischen Geometrie. Berlin-Göttingen-Heidelberg: Springer 1956.

[*8*] Lane, E. P.: Projective Differential Geometry of Curves and Surfaces. University of Chicago Press 1932, p. 274. (Ed. of 1942, p. 438.)

[*9*] Nakano, S.: Tangential vector bundle and Todd canonical systems of an algebraic variety. Memoirs of the College of Science, University of Kyoto, Series A, 29, Mathematics No. 2, 1955, 145—149.

[*10*] Pohl, W. F.: Differential geometry of higher order. Topology **1**, 169—211 (1962).

[*11*] Porteous, I. R.: Blowing up Chern classes. Proc. Cambridge Phil. Soc. **56**, part 2, 118—124 (1960).

[*12*] — Simple singularities of maps. Technical report, Columbia Univ., 1962.

[*13*] Santaló, L. A.: Integral geometry in hermitian spaces. Amer. J. Math. **74**, 423—434 (1952).

[*14*] Wirtinger, W.: Eine Determinantenidentität und ihre Anwendung auf analytische Gebilde in euclidischer und hermitischer Maßbestimmung. Monatshefte Math. u. Physik **44**, 343—365 (1936).

Department of Mathematics
Stanford University
Stanford, California

Department of Mathematics
University of Minnesota
Minneapolis, Minnesota

Some Properties of Pseudo-conformal Images of Circular Domains in the Theory of Two Complex Variables*

By

S. BERGMAN[1]

With 2 Diagrams

1. Distinguished boundary sets

A biholomorphic mapping

$$\boldsymbol{W}: \qquad z_k^* = z_k^*(z_1, z_2), \qquad k = 1, 2, \tag{1}$$

of a domain $\mathfrak{B}$, which is one-to-one and continuous in $\overline{\mathfrak{B}}$, will be called a PCT (pseudo-conformal transformation) of $\overline{\mathfrak{B}}$.

Two bounded simply connected domains of the space of two complex variables, in general, cannot be mapped pseudo-conformally onto each other.

If the domain $\mathfrak{B}$ and the class $\boldsymbol{q} \equiv \boldsymbol{q}(\mathfrak{B}^*)$ of domains $\mathfrak{B}^*$ are given, there arises the problem to decide whether $\mathfrak{B}$ can be mapped onto $\mathfrak{B}^* \in \boldsymbol{q}$ and to determine the function pair (1) transforming $\mathfrak{B}$ onto $\mathfrak{B}^*$ if the mapping $\boldsymbol{W}(\mathfrak{B}) = \mathfrak{B}^*$ exists.

Remark. In the case of doubly-connected domains in the space of one complex variable, an analogous problem is to determine the condition for the given domain[2] $\mathfrak{B}^2$ to be mapped onto the ring domain $\mathfrak{B}^{*2}$ and to compute the mapping function, if the mapping is possible. In this case $\boldsymbol{q} \equiv \boldsymbol{q}(\mathfrak{B}^{*2})$ is the class of domains $[0 < a_1 < |z^*| < a_2 \leqq 1]$, see [Z. *1*]. In general, the domains $\mathfrak{B}^* \in \boldsymbol{q}$ cannot be mapped onto each other by a PCT.

Using the theory of the kernel function, this question was treated under some additional assumptions in [B. *9*], p. 48, in the case where $\boldsymbol{q} = \boldsymbol{q}_0$ is the class of REINHARDT circular domains (see the Definition on p. 33/34). The approach was further developed and generalized by SPRINGER [S. *2*] and in [B. *10*].

The case where $\boldsymbol{q} = \boldsymbol{q}_1$ is the class of general circular domains was considered in [B. *13*] ($\boldsymbol{q}_0$ is a subclass of $\boldsymbol{q}_1$).

---

* Received May 29, 1964. This work was supported in part by the National Science Foundation Grant 2735.

[1] The author wishes to express his thanks to Dr. K. T. HAHN for his help and valuable suggestions in connection with this paper.

[2] The upper index n, $n = 1, 2, 3$, indicates the dimension of the set. If $n = 0$ or 4, or if the dimension of the set in the context is of no importance, the index n is omitted.

In the present paper a new method for answering the above question in the case of the class q_1 is presented. This new method can be applied to cases not solved by the considerations of [B. *13*].

The introduction and the study of distinguished sets is one of the useful tools for solving our problem.

Let $\{f\}$ be a class of functions defined in $\overline{\mathfrak{B}}$ and possessing a property $\mathscr{P}$ which is preserved under PCT's of $\overline{\mathfrak{B}}$. We call the set $\mathfrak{S}$ of $\overline{\mathfrak{B}}$ a distinguished set with respect to the property $\mathscr{P}$ if $\mathscr{P}$ holds for all domains $\boldsymbol{W}(\mathfrak{B})$, with $\mathfrak{S}$ replaced by $\boldsymbol{W}(\mathfrak{S})$.

Distinguished boundary sets have been defined using either $\boldsymbol{C}(\mathfrak{B})$ or $\boldsymbol{L}^2(\mathfrak{B})$.

$\boldsymbol{C}(\mathfrak{B})$ is the class of functions holomorphic in $\mathfrak{B}$ and continuous in $\overline{\mathfrak{B}}$.

$\boldsymbol{L}^2(\mathfrak{B})$ is the class of functions $f(z_1, z_2)$ holomorphic in $\mathfrak{B}$ for which

$$\int_{\mathfrak{B}} |f|^2 d\omega < \infty . \tag{2}$$

$d\omega$ is the volume element, $\int$ is the Lebesgue integral.

The distinguished sets obtained using the class $\boldsymbol{C}(\mathfrak{B})$ have been introduced already in [B. *2*], where the smallest maximum surface has been considered. The smallest maximum boundary is the (smallest) set $\mathfrak{m}$ consisting of boundary points with the properties a) and b).

a) For every function $f \in \boldsymbol{C}(\mathfrak{B})$

$$\max_{\zeta \in \mathfrak{m}} |f(\zeta)| \geqq |f(z)|, \quad z = (z_1, z_2) \in \overline{\mathfrak{B}}. \tag{3a}$$

b) To every point ζ_0 of $\mathfrak{m}$ there exists a function $f \in \boldsymbol{C}(\mathfrak{B})$, such that

$$|f(\zeta_0)| > |f(z)|, \quad z \in \overline{\mathfrak{B}} - \zeta_0 . \tag{3b}$$

The introduction of the smallest maximum boundary, which represents only a part of the boundary $\mathfrak{b}^3$ of $\mathfrak{B}$ (i.e., in the case where $\mathfrak{b}^3 - \mathfrak{m}$ is not empty) enables us to derive an interesting generalization of the Schwarz lemma. See [B. *4*], p. 76, [B. *11*], Sec. 2.

An analytic polyhedron $\mathfrak{P}$, i.e., a domain which is bounded by n segments, $n < \infty$,

$$\bar{\mathfrak{e}}^3_\varkappa = \bigcup_{\varphi_\varkappa = \varphi_{\varkappa 1}}^{\varphi_{\varkappa 4}} \mathfrak{J}^2_\varkappa(s_\varkappa \exp((i\varphi_\varkappa))), \tag{4}$$

$$\mathfrak{J}^2_\varkappa(t_\varkappa) = \{z_k = h_{k\varkappa}(\hat{Z}_\varkappa, t_\varkappa)\}, \quad h_{k\varkappa}(\hat{Z}_\varkappa, t_\varkappa) \equiv h_{k\varkappa}(\hat{Z}_\varkappa, t_\varkappa, \bar{t}_\varkappa), \quad t_\varkappa = s_\varkappa e^{i\varphi_\varkappa},$$

$$k = 1, 2, \quad \varkappa = 1, 2, \ldots, n,$$

of analytic hypersurfaces, has the smallest maximum boundary

$$\mathfrak{J}^2 = \bigcup_{s=1}^{n} \bigcup_{\substack{\varkappa=1 \\ s \neq \varkappa}}^{n} \mathfrak{J}^2_{s\varkappa}, \quad \mathfrak{J}^2_s = \bar{\mathfrak{e}}^3_s \cap \bar{\mathfrak{e}}^3_\varkappa . \tag{5}$$

When considering segments $\mathfrak{e}_\varkappa^3$ among others, we can distinguish the following two cases:

$$\begin{aligned} &\text{I.}\quad \mathfrak{J}_\varkappa^2(s_\varkappa \exp(i\psi_\varkappa^{(0)})) \cap \mathfrak{J}_\varkappa^2(s_\varkappa \exp(i\psi_\varkappa^{(1)})) = \emptyset \quad \text{for} \quad \psi_\varkappa^{(0)} \neq \psi_\varkappa^{(1)}. \\ &\text{II.}\quad \mathfrak{J}_\varkappa^2(s_\varkappa \exp(i\psi_\varkappa)),\ 0 \leqq \psi_\varkappa \leqq 2\pi, \text{ have a common point } P_\varkappa. \end{aligned} \tag{6}$$

It is of interest to consider functions $f \in C(\mathfrak{P})$ which omit laminarly in $(n-1)$ segments $\bar{\mathfrak{e}}_\varkappa^3$, $\varkappa = 1, 2, \ldots, n-1$, two values $A_{\varkappa 1}(\psi_\varkappa)$ and $A_{\varkappa 2}(\psi_\varkappa)$, $A_{\varkappa 1}(\psi_\varkappa) \neq A_{\varkappa 2}(\psi_\varkappa)$. In this case one obtains an upper bound for $|f(z_1, z_2)|$ in terms of $A_{\varkappa k}(\psi_\varkappa)$, $k = 1, 2$, and the values of f along $(n-1)$ curves $\mathfrak{c}_\varkappa^1$ lying in $\bar{\mathfrak{e}}_\varkappa^3$, $\varkappa = 1, 2, \ldots, n-1$.

Remark. "Laminarly" means that the function f omits in every lamina $\mathfrak{J}_\varkappa^2(s_\varkappa \exp(i\psi_\varkappa))$ of $\bar{\mathfrak{e}}_\varkappa^3$ the values $A_{\varkappa 1}(\psi_\varkappa)$ and $A_{\varkappa 2}(\psi_\varkappa)$. This property will be called the property $\mathscr{P}_1$ of the functions $f \in C(\mathfrak{P})$.

An interesting simplification arises in the case II. In this case one can replace the values of f along $\mathfrak{c}_\varkappa^1$ by the value of $f \in C(\mathfrak{P})$ at the point $P_\varkappa$. The points $P_\varkappa$ will be called the distinguished boundary points with respect to the property $\mathscr{P}_1$, i.e., to the property of f to omit laminarly two values $A_{\varkappa 1}(\psi_\varkappa)$ and $A_{\varkappa 2}(\psi_\varkappa)$ in $\bar{\mathfrak{e}}_\varkappa^3$, $\varkappa = 1, 2, \ldots, n-1$. For details, see for instance [B. *12*], pp. 15—21, where a polyhedron bounded by two analytic hypersurfaces is considered.

Remark. Instead of using the property $\mathscr{P}_1$, one can also use the property $\mathscr{P}_2$, namely, we assume that in every lamina $\mathfrak{J}_\varkappa^2(s_\varkappa \exp(i\psi_\varkappa))$ the function $f \in C(\mathfrak{P})$ is schlicht, i.e., assumes every value at most once. See also [B. *5*], [B. *6*] § 3, CHARZYNSKI [C. *1*] and ŚLADKOWSKA [S. *1*].

When using the class $L^2(\mathfrak{B})$ in the study of PCT's, it is useful to introduce the kernel function $K_\mathfrak{B}$ of the domain $\mathfrak{B}$, the metric

$$ds_\mathfrak{B}^2 = \sum T_{m\bar{n}}\, dz_m\, d\bar{z}_n, \quad T_{m\bar{n}} = \frac{\partial^2 \log K}{\partial z_m\, \partial \bar{z}_n}, \quad K \equiv K_\mathfrak{B}, \tag{7}$$

which is invariant with respect to PCT's and various invariants, e.g.,

$$\begin{gathered} J^{(1)}(z_1, z_2) = \frac{\begin{vmatrix} T_{1\bar{1}} & T_{1\bar{2}} \\ T_{2\bar{1}} & T_{2\bar{2}} \end{vmatrix}}{K}, \quad J^{(2)}(z_1, z_2) = \frac{\begin{vmatrix} S_{1\bar{1}}^{(1)} & S_{1\bar{2}}^{(1)} \\ S_{2\bar{1}}^{(1)} & S_{2\bar{2}}^{(1)} \end{vmatrix}}{K}, \quad S_{m\bar{n}}^{(1)} = \frac{\partial^2 J^{(1)}}{\partial z_m\, \partial \bar{z}_n}, \\ \ldots J^{(p+1)}(z_1, z_2) = \frac{\begin{vmatrix} S_{1\bar{1}}^{(p)} & S_{1\bar{2}}^{(p)} \\ S_{2\bar{1}}^{(p)} & S_{2\bar{2}}^{(p)} \end{vmatrix}}{K}, \quad S_{m\bar{n}}^{(p)} = \frac{\partial^2 J^{(p)}}{\partial z_m\, \partial \bar{z}_n}, \ldots \end{gathered} \tag{8}$$

The behavior of K, $ds_\mathfrak{B}^2$ and the invariants can be used for the definition and classification of boundary points. In the case of an existence domain, the kernel function becomes in general infinite at the boundary. In the case of an analytic polyhedra, the kernel function becomes infinite of the fourth order at the points of the smallest maximum surface $\mathfrak{m} = \mathfrak{F}^2$, and of the second order at the remaining part of the boundary

$$\mathfrak{b}^3 - \mathfrak{F}^2, \quad \mathfrak{b}^3 = \bigcup_{\nu=1}^{n} \bar{\mathfrak{e}}_\nu^3. \tag{9}$$

Similar results hold for the invariant metric, the invariants, etc. See [B. *3*], [B. *8*], [B. *9*]. It is natural to study also *interior* distinguished sets, i.e., *interior* sets which can be characterized by a property which remains preserved under PCT's $\boldsymbol{W}$. The critical sets of invariants $J_{\mathfrak{B}}^{(p)}$ are sets of this kind.

The center[3] O of a circular domain $\mathfrak{C}$ is a distinguished point and the problem arises to characterize this point in a way invariant under PCT's $\boldsymbol{W}$. Such a characterization makes it possible to determine $\boldsymbol{W}(O)$ in a domain $\boldsymbol{W}(\mathfrak{C})$. In the next section we shall discuss this problem.

2. Interior distinguished sets in $\boldsymbol{W}(\mathfrak{C})$

The critical sets of the invariants $J_{\mathfrak{B}}^{(p)}(z_1, z_2)$, $p = 1, 2, \ldots$, see (1.8)[4], represent an important class of interior distinguished sets of $\mathfrak{B}$. When studying these sets, we can distinguish a number of special cases.

I. The invariants $J_{\mathfrak{B}}^{(p)}(z_1, z_2)$, $p = 1, 2, \ldots$, are constants. This is obviously the case if the domain is homogeneous, i.e., admits the group of automorphism (PCT onto itself), transforming an arbitrary point P of $\mathfrak{B}$ into another point P' of $\mathfrak{B}$. The question, whether the domain $\mathfrak{B}$ must be a homogeneous one if $J_{\mathfrak{B}}^{(1)}(z_1, z_2)$ is a constant, is not clarified. In this case $K_{\mathfrak{B}}$ satisfies the equation (1.3) of [B. *13*].

II. $J^{(1)} \equiv J_{\mathfrak{B}}^{(1)}(z_1, z_2)$ is not a constant, but $J^{(1)}$ and $J^{(2)}$ are *not* independent.

III. $J_{\mathfrak{B}}^{(p)}(z_1, z_2)$, $p = 1$ and 2, are nonconstant and are independent from each other. $J_{\mathfrak{B}}^{(p)}(z_1, z_2)$, $p = 1$ and 2, are analytic functions of four real variables and for our purposes it is of interest to consider the cases where the $J^{(p)}(z_1, z_2)$ are independent in the neighborhood of some interior distinguished set.

Remark. We note that $J_{\mathfrak{B}}^{(1)}(z_1, z_2) = 1/\boldsymbol{J}(z_1, z_2)$ where $\boldsymbol{J}(z_1, z_2)$ is the invariant considered in [B. *7*] p. 54 [B. *9*] p. 19 and in [B. *13*].

We shall here consider a special class of domains $\mathfrak{B}$, namely those which are pseudo-conformally equivalent to a circular domain $\mathfrak{C}$, $(\mathfrak{B} \approx \mathfrak{C})$. A simply connected domain which admits the group of transformations

$$z_k^* = z_k e^{i\varphi}, \quad 0 \leqq \varphi \leqq 2\pi, \quad k = 1, 2, \tag{1}$$

is called a circular domain with the center at the origin $O = (0, 0)$. The simply-connected domains which admit the (two-parameter) group

[3] In the present paper we assume that the origin is the center O of $\mathfrak{C}$.

[4] (1.8) = Eq. (8) of Sec. 1.

of transformations

$$z_k^* = z_k e^{i\varphi_k}, \quad 0 \leqq \varphi_k \leqq 2\pi, \quad k = 1, 2, \tag{2}$$

is called a REINHARDT circular domain with the center at O.

Example. The REINHARDT circular domains

$$\mathfrak{R} = [|z_2|^{2/p} + |z_1|^2 < 1], \quad p \neq 1, \quad p > 0, \tag{2a}$$

represent an example of domains of the class II, for which $J_{\mathfrak{R}}^{(p)}$, $p = 1, 2$, are not independent.

A formal computation (see [B. *7*] p. *21*, [B. *13*]) shows that

$$K_{\mathfrak{R}}(z_1, z_2; \bar{z}_1, \bar{z}_2) = \frac{(1 - |z_1|^2)^{p-2} D}{\pi^3 C^3}, \quad p > 0, \tag{3}$$

where

$$C = (1 - |z_1|^2)^p - |z_2|^2, \quad D = (p + 1)(1 - |z_1|^2)^p + (p - 1)|z_2|^2,$$

and

$$J_{\mathfrak{R}}^{(1)}(z_1, z_2) = \frac{9\pi^2}{2}\left[1 - \left(\frac{kC}{D}\right)^4\right], k = \left(\frac{p^2 - 1}{3}\right)^{1/2}. \tag{4}$$

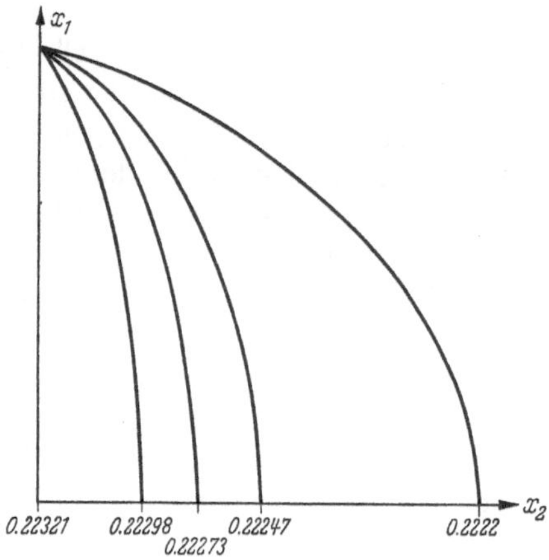

Fig. 1. The level lines of the invariant $\pi^2/J_{\mathfrak{R}}^{(1)}(z_1, z_2)$, $y_1 = y_2 = 0$.

In Fig. 1 the intersection of the level hypersurfaces $\pi^2/J_{\mathfrak{R}}^{(1)}(z_1, z_2) = c$, $x_1 \geqq 0, x_2 \geqq 0$, with the plane $y_1 = y_2 = 0$, $p = 1.5$ are indicated.

If the coefficients a_n do not satisfy certain relations, the REINHARDT circular domain

$$|z_2|^2 < G(|z_1|^2) \equiv \sum_{n=0}^{\infty} a_n |z_1|^{2n},$$

$$a_0 = 1, \quad \sum_{n=0}^{\infty} a_n = 0 \tag{5}$$

represents a class of domains for which $J_{\mathfrak{R}}^{(p)}$, $p = 1$ and 2, are nonconstant and independent from each other.

A formal computation yields the series development

$$J_{\mathfrak{R}}^{(1)}(z_1, z_2) = \sum_{n=0}^{\infty} \sum_{M+N=n} E_{MN}^{(1)} |z_1|^{2M} |z_2|^{2N},$$

$$E_{MN}^{(1)} = B_{00}^{-3} \sum \prod_{\nu=1}^{2} B_{m_\nu n_\nu} L_{m_1 n_1 m_2 n_2}^{(MN)} + B_{00}^{-4} \sum \prod_{\nu=1}^{3} B_{m_\nu n_\nu} L_{m_1 n_1 m_2 n_2 m_3 n_3}^{(MN)}$$

$$+ \cdots + B_{00}^{-(3+M+N)} \sum \prod_{\nu=1}^{M+N+2} B_{m_\nu n_\nu} L_{m_1 n_1 m_2 n_2 \ldots m_{M+N+2}\, n_{M+N+2}}^{(MN)}. \tag{6}$$

Here

$$\sum \equiv \sum_{\substack{\Sigma m_\nu = M+1 \\ \Sigma n_\nu = N+1}}, \quad B_{mp}^{-1} = \sum_{\alpha_1, \alpha_2, \ldots, \alpha_{p+1} = 0}^{N} \frac{\prod_{\nu=1}^{p+1} a_{\alpha_n}}{\sum_{n=1}^{p+1} \alpha_n + m + 1}$$

and $L_{m_1 n_1 \ldots m_\mu n_\mu}^{(MN)}$ are numerical constants.

In the case of REINHARDT circular domains (5) it is of interest to consider the cases where

$$\begin{vmatrix} \dfrac{\partial \tilde{J}^{(1)}}{\partial \varrho_1} & \dfrac{\partial \tilde{J}^{(1)}}{\partial \varrho_2} \\ \dfrac{\partial \tilde{J}^{(2)}}{\partial \varrho_1} & \dfrac{\partial \tilde{J}^{(2)}}{\partial \varrho_2} \end{vmatrix} \neq 0, \tag{7}$$

$$\tilde{J}^{(p)} \equiv \sum_{n=0}^{\infty} \sum_{M+N=n} E_{MN}^{(p)} \varrho_1^M \varrho_2^N, \quad J_{\mathfrak{R}}^{(p)}(z_1, z_2) = \sum_{M=0}^{\infty} \sum_{N=0}^{\infty} E_{MN}^{(p)} |z_1|^{2M} |z_2|^{2N},$$

in a sufficiently small neighborhood of the origin. As indicated in [B. *13*], an isolated maximum or minimum of $J_{\mathfrak{B}}^{(1)}(z_1, z_2)$ in a domain $\mathfrak{B} = \boldsymbol{W}(\mathfrak{R})$ is the image $\boldsymbol{W}(O)$ of the center O.

If the intersection of the level hypersurfaces $J_{\mathfrak{R}}^{(p)}(z_1, z_2) = \text{const}$, $p = 1, 2$, passes through the center O of $\mathfrak{R}$ and the functional determinant in (7) does not vanish at O, then the intersection of the above hypersurfaces is an isolated point. In the neighborhood of every other point $Q \neq O$, the above hypersurfaces intersect along a line or along a torus. Therefore, considering both invariants, we obtain in this case a new procedure to determine the image of the center O in the domain $\boldsymbol{W}(\mathfrak{R})$ also in the case where the level hypersurfaces of $J_{\mathfrak{R}}^{(p)}(z_1, z_2) = \text{const}$, $p = 1, 2$, have a minimax at the center.

One of the problems of the theory of PCT's is to decide whether a given domain $\mathfrak{B}$ is pseudo-conformally equivalent to a circular domain. To answer these questions, it is sufficient to determine in $\mathfrak{B}$ the image $\boldsymbol{W}(O)$ of the center O of $\mathfrak{C}$. Indeed, if the image $P = \boldsymbol{W}(O)$ is known, we determine, using the mapping

$$v_1(z_1, z_2) = \frac{M^{010}(z_1, z_2; t_1, t_2)}{M^1(z_1, z_2; t_1, t_2)}, \quad v_2(z_1, z_2) = \frac{M^{001}(z_1, z_2; t_1, t_2)}{M^1(z_1, z_2; t_1, t_2)}, \tag{8}$$

the representative domain of $\mathfrak{B}$ with respect to the point $P = \boldsymbol{W}(O) = (t_1, t_2)$. This representative domain must be a circular domain if $\mathfrak{B}$ can be mapped pseudo-conformally onto a circular domain [B. *1*], [W. *1*], [B. *9*]. Here

$$M^{X_{00} X_{10} X_{01}}(z_1, z_2; t_1, t_2) = - \frac{\begin{vmatrix} 0 & K(z,\bar{t}) & K_{00 1\bar{0}}(z,\bar{t}) & K_{000\bar{1}}(z,\bar{t}) \\ X_{00} & K & K_{00\bar{1}\bar{0}} & K_{000\bar{1}} \\ X_{10} & K_{10\bar{0}\bar{0}} & K_{10\bar{1}\bar{0}} & K_{000\bar{1}} \\ X_{01} & K_{01\bar{0}\bar{0}} & K_{01\bar{1}\bar{0}} & K_{010\bar{1}} \end{vmatrix}}{\begin{vmatrix} K & K_{00\bar{1}\bar{0}} & K_{000\bar{1}} \\ K_{10\bar{0}\bar{0}} & K_{10\bar{1}\bar{0}} & K_{100\bar{1}} \\ K_{01\bar{0}\bar{0}} & K_{01\bar{1}\bar{0}} & K_{010\bar{1}} \end{vmatrix}}, \tag{9}$$

$$K_{mn\overline{M}\overline{N}} = \frac{\partial^{m+n+M+N} K(t_1, t_2; \bar{t}_1, \bar{t}_2)}{\partial t_1^m \, \partial t_2^n \, \partial \bar{t}_1^M \, \partial \bar{t}_2^N}, (X_{00}, X_{10}, X_{01}) = (0,1,0) \text{ and } = (0,0,1),$$

$$M^1(z_1, z_2; t_1, t_2) = \frac{K(z_1, z_2; \bar{t}_1, \bar{t}_2)}{K}.$$

See [B. *7*], p. 45, [B. *9*], p. 29, [F. *1*], p. 359.

Remark. The kernel function yields invariants of other types. In the case of (multiply-connected) domains $\mathfrak{B}$ including closed surfaces $\mathfrak{S}^2 = [z_k = z_k(u_1, u_2)]$, $k = 1, 2$, which cannot be reduced to a point in $\mathfrak{B}$, one obtains invariants (numbers) by evaluating the integrals

$$\omega(\mathfrak{S}^2_{1\bar{\zeta}}, \mathfrak{S}^2_{2z}) =$$

$$\iint\limits_{\mathfrak{S}^2_{1\bar{\zeta}}} \iint\limits_{\mathfrak{S}^2_{2z}} K(z_1, z_2; \bar{\zeta}_1, \bar{\zeta}_2) \frac{\partial(z_1, z_2)}{\partial(u_1^{(1)}, u_2^{(1)})} \frac{\partial(\bar{\zeta}_1, \bar{\zeta}_2)}{\partial(u_1^{(2)}, u_2^{(2)})} du_1^{(1)} du_2^{(1)} du_1^{(2)} du_2^{(2)}. \quad (10)$$

Here $\mathfrak{S}^2_{1\bar{\zeta}}$ is a surface in the $\bar{\zeta}_1, \bar{\zeta}_2$-space, $\mathfrak{S}^2_{2z}$ is a surface in the z_1, z_2-space.

3. A procedure for the determination of the image $W(O)$ of the center of a circular domain

In this section a procedure will be discussed to determine in certain cases the image $W(O)$ of the center O of $\mathfrak{C}_1 \in q_1$ in a domain $\mathfrak{B} = W(\mathfrak{C}_1)$.

We make at first a number of hypotheses about $\mathfrak{C}_1$.

1. $\mathfrak{C}_1$ is a bounded circular domain. The invariant $J^{(1)} \equiv J^{(1)}_{\mathfrak{C}_1}(z_1, z_2)$ is not constant in $\mathfrak{C}_1$ and assumes on the boundary $\mathfrak{c}_1^3$ of $\mathfrak{C}_1$ the constant limit values c_0. The metric (1.7) of $\mathfrak{C}_1$ is complete. See [H. *1*].

Remark. The conditions in order that $J^{(1)}_{\mathfrak{B}}(z_1, z_2)$ has the boundary values $c_0 = \dfrac{9\pi^2}{2}$ at the boundary point P have been given in [B. *3*], [B. *9*], p. 12, [H. *2*]. The condition for $c_0 = 4\pi^2$ has been established in [B. *3*], [B. *8*], [B. *9*], p. 25.

2. There exists a hypersurface $\mathfrak{c}_2^3 = [J^{(1)}_{\mathfrak{C}_1}(z_1, z_2) = c_2]$, $c_2 < c_0$, which is the boundary of the (simply connected) circular domain $\mathfrak{C}_2$, $\overline{\mathfrak{C}}_2 \subset \mathfrak{C}_1$.

We shall now describe a geometrical construction to be applied in the following. Using the non-Euclidean metric whose line element is given by

$$ds_{\mathfrak{C}_1} = [\sum T_{m\bar{n}} \, dz_m \, d\bar{z}_n]^{1/2}, \quad T_{m\bar{n}} \equiv \frac{\partial^2 \log K}{\partial z_m \, \partial \bar{z}_n}, \quad K = K_{\mathfrak{C}_1},$$

we draw around every point $P_\nu \in \mathfrak{c}_2^3$ the non-Euclidean hypersphere $\mathfrak{h}(\varrho, P_\nu)$ of radius ϱ. If ϱ is sufficiently large, there will exist a subdomain $\mathfrak{d}(\varrho, \mathfrak{c}_2^3)$ of $\mathfrak{C}_2$ for every ϱ, such that every point Q of $\mathfrak{d}(\varrho, \mathfrak{c}_2^3)$ is covered by *all* $\mathfrak{h}(\varrho, P_\nu)$, $P_\nu \in \mathfrak{c}_2^3$. Further, we associate with every ϱ_1 the set $\mathfrak{l}(\varrho_1, \mathfrak{c}_2^3)$ of points which belong to $\mathfrak{d}(\varrho_1, \mathfrak{c}_2^3)$ but not to $\mathfrak{d}(\varrho, \mathfrak{c}_2^3)$, $\varrho < \varrho_1$.

Remark. $\mathfrak{d}(\varrho, \mathfrak{c}_2^3)$ is the set of points $Q \in \mathfrak{C}_2$ possessing the property that $D_{\mathfrak{C}_1}(Q, P_\nu) \leqq \varrho$ for all $P_\nu \in \mathfrak{c}_2^3$. Here $D_{\mathfrak{C}_1}(Q, P_\nu)$ is the non-Euclidian distance between Q and P_ν.

Let ϱ_0 be the maximum[5] of the (non-Euclidian) distances between O and the points of $\mathfrak{c}_2^3$.

Theorem. $\mathfrak{l}(\varrho_0, \boldsymbol{W}(\mathfrak{c}_3^2))$ *includes the isolated point P and $P = \boldsymbol{W}(O)$.*

Proof. We consider at first the situation in $\mathfrak{C}_1$. Since ϱ_0 is the maximum distance between O and points $P_\nu \in \mathfrak{c}_2^3$, O will belong to every $\mathfrak{h}(\varrho_0, P_\nu), P_\nu \in \mathfrak{c}_2^3$. The (non-Euclidian) distance between the points $P_\nu \in \mathfrak{c}_2^3$ and a fixed point $T \in \mathfrak{C}_2$ is a continuous function of P_ν.

For every $\varrho < \varrho_0$ there must exist at least one point, say P_1 of $\mathfrak{c}_2^3$, whose distance from O is larger than ϱ and therefore O will not be covered by $\mathfrak{h}(\varrho, P_1)$.

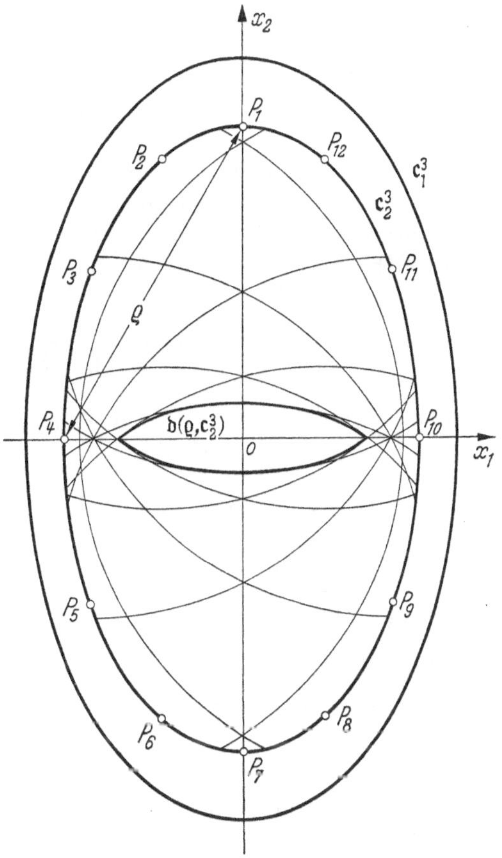

Fig. 2. The hypersurfaces[6] $\mathfrak{c}_1^3$, $\mathfrak{c}_2^3$ and the domain $\mathfrak{d}(\varrho, \mathfrak{c}_2^3)$, $\varrho > \varrho_0$.

Thus ϱ_0 is the smallest value of ϱ for which $\mathfrak{l}(\varrho,\mathfrak{c}_2^3)$ includes the point O. The orbit $\mathfrak{T}(O)$ is the *point* O. $\mathfrak{l}(\varrho, \mathfrak{c}_2^3)$ for $\varrho \neq \varrho_0$ does not include an isolated point since $\mathfrak{l}(\varrho, \mathfrak{c}_2^3)$ does not include the origin O and the orbit $\mathfrak{T}(Q)$ of every $Q \neq O$ is a closed *curve*.

If starting from $\varrho = 0$ we increase ϱ continuously, there will exist the smallest $\varrho = \varrho_0$ such that $\mathfrak{l}(\varrho_0, \mathfrak{c}_2^3)$ includes an isolated point. This

[5] Since $J^{(1)}_{\mathfrak{C}_1}(z_1, z_2) = c_2$ lies inside of $\mathfrak{C}_1$, the non-Euclidean distance between O and P, $P \in \mathfrak{c}_2^3$, is a continuous function of P. The maximum of non-Euclidean distances $D_{\mathfrak{C}_1}(O, P)$, $P \in \mathfrak{c}_2^3$, exists since $\mathfrak{c}_2^3$ is compact.

[6] Segments of the boundaries of non-Euclidean hyperspheres with the centers at P_ν are replaced by segments of Euclidean circles in Fig. 2. Also, instead of the three parameter family of domains $\mathfrak{h}(\varrho\ P_\nu)$, only a finite number of segments of circles are drawn. Note that the totality of points P_ν forms the hypersurface $\mathfrak{c}_3^2$. $\mathfrak{d}(\varrho, \mathfrak{c}_3^2)$ appears as a two-dimensional domain in Fig. 2.

point is the center O. Since our construction is invariant under PCT's $\mathfrak{W}$, and the dimension of the set $\mathfrak{T}(Q)$ is preserved under PCT's $\mathfrak{W}$, ϱ_0 will be the smallest value of ϱ for which $\mathfrak{l}(\varrho, \mathfrak{W}(\mathfrak{c}_2^3))$ includes an isolated point P and $P = \mathfrak{W}(O)$.

4. A remark on the behavior of $J_{\mathfrak{B}}^{(1)}$ at boundary points

In Section 3, p. 36, we made an assumption about the behavior of the invariant $J_{\mathfrak{B}}^{(1)}$ in the neighborhood of the boundary. It is of interest to indicate a procedure which enables us to investigate these properties. One of the methods is based on the use of the relation

$$J_{\mathfrak{B}}^{(1)} = \frac{[\lambda_{\mathfrak{B}}^{1}(z)]^3}{\lambda_{\mathfrak{B}}^{01}(z)\,\lambda_{\mathfrak{B}}^{001}(z)}, \quad z \equiv (z_1, z_2) \in \mathfrak{B} \tag{1}$$

where $\lambda_{\mathfrak{B}}^{X_{00}X_{10}X_{01}}$ is the minimum value of the integral (1.2), see [B. *3*], [B. *7*], p. 55, under the conditions

$$f(z_1, z_2) = X_{00}, \quad f_{z_1}(z_1, z_2) = X_{10}, \quad f_{z_2}(z_1, z_2) = X_{01}. \tag{2}$$

Since the inequalities[7]

$$\lambda_{\mathfrak{J}}^{X_{00}X_{10}X_{01}}(z) \leqq \lambda_{\mathfrak{B}}^{X_{00}X_{10}X_{01}}(z) \leqq \lambda_{\mathfrak{A}}^{X_{00}X_{10}X_{01}}(z), \qquad z \in \mathfrak{J} \subset \mathfrak{B} \subset \mathfrak{A}, \tag{3}$$

hold, one introduces the domains of comparison $\mathfrak{J}$ and $\mathfrak{A}$, i.e., domains possessing the property that

$$\mathfrak{J} \subset \mathfrak{B} \subset \mathfrak{A} \tag{4}$$

and for which the kernel functions $K_{\mathfrak{J}}$ and $K_{\mathfrak{A}}$ can be expressed in closed form.

Applying to the hyperspheres $|z_1 - \sigma_k^{-1}|^2 + |z_2|^2 < \sigma_k^{-2}$, $k = 1, 2$, the transformations

$$z_1^* = z_1(1 + \alpha_1 z_1)^{-1}, \quad z_2^* = \frac{z_2(1 + \alpha_1 z_1)}{1 + (\alpha_1 + \beta_1) z_1} \tag{5a}$$

and

$$z_1^* = \frac{z_1}{(1 - \alpha_2 z_1)}, \quad z_2^* = z_2 \left[\frac{1 + (\beta_2 - \alpha_2) z_1}{1 - \alpha_2 z_1}\right] \tag{5b}$$

we obtain in the case of points of third order the domains of comparison $\mathfrak{J}$ and $\mathfrak{A}$, respectively. Here $\alpha_k, \beta_k, \sigma_k$, $k = 1, 2$, are conveniently chosen

---

[7] Naturally, for $\lambda_{\mathfrak{B}}^{X_{00}}(z)$ and $\lambda_{\mathfrak{B}}^{X_{00}X_{10}}(z)$ inequalities analogous to (3) hold.

constants. Using (1) and (3) we get the bounds

$$\begin{aligned}
&\frac{2}{9\pi^2}\left(\frac{\sigma_2}{\sigma_1}\right)^3\Big\{1+3[-\beta_2-\beta_1+\alpha_2+\alpha_1](z_1+\bar{z}_1)+\\
&\qquad+(-18\alpha_2-18\alpha_1+9\sigma_2-9\sigma_1)\frac{|z_1|^2}{z_1+\bar{z}_1}+\\
&\qquad+(9\sigma_2-9\sigma_1)\frac{|z_2|^2}{z_1+\bar{z}_1}+\Omega\Big\}\leqq\frac{1}{J_{\mathfrak{B}}^{(1)}(z_1,z_2)} \qquad (6)\\
&\qquad\leqq\frac{2}{9\pi^2}\left(\frac{\sigma_1}{\sigma_2}\right)^3\Big\{1+3[-\alpha_2-\alpha_1+\beta_2+\beta_1](z_1+\bar{z}_1)+\\
&\qquad+(18\alpha_1+18\alpha_2+9\sigma_1-9\sigma_2)\frac{|z_1|^2}{z_1+\bar{z}_1}+\\
&\qquad+(9\sigma_1-9\sigma_2)\frac{|z_2|^2}{z_1+\bar{z}_1}+\Omega\Big\},\quad \lim_{(z_1 z_2)\to O}[(z_1+\bar{z}_1)\Omega]=0
\end{aligned}$$

As in the case of the hypersphere, one can apply the transformations (5a) and (5b) to the domains $\mathfrak{R}$ introduced in (2.2a) to obtain useful new domains of comparison $\mathfrak{Z}$ and $\mathfrak{A}$.

Using (1) and the latter domains of comparison, we can obtain conditions for the boundary points P ensuring that $J_{\mathfrak{B}}^{(1)}(z_1, z_2)$ decreases if we move inside $\mathfrak{C}_1$ along the interior normal at P.

These considerations will be discussed in more detail in a later paper. See [B, *14*].

References

[B. *1*] Bergman, S.: Über die Existenz von Repräsentantenbereichen. Math. Ann. **103**, 430—446 (1929).

[B. *2*] — Über die Abbildung durch Paare von Funktionen von zwei komplexen Veränderlichen. Jber. Deutsch. Math. Verein. **39**, 24—27 (1930).

[B. *3*] — Über die Kernfunktion eines Bereiches und ihr Verhalten am Rande. J. Reine Angew. Math. **169**, 1—42 (1933); **172**, 89—128 (1934).

[B. *4*] — Über eine in gewissen Bereichen mit Maximumfläche gültige Integraldarstellung der Funktionen zweier komplexer Variabler, I, II. Math. Z. **39**, 76—94; 605—608 (1934).

[B. *5*] — Über eine Abschätzung von meromorphen Funktionen zweier komplexer Veränderlicher in Bereichen mit ausgezeichneter Randfläche. Trav. Inst. Math. Tbilissi **1**, 187—204 (1937).

[B. *6*] — Über das Verhalten der Funktionen von zwei komplexen Veränderlichen in Gebieten mit einer ausgezeichneten Randfläche. Mathematica **14**, 107—123 (1938).

[B. *7*] — Sur les fonctions orthogonales de plusieur variables complexes avec les applications à la théorie des fonctions analytiques. Interscience Publishers 1941 and Mémorial des Sciences Mathématique 106, Paris 1947.

[B. *8*] — The behavior of the kernel function at boundary points of the second order. Amer. J. Math. **65**, 679—700 (1943).

[B. *9*] — Sur la fonction-noyau d'un domaine et ses applications dans la théorie des transformations pseudo-conformes. Mémor. Sci. Math. **108**, Paris (1948).

[B. 10] Bergman, S.: Distinguished boundary sets in the theory of functions of two complex variables. General topology and its relations to modern analysis and algebra. Proc. of the Symposium, Prague, 1961, 79—86.
[B. 11] — The kernel function and conformal mapping. Math. Surveys **5** (1950).
[B. 12] — The distinguished boundary sets and value distribution of functions of two complex variables. Ann. Acad. Sci. Fenn. Ser. A. I **336/12**, 1—28 (1963).
[B. 13] — Distinguished sets of invariants in the theory of pseudo-conformal transformations. J. Analyse Math. **13** (1964) (to appear).
[B. 14] — On boundary behavior of some domain functionals in the theory of functions of two complex variables. Bull. Soc. Sci. Lettres Łódź (to appear).
[C. 1] Charzyński, Z.: Bounds for analytic functions of two complex variables. Math. Z. **75**, 29—35 (1961).
[F. 1] Fuks, B. A.: Special chapters in the theory of analytic functions of several complex variables. (Russian) Moscow 1963.
[H. 1] Hahn, K. T.: Minimum problems of the Plateau type in the Bergman metric space. Pacific Journal (1964 or 1965) (to appear).
[H. 2] Hörmander, Lars: Existence theorems for the ∂ operator by L^2 methods. Acta mathematica (to appear).
[S. 1] Śladkowska, J.: Bounds of analytic functions of two complex variables in domains with the Bergman-Shilov boundary. Pacific J. Math. **12**, 1435—1451 (1962).
[S. 2] Springer, G.: Pseudo-conformal transformations onto circular domains. Duke Math. J. **18**, 411—424 (1951).
[W. 1] Welke, H.: Über die analytischen Abbildungen von Kreiskörpern und Hartogsschen Bereichen. Math. Ann. **103**, 437—449 (1930).
[Z. 1] Zarankiewicz, K.: Über ein numerisches Verfahren zur konformen Abbildung zweifach zusammenhängender Gebiete. Z. Angew. Math. Mech. **14**, 97—104 (1934).

Department of Mathematics
Stanford University
Stanford, California

Holomorphic Imbeddings of Symmetric Domains into a Siegel Space*

By

I. Satake

1. A symmetric domain is a bounded domain in $\boldsymbol{C}^N$ which becomes a symmetric Riemannian space with respect to its Bergman metric. For a symmetric domain $\mathcal{D}$, we denote by G the connected component of the group of all analytic automorphisms of $\mathcal{D}$ (or, what is the same, that of the group of all isometries of $\mathcal{D}$ onto itself) with its natural topology and by K the isotropy subgroup of G at an (arbitrary) point $z_0 \in \mathcal{D}$. Then, as is well-known ([1], [2]), G is a (connected) semi-simple

* Received June 9, 1964.

Lie group of non-compact type with center reduced to the identity, K is a maximal compact subgroup of G and $\mathscr{D}$ is identified with the coset-space G/K. A symmetric domain $\mathscr{D}$ is decomposed uniquely into the direct product $\mathscr{D}_1 \times \cdots \times \mathscr{D}_s$ of irreducible symmetric domains $\mathscr{D}_i$ (i. e. the domains which cannot be decomposed any further) corresponding to the direct decomposition $G_1 \times \cdots \times G_s$ of G into simple components. Irreducible symmetric domains have been classified completely by E. CARTAN [*1*] into four main series (I), (II), (III) and (IV) of classical domains (see no 3 below) and two exceptional domains corresponding to the simple Lie groups of type (E_6) and (E_7). Among them, an irreducible symmetric domain of type (III), called a "Siegel space", is of particular interest, because it can be realized as the space $\mathscr{H}_n$ of all symmetric $n \times n$ complex matrices $Z = X + iY$ with the positive-definite imaginary part Y, and thus considered as a parameter-space of the family of (polarized) abelian varieties A_Z with the period matrix $(1_n, Z)$.

The purpose of the present note is to answer the following question raised recently by KUGA [*3*]. Here we shall restrict ourselves to stating only the results, for the proofs are to be published elsewhere [*4*].

Problem 1. For a given symmetric domain $\mathscr{D}$, determine all holomorphic isometries ϱ of $\mathscr{D}$ into a Siegel space $\mathscr{D}' = \mathscr{H}_n$ such that $\varrho(\mathscr{D})$ is totally geodesic in $\mathscr{D}'$.

Let us explain briefly the meaning of the problem. Let $G' = \mathrm{Sp}(2n, \boldsymbol{R})$ be the real symplectic group operating on $\mathscr{D}' = \mathscr{H}_n$ in the usual manner, K' the isotropy subgroup of G' at z_0' (say $i\,1_n$) and $\Gamma' = \mathrm{Sp}(2n, \boldsymbol{Z})$ the Siegel modular group. Then, two (polarized) abelian varieties A_{Z_1} and A_{Z_2} with $Z_1, Z_2 \in \mathscr{H}_n$ being isomorphic if and only if Z_1 and Z_2 are equivalent under Γ', we have actually a family of abelian varieties on the quotient space $\Gamma' \backslash \mathscr{D}' = \Gamma' \backslash G'/K'$. (Precisely speaking, what we have here is a quotient variety of such a family by a finite group of automorphisms.) Now, suppose that we have a faithful representation ϱ of G into G' such that $\varrho(K)$ is contained in K'. Then ϱ induces in a natural manner an isometry, denoted again by ϱ, of $\mathscr{D} = G/K$ into $\mathscr{D}' = G'/K'$ such that $\varrho(\mathscr{D})$ is totally geodesic in $\mathscr{D}'$; and, conversely, it is not hard to see that any isometry ϱ of $\mathscr{D}$ into $\mathscr{D}'$ with $\varrho(z_0) = z_0'$ and having this property can be obtained in this manner if we replace G by a suitable covering group with finite center. Suppose further we have a discrete subgroup Γ of G with compact quotient $\Gamma \backslash G$ such that $\varrho(\Gamma)$ is contained in Γ'. Then by means of the mapping from $\Gamma \backslash \mathscr{D}$ into $\Gamma' \backslash \mathscr{D}'$, induced by ϱ, we obtain, by pulling back the family of abelian varieties on $\Gamma' \backslash \mathscr{D}'$, one on $\Gamma \backslash \mathscr{D}$. Under these circumstances, KUGA has proved, among other things, that the family of abelian varieties thus obtained is *algebraic*, if and only if the injection ϱ of $\mathscr{D}$ into $\mathscr{D}'$ is holomorphic. Thus, to solve the above-mentioned problem would be the first step to the determination of all algebraic families of abelian varieties on the quotient space $\Gamma\backslash\mathscr{D}$ obtained in the manner described above.

2. It is now convenient to translate the problem into terms of Lie algebras. Let $\mathfrak{g}$ and $\mathfrak{k}$ be the Lie algebras of G and K, respectively, and let $\mathfrak{g} = \mathfrak{k} + \mathfrak{p}$ be the corresponding Cartan decomposition of $\mathfrak{g}$, where $\mathfrak{p}$ is the orthogonal complement of $\mathfrak{k}$ in $\mathfrak{g}$ with respect to the Killing form. Then, $\mathfrak{p}$, being identified with the tangent space of $\mathscr{D}$ at z_0, has a complex structure determined by that on $\mathscr{D}$, and it is wellknown that there exists uniquely an element H_0 in the center of $\mathfrak{k}$ such that $\mathrm{ad}(H_0)$ (adjoint representation of H_0) induces on $\mathfrak{p}$ this complex structure. (Incidentally among semi-simple Lie algebras of non-compact type, the existence of such an element H_0 characterizes those corresponding to symmetric domains.) Similarly, let $\mathfrak{g}' = \mathfrak{k}' + \mathfrak{p}'$ be a Cartan decomposition of $\mathfrak{g}'$, the Lie algebra of G', taken at z_0' and let H_0' be the element in the center of $\mathfrak{k}'$ giving the complex structure on $\mathfrak{p}'$. Then it is clear, from what we have mentioned in no 1, that an isometry ϱ of $\mathscr{D}$ into $\mathscr{D}' = \mathscr{H}_n$ with $\varrho(z_0) = z_0'$ and satisfying the conditions in Problem 1 gives rise, in a natural and one-to-one way, to a monomorphism, denoted again by ϱ, of $\mathfrak{g}$ into $\mathfrak{g}'$ satisfying the condition

$$(H_1) \qquad \varrho \circ \mathrm{ad}(H_0) = \mathrm{ad}(H_0') \circ \varrho ,$$

which clearly implies that $\varrho(\mathfrak{k}) \subset \mathfrak{k}'$, $\varrho(\mathfrak{p}) \subset \mathfrak{p}'$. Thus our problem is equivalent to the following one:

Problem 1′. For a given semi-simple Lie algebra of non-compact type $\mathfrak{g}$, corresponding to a symmetric domain, determine all faithful representations ϱ of $\mathfrak{g}$ into $\mathfrak{g}'$ (the Lie lagebra of $\mathrm{Sp}(2n, \boldsymbol{R})$) satisfying the condition (H_1).

For convenience in applications, we shall consider the problem in a slightly more general form. Namely, let us call a semi-simple Lie algebra (over $\boldsymbol{R}$) "of hermitian type" if all the non-compact simple components of it correspond to symmetric domains, and consider the following Problem 2. (In fact, for such a Lie algebra $\mathfrak{g}$, one can again define the element H_0 in the same way as above in taking as $\mathfrak{k}$ a maximal compact subalgebra of $\mathfrak{g}$.)

Problem 2. For a given semi-simple Lie algebra of hermitian type $\mathfrak{g}$, determine all representations ϱ of $\mathfrak{g}$ into $\mathfrak{g}'$ satisfying the condition (H_1).

It will be also useful to consider the following condition (H_2) which is stronger than (H_1):

$$(H_2) \qquad \varrho(H_0) = H_0' .$$

Actually these conditions can be formulated for any two semi-simple Lie algebras of hermitian type $\mathfrak{g}$ and $\mathfrak{g}'$. We call two representations ϱ_1 and ϱ_2

of $\mathfrak{g}$ into $\mathfrak{g}'$ "(k)-equivalent", if there exists an element k' in K' such that $\varrho_2 = \mathrm{ad}(k') \circ \varrho_1$. The conditions (H_1) and (H_2) being invariant under (k)-equivalence, our problem is actually to determine the (k)-equivalence-classes of representations satisfying (H_1).

3. In order to fix notations, we give here a list of irreducible symmetric domains and the corresponding simple Lie algebras of the four main series in their usual matricial expressions. The notation $Z^{(p,q)}$ will mean that Z is a $p \times q$ (complex) matrix, and $Z^{(p)}$ will stand for $Z^{(p,p)}$. As usual, 1_p denotes the identity-matrix of degree p.

$$(\mathrm{I})_{p,q}\,(p, q \geqq 1) \quad \mathscr{D} = \{Z^{(p,q)} \mid 1_q - {}^t\bar{Z}Z \gg 0\},$$

$$\mathfrak{g} = \left\{ \begin{pmatrix} X_1^{(p)} & X_{12}^{(p,q)} \\ {}^t\bar{X}_{12} & X_2^{(q)} \end{pmatrix} \middle| \; {}^t\bar{X}_i = -X_i \,(i = 1, 2), \mathrm{tr}(X_1) + \mathrm{tr}(X_2) = 0 \right\},$$

$$\mathfrak{k} = \left\{ \begin{pmatrix} X_1 & 0 \\ 0 & X_2 \end{pmatrix} \in \mathfrak{g} \right\},$$

$$H_0 = \mathrm{diag.}\left(\frac{q}{p+q} \sqrt{-1}\, 1_p, -\frac{p}{p+q} \sqrt{-1}\, 1_q \right).$$

The group G is the projective unitary group of a hermitian form of signature (p, q). (In accordance with our generalization of the problem, it will be convenient to include here the compact case where p or $q = 0$.)

$$(\mathrm{II})_p\,(p \geqq 3) \quad \mathscr{D} = \{Z^{(p)} \mid {}^tZ = -Z, 1_p + \bar{Z}Z \gg 0\},$$

$$\mathfrak{g} = \left\{ \begin{pmatrix} X_1^{(p)} & X_{12}^{(p)} \\ -\bar{X}_{12} & \bar{X}_1 \end{pmatrix} \middle| \; {}^t\bar{X}_1 = -X_1, \quad {}^tX_{12} = -X_{12} \right\},$$

$$\mathfrak{k} = \left\{ \begin{pmatrix} X_1 & 0 \\ 0 & \bar{X}_1 \end{pmatrix} \in \mathfrak{g} \right\},$$

$$H_0 = \mathrm{diag.}\left(\frac{\sqrt{-1}}{2} 1_p, \quad -\frac{\sqrt{-1}}{2} 1_p \right).$$

G is the projective unitary group of a quaternionic skew-hermitian form of p variables.

$$(\mathrm{III})_p\,(p \geqq 1) \quad \mathscr{D} = \{Z^{(p)} \mid {}^tZ = Z, \quad 1_p - \bar{Z}Z \gg 0\},$$

$$\mathfrak{g} = \left\{ \begin{pmatrix} X_1^{(p)} & X_{12}^{(p)} \\ \bar{X}_{12} & \bar{X}_1 \end{pmatrix} \middle| \; {}^t\bar{X}_1 = -X_1, \quad {}^tX_{12} = X_{12} \right\},$$

$\mathfrak{k}$ and H_0 are the same as in $(\mathrm{II})_p$. G is the projective symplectic group of a real skew-symmetric form of $2p$ variables.

(IV)$_p$ ($p \geqq 1, p \neq 2$) $\mathscr{D} = \{\mathfrak{z} = (z_i) \in \boldsymbol{C}^p \mid \sum |z_i|^2 < \frac{1}{2}(1 + |\sum z_i^2|^2) < 1\}$,

$$\mathfrak{g} = \left\{ \begin{pmatrix} X_1^{(p)} & X_{12}^{(p,2)} \\ {}^tX_{12} & X_2^{(2)} \end{pmatrix} \middle| X_1, X_{12}, X_2 \text{:real}, {}^tX_i = -X_i (i = 1, 2) \right\},$$

$$\mathfrak{k} = \left\{ \begin{pmatrix} X_1 & 0 \\ 0 & X_2 \end{pmatrix} \in \mathfrak{g} \right\},$$

$$H_0 = \text{diag.}\left(0, \dots, 0, \begin{pmatrix} 0 & -1 \\ 1 & 0 \end{pmatrix}\right).$$

G is the connected component of the projective orthogonal group of a real quadratic form of signature $(p, 2)$.

There are the following isomorphisms between these simple Lie algebras. (All these isomorphisms, being taken to satisfy the condition (H_2), are naturally contained in the list given in no 5.)

$$(\mathrm{I})_{p,q} \cong (\mathrm{I})_{q,p}, \quad (\mathrm{I})_{1,1} \cong (\mathrm{III})_1 \cong (\mathrm{IV})_1, \quad (\mathrm{I})_{3,1} \cong (\mathrm{II})_3,$$
$$(\mathrm{III})_2 \cong (\mathrm{IV})_3, \quad (\mathrm{I})_{2,2} \cong (\mathrm{IV})_4, \quad (\mathrm{II})_4 \cong (\mathrm{IV})_6.$$

Therefore, if one wants to avoid overlappings, one may restrict the parameters as follows:

$$(\mathrm{I})_{p,q} (p \geqq q \geqq 1, p + q \geqq 3), \quad (\mathrm{II})_p (p \geqq 5),$$
$$(\mathrm{III})_p (p \geqq 1), \quad (\mathrm{IV})_p (p \geqq 5).$$

We note also that there are natural injections of $(\mathrm{II})_p$ and $(\mathrm{III})_p$ into $(\mathrm{I})_{p,p}$, satisfying (H_2), given by the identical mappings. On the other hand, there is a canonical injection of the symmetric domain $(\mathrm{I})_{p,q}$ into $(\mathrm{III})_{p+q}$ defined by the mapping

$$Z \to \begin{pmatrix} 0 & {}^tZ \\ Z & 0 \end{pmatrix}.$$

The corresponding homomorphism of the Lie algebra, which we denote by $\iota_{p,q}$, is given as follows:

$$(\mathrm{I})_{p,q} \ni X = \begin{pmatrix} X_1 & X_{12} \\ {}^t\bar{X}_{12} & X_2 \end{pmatrix} \to \begin{pmatrix} \bar{X}_2 & 0 & 0 & {}^tX_{12} \\ 0 & X_1 & X_{12} & 0 \\ 0 & {}^t\bar{X}_{12} & X_2 & 0 \\ \bar{X}_{12} & 0 & 0 & \bar{X}_1 \end{pmatrix} \in (\mathrm{III})_{p+q},$$

which, of course, satisfies (H_1) (but not (H_2) in general). Considered as a representation in the usual sense, $\iota_{p,q}$ is equivalent to $id \dotplus \overline{id}$.

4. To state the results, we make the following conventions. If $\varrho_i (1 \leqq i \leqq r)$ are representations of $\mathfrak{g}$ into $(\mathrm{I})_{p_i,q_i}$ and $\varrho_i(X) = \begin{pmatrix} {}_iX_1 & {}_iX_{12} \\ {}_i\bar{X}_{12} & {}_iX_2 \end{pmatrix}$

($X \in \mathfrak{g}$), we denote by $\varrho_1 \dot{+} \cdots \dot{+} \varrho_r$ the representation of $\mathfrak{g}$ into $(\mathrm{I})_{p,q}$ with $p = \sum p_i$, $q = \sum q_i$ given by

$$\mathfrak{g} \ni X \to \left(\begin{array}{ccc:ccc} {}_1X_1 & & & {}_1X_{12} & & \\ & \ddots & & & \ddots & \\ & & {}_rX_1 & & & {}_rX_{12} \\ \hdashline {}^t_1\bar{X}_{12} & & & {}_1X_2 & & \\ & \ddots & & & \ddots & \\ & & {}^t_r\bar{X}_{12} & & & {}_rX_2 \end{array}\right) \in (\mathrm{I})_{p,q}.$$

By a trivial representation (abbreviated as triv.) we mean a representation (of any dimension) such that $\varrho(X) = 0$ for all $X \in \mathfrak{g}$.

In this notation, we obtain the following results.

Theorem 1. *Let $\mathfrak{g}$ be a semi-simple Lie algebra of hermitian type and let ϱ be a representation of $\mathfrak{g}$ into $(\mathrm{III})_n$ satisfying the condition (H_1). Then there exist absolutely irreducible representations ϱ_i $(1 \leqq i \leqq r_1)$ of $\mathfrak{g}$ into $(\mathrm{III})_{n_i}$ satisfying (H_2) and absolutely irreducible representations ϱ_i $(r_1+1 \leqq i \leqq r_1+r_2)$ of $\mathfrak{g}$ into $(\mathrm{I})_{p_i,q_i}$ $(p_i, q_i \geqq 0,\ p_i+q_i > 0)$ satisfying (H_2) such that ϱ is (k)-equivalent (in $(\mathrm{III})_n$) to a direct sum of the following form:*

$$\varrho \underset{(k)}{\sim} \sum_{i=1}^{r_1} \varrho_i \dot{+} \sum_{i=r_1+1}^{r_1+r_2} \iota_{p_i,q_i} \circ \varrho_i \dot{+} (\text{triv.})\,.$$

If, moreover, we assume that ϱ_i $(r_1+1 \leqq i \leqq r_1+r_2)$ with $p_i = q_i$ are not (k)-equivalent to representations contained in $(\mathrm{III})_{p_i}$, then the above decomposition is unique up to the order and (k)-equivalence. Conversely, any representation given in this form is a representation of $\mathfrak{g}$ into $(\mathrm{III})_n$ satisfying (H_1).

It should be remarked that, for absolutely irreducible representations satisfying (H_2), the (k)-equivalence coincides with the usual equivalence. More precisely, let ϱ_1 and ϱ_2 be absolutely irreducible representations of $\mathfrak{g}$ into $(\mathrm{I})_{p,q}$ and $(\mathrm{I})_{p',q'}$, respectively, satisfying (H_2). If $p+q = p'+q'$ and if ϱ_1 and ϱ_2 are equivalent in the usual sense, then we have $p = p'$, $q = q'$ and ϱ_1 and ϱ_2 are (k)-equivalent. If, moreover, $p = q$ and $\varrho_1(\mathfrak{g})$ and $\varrho_2(\mathfrak{g})$ are both contained in $(\mathrm{III})_p$ or $(\mathrm{II})_p$, then they are (k)-equivalent in $(\mathrm{III})_p$ or $(\mathrm{II})_p$, respectively. It then follows from Theorem 1 that, for representations of $\mathfrak{g}$ into $(\mathrm{III})_n$ satisfying (H_1), the (k)-equivalence coincides with the usual equivalence.

Theorem 2. *Let $\mathfrak{g}$ be a semi-simple Lie algebra of hermitian type and let*

$$\mathfrak{g} = \mathfrak{g}_0 + \sum_{i=1}^{s} \mathfrak{g}_i$$

be a decomposition of $\mathfrak{g}$ into the direct sum of ideals, where $\mathfrak{g}_0$ is compact and $\mathfrak{g}_i$ $(1 \leqq i \leqq s)$ are simple and non-compact. Let ϱ be an absolutely irreducible *representation of* $\mathfrak{g}$ *into* $(\mathrm{I})_{p,q}$ *satisfying* (H_2). *Then there exist an absolutely irreducible representation* ϱ_0 *of* $\mathfrak{g}_0$ *into* $(\mathrm{I})_{n_0,0}$ *and, for some* i_0 $(1 \leqq i_0 \leqq s)$, *an absolutely irreducible representation* ϱ_{i_0} *of* $\mathfrak{g}_{i_0}$ *into* $(\mathrm{I})_{p_0,q_0}$ *with* $p = n_0 p_0$, $q = n_0 q_0$ *satisfying* (H_2) *such that* ϱ *is* (k)*-equivalent to a representation of the following form*:

$$\varrho\left(\sum_{i=0}^{s} X_i\right) = \varrho_0(X_0) \otimes 1_{p_0+q_0} + 1_{n_0} \otimes \varrho_{i_0}(X_{i_0})$$

where $X_i \in \mathfrak{g}_i (0 \leqq i \leqq s)$. *Conversely, any representation of this form is an absolutely irreducible representation of* $\mathfrak{g}$ *into* $(\mathrm{I})_{p,q}$ *satisfying* (H_2). *Moreover, in case* $p = q$, *the representation* ϱ *is* (k)*-equivalent to a representation contained in* $(\mathrm{III})_p$, *if and only if*

1°. ϱ_0 *is an absolutely irreducible representation of* $\mathfrak{g}_0$ *into the subalgebra of* $(\mathrm{I})_{n_0,0}$ *corresponding to the orthogonal group of a real positive-definite quadratic form of* n_0 *variables and* ϱ_{i_0} *is an absolutely irreducible representation of* $\mathfrak{g}_{i_0}$ *into* $(\mathrm{III})_{p_0}$ *satisfying* (H_2), *or*

2°. n_0 *is even, and* ϱ_0 *is an absolutely irreducible representation of* $\mathfrak{g}_0$ *into the subalgebra of* $(\mathrm{I})_{n_0,0}$ *corresponding to the unitary group of a quaternionic positive-definite hermitian form of* $\frac{n_0}{2}$ *variables and* ϱ_{i_0} *is an absolutely irreducible representation of* $\mathfrak{g}_{i_0}$ *into* $(\mathrm{II})_{p_0}$ *satisfying* (H_2).

By means of these theorems, our problem is reduced to the following one, if we leave aside the easier problem of determining the representations ϱ_0 of the compact factor $\mathfrak{g}_0$ as described in Theorem 2.

Problem 3. For a given non-compact simple Lie algebra $\mathfrak{g}$ corresponding to an irreducible symmetric domain, determine all absolutely irreducible representations ϱ of $\mathfrak{g}$ into $\mathfrak{g}' = (\mathrm{I})_{p',q'}$ satisfying (H_2). Moreover, in case $p' = q'$, determine whether or not ϱ is equivalent to a representation contained in $(\mathrm{III})_{p'}$ or $(\mathrm{II})_{p'}$.

5. Finally we shall list all the solutions of Problem 3.

For two simple Lie algebras of exceptional type corresponding to symmetric domains, there is no *solution of Problem 3.*

We add some comments to the following table.

(a) Precisely speaking, the representation takes the following form:

$$(\mathrm{I})_{p,q} \ni X = \begin{pmatrix} X_1 & X_{12} \\ {}^t\bar{X}_{12} & X_2 \end{pmatrix} \to \begin{pmatrix} 0 & 1_q \\ 1_p & 0 \end{pmatrix} \bar{X} \begin{pmatrix} 0 & 1_p \\ 1_q & 0 \end{pmatrix} = \begin{pmatrix} \bar{X}_2 & {}^tX_{12} \\ \bar{X}_{12} & \bar{X}_1 \end{pmatrix} \in (\mathrm{I})_{q,p}\,.$$

This gives the isomorphism $(\mathrm{I})_{p,q} \cong (\mathrm{I})_{q,p}$.

(b) The hermitian form on the representation-space is defined as follows. Let V be a $(p+1)$-dimensional vector-space over $\boldsymbol{C}$ provided with a hermitian form F of signature $(p, 1)$, for which $\mathfrak{g} = (\mathrm{I})_{p,1}$ is the Lie algebra of its special unitary group. Let $\Lambda(V) = \sum_{m=0}^{p+1} \Lambda^m(V)$ be the

$\mathfrak{g}$		ϱ	p'	q'	Cases for which $\varrho(\mathfrak{g}) \subset (\mathrm{III})_{p'}$	Cases for which $\varrho(\mathfrak{g}) \subset (\mathrm{II})_{p'}$
$(\mathrm{I})_{p,q}$	$p \geqq q \geqq 2$	id.	p	q	—	—
		$\overline{\mathrm{id.}}$ (a)	q	p	—	—
	$p \geqq q = 1$	skew-symmetric tensor representations of degree m $(1 \leqq m \leqq p)$ (b)	$\binom{p}{m}$	$\binom{p}{m-1}$	$p \equiv 1\ (4)$ $m = \frac{p+1}{2}$	$p \equiv 3\ (4)$ $m = \frac{p+1}{2}$
$(\mathrm{II})_p$	$p \geqq 5$	id.	p	p	—	always
$(\mathrm{III})_p$	$p \geqq 1$	id.	p	p	always	—
$(\mathrm{IV})_p$	$p \geqq 4$, even	(two) spin representations (c)	$2^{\frac{p}{2}-1}$	$2^{\frac{p}{2}-1}$	$p \equiv 2\ (8)$	$p \equiv 6\ (8)$
	$p \geqq 1$, odd	spin representation (c)	$2^{\frac{p-1}{2}}$	$2^{\frac{p-1}{2}}$	$p \equiv 1,3\ (8)$	$p \equiv 5,7\ (8)$

exterior algebra of V, $\Lambda^m(V)$ denoting the space of skew-symmetric tensors of degree m. Then we can define a hermitian form $F^{(m)}$ on $\Lambda^m(V)$ by putting

$$F^{(m)}(x_1 \wedge \cdots \wedge x_m,\ \ y_1 \wedge \cdots \wedge y_m) = \det(F(x_i, y_j))$$

where $x_i, y_i \in V (1 \leqq i \leqq m)$, which is of signature $(\binom{p}{m}, \binom{p}{m-1})$. It is easy to verify that the skew-symmetric tensor representation is actually a representation of $\mathfrak{g}$ into the Lie algebra of the special unitary group of this hermitian form, which for $1 \leqq m \leqq p$, satisfies the condition (H_2).

(c) Let $V_{\boldsymbol{R}}$ be a $(p+2)$-dimensional vector-space over $\boldsymbol{R}$ provided with a quadratic form (or symmetric bilinear form) S of signature $(p, 2)$, for which $\mathfrak{g} = (\mathrm{IV})_p$ is the Lie algebra of its orthogonal group. We denote by $V = V_{\boldsymbol{C}}$ the complexification of $V_{\boldsymbol{R}}$ and define a hermitian form F on V by

$$F(x, y) = 2\,S(\bar{x}, y) \qquad \text{for} \quad x, y \in V,$$

which is also of signature $(p, 2)$. Put $p_1 = \left[\frac{p}{2}\right]$. Then we can find a (p_1+1)-dimensional (complex) subspace W of V such that the restriction of S on W is identically zero (i. e., W is a maximal totally isotropic subspace of V) and that that of F on W is of signature $(p_1, 1)$. Let $\Lambda(W)$ be the exterior algebra of W, and denote by $\Lambda^{\pm}(W)$ the subspaces of it consisting of all skew-symmetric tensors of even and odd degree, respectively. Then, according to the theory of spinors, the representation-spaces of spin representations of $\mathfrak{g}$ are given by $\Lambda^{\pm}(W)$, if p is even, and by $\Lambda(W)$, if p is odd. In case p is even (resp. odd), extending the hermitian form F (resp. $-F$) on W to those on $\Lambda^{\pm}(W)$ (resp. that on $\Lambda(W)$) in the manner explained in (b), we can prove that the spin representations are actually representations of $\mathfrak{g}$ into the Lie algebras of the special unitary groups of those hermitian forms, which satisfy the condition (H_2).

References

[1] Cartan, E.: Sur les domaines bornés homogènes de l'espace de n variables complexes. Abh. Math. Sem. Univ. Hamburg 11, 116—162 (1936).

[2] Helgason, S.: Differential geometry and symmetric spaces. New York and London: Academic Press, 1962.

[3] Kuga, M.: Fiber varieties over a symmetric space whose fibers are abelian varieties. Lecture Notes, Univ. of Chicago. 1963—64.

[4] Satake, I.: Holomorphic imbeddings of symmetric domains into a Siegel space. To appear in Amer. J. Math.

Department of Mathematics
University of Chicago
Chicago, Illinois

On Determining Sets in a Stein Manifold*,**

By

A. Aeppli

Dedicated to Professor Heinz Hopf on his 70th birthday, in gratitude for his inspiration and guidance

1. Definitions and some introductory remarks

a) A set D in a Stein manifold[1] X is called a determining set in X if the following "unique extendability property" holds:

(e) every holomorphic function on D can be extended to a holomorphic function on X, and the extension is unique[2].

More generally for $D \subset X$, D is form-determining in X if

(e_f) every holomorphic form on D has a unique holomorphic extension over X.

* Research supported by NSF G-24336.

** Received April 27, 1964.

[1] "Stein manifold" in the sense of [2], Exposé IX.

[2] Minimal determining sets might be called distinguished sets, cf. [6].

(e_f) includes (e) trivially, and if $X = X^{(n)}$ is a Stein manifold with $\dim_{\mathbf{C}} X = n$ which is (holomorphically) immersed in $\mathbf{C}^n$ then a determining set D in X is form-determining since there is a global complex coordinate system in X. (e) implies (e_f) already in case there is a nowhere vanishing holomorphic n-form on $X^{(n)}$, and such a form exists e.g. if X is simply connected and $c_1(X) = 0$ ($c_1 =$ first Chern class given by $\sqrt{-1} \cdot \nabla \log g$, $g =$ nowhere vanishing density), cf. [*10*], Sec. 7.

b) If X is a Stein manifold such that $\bar{X}$ is a finite manifold with boundary $\partial X = \bar{X} - X$, we say the "boundary property" for $D \subset X$ holds if

(b) there is a neighborhood of the boundary ∂X (in X) which is contained in D [3].

(b) implies the weaker property $\bar{D} \supset \partial X$. If a determining or form-determining set D in X has the boundary property, then D contains an open set of X, and hence the uniqueness part in (e) and (e_f) follows. The condition (b) will enter in Theorem 1 below. (b) must hold e.g. in case D is a determining domain in a complex manifold $X^{(n)}$ which is immersed in $\mathbf{C}^n$ such that X has a smooth (i.e. sufficiently differentiable) everywhere strongly pseudo-convex boundary.

c) A well known and often used method to get a form-determining set D in X consists of taking a domain $D \subset \mathbf{C}^n$ together with $X =$ holomorphy envelope of D in $\mathbf{C}^n$ [4]; X is a complex n-manifold immersed in $\mathbf{C}^n$, cf. [*2*], Exp. VII.

Given a form-determining set D in X, it is natural to ask how far and in what way the "geometric structure" of X will be determined by suitable properties of D; in particular the cohomology groups of D and of X (with coefficients in $\mathbf{C}$) are related by the homomorphism

$H^q(X; \mathbf{C}) \xrightarrow{i^*} H^q(D; \mathbf{C})$ induced by the imbedding $i: D \to X$. In the sequel we are interested in the question whether i^* is a monomorphism or not. Theorems 1 and 2 deal with sufficient conditions for i^* to be a monomorphism. Afterwards examples are constructed where i^* is not injective. As a side result we prove a non-imbeddability theorem for Stein manifolds. Finally a uniqueness theorem is given followed by some remarks on the functor $D \to H^q(X; \mathbf{C})$.

---

[3] If $\bar{X}$ is not finite, the boundary ∂X in (b) has to include the "ideal" boundary (the "ends", cf. [*7*]). The following Theorems 1, $\bar{1}$ and 2, $\bar{2}$ hold — properly formulated — in this case too.

[4] Similarly, the holomorphy envelope of a domain D in a complex manifold M may produce an example of the required type, or more generally the set $X =$ "form-completion" of D in M, i. e. $X =$ maximal set immersed in M such that all holomorphic forms on D can be holomorphically and uniquely extended over X.

2. Situations where i* is injective

a) Let $X = X^{(n)}$ be a Stein manifold such that $\bar{X}$ is finite with "combinatorially smooth" boundary[5] $\partial X = \bar{X} - X$. Consequently the Poincaré-Lefschetz duality theorem holds:

$$H^q(\bar{X}, \partial X) \cong H_{2n-q}(X), \tag{1}$$

and furthermore $H^q(X) \cong H^q(\bar{X})$. We can restrict q by $0 \leqq q \leqq 2n = \dim_{\mathbf{R}} X$. D is assumed to be an open set in X, and property (b) should hold (for D in X).

For a complex manifold M, $H^q_h(M)$ denotes the q-th cohomology group for holomorphic forms on M, and $H^q_d(M)$ the q-th cohomology group for differential forms on M with coefficients in $\boldsymbol{C}$ (= q-th de Rham group). Then

$$H^q_d(M) \cong H^q(M; \boldsymbol{C}) \tag{2}$$

for any differentiable manifold M by de Rham; moreover

$$H^q_h(X) \cong H^q_d(X) \tag{3}$$

for a Stein manifold X by Serre [*3*], and the isomorphism (3) is naturally induced by the imbedding of the holomorphic in the differentiable closed forms. We use in general the notation $\lambda: H^q_h \to H^q_d$ for the homomorphism defined by sending the holomorphic into the differentiable closed forms on a complex manifold. (2), (3) yield for a Stein manifold $X^{(n)}$

$$H^q(X; \boldsymbol{C}) = 0 \quad \text{for} \quad q \geqq n+1. \tag{4}$$

Theorem 1. *Let $X^{(n)}$ be a Stein manifold such that $\bar{X}$ is finite with combinatorially smooth boundary, and let D be a form-determining open set in X. Assume the boundary property for $D \subset X$. Then*

$$0 \to H^q(X; \boldsymbol{C}) \xrightarrow{i^*} H^q(D; \boldsymbol{C}) \tag{5}$$

and

$$0 \to H^q_h(D) \xrightarrow{\lambda} H^q_d(D) \tag{6}$$

are exact for $0 \leqq q \leqq n-1$.

---

[5] "X is finite with combinatorially smooth boundary" means: the manifold X is the interior of the finite (closed) polyhedron $\bar{X}$ such that the intersection of an open neighborhood of a point $x \in \bar{X} - X$ with $\bar{X}$ contains a neighborhood of x which is homeomorphic with a closed euclidean half-space (and this should hold for all boundary points x). Such a smoothness condition is a convenient assumption which is often not entirely needed: using singular homology theory and cochains with compact carriers etc. we get theorems similar to the ones given in the sequel covering more general situations. Cf. also footnote 4.

Proof. The coefficient domain is $\boldsymbol{C}$ for the considered cohomology groups. (1), (4) and the exact cohomology sequence for the imbedding $j:\partial X \to \bar{X}$ imply

$$
\begin{array}{lll}
j^*: H^q(\bar{X}) \to H^q(\partial X) & \text{is an isomorphism for} & 0 \leqq q \leqq n-2 \\
 & \text{and a monomorphism for} & 0 \leqq q \leqq n-1 .
\end{array} \qquad (7)
$$

Since ∂X is combinatorially smooth and since (b) holds there is an open neighborhood $\widetilde{\partial X}$ of ∂X in X with the imbedding $k:\widetilde{\partial X} \to D$ such that the inclusion $\tilde{j}:\widetilde{\partial X} \to X$ induces by virtue of (7) $\tilde{j}^*$ with:

$$
\begin{array}{lll}
\tilde{j}^*: H^q(X) \to H^q(\widetilde{\partial X}) & \text{is an isomorphism for} & 0 \leqq q \leqq n-2 \\
 & \text{and a monomorphism for} & 0 \leqq q \leqq n-1 .
\end{array} \qquad (\tilde{7})
$$

Furthermore, $i:D \to X$ induces an isomorphism

$$H^q_h(X) \cong H^q_h(D) \qquad (8)$$

for all $q = 0, 1, \ldots$ since D is form-determining in X. Consider now the following commutative diagram for $0 \leqq q \leqq n-1$:

$$
\left.
\begin{array}{ccc}
 & \overset{(\tilde{7})}{\Downarrow} & \\
H^q(X) & \xrightarrow[\text{monom.}]{\tilde{j}^*} & H^q(\widetilde{\partial X}) \\
(2), (3) \Longrightarrow \Big\downarrow \text{isom.} & \searrow i^* & \Big\uparrow k^* \\
H^q_h(X) & & H^q(D) \\
(8) \Longrightarrow \Big\downarrow \text{isom.} & & \Big\uparrow \text{isom.} \Longleftarrow (2) \\
H^q_h(D) & \xrightarrow{\lambda} & H^q_d(D)
\end{array}
\right\} \qquad (9)
$$

The conclusions (5), (6) in Theorem 1 follow from (9), q.e.d.

The boundary property for $D \subset X$ in Theorem 1 can be replaced by the condition $\bar{D} \supset \partial X$, i.e. we claim.

Theorem $\bar{1}$. *Let $X^{(n)}$ be as above and D a form-determining open set in X with $\bar{D} \supset \partial X$. Then the conclusions* (5), (6) *hold.*

Proof. Since D is open and $\bar{D} \supset \partial X$, an open neighborhood of ∂X in X intersected with D yields an open neighborhood of ∂X contained in D, i.e. (b) holds. Hence all the hypotheses of Theorem 1 are fulfilled, therefore Theorem $\bar{1}$ follows from Theorem 1.

b) We formulate a slightly more general monomorphism theorem:

Theorem $\bar{2}$. *Let $X^{(n)}$ be as in Theorem 1 and D a form-determining open set in X with combinatorially smooth boundary ∂D*[6] *such that the*

[6] I.e. $\bar{D}$ = subpolyhedron of the finite polyhedron $\bar{X}$ with combinatorially smooth ∂D in $\bar{X}$.

inclusion $l: \partial X \cap \bar{D} \to \partial X$ *induces a monomorphism* $l^*: H^q(\partial X; \boldsymbol{C}) \to H^q(\partial X \cap \bar{D}; \boldsymbol{C})$ *for* $0 \leqq q \leqq n-1$. *Then the conclusions* (5), (6) *hold.*

Proof. Since $X^{(n)}$ is as in Theorem 1 and since ∂D is combinatorially smooth we can find $\widetilde{\partial X} \subset X$ as above such that $(\tilde{7})$ holds and such that the monomorphism l^* induces

$$\tilde{l}^*: H^q(\widetilde{\partial X}) \to H^q(\widetilde{\partial X} \cap D) \text{ is a monomorphism for } 0 \leqq q \leqq n-1. \quad (10)$$

Then we get a commutative diagram analogous to (9):

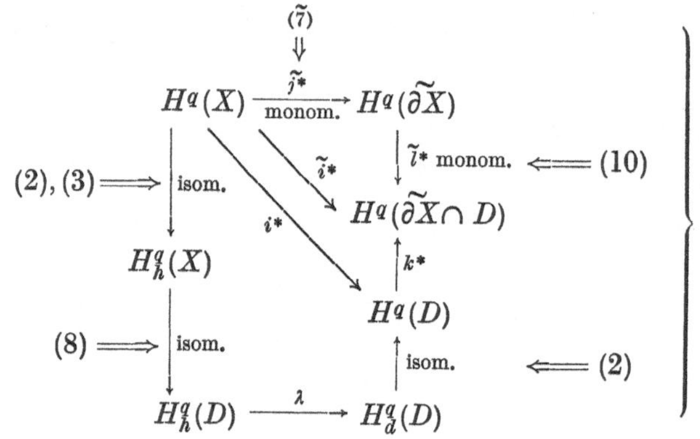

(11)

(11) implies (5), (6), q.e.d.

Remark 1. Theorem $\bar{2}$ still holds if ∂D is combinatorially smooth only in a neighborhood of ∂X. This version of Theorem $\bar{2}$ is then a strengthening of Theorem $\bar{1}$.

Remark 2. There is a theorem — which one would call Theorem 2 — strengthening Theorem 1 in a similar manner as Theorem $\bar{2}$ contains Theorem $\bar{1}$: one requires conditions which imply directly (10), hence injectivity of $\tilde{i}^*$ in (11).

Remark 3. In the proofs above we do not use the full strength of de Rham's theorem for D, only the existence of the natural homomorphism $H^q_d(D) \xrightarrow{\varrho} H^q(D)$ (by Stokes' theorem) has to be known.

Remark 4. From $(\tilde{7})$ and (11) is easily seen: for $0 \leqq q \leqq n-1$
$\tilde{l}^*$ monomorphism $\Rightarrow \tilde{i}^*, i^*, \lambda$ monomorphisms;
$k^* \varrho \lambda$ isomorphism, epimorphism, monomorphism
$\Leftrightarrow \tilde{i}^*$ isomorphism, epimorphism, monomorphism resp.;
and for $0 \leqq q \leqq n-2$
$k^* \varrho \lambda$ isomorphism, epimorphism, monomorphism
$\Leftrightarrow \tilde{l}^*$ isomorphism, epimorphism, monomorphism resp.

Remark 5. ROYDEN proved in [4] $H^1(X, D; \boldsymbol{Z}) = 0$ for an open determining set D in the Stein manifold X without assuming (b) or any condition on l or $\tilde{l}$. Hence $i^*: H^1(X; \boldsymbol{Z}) \to H^1(D; \boldsymbol{Z})$ is a monomorphism. So the given theorems appear as rather weak counterparts of Royden's result: torsion is neglected, and an additional condition (essentially on ∂D) is made. However, the next section shows that we have to assume some property like (b) or (10) in order to get the conclusions (5), (6). Of course the conclusion (5) in Theorems 1 and $\bar{1}$ holds over $\boldsymbol{Z}$ if $H_k(X; \boldsymbol{Z})$ does not have torsion for $k \geq n+1$ (which might always be true ?). An analogous statement holds in the case of Theorem $\bar{2}$ (and 2), and also (7) over $\boldsymbol{Z}$ follows from the indicated torsion conditions (implying homotopy relationships between X and ∂X by arguments of J. H. C. WHITEHEAD if X and ∂X are simply connected or more generally if $j_*: \pi_1(\partial X) \xrightarrow{\cong} \pi_1(\bar{X})$ and under appropriate compatibility assumptions). Finally, $i_*: \pi_1(D) \to$ $\to \pi_1(X)$ is an epimorphism as shown in [10] (Theorem 7 in [10]; also $H_1(X, D; \boldsymbol{Z}) = 0$ follows from [10], Sec. 7; D is again assumed to be open).

3. Examples for non-injective homomorphisms i*

a) Let M be a compact, oriented[7], real analytic manifold with $\chi = \chi(M) \neq 0$ (χ = Euler characteristic; our assumption implies: $\dim_{\mathbf{R}} M = n$ is even). Construct a "small enough" Grauert tube $T^{(n)} = T(M)$ according to [1], i. e. an open neighborhood of M in the complexification $\tilde{M}$ of M such that T is a Stein manifold, $\bar{T}$ is a finite polyhedron with combinatorially smooth (or even differentiable as can be assumed without difficulty) boundary ∂T, and ∂T is fibered over M by spheres S^{n-1} since $\bar{T}$ is fibered by closed n-cells, this fibration being induced by the fiber bundle of the purely imaginary tangent n-planes along M (in $\tilde{M}$). We denote the S^{n-1}-bundle over M with total space ∂T by $\mathfrak{B}(M) = \{\partial T, S^{n-1}, M\}$ and observe that $\mathfrak{B}(M)$ is isomorphic with the tangent bundle $\mathfrak{T}(M)$ (= bundle of unit tangent vectors on M, M considered as a Riemannian manifold). We recall Gysin's exact sequence for $\mathfrak{B}(M) \cong \mathfrak{T}(M)$:

$$\cdots \to H^0(M; \boldsymbol{Z}) \xrightarrow{\chi^*} H^n(M; \boldsymbol{Z}) \xrightarrow{p^*} H^n(\partial T; \boldsymbol{Z}) \to H^1(M; \boldsymbol{Z}) \to \cdots \tag{12}$$

where χ^* = multiplication by χz^n, z^n = fundamental n-cocycle of M and p^* is induced by the projection $p: \partial T \to M$.

Assume now $H^1(M; \boldsymbol{Z}) = 0$. Then (12) together with $H^0(M; \boldsymbol{Z}) \cong H^n(M; \boldsymbol{Z}) \cong \boldsymbol{Z}$ implies

$$H^n(\partial T; \boldsymbol{Z}) \cong \boldsymbol{Z}/\chi\boldsymbol{Z} = \boldsymbol{Z}_\chi. \tag{13}$$

---

[7] Orientability enables us to work with the coefficient domain $\mathbf{Z}$. If M is not orientable one uses cohomology with coefficients in a non-trivial sheaf (twisted coefficients), and then similar considerations are possible.

Clearly

$$H^k(\bar{T}) \cong H^k(T) \cong H^k(M), \qquad k = 0, 1, \ldots, \quad (14)$$

for any coefficient domain since M is a homotopy retract of T (or of $\bar{T}$). In particular, (14) yields $H^n(T; \boldsymbol{Z}) \cong H^n(\bar{T}; \boldsymbol{Z}) \cong \boldsymbol{Z}$, and from this and (13) we conclude that $j^* : H^n(\bar{T}; J) \to H^n(\partial T; J)$ is not injective for $J = \boldsymbol{Z}$, and hence the same remains true for $J = \boldsymbol{C}$, i. e. the second part in (7) and ($\tilde{7}$) holds no longer for $q = n$.

b) $D = \widetilde{\partial T}$ = open neighborhood of ∂T in T is a form-determining set in T (follows from [1]), and we can choose D so that ∂T is a homotopy retract of $\partial T \cup D$, i. e. so that $H(\partial T) \cong H(D)$. Then for $X = T^{(n)} = T(M)$ with M as above together with $D = \widetilde{\partial T}$ as just described, the homomorphism $i^* : \boldsymbol{C} \cong H^n(X; \boldsymbol{C}) \to H^n(D; \boldsymbol{C}) = 0$ is not injective as well as $\lambda : \boldsymbol{C} \cong H^n_h(D) \to H^n_d(D) = 0$. A simple example is given by $M = S^2 \simeq T(S^2) = X$ together with $D = \widetilde{\partial X} \simeq \partial T = P^3$ = 3-dimensional real projective space where $H^2(X; \boldsymbol{Z}) \cong \boldsymbol{Z}$ and $H^2(D; \boldsymbol{Z}) \cong \boldsymbol{Z}_2$.

c) To get examples which show that (5), (6) do not hold in general even for $q < n$ we form cartesian products and use the fact that $D_1 \times D_2$ is form-determining resp. determining in $X_1 \times X_2$ if D_t is form-determining resp. determining in X_t for $t = 1, 2$. Namely: take two Stein manifolds $X^{(n)}$, $Y^{(m)}$ with form-determining open sets D in X and E in Y such that $i^* : H^n(X; \boldsymbol{C}) \to H^n(D; \boldsymbol{C})$ and $\lambda : H^n_h(D) \to H^n_d(D)$ are not monomorphic (e. g. $X = T(M)$ as above and $D = \widetilde{\partial X}$); then

$$i^* : H^n(X \times Y; \boldsymbol{C}) \to H^n(D \times E; \boldsymbol{C}) \quad \text{and} \quad \lambda : H^n_h(D \times E) \to H^n_d(D \times E)$$

are not monomorphic either as is easily seen by Künneth type arguments.

4. A non-imbeddability theorem

a) If a combinatorial manifold V^N is imbedded in $\boldsymbol{R}^{N+1}$ then $\boldsymbol{R}^{N+1} = A \cup B$ with $A \cap B = V$, and the Mayer-Vietoris exact sequence for A, B yields

$$H^q(A) + H^q(B) \xrightarrow{\cong} H^q(V) \quad \text{for} \quad q \geqq 1$$

and

$$0 \to H^0(R^{N+1}) \to H^0(A) + H^0(B) \to H^0(V) \to 0,$$

hence

$$\left.\begin{array}{l} j^* : H^q(A) \to H^q(V) \quad \text{and} \quad j^* : H^q(B) \to H^q(V) \\ \text{are monomorphisms for } q = 0, 1, 2, \ldots, \end{array}\right\} \quad (15)$$

j^* induced by the imbeddings $V \subset A$ and $V \subset B$ (for $q \geq 1$ immediate; for $q = 0$: use the definition of the homomorphisms in the Mayer-Vietoris sequence).

b) Assume now $X^{(n)} \subset \boldsymbol{C}^n$, i.e. we consider a Stein manifold $X^{(n)}$ imbedded in $\boldsymbol{C}^n = \boldsymbol{R}^{2n}$, X being combinatorially smooth in $\boldsymbol{C}^n$. Putting $A = \bar{X}$ and $B = \boldsymbol{C}^n - X$ we infer from (15):

$$j^* : H^q(\bar{X}) \to H^q(\partial X) \text{ is a monomorphism for } q = 0, 1, 2, \ldots . \tag{16}$$

(15), (16) hold for any coefficient domain. On the other hand, we have seen in 3a) that $j^* : H^n(\bar{T}; \boldsymbol{Z}) \to H^n(\partial T; \boldsymbol{Z})$ is not injective for a T as described there, against (16) for $q = n$. We conclude[8]:

Theorem 3. *A Grauert tube $T^{(n)}$ (which is a Stein manifold) of a compact, orientable, real analytic manifold M^n with $\chi(M) \neq 0$ and $H^1(M; \boldsymbol{Z}) = 0$ cannot be imbedded in $\boldsymbol{C}^n$.*

Remark 1. In Theorem 3, imbeddings of the compact ("small enough") tube $\bar{T}^{(n)}$ in $\boldsymbol{C}^n$ are ruled out to begin with; since an imbedding $T \subset \boldsymbol{C}^n$ induces an imbedding of a compact Grauert tube $\bar{T}_1$ about M, $\bar{T}_1 \subset T$, $\bar{T}_1$ being fibered by closed n-cells as indicated above, Theorem 3 holds as formulated.

Examples. (i) Take $X = T^{(2)} = T(S^2) : X \simeq S^2 = 2$-sphere and therefore $H(X) \cong H(S^2)$, and $\partial X = P^3$ so that $H(\partial X) = H(P^3)$; hence X cannot be imbedded in $\boldsymbol{C}^2$. (ii) The Mayer-Vietoris sequence for $\partial X \times X, X \times \partial X, \partial(X \times X) = (\partial X \times X) \cup (X \times \partial X), X = T(S^2)$, yields $H^4(\partial(X \times X); \boldsymbol{C}) \cong H^4(\partial X \times X; \boldsymbol{C}) + H^4(X \times \partial X; \boldsymbol{C}) = 0$; hence $j^* : \boldsymbol{C} \cong H^4(X \times X; \boldsymbol{C}) \to H^4(\partial(X \times X); \boldsymbol{C}) = 0$ is not injective, and therefore $X \times X$ is not imbeddable in $\boldsymbol{C}^4$. (iii) Similarly $X \times X \times \cdots \times X = r$-fold cartesian product of X, $X = T(S^2)$, cannot be imbedded in $\boldsymbol{C}^{2r}$. (iv) In many other cases, an analogous argument leads to cartesian products of Stein manifolds $X_1^{(n_1)}, \ldots, X_r^{(n_r)}$ which cannot be imbedded in $\boldsymbol{C}^N$, $N = n_1 + n_2 + \cdots + n_r$.

Remark 2. (15) shows: a non-compact Riemann surface $X^{(1)}$ (Stein manifold X with $\dim_{\boldsymbol{C}} X = 1$) such that $H^1(\bar{X}^{(1)}) \to H^1(\partial X^{(1)})$ is not monomorphic cannot be imbedded in $\boldsymbol{C}$. Hence: if $X^{(1)}$ has a handle it is not imbeddable in $\boldsymbol{C}$.

Remark 3. A Grauert tube $T^{(n)}$ of a compact real analytic manifold M^n with non-trivial Pontrjagin class cannot be immersed in $\boldsymbol{C}^n$ since such an immersion would imply complex parallelizability of the complexified tangent bundle of M^n (such that the Pontrjagin classes vanish). Example: $M = P^{(n)}$ = complex projective n-space, $n = 2m \geq 2$. Thus a Stein manifold $T^{(n)}$ of this type does not have n everywhere independent holomorphic functions.

---

[8] Compare Theorem 3 and its proof with [5], p. 298 where the same tools of algebraic topology are used.

5. Uniqueness of X for a given D

a) Although Theorem 4 below or the facts which immediately imply it are well known to Remmert and others (cf. [8], [9]) Theorem 4 is important enough (in connection with the previous considerations) to justify the following discussion. Suppose: D_1 is a determining open set in $X_1^{(n)}$, D_2 a determining open set in $X_2^{(n)}$, X_1 and X_2 are Stein manifolds, and there is a biholomorphic map $\varphi : D_1 \to D_2$. Every Stein manifold can be holomorphically imbedded in $\boldsymbol{C}^N$ (for some N), thus let $\psi : X_2 \to \boldsymbol{C}^N$ be a holomorphic imbedding (ψ can even be assumed compact, i. e. proper, according to Remmert). Then $\psi\varphi : D_1 \to \boldsymbol{C}^N$ imbeds D_1 holomorphically in C^N, and $\psi\varphi$ is given by N holomorphic functions $g_1, g_2, \ldots, g_N$ in D_1. Since D_1 is determining in X_1, the functions $g_1, \ldots, g_N$ are the restrictions of holomorphic functions $f_1, \ldots, f_N$ in $X_1 : g_t = f_t | D_1$ for $t = 1, \ldots, N$. $f_1, \ldots, f_N$ describe a holomorphic mapping $\mu : X_1 \to \boldsymbol{C}^N$ with $\mu | D_1 = \psi\varphi$. μ has maximal rank everywhere: the singularity set of μ is an analytic set A in X_1, A being the zero-set of global holomorphic functions $h_1, h_2, \ldots$ (since every analytic set in a Stein manifold is defined by global holomorphic functions, cf. [3]); if A is non-empty then $A \cap D_1$ is non-empty since otherwise there exists a holomorphic function g in D_1, e. g. $g = 1/h_1 | D_1$, such that g does not have a holomorphic extension over X_1; since μ is regular in D_1 it follows that A is empty, hence μ is regular everywhere, i. e. μ immerses X_1 in $\boldsymbol{C}^N$. Therefore we end up with the following situation: $\tilde{X}_1^{(n)} = \mu X_1$ and $\tilde{X}_2^{(n)} = \psi X_2$ are two Stein manifolds, holomorphically immersed in $\boldsymbol{C}^N$ with $\tilde{X}_1 \cap \tilde{X}_2 \supset \tilde{D}_1 = \tilde{D}_2$ where $\tilde{D}_1 = \mu D_1$ and $\tilde{D}_2 = \psi D_2$. If $\tilde{X}_1 - \tilde{X}_1 \cap \tilde{X}_2$ is not empty, then there exists a holomorphic function in $\tilde{X}_1 \cap \tilde{X}_2$ (and hence in $\tilde{D}_1$) which cannot be holomorphically extended over $\tilde{X}_1$ (since $\tilde{X}_1 \cap \tilde{X}_2$ is a Stein manifold; given an infinite sequence of points x_r, $r = 1, 2, \ldots$, without interior accumulation point in the Stein manifold X, there is a holomorphic function f in X with $\lim_{r\to\infty} f(x_r) = +\infty$, cf. [3]); therefore $\tilde{X}_1 = \tilde{X}_1 \cap \tilde{X}_2$, and similarly $\tilde{X}_2 = \tilde{X}_1 \cap \tilde{X}_2$, hence $\tilde{X}_1 = \tilde{X}_2$. Thus we proved[9]:

Theorem 4. *If a fixed D (with a complex analytic structure) is an open determining set in the Stein manifold X, then X is unique up to a biholomorphic mapping.*

If D is not open, but if a structural sheaf over D is given defining a complex structure, then the Stein manifold X (such that D is deter-

[9] Theorem 4, considered in the framework of function algebras, is quickly proved using results of [8], [9]: two Stein manifolds X_1, X_2 with isomorphic function algebras of holomorphic functions are biholomorphically equivalent ($X_1 \cong X_2$), hence $D_1 \cong D_2$ implies $X_1 \cong X_2$ for D_t determining in X_t, $t = 1, 2$.

mining in X) is still unique up to a biholomorphic mapping provided X exists. The proof is essentially the same as above. There is also a version of Theorem 4 in the case of complex spaces instead of manifolds.

b) Theorem 4 asserts: the determining set D (with its complex structure) in X is "responsible" for the Stein manifold X. In particular: in this paper we considered the functor $D \to H^q(X; \boldsymbol{C})$ which in case of a form-determining set D is given by $D \to H^q_h(D) \cong H^q_h(X) \cong H^q(X; \boldsymbol{C})$. The results in 2. might suggest that in some cases this mechanism leading from D to $H^q(X; \boldsymbol{C})$ can be described by cohomology properties of D alone; however, the examples in 3b), c) of non-injective homomorphisms $i^*: H^q(X; \boldsymbol{C}) \to H^q(D; \boldsymbol{C})$ show that the knowledge of $H(D; \boldsymbol{C})$ is not enough to get $H(X; \boldsymbol{C})$ ($H^q(D; \boldsymbol{C}) = 0$ for $1 \leqq q \leqq n$ can occur together with $H^q(X; \boldsymbol{C}) \neq 0$ for some $q \geqq 1$).

Remark. Given the homotopy type of a finite dimensional, countable polyhedron P, there exists a Stein manifold X realizing this homotopy type. Proof. P can be imbedded in some $\boldsymbol{R}^n$; there is an open neighborhood V of P in $\boldsymbol{R}^n$ such that $V \simeq P$; take a Grauert tube $W^{(n)} = T(V^n)$ about V in the complexification $\tilde{V}$ of V with $W \simeq V \simeq P$. Hence every cohomology ring $H^*(P; J)$, $J =$ coefficient ring, can be realized as cohomology ring of a Stein manifold, and the range of the functor $D \to H^*(X; J)$, D determining in X, is not restricted by the condition that X has to stay inside the category of Stein manifolds.

c) As the proof of Theorem 4 shows, a little more than in a) can be said, namely[10]: a holomorphic mapping $\varphi: D_1 \to X_2$ has a unique holomorphic extension $\hat{\varphi}: X_1 \to X_2$ (for D_1 determining in X_1; X_1, X_2 Stein manifolds), and $\hat{\varphi}$ is regular if φ is regular; if $\varphi D_1 = D_2 =$ determining set in X_2 and if φ is biholomorphic then $\hat{\varphi}$ is also biholomorphic. Consequently $\mathrm{Aut}(D)$ is a subgroup of $\mathrm{Aut}(X)$ for D determining in X where $\mathrm{Aut}(M) =$ automorphism group of the complex manifold (or space) M, automorphism = biholomorphic self-mapping. Corollary: every determining set in a rigid Stein manifold X (i. e. $\mathrm{Aut}(X) = \{\text{Identity}\}$) is rigid. — In cohomology theory the question arises how $\hat{\varphi}^*: H^q(X_2) \to H^q(X_1)$ is determined by $\varphi: D_1 \to X_2$. If the imbedding $i_1: D_1 \to X_1$ induces a monomorphism $i_1^*: H^q(X_1) \to H^q(D_1)$ then $\hat{\varphi}^*$ is given by $\hat{\varphi}^* = i_1^{*-1} \varphi^*$. This illustrates the significance of Theorems 1 and 2 and the related considerations in 2.

References

[1] Grauert, H.: On Levi's problem and the imbedding of real analytic manifolds, Ann. Math. 68, 460—472 (1958).
[2] Cartan, H.: Séminaire 1951—52. ENS Paris.
[3] Serre, J.-P.: H. Cartan Séminaire 1951—52. ENS Paris. Exposé XX.

[10] This again is contained in [8], [9].

[4] Royden, H. L.: One-dimensional cohomology in domains of holomorphy. Ann. Math. 78, 197—200 (1963).

[5] Morimoto, A., and T. Nagano: On pseudo-conformal transformations of hypersurfaces, J. Math. Soc. Japan 15, 289—300 (1963).

[6] Bergman, S.: Les ensembles distingués dans la théorie des fonctions de deux variables complexes, Rend. di Mat. 21, 240—244 (1962).

[7] Specker, E.: Endenverbände von Räumen und Gruppen. Math. Ann. 122, 167—174 (1950).

[8] Royden, H. L.: Function algebras. Bull. A.M.S. 69, 281—298 (1963).

[9] Edwards, R. E.: Algebras of holomorphic functions. Proc. London Math. Soc. (3) 7, 510—517 (1957).

[10] Aeppli, A.: On the cohomology structure of Stein manifolds. These Proceedings 58—70.

School of Mathematics
Institute of Technology
University of Minnesota
Minneapolis, Minnesota

On the Cohomology Structure of Stein Manifolds*, **

By

A. Aeppli

Introduction. Summary of results

W. V. D. Hodge considered in [5] differential forms on a compact Kähler manifold and proved certain "natural" isomorphisms between mixed cohomology groups and modules of harmonic forms. Here we study these mixed groups on a Stein manifold X (for the definition and properties of Stein manifolds see [1], [11]) and get the isomorphisms $H^{p,q}_{d/\nabla}(X) \cong H^{p+q}(X;\mathbf{C})$, $H^{p,q}_{\nabla/d',d''}(X) \cong H^{p+q+1}(X;\mathbf{C})$ for $p, q \geqq 1$ and $H^{p,s}_{(d/d')_{d''}}(X) \cong H^{p+s}(X;\mathbf{C})$ for $p \geqq 1$ (Theorems 1 and 2 in Section 4). This last isomorphism generalizes Serre's isomorphism given in [2]. We discuss naturality in Section 5: the mentioned isomorphisms are induced by the obvious imbeddings of forms and with the help of an isomorphism $d': H^{p,q}_{\nabla/d',d''}(X) \to H^{p+1,q}_{(d/d')_{d''}}(X)$ $(p, q \geqq 1)$. As a result we state in Corollaries 1 and 2: every d-exact $(p+q)$-form, $p + q \geqq 1$, on X is d-cohomologous to a pure type (p, q)-form, and a d-total (p, q)-form, $p, q \geqq 1$, on X is ∇-total. In Section 6 the relative d/∇ cohomology groups are treated: Theorem 4 asserts $H^{p,q}_{d/\nabla}(K, L) \cong H^{p+q}(K, L;\mathbf{C})$ if $p, q \geqq 2$ for a pair of Stein manifolds, and in Theorem 5 a short exact sequence is given relating mixed groups with relative mixed groups in case of a pair $(X, \partial\tilde{X})$ where $\partial\tilde{X}$ is a suitable open neighborhood (in X) of

* Research supported by NSF G-24336.
** Received May 1, 1964.

the boundary of X. Finally, Section 7 contains two results on determining and form-determining sets in X: (i) a determining open set in a Stein manifold X with vanishing first Chern class $c_1(X)$ is form-determining (Theorem 6); (ii) $i_*: \pi_1(D) \to \pi_1(X)$ is an epimorphism induced by the imbedding $i: D \to X$, D determining and open in X (Theorem 7).

Preliminary considerations are made in the first three sections; in particular it is shown that for $p, q \geqq 1$ every d-exact (p, q)-form is locally ∇-total, and every ∇-exact (p, q)-form is locally the sum of a d'-total and a d''-total form. A special result of this type is well known and can be found e.g. in [6], p. 72: a Kähler metric has locally a potential.

The methods are sheaf theoretic at the beginning of the paper, following the lines of CARTAN's and SERRE's work [1], [2], [3]. A glossary of notations is given at the end of this introduction. The indexes r, s range preferably, but not necessarily, over the integers between 0 and n (including 0 and n), and the same holds true for the indexes p, q except for further restrictions indicated at the appropriate places. Sequences of groups (mostly of $\boldsymbol{C}$-modules) or sheaves and homomorphisms are understood to be exact. In cohomology, we work with the complex coefficients $\boldsymbol{C}$ and with $\boldsymbol{C}$-sheaves unless stated otherwise. For convenience, all forms are of class C^∞. Complex manifolds are treated here, but often generalizations of the formulas and results in the sequel can be made for complex spaces. Moreover, it appears that the only genuine Stein manifold property used in the first five sections is $H^q(X; \Phi_z^r) = 0$ for $q \geqq 1$.

Glossary of notations

$M = M^{(n)} = M^{2n}$: connected complex manifold of complex dimension n

$X = X^{(n)}$: (connected) Stein manifold of complex dimension n

$d = d' + d''$: exterior differentiation operator, d' of type $(-1, 0)$, d'' of type $(0, -1)$

$\nabla = d'd'' = -d''d'$

$\Omega^{r,s}$: sheaf of germs of (r, s)-forms (on M or X; $\Omega^{r,s} \neq 0$ for $0 \leqq r, s \leqq n$)

$\Omega^k = \sum_{r+s=k} \Omega^{r,s}$: sheaf of germs of k-forms

$Z_{d'}^{r,s} = \{\alpha \mid \alpha \in \Omega^{r,s},\ d'\alpha = 0\}$: sheaf of germs of d'-exact (i. e. d'-closed) (r, s)-forms

$\Phi_{\bar z}^s = Z_{d'}^{0,s}$: sheaf of germs of $\bar z$-holomorphic s-forms

$Z_{d''}^{r,s} = \{\alpha \mid \alpha \in \Omega^{r,s}, d''\alpha = 0\}$: sheaf of germs of d''-exact (r, s)-forms

$\Phi_z^r = Z_{d''}^{r,0}$: sheaf of germs of holomorphic r-forms

$Z_d^{r,s} = \{\alpha \mid \alpha \in \Omega^{r,s},\ d\alpha = 0\} = Z_{d'}^{r,s} \cap Z_{d''}^{r,s}$

$Z^{r,s}_{\nabla} = \{\alpha \mid \alpha \in \Omega^{r,s},\ \nabla\alpha = 0\}$

$Z^{r}_{z} = \{\alpha \mid \alpha \in \Phi^{r}_{z},\ d'\alpha = 0\} = Z^{r,0}_{d}$

$Z^{s}_{\bar z} = \{\alpha \mid \alpha \in \Phi^{s}_{\bar z},\ d''\alpha = 0\} = Z^{0,s}_{d}$

$C = Z^{0}_{z} = Z^{0}_{\bar z}$: trivial sheaf of complex numbers

$P^{r,s}_{d'} = d'\Omega^{r-1,s} = \{\alpha \mid \alpha \in \Omega^{r,s},\ \alpha = d'\beta$ for some $\beta\}$: sheaf of germs of d'-total (r, s)-forms

$P^{r,s}_{d''} = d''\Omega^{r,s-1}$, $P^{r,s}_{\nabla} = \nabla\Omega^{r-1,s-1}$

$\Gamma(M; \Theta) = H^0(M; \Theta)$: module of sections in the sheaf Θ over M

1. The Dolbeault isomorphisms and some related results

$d'd' = 0$, $d''d'' = 0$ and the Poincaré lemma for d' resp. d'' (cf. [3] and [4]) imply $P^{r+1,s}_{d'} = Z^{r+1,s}_{d'}$ resp. $P^{r,s+1}_{d''} = Z^{r,s+1}_{d''}$, hence

$$0 \to Z^{r,s}_{d'} \to \Omega^{r,s} \xrightarrow{d'} Z^{r+1,s}_{d'} \to 0, \tag{1'}$$

$$0 \to Z^{r,s}_{d''} \to \Omega^{r,s} \xrightarrow{d''} Z^{r,s+1}_{d''} \to 0. \tag{1''}$$

Well known arguments lead from (1′), (1″) to the Dolbeault isomorphisms

$$H^{r,s}_{d'}(M) = \Gamma(M; Z^{r,s}_{d'})/d'\Gamma(M; \Omega^{r-1,s}) \cong H^r(M; \Phi^{s}_{\bar z}),$$
$$H^{r,s}_{d''}(M) = \Gamma(M; Z^{r,s}_{d''})/d''\Gamma(M; \Omega^{r,s-1}) \cong H^s(M; \Phi^{r}_{z}).$$

This includes in particular

$$H^q(M; \Phi^{r}_{z}) = 0 \quad \text{for} \quad q > n, \tag{2'}$$

$$H^q(M; \Phi^{s}_{\bar z}) = 0 \quad \text{for} \quad q > n. \tag{2''}$$

The Poincaré lemma holds for holomorphic forms too (and for $\bar z$-holomorphic forms); hence

$$0 \to Z^{r}_{z} \to \Phi^{r}_{z} \xrightarrow{d'} Z^{r+1}_{z} \to 0, \tag{3'}$$

$$0 \to Z^{s}_{\bar z} \to \Phi^{s}_{\bar z} \xrightarrow{d''} Z^{s+1}_{\bar z} \to 0. \tag{3''}$$

(2′), (3′) resp. (2″), (3″) imply

$$H^q(M; Z^{r}_{z}) \cong H^{q+r}(M; C) \quad \text{for} \quad q > n, \tag{4'}$$

$$H^q(M; Z^{s}_{\bar z}) \cong H^{q+s}(M; C) \quad \text{for} \quad q > n. \tag{4''}$$

If $M = X$ is a Stein manifold then $H^q(X; \Phi^{r}_{z}) = 0$ for $q \geqq 1$ (and similarly in the $\bar z$-holomorphic case), and (3′) resp. (3″) yields

$$H^q(X; Z^{r}_{z}) \cong H^{q+r}(X; C) \quad \text{for} \quad q \geqq 1, \tag{$4'_h$}$$

$$H^q(X; Z^{s}_{\bar z}) \cong H^{q+s}(X; C) \quad \text{for} \quad q \geqq 1. \tag{$4''_h$}$$

2. Local considerations involving the operator ∇

The rows and columns in the following diagram are exact by (1'), (1''), (3'), (3''):

$$\left.\begin{array}{ccccccccc}
 & & 0 & & 0 & & 0 & & 0 \\
 & & \downarrow & & \downarrow & & \downarrow & & \downarrow \\
0 \to & \boldsymbol{C} & \to & \Phi^0_z & \overset{d'}{\to} & \Phi^1_z & \overset{d'}{\to} & \Phi^2_z & \overset{d'}{\to} \cdots \\
 & \downarrow & & \downarrow & & \downarrow & & \downarrow & \\
0 \to & \Phi^0_{\bar z} & \to & \Omega^{0,0} & \overset{d'}{\to} & \Omega^{1,0} & \overset{d'}{\to} & \Omega^{2,0} & \overset{d'}{\to} \cdots \\
 & \downarrow d'' & & \downarrow d'' & & \downarrow d'' & & \downarrow d'' & \\
0 \to & \Phi^1_{\bar z} & \to & \Omega^{0,1} & \overset{d'}{\to} & \Omega^{1,1} & \overset{d'}{\to} & \Omega^{2,1} & \overset{d'}{\to} \cdots \\
 & \downarrow d'' & & \downarrow d'' & & \downarrow d'' & & \downarrow d'' & \\
0 \to & \Phi^2_{\bar z} & \to & \Omega^{0,2} & \overset{d'}{\to} & \Omega^{1,2} & \overset{d'}{\to} & \Omega^{2,2} & \overset{d'}{\to} \cdots \\
 & \downarrow d'' & & \downarrow d'' & & \downarrow d'' & & \downarrow d'' & \\
 & \vdots & & \vdots & & \vdots & & \vdots &
\end{array}\right\} \qquad (5)$$

The diagram (5) is commutative along the boundary and anti-commutative inside since $d'd'' + d''d' = 0$. Define $\Lambda^{p,q} = Z^{p,q}_d / P^{p,q}_\nabla$, $\Lambda^{p,0} = Z^p_z / d'\Phi^{p-1}_z$, $\Lambda^{0,q} = Z^q_{\bar z} / d''\Phi^{q-1}_{\bar z}$ for $p, q \geqq 1$. Standard arguments on (5) yield the isomorphisms

$$\Lambda^{r+1,s} \cong \Lambda^{r,s+1}$$

(for $r, s \geqq 0$), and (3'), (3'') contain

$$\Lambda^{p,0} \cong \Lambda^{0,q} = 0 \quad \text{for} \quad p, q \geqq 1 .$$

Thus we get $\Lambda^{r,s} \cong \Lambda^{r+s,0} \cong \Lambda^{0,r+s} = 0$ for $r + s \geqq 1$, in particular $\Lambda^{p,q} = 0$ for $p, q \geqq 1$: *every d-exact (p, q)-form is locally ∇-total for $p, q \geqq 1$*. Therefore:

$$0 \to Z^{r,s}_\nabla \to \Omega^{r,s} \overset{\nabla}{\to} Z^{r+1,s+1}_d \to 0 . \qquad (6)$$

Take an element $\alpha \in Z^{p,q}_\nabla$, i. e. $\nabla\alpha = d'd''\alpha = 0$; if $p, q \geqq 1$ then $\xi = d''\alpha \in Z^{p,q+1}_d$ and by (6) there is a β such that $\xi = -\nabla\beta = d''d'\beta = d''\alpha$, hence $d''(\alpha - d'\beta) = 0$ and by (1'') $\alpha - d'\beta = d''\gamma$ for some γ, i. e. $\alpha = d'\beta + d''\gamma$. This shows[1]

$$Z^{p,q}_\nabla = P^{p,q}_{d'} \oplus P^{p,q}_{d''} = Z^{p,q}_{d'} \oplus Z^{p,q}_{d''} \quad \text{for} \quad p, q \geqq 1 . \qquad (7)$$

[1] $\oplus$ has the following meaning: if A and B are submodules (here we are concerned about $\boldsymbol{C}$-modules) of D, then $A \oplus B$ is the submodule in D generated by $A \cup B$, hence $A \oplus B \cong A + B/B \cap A \cong B + A/A \cap B$ if we consider modules

Thus: *for $p, q \geqq 1$, every ∇-exact (p, q)-form is locally the sum of a d'-total and a d''-total form.*

3. Cohomology with coefficients in $Z_d^{r,s}$ and in $Z_\nabla^{r,s}$

It follows from (6) and (7) that

$$H^q(M; Z_d^{r+1,s+1}) = H^q(M; Z_{d'}^{r+1,s+1} \cap Z_{d''}^{r+1,s+1}) \cong H^{q+1}(M; Z_{d'}^{r,s} \oplus Z_{d''}^{r,s})$$
$$= H^{q+1}(M; Z_\nabla^{r,s}) \text{ for } q \geqq 1, \text{ i. e.}$$

$$H^q(M; Z_d^{r+1,s+1}) \cong H^{q+1}(M; Z_\nabla^{r,s}) \quad \text{for} \quad q \geqq 1. \tag{8}$$

Furthermore, (6) implies

$$0 \to Z_d^{r,s} \to Z_{d''}^{r,s} \xrightarrow{d'} Z_d^{r+1,s} \to 0, \tag{9}$$

and therefore

$$\cdots \to H^q(M; Z_{d''}^{r,s}) \to H^q(M; Z_d^{r+1,s}) \to$$
$$\to H^{q+1}(M; Z_d^{r,s}) \to H^{q+1}(M; Z_{d''}^{r,s}) \to \cdots.$$

(1″) and the Dolbeault isomorphisms yield $H^q(M; Z_{d''}^{r,s}) \cong H^{q+s}(M; \Phi_z^r) \cong H_{d''}^{r,q+s}(M)$ for $q \geqq 1$ and hence $H^q(M; Z_{d''}^{r,s}) = 0$ for $q + s > n$. Thus we get from (9) $H^q(M; Z_d^{r+1,s}) \cong H^{q+1}(M; Z_d^{r,s}) \cong \cdots \cong H^{q+r+1}(M; Z_d^{0,s}) = H^{q+r+1}(M; Z_{\bar z}^s)$ if $q + s > n$, and by (4″)

$$H^q(M; Z_d^{r+1,s}) \cong H^{q+r+1+s}(M; \mathbf{C})$$

for $q + r + 1 > n$ and $q + s > n$, hence[2] (using (4″) again)

$$H^q(M; Z_d^{r,s}) \cong H^{q+r+s}(M; \mathbf{C}) \quad \text{for} \quad q + r > n \quad \text{and} \quad q + s > n. \tag{10}$$

(10) contains (4′) and (4″). (8) and (10) imply

$$H^{q+1}(M; Z_\nabla^{r,s}) \cong H^{q+r+s+2}(M; \mathbf{C})$$
$$\text{for} \quad q \geqq 1, \quad q + r \geqq n, \quad q + s \geqq n. \tag{11}$$

If $M = X$ is a Stein manifold then $H^q(X; Z_{d''}^{r,s}) \cong H^{q+s}(X; \Phi_z^r) = 0$ for $q \geqq 1$, and (9), ($4_h''$) imply

$$H^q(X; Z_d^{r,s}) \cong H^{q+r+s}(X; \mathbf{C}) \quad \text{for} \quad q \geqq 1 \tag{10_h}$$

---

over a field; + stands for taking the direct sum. The $\oplus$-sum $\Theta \oplus \Xi$ of two sheaves Θ, Ξ (Θ, Ξ are subsheaves of Π) is the sheaf in Π whose stalk over a point $x \in M$ is the $\oplus$-sum of the stalks of Θ and of Ξ over x, $\Theta \oplus \Xi$ having the topology induced by the imbedding $\Theta \oplus \Xi \subset \Pi$. We use the operation $\oplus$ in (7) for $\Theta = Z_{d'}^{p,q}$, $\Xi = Z_{d''}^{p,q}$, $\Pi = \Omega^{p,q}$. — *Remark.* $\oplus$ indicates in [7], p. 792 the operation as described here, but unfortunately it appears in the same paper on p. 796—797 as direct sum sign.

[2] Notice, (10) is trivially true for $q = 0$ since then $r, s > n$.

containing $(4_h')$ and $(4_h'')$. (10_h) yields together with (8)

$$H^{q+1}(X; Z_\nabla^{r,s}) \cong H^{q+r+s+2}(X; \mathbf{C}) \quad \text{for} \quad q \geqq 1. \qquad (11_h)$$

We get from (9) and $H^1(X; Z_{d''}^{q-1,s}) = 0$ (for $q \geqq 1$)

$$H^1(X; Z_d^{q-1,s}) \cong \Gamma(X; Z_d^{q,s})/d'\Gamma(X; Z_{d''}^{q-1,s}),$$

hence by (10_h)

$$H^{q+s}(X; \mathbf{C}) \cong \Gamma(X; Z_d^{q,s})/d'\Gamma(X; Z_{d''}^{q-1,s}) \quad \text{for} \quad q \geqq 1, \qquad (12)$$

and (12) contains for $s = 0$ the Serre isomorphisms (cf. [2])

$$H^q(X; \mathbf{C}) \cong \Gamma(X; Z_z^q)/d'\Gamma(X; \Phi_z^{q-1}) = H_h^q(X) \qquad (12_0)$$

which hold for all $q \geqq 0$.

4. Mixed cohomology groups for a Stein manifold

(7) and (9) yield

$$0 \to Z_{d'}^{r,s} \to Z_\nabla^{r,s} \xrightarrow{d'} Z_d^{r+1,s} \to 0 \qquad (13)$$

which induces for a Stein manifold X (using $H^q(X; Z_{d'}^{r,s}) = 0$ for $q \geqq 1$)

$$0 \to H^q(X; Z_\nabla^{r,s}) \xrightarrow{d'} H^q(X; Z_d^{r+1,s}) \to 0 \quad \text{for} \quad q \geqq 1$$

establishing a natural relation between (10_h) and (11_h) and showing that (11_h) is also true for $q = 0$, i. e.

$$H^1(X; Z_\nabla^{r,s}) \cong H^1(X; Z_d^{r+1,s}) \cong H^{r+s+2}(X; \mathbf{C}). \qquad (11_h^0)$$

On the other hand, (6) implies

$$H^1(M; Z_\nabla^{r,s}) \cong \Gamma(M; Z_d^{r+1,s+1})/\nabla\Gamma(M; \Omega^{r,s}).$$

If we combine this isomorphism with (11_h^0) we get

$$H^{p+q}(X; \mathbf{C}) \cong \Gamma(X; Z_d^{p,q})/\nabla\Gamma(X; \Omega^{p-1,q-1}) \quad \text{for} \quad p, q \geqq 1. \qquad (14)$$

Remark. (13) relates (10) with (11) for any complex manifold M just as well as (10_h) with (11_h) in the case of $M = X$.

Defining $H_{d/\nabla}^{p,q}(M) = \Gamma(M; Z_d^{p,q})/\nabla\Gamma(M; \Omega^{p-1,q-1})$ and $H_{(d/d')_{d''}}^{p,s}(X) = \Gamma(M; Z_d^{p,s})/d'\Gamma(M; Z_{d''}^{p-1,s})$ we summarize the results (12) and (14) in

Theorem 1. *For a Stein manifold X, the d/∇ and the $(d/d')_{d''}$ cohomology groups are related to the ordinary cohomology groups by*

$$H_{d/\nabla}^{p,q}(X) \cong H^{p+q}(X; \mathbf{C}) \quad \textit{for} \quad p, q \geqq 1, \qquad (14)$$

$$H_{(d/d')_{d''}}^{p,s}(X) \cong H^{p+s}(X; \mathbf{C}) \quad \textit{for} \quad p \geqq 1. \qquad (12)$$

(13) and $H^1(X; Z_{d'}^{r,s}) = 0$ imply $\Gamma(X; Z_d^{r+1,s}) = d'\Gamma(X; Z_\nabla^{r,s}) \cong \Gamma(X; Z_\nabla^{r,s})/\Gamma(X; Z_{d'}^{r,s})$, and since by (1') $\Gamma(X; Z_{d'}^{p,s}) = d'\Gamma(X; \Omega^{p-1,s})$

for $p \geqq 1$ we also get

$$\Gamma(X; Z_d^{p+1,s}) \cong \Gamma(X; Z_\nabla^{p,s})/d'\Gamma(X; \Omega^{p-1,s}) \quad \text{for} \quad p \geqq 1 . \tag{15}$$

We have (because of (1′), (1″)) for $p, q \geqq 1$

$$\Gamma(X; Z_d^{p,q}) = d'\Gamma(X; \Omega^{p-1,q}) \cap d''\Gamma(X; \Omega^{p,q-1}),$$

and therefore, using (9),

$$\begin{aligned}
& d'\Gamma(X; \Omega^{p-1,q}) \oplus d''\Gamma(X; \Omega^{p,q-1}) \\
& \cong d'\Gamma(X; \Omega^{p-1,q}) + d''\Gamma(X; \Omega^{p,q-1})/(d'\Gamma(X; \Omega^{p-1,q}) \cap d''\Gamma(X; \Omega^{p,q-1})) \\
& \cong d'\Gamma(X; \Omega^{p-1,q}) + d''\Gamma(X; \Omega^{p,q-1})/\Gamma(X; Z_d^{p,q}) \\
& \cong d'\Gamma(X; \Omega^{p-1,q}) + \Gamma(X; Z_{d''}^{p,q})/\Gamma(X; Z_d^{p,q}) \\
& \cong d'\Gamma(X; \Omega^{p-1,q}) + d'\Gamma(X; Z_{d''}^{p,q}) .
\end{aligned}$$

Now we get with the help of (15) and (12)

$$\begin{aligned}
H_{\nabla/d',d''}^{p,q}(X) &= \Gamma(X; Z_\nabla^{p,q})/(d'\Gamma(X; \Omega^{p-1,q}) \oplus d''\Gamma(X; \Omega^{p,q-1})) \\
&\cong \Gamma(X; Z_\nabla^{p,q})/(d'\Gamma(X; \Omega^{p-1,q}) + \Gamma(X; Z_{d''}^{p,q})/\Gamma(X; Z_d^{p,q})) \\
&\cong \Gamma(X; Z_d^{p+1,q})/d'\Gamma(X; Z_{d''}^{p,q}) \\
&= H_{(d/d')_{d''}}^{p+1,q}(X) \cong H^{p+q+1}(X; \mathbf{C});
\end{aligned}$$

thus we proved

Theorem 2. *For a Stein manifold X, the $\nabla/d', d''$ cohomology groups fulfil the isomorphisms*

$$H_{\nabla/d',d''}^{p,q}(X) \cong H^{p+q+1}(X; \mathbf{C}) \quad \textit{for} \quad p, q \geqq 1 . \tag{16}$$

(12_0), (12), (14), (16) imply $H^p(X; \mathbf{C}) = 0$ for $p > n$, $H_{d/\nabla}^{p,q}(X) = 0$ for $p, q \geqq 1$ and $p + q > n$, $H_{(d/d')_{d''}}^{p,s}(X) = 0$ for $p \geqq 1$ and $p + s > n$, $H_{\nabla/d',d''}^{p,q}(X) = 0$ for $p, q \geqq 1$ and $p + q \geqq n$.

5. Naturality

We put $H_d^k(M) = \Gamma(M; Z_d^k)/d\Gamma(M; \Omega^{k-1})$ for a differentiable manifold M and $Z_d^k = \{\alpha \mid \alpha \in \Omega^k, d\alpha = 0\}$. De Rham's theorem says

$$H_d^k(M) \cong H^k(M; \mathbf{C}) \tag{17}$$

which is proved in a similar way as (12) or (14). The isomorphisms in Theorems 1 and 2 will be regarded as naturally induced by homomorphisms into $H_d^k(X)$, namely:

$$h_1 : H_{d/\nabla}^{p,q}(X) \to H_d^{p+q}(X) , \qquad p, q \geqq 1 ,$$

and

$$h_2 : H_{(d/d')_{d''}}^{p,q}(X) \to H_d^{p+q}(X) , \qquad p \geqq 1 ,$$

are the naturally induced homomorphisms which we get by considering a d-exact (p, q)-form as a d-exact $(p+q)$-form and passing from the d/∇ resp. $(d/d')_{d''}$ cohomology to the d cohomology. Starting with the commutative diagram

$$\begin{array}{ccccccccc} 0 & \to & Z_d^{r,s} & \to & Z_{d''}^{r,s} & \xrightarrow{d'} & Z_d^{r+1,s} & \to & 0 \\ & & \downarrow & & \downarrow & & \downarrow & & \\ 0 & \to & Z_d^{r+1} & \to & \Omega^{r+s} & \xrightarrow{d} & Z_d^{r+s+1} & \to & 0 \end{array}$$

(the vertical mappings are imbeddings) and going through the steps of the proof of (12) and (14) we "map" all the arguments into the corresponding ones of the proof for (17), and we end up with recognizing h_1 and h_2 as isomorphisms[3] corresponding to (14) and (12).

Next we define the homomorphism

$$d' : H^{p,q}_{\nabla/d',d''}(X) \to H^{p+1,q}_{(d/d')_{d''}}(X)\,, \qquad p, q \geqq 1\,, \tag{18}$$

by sending a ∇-exact (p, q)-form α (representing an element in $H^{p,q}_{\nabla/d',d'}(X)$) into $d'\alpha$ which is d-exact and determines an element in $H^{p+1,q}_{(d/d')_{d''}}(X)$.

Lemma. *d' is an isomorphism.*

Proof. (i) d' is monomorphic: suppose $d'\alpha = d'\beta$ for the ∇-exact (p, q)-form α and the d''-exact β. Since $H^{p,q}_{d''}(X) = 0$ for $q \geqq 1$ there exists a $(p, q-1)$-form γ with $\beta = d''\gamma$. Hence $d'\alpha = d'd''\gamma$, i.e. $d'(\alpha - d''\gamma) = 0$. $H^{p,q}_{d'}(X) = 0$ for $p \geqq 1$ yields the existence of a form ε such that $\alpha - d''\gamma = d'\varepsilon$, hence $\alpha = d'\varepsilon + d''\gamma$ determines the element 0 in $H^{p,q}_{\nabla/d',d''}(X)$.

(ii) d' is epimorphic: if α is a $(p+1, q)$-form representing an element in $H^{p+1,q}_{(d/d')_{d''}}(X)$ then $d'\alpha = d''\alpha = 0$, and since $H^{p+1,q}_{d'}(X) = 0$ there is a β with $d'\beta = \alpha$ and $\nabla\beta = 0$ determining an element in $H^{p,q}_{\nabla/d',d''}(X)$.

We define now the homomorphism

$$h_3 = h_2 d' : H^{p,q}_{\nabla/d',d''}(X) \to H^{p+q+1}_d(X)\,, \qquad p, q \geqq 1\,,$$

which is an isomorphism by the Lemma and since h_2 is bijective, in fact h_3 is the isomorphism (16) if we replace $H(X; \mathbf{C})$ by $H_d(X)$. Therefore, we state

Theorem 3. *Using de Rham's theorem* (17) *we get the isomorphisms* (12), (14), (16) *by the natural imbeddings of the forms and by composition with the isomorphism d' in* (18).

---

[3] The isomorphism $H^{p,q}_{d/\nabla}(X) \cong H^{p,p}_{(d/d')_{d''}}(X)$ (for $p, q \geqq 1$) and injectivity of the homomorphisms h_1, h_2 above can be directly proved by using repeatedly $H^{p,q}_{d'}(X) = 0$ for $p \geqq 1$, $H^{p,q}_{d''}(X) = 0$ for $q \geqq 1$, and type arguments. (14) and (12) imply then bijectivity of h_1 and h_2 in case of finite Betti numbers. Moreover, a diagram like (5) for the modules $\Gamma(\Omega^{r,s})$, $\Gamma(\Phi^r_z)$, $\Gamma(\Phi^s_{\bar z})$ leads to $H^{p,q}_{d/\nabla}(X) \cong H^{p+q}_h(X)$, $p, q \geqq 1$, and by the naturality of the Serre isomorphism (12_0) h_1 is proved to be the naturally induced isomorphism corresponding to (14).

We deduce from the bijectivity of h_1

Corollary 1. *A d-total form of pure type (p, q), $p, q \geqq 1$, on a Stein manifold is ∇-total.*

Since h_2 is bijective and also the analogous homomorphism from $H^{p,q}_{(d/d'')_{d'}}(X)$ to $H^{p+q}_{d}(X)$ for $q \geqq 1$, we get

Corollary 2. *A d-exact $(p+q)$-form, $p + q \geqq 1$, on a Stein manifold is d-cohomologous to a form of pure type (p, q).*

Remark. The isomorphisms h_1, h_2 enable us to relate the cohomology rings[4] $H^{*}_{d', d''/\nabla}(X)$, $H^{*}_{(d/d')_{d''}}(X)$ with $H^{*}_{d}(X)$: h_1 and h_2 respect the product induced by the exterior multiplication of forms. — Of course, the $\nabla/d', d''$ cohomology groups do not have a natural ring structure, and a multiplicative homomorphism h_3 does not make sense.

6. The relative d/∇ cohomology groups

Let (K, L) be a pair of complex manifolds such that either (i) dim K = dim L, L = open subset of K, or (ii) dim $K >$ dim L, L = complex submanifold of K, as a point set a closed subset of K. For these situations there is an exact sequence described in [7], 4. (a), (b), (c) which combined with the de Rham exact sequence for (K, L) leads to the following diagram:

$$\left.\begin{array}{l}
H^{p,q}_{\nabla/d', d''}(K) \xrightarrow{\bar{i}} H^{p,q}_{\nabla/d', d''}(L) \xrightarrow{\nabla} H^{p+1,q+1}_{d/\nabla}(K, L) \xrightarrow{\bar{j}} H^{p+1,q+1}_{d/\nabla}(K) \xrightarrow{\bar{\bar{i}}} \\
\quad\downarrow -h_3 \qquad\qquad\quad \downarrow -h_3 \qquad\qquad\quad \downarrow \tilde{h}_1 \qquad\qquad\qquad \downarrow h_1 \\
H^{p+q+1}_{d}(K) \xrightarrow{i} H^{p+q+1}_{d}(L) \xrightarrow{d} H^{p+q+2}_{d}(K, L) \xrightarrow{j} H^{p+q+2}_{d}(K) \xrightarrow{i} \\
\qquad\qquad \xrightarrow{\bar{\bar{i}}} H^{p+1,q+1}_{d/\nabla}(L) \\
\qquad\qquad\qquad\quad \downarrow h_1 \\
\qquad\qquad \xrightarrow{i} H^{p+q+2}_{d}(L)
\end{array}\right\} \quad (19)$$

Naturality of the homomorphisms h_1, h_3 implies commutativity in the diagram (19); since $\nabla = d'd'' = -d''d'$ we have to choose $-h_3 = -h_2 d'$ instead of h_3.

a) Suppose K and L are Stein manifolds. If $p,q \geqq 1$ then Theorem 3 asserts that the homomorphisms $-h_3$ and h_1 in (19) are isomorphisms, hence $\tilde{h}_1$ is an isomorphism by the five lemma; moreover, h_3 is epimorphic for all $p,q \geqq 0$ (since d' is always epimorphic and h_2 is an isomorphism for $p \geqq 1$) and the h_1's are still isomorphisms in this case whence $\tilde{h}_1$ is an epimorphism for $p,q \geqq 0$ by one part of the five lemma (cf. [9], p. 14). We proved

[4] Notice, here we have to write $H_{d', d''/\nabla}(M)$ whereas $H^{p,q}_{d', d''/\nabla}(M) = H^{p,q}_{d/\nabla}(M)$ by type arguments.

Theorem 4. *For a pair (K, L) of Stein manifolds, $\tilde{h}_1: H^{p,q}_{d/\nabla}(K, L) \to H^{p+q}_d(K, L)$ is an isomorphism for $p, q \geqq 2$ and an epimorphism for $p, q \geqq 1$.*

b) Let $K = X^{(n)}$ be a Stein manifold and $L = \widetilde{\partial X}$ an open neighborhood of the boundary ∂X (∂X includes the "ends") in X such that $X \simeq X - \widetilde{\partial X}$. Then $H^k_d(X, \widetilde{\partial X}) \cong H^k(X, \widetilde{\partial X}; \boldsymbol{C}) \cong H_{2n-k}(X - \widetilde{\partial X}; \boldsymbol{C}) \cong H_{2n-k}(X; \boldsymbol{C}) = 0$ for $k < n$ which implies[5]

$$\left.\begin{array}{l} i: H^k_d(X) \to H^k_d(\widetilde{\partial X}) \text{ is an isomorphism for } 0 \leqq k \leqq n-2 \\ \qquad\qquad\qquad \text{and a monomorphism for } 0 \leqq k \leqq n-1 \end{array}\right\} \tag{20}$$

(cf. also [8]). Since h_1 and h_3 in dimension (p, q) are isomorphisms for $K = X$ and for $p, q \geqq 1$ (by Theorem 3), (19) and (20) yield

Theorem 5. *Let $\widetilde{\partial X}$ be an open boundary neighborhood in $X^{(n)}$ such that $X \simeq X - \widetilde{\partial X}$. Then*

$$\left.\begin{array}{l} 0 \to H^{p,q}_{\nabla/d',d''}(X) \xrightarrow{\bar{i}} H^{p,q}_{\nabla/d',d''}(\widetilde{\partial X}) \xrightarrow{\nabla} H^{p+1,q+1}_{d/\nabla}(X, \widetilde{\partial X}) \to 0 \\ \text{for} \quad p, q \geqq 1 \quad \text{and} \quad p + q \leqq n - 3, \end{array}\right\} \tag{21}$$

$$0 \to H^{p,q}_{\nabla/d',d''}(X) \xrightarrow{\bar{i}} H^{p,q}_{\nabla/d',d''}(\widetilde{\partial X}) \quad \text{for} \quad p, q \geqq 1, \quad p + q \leqq n - 2, \tag{22}$$

$$0 \to H^{p,q}_{d/\nabla}(X) \xrightarrow{\bar{\bar{i}}} H^{p,q}_{d/\nabla}(\widetilde{\partial X}) \quad \text{for} \quad p, q \geqq 1 \quad \text{and} \quad p + q \leqq n - 1. \tag{23}$$

(23) (resp. (22)) asserts that a d- (resp. ∇-) exact (p, q)-form in X, $p, q \geqq 1$ and $p + q \leqq n - 1$ (resp. $\leqq n - 2$), which is ∇- (resp. (d', d'')-) total in $\widetilde{\partial X}$ is ∇- (resp. (d', d'')-) total in X. (21) combines (22) and (23). Thus Theorem 5 deals with the passage from the boundary behavior to the global behavior for forms on Stein manifolds, in analogy to well known circumstances in the theory of holomorphic functions.

7. Determining and form-determining sets

As an application of Corollary 1 in Section 5 we prove:

Proposition. *If D is a determining set in a simply connected Stein manifold X with vanishing first Chern class $c_1(X)$ then D is form-determining in X.*

This proposition will be improved below in case D is open. The definitions and some properties of determining and form-determining sets are given in [8]. $c_1(X)$ is defined to be the (unique) 2-dimensional real cohomology class determined by $\sqrt{-1}\, \nabla \log g$, g a nowhere vanishing density on $X^{(n)}$ (g real, $g > 0$, $g\, dz^1 \wedge \ldots \wedge dz^n \wedge d\bar{z}^1 \wedge \ldots \wedge d\bar{z}^n$ is

[5] This implies in particular: every boundary neighborhood $\widetilde{\partial X}$ in the (connected) Stein manifold $X^{(n)}$, $n \geqq 2$, is connected.

a (n, n)-form on X), e.g. $g = \det g_{ij}$ for some Hermitian metric

$$\sum_{i,j=1}^{n} g_{ij}\, dz^i\, d\bar{z}^j \quad \text{on} \quad X.$$

Proof of the Proposition. $c_1(X) = 0$ means: the (1,1)-form $\omega = \sqrt{-1}\, \nabla \log g$ is d-total, hence by Corollary 1 in 5. ∇-total, i.e. there exists a function f on X such that $\omega = -\sqrt{-1}\, \nabla f$. Since g is real we get $\omega = \bar{\omega} = -\sqrt{-1}\, \nabla \bar{f}$ (using $\bar{\nabla} = d''d' = -\nabla$) whence $\nabla(f - \bar{f}) = 0$ so that f can be chosen as a real function. Thus $\nabla \log(g \cdot e^f) = 0$ for some real function f. Therefore $g \cdot e^f = e^h \cdot e^{\bar{h}}$ locally for a holomorphic h where e^h is unique up to a constant factor of absolute value 1; if X is simply connected then $e^h dz^1 \wedge \ldots \wedge dz^n$ represents a global nowhere vanishing holomorphic n-form ζ on X starting with a specific choice of e^h at some point.

The Proposition is now a consequence of the following statement: if there exists on $X^{(n)}$ a nowhere vanishing holomorphic n-form ζ then a determining set D is form-determining. Proof: assume there is a holomorphic k-form α in D which does not have a unique holomorphic extension over X; then we find (since X is a Stein manifold) a holomorphic $(n-k)$-form β on X such that $\alpha \wedge \beta$ is a holomorphic n-form in D and $\alpha \wedge \beta$ is not uniquely extendable over X; hence $\frac{\alpha \wedge \beta}{\zeta} = \frac{u}{v}$ (where $\alpha \wedge \beta = u\, dz^1 \wedge \ldots \wedge dz^n$ and $\zeta = v\, dz^1 \wedge \ldots \wedge dz^n$ in local coordinates) is a holomorphic function in D, non-extendable (in a unique way) over X; this contradicts the assumption that D is determining in X, q.e.d.

The simple connectedness condition for X in the Proposition can be left out:

Theorem 6. *Suppose X is a Stein manifold with vanishing first Chern class, then a determining open set D in X is form-determining.*

Corollary. *If X is a Stein manifold for which the second Betti number is zero then a determining open set in X is form-determining.*

Theorem 6 is proved with the help of the following two lemmas. We recall that a covering $\hat{X}$ of a Stein manifold X is a Stein manifold by [*10*]. "Covering" means unbounded, unramified covering.

Lemma 1. *If X is a Stein manifold, $\hat{X}$ a covering of X, $p: \hat{X} \to X$ the projection, $\hat{D} = p^{-1} D$ a determining resp. form-determining set in $\hat{X}$ (for $D \subset X$), then D is determining resp. form-determining in X.*

Lemma 2. *If $p: \hat{X} \to X$ as in Lemma 1, and if D is determining and open in X, then $\hat{D} = p^{-1} D$ is determining in $\hat{X}$.*

Proof of Theorem 6 (assuming Lemmas 1 and 2). Consider the universal covering $\hat{X}$ of X, $p: \hat{X} \to X$. $\hat{D} = p^{-1} D$ is determining in $\hat{X}$ according to Lemma 2. $c_1(\hat{X}) = p^* c_1(X) = 0$ whence $\hat{D}$ is form-determining in $\hat{X}$ by the Proposition. Then Lemma 1 says that D is form-determining in X.

Proof of Lemma 1. Every holomorphic function or form α in D induces a lifted holomorphic function or form $\hat{\alpha}$ in $\hat{D}$ compatible with the projection p, i.e. if $p\hat{x}_1 = p\hat{x}_2$ then the induced biholomorphic mapping q of a neighborhood N_1 of $\hat{x}_1$ onto a neighborhood N_2 of $\hat{x}_2$ relates $\hat{\alpha}_1 = \hat{\alpha}\,|\,N_1$ with $\hat{\alpha}_2 = \hat{\alpha}\,|\,N_2$: $\hat{\alpha}_1 = q^*\hat{\alpha}_2$. $\hat{\alpha}$ has a holomorphic extension $\hat{\beta}$ over $\hat{X}$ which is again compatible with p; hence $\hat{\beta} = p^*\beta$ for a unique holomorphic β extending α over X; thus D is determining or form-determining in X.

Proof of Lemma 2. Since every covering is covered by the universal covering it is enough by Lemma 1 to prove Lemma 2 for $\hat{X}$ = universal covering of X. Construct the holomorphy envelope $\hat{Y}$ for a connectivity component $\hat{D}_1$ of $\hat{D} = p^{-1}D$ in $\hat{X}$ (holomorphy envelope in the sense of [*1*], Exp. VII: $\hat{D}_1$ is a domain in $\hat{X}$ and $\hat{Y}$ is a complex manifold containing $\hat{D}_1$, $n = \dim_{\mathbb{C}} \hat{Y} = \dim_{\mathbb{C}} \hat{X}$). $\hat{Y}$ is a Stein manifold immersed in $\hat{X}$ by a holomorphic mapping $i\colon \hat{Y} \to \hat{X}$, and $\hat{D}_1$ is determining in $\hat{Y}$. $i\,|\,\hat{D}_1$ is an imbedding. A subgroup G of $\pi_1(X)$ acts on $\hat{D}_1$ (and on $\hat{X}$) as fixed point free automorphism group such that $\hat{D}_1/G = D$, and therefore G acts also freely on $\hat{Y}$ as an automorphism group (since every automorphism of $\hat{D}_1$ is the restriction of a well determined automorphism of $\hat{Y}$, cf. [*8*], Sec. 5 c); the fixed point set of an automorphism is an analytic set which has to intersect $\hat{D}_1$ if it is not empty, hence a free automorphism of $\hat{D}_1$ comes from a free automorphism of $\hat{Y}$). We conclude: $\hat{D}_1/G = p\hat{D}_1 = D \subset \hat{Y}/G = p\hat{Y} = Y$ immersed in $\hat{X}/\pi_1(X) = p\hat{X} = X$. Y is an open set immersed in the Stein manifold X, Y is pseudo-convex. Y is even a Stein manifold (because Y can be considered as a fundamental domain of G in $\hat{Y}$, hence as a pseudo-convex domain in a Stein manifold; cf. [*10*], Introduction and [*11*]), and D is determining in both Stein manifolds X and Y. Therefore $X = Y$ by the uniqueness theorem in [*8*], Sec. 5 whence $\hat{X} = \hat{Y}$, i.e. $\hat{D}_1$ is determining in $\hat{X}$. Hence $\hat{D}_1 = \hat{D}$ is connected, and $\hat{D}$ is determining in $\hat{X}$, q.e.d.

The connectivity of $\hat{D} = p^{-1}D$ in the situation just described implies:

Theorem 7. $\pi_1(D) \xrightarrow{i_*} \pi_1(X) \to 0$ *for a determining open set D in the Stein manifold X, D imbedded by i.*

Remark. Lemma 2 is proved in a simple way for any determining D (open or not) if $\hat{X}$ is a finite covering of X, with the help of the elementary symmetric functions.

References

[1] H. Cartan Séminaire 1951—52, ENS Paris.
[2] SERRE, J.-P.: H. Cartan Séminaire 1951—52, ENS Paris, Exposé XX.
[3] — H. Cartan Séminaire 1953—54, ENS Paris, Exposé XVIII.
[4] NICKERSON, H. K.: On the complex form of the Poincaré lemma. Proc. Amer. Math. Soc. **9**, 182—188 (1958).
[5] HODGE, W.V.D.: Differential forms on a Kähler manifold. Proc. Cambridge Phil. Soc. **47**, 504—517 (1951).
[6] WEIL, A.: Introduction à l'étude des variétés kählériennes. Actualités scientifiques et industrielles 1267. Paris: Hermann 1958.
[7] AEPPLI, A.: Some exact sequences in cohomology theory for Kähler manifolds. Pac. J. Math. **12**, 791—799 (1962).
[8] — On determining sets in a Stein manifold. These Proceedings 48—58.
[9] MACLANE, S.: Homology. Springer — Academic Press 1963.
[10] STEIN, K.: Überlagerungen holomorph-vollständiger komplexer Räume. Arch. Math. VII 354—361, (1956).
[11] GRAUERT, H.: Charakterisierung der holomorph-vollständigen komplexen Räume. Math. Ann. **129**, 233—259 (1955).

School of Mathematics
Institute of Technology
University of Minnesota
Minneapolis, Minnesota

Normal Families of Non-negative Divisors*, **

By

W. STOLL

As early as 1934, OKA [7] mentioned, without proof, that a family of non-negative divisors is normal if and only if the area of the divisors is uniformly bounded on compact subsets. This report is concerned with recent results obtained on this problem. At first some pertinent definitions and facts shall be reviewed.

Let G be a complex manifold[1] with the structure sheaf $\mathfrak{O}$. For $a \in G$, let $\mathfrak{O}_a$ be the stalk over a. Let $\mathfrak{m}_a$ be the maximal ideal in $\mathfrak{O}_a$. If $0 \neq f_a \in \mathfrak{O}_a$, one and only one number $\nu = \nu_{f_a}$ exists such that

$$f_a \in \mathfrak{m}_a^{\nu} \quad \text{but} \quad f_a \notin \mathfrak{m}_a^{\nu+1}.$$

* This is an abstract of a talk which was given at the Conference on Complex Analysis at the University of Minnesota, March 16—March 22, 1963. The results were obtained in a paper (STOLL [13]) which is being published in the Mathematische Zeitschrift. This research was partially supported by the National Science Foundation under Grant G-22126.

** Received April 22, 1964.

1 Here, all complex manifolds are pure-dimensional and have a countable base of open sets.

A function $\nu: G \to \mathbf{Z}$ is said to be a *divisor* on G, if and only if for every $a \in G$ an open connected neighborhood U of a and holomorphic functions $g \not\equiv 0$ and $h \not\equiv 0$ exist such that

$$\nu(z) = \nu_{g_z} - \nu_{h_z} \qquad \text{for} \quad z \in U,$$

where g_z and h_z are the germs at z of g and h respectively. The set $\mathfrak{D}(G)$ of all divisors on G forms a partially ordered abelian group under function addition. A *divisor* ν is said to be *non-negative* if $\nu(z) \geqq 0$ for all $z \in G$. The set $\mathfrak{D}^+(G)$ of all non-negative divisors is a semi-group. The *support* of a divisor ν is defined by

$$\mathfrak{M}(\nu) = \overline{\{z \mid \nu(z) \neq 0\}}.$$

The support is empty or an analytic set of pure codimension 1. If ν is non-negative, then

$$\mathfrak{M}(\nu) = \{z \mid \nu(z) > 0\}.$$

Each meromorphic function f, which is not identically zero on each connectivity component of G, defines a divisor ν_f, which is called the *principal divisor of* f. The meromorphic function f is holomorphic if and only if ν_f is non-negative. An open subset U of G is called a COUSIN-II-domain if and only if each divisor on U is a principal divisor. The COUSIN-II-domains on G form a base of open subsets of G.

The concept of a normal family of non-negative divisors requires a convergence concept on $\mathfrak{D}^+(G)$. KELLEY [4] gives four conditions which assert that the convergence is a topological convergence. These conditions involve MOORE-SMITH sequences which are also called *nets*. Therefore, a concept of convergence shall now be introduced for nets of non-negative divisors.

A net $\mathfrak{N} = \{\nu_\lambda\}_{\lambda \in \Lambda}$ of non-negative divisors ν on G is said to be *convergent* on G if and only if *for every* $a \in G$ *an open connected neighborhood* U *of* a *and a net* $\{f_\lambda\}_{\lambda \in \Lambda}$ *exists such that*

1. *The function* f_λ *is holomorphic and not identically zero on* U *for each* $\lambda \in \Lambda$.

2. *For each* $\lambda \in \Lambda : \nu_{f_\lambda} = \nu_\lambda \mid U$.

3. *The net* $\{f_\lambda\}_{\lambda \in \Lambda}$ *converges:*

$$f_\lambda \to f \not\equiv 0 \quad \text{for} \quad \lambda \to \Lambda.$$

The convergence is uniform on every compact subset of U. *The limit* f *is not identically zero on* U. *(Of course,* f *is holomorphic on* U.*)*

If $\mathfrak{N}$ converges, one and only one non-negative divisor ν exists such that, for every family $\{f_\lambda\}_{\lambda \in \Lambda}$ satisfying these three conditions, $\nu \mid U = \nu_f$ holds. Define ν to be the *limit of* $\mathfrak{N}$, that is:

$$\nu = \lim_{\lambda \to \Lambda} \nu_\lambda \quad \text{or} \quad \nu_\lambda \to \nu \quad \text{for} \quad \lambda \to \Lambda.$$

A family $\mathfrak{N}$ of non-negative divisors on G is said to be *normal* on G if every net in $\mathfrak{N}$ has a convergent subnet.

This concept of convergence of non-negative divisors satisfies the four conditions of Kelley [4]. However the proof requires some preparations which shall be reported now.

Let G be a complex manifold of dimension $m > 1$. Let χ be the exterior form of bidegree (1, 1) associated with a continuous Hermitian metric on G. Define

$$\chi_p = \frac{1}{p} \chi \wedge \cdots \wedge \chi \qquad (p \text{ times}).$$

Let N be a pure p-dimensional analytic subset of G. Let K be a subset of N such that $N \cap K$ is measurable on N. Then, the *area* (*volume*) of N on K is

$$V_N(K, \chi) = \int_{N \cap K} \chi_p \geqq 0 .$$

If $\bar{K}$ is compact, then $V_N(K, \chi) < \infty$. Define

$$V_\varnothing(K, \chi) = 0 .$$

A family $\mathfrak{N} = \{N_\lambda\}_{\lambda \in \Lambda}$ of analytic sets N_λ where N_λ is empty or has pure dimension p is said to be *bounded on the compact set* K in G if and only if a constant $L > 0$ exists such that

$$V_{N_\lambda}(K, \chi) \leqq L \quad \text{for all} \quad \lambda \in \Lambda .$$

If $\mathfrak{N}$ is bounded on K with respect to one continuous Hermitian metric then it is also bounded on K with respect to any other continuous Hermitian metric on G.

Let ν be a divisor on the complex manifold G. Let K be a compact subset of G. Define

$$V_\nu(K, \chi) = \int_{K \cap \mathfrak{M}(\nu)} \nu(z) \, \chi_{m-1} .$$

If ν is non-negative, then $V_\nu(K, \chi) \geqq 0$. A family $\mathfrak{N} = \{\nu_\lambda\}_{\lambda \in \Lambda}$ of non-negative divisors is said to be *bounded on the compact subset* K of G if and only if a constant $L > 0$ exists such that

$$V(K, \chi) \leqq L \quad \text{for all} \quad \lambda \in \Lambda .$$

Again, if $\mathfrak{N}$ is bounded on K with respect to one continuous Hermitian metric then it is bounded on K with respect to any other continuous Hermitian metric on G.

Now, the special case of open subsets of the m-dimensional complex vector space $\boldsymbol{C}^m$ shall be considered.

For

$$\mathfrak{z} = (z_1, \ldots, z_m) \in \boldsymbol{C}^m , \qquad \mathfrak{w} = (w_1, \ldots, w_m) \in \boldsymbol{C}^m$$

define

$$(\mathfrak{z} \mid \mathfrak{w}) = \sum_{\mu=1}^{m} z_\mu \bar{w}_\mu , \qquad |\mathfrak{z}| = \sqrt{(\mathfrak{z} \mid \mathfrak{z})} ,$$

$$(d\mathfrak{z} \mid \mathfrak{z}) = \sum_{\mu=1}^{m} \bar{z}_\mu dz_\mu , \qquad (\mathfrak{z} \mid d\mathfrak{z}) = \sum_{\mu=1}^{m} z_\mu d\bar{z}_\mu ,$$

$$(d\mathfrak{z} \mid d\mathfrak{z}) = \sum_{\mu=1}^{m} dz_\mu \wedge d\bar{z}_\mu .$$

The exterior form associated with the Hermitian metric of Euclidean space is

$$v(\mathfrak{z}) = \frac{i}{2} (d\mathfrak{z} \mid d\mathfrak{z});$$

put

$$v_p(\mathfrak{z}) = \frac{1}{p!} v(\mathfrak{z}) \wedge \cdots \wedge v(\mathfrak{z}) \qquad (p \text{ times}) .$$

If $\mathfrak{z} \neq 0$, define

$$\omega(\mathfrak{z}) = v\left(\frac{\mathfrak{z}}{|\mathfrak{z}|}\right) = \frac{i}{2} |\mathfrak{z}|^{-4} [|\mathfrak{z}|^2 (d\mathfrak{z} \mid d\mathfrak{z}) - (d\mathfrak{z} \mid \mathfrak{z}) \wedge (\mathfrak{z} \mid d\mathfrak{z})],$$

$$\omega_p(\mathfrak{z}) = v_p\left(\frac{\mathfrak{z}}{|\mathfrak{z}|}\right) = \frac{1}{p!} \omega(\mathfrak{z}) \wedge \cdots \wedge \omega(\mathfrak{z}) .$$

The form ω_p has a non-negative density in every pure p-dimensional analytic subset of $\boldsymbol{C}^m - \{0\}$. Define

$$B(\mathfrak{a}, r) = \{\mathfrak{z} \mid |\mathfrak{z} - \mathfrak{a}| < r\} ,$$

$$B(r) = B(0, r) ,$$

$$S(\mathfrak{a}, r) = \{\mathfrak{z} \mid |\mathfrak{z} - \mathfrak{a}| = r\} ,$$

$$S(r) = S(0, r) ,$$

$$W_p = \frac{\pi^p}{p!} , \qquad \Phi(r) = \frac{2\pi^m}{(m-1)!} r^{2m-1} .$$

Denote by σ_r the volume element of $S(\mathfrak{a}, r)$ which is oriented towards the exterior.

If ν is a divisor on the open subset G of $\boldsymbol{C}^m$ and if K is compact and contained in G, define as before

$$V_\nu(K) = V_\nu(K, v) = \int_{K \cap \mathfrak{M}(\nu)} \nu(\mathfrak{z})\, v_{m-1}(\mathfrak{z}) .$$

If $B(\mathfrak{a}, R) \subseteqq G$ and $0 < s < r < R$ define

$$n_\nu(\mathfrak{a}, r) = \frac{V_\nu(\overline{B(\mathfrak{a}, r)})}{W_{m-1}\, r^{2m-2}} ,$$

$$N_\nu(\mathfrak{a}, r, s) = \int_s^r n_\nu(\mathfrak{a}, t) \frac{dt}{t} .$$

If ν is non-negative, then the functions $n_\nu(\mathfrak{a}, r)$ (for $R > r > 0$) and $N_\nu(\mathfrak{a}, r, s)$ (for $R > r > s$) are non-negative and monotonically increasing. If $\mathfrak{a} \notin \mathfrak{M}(v)$ the integral $N_\nu(\mathfrak{a}, r, s)$ exists also for $s = 0$. Define

$$N_\nu(\mathfrak{a}, r) = N_\nu(\mathfrak{a}, r, 0) \quad \text{if} \quad \mathfrak{a} \notin \mathfrak{M}(\nu).$$

If $\mathfrak{a} = 0$, define

$$n_\nu(r) = n_\nu(0, r), \quad N_\nu(r, s) = N_\nu(0, r, s), \quad N_\nu(r) = N_\nu(0, r).$$

For $|z| < 1$ define

$$l(z) = \frac{1}{(m-1)!} \frac{d^{m-1}}{dz^{m-1}} [z^{m-1} \operatorname{Log}(1 - z)]$$

where Log means the principal value of the logarithm. Then the following Jensen formula can be proved:

Jensen formula[2]. *Let $f \not\equiv 0$ be a meromorphic function on the open connected neighborhood G of $\overline{B(r)}$. Take s in $0 < s < r$. Then*

$$N_{\nu_f}(r, s) = \frac{1}{\Phi(r)} \int_{S(r)} \log|f|\, \sigma_r - \frac{1}{\Phi(s)} \int_{S(s)} \log|f|\, \sigma_s .$$

If $0 \notin \mathfrak{M}(\nu_f)$ then

$$N_{\nu_f}(r) = \frac{1}{\Phi(r)} \int_{S(r)} \log|f|\, \sigma_r - \log|f(0)| .$$

Jensen-Poisson formula[2]. *Let $f \not\equiv 0$ be a meromorphic function on the open connected neighborhood G of $\overline{B(r)}$. Suppose that $0 < s < r$ and $\mathfrak{M}(\nu_f) \cap \overline{B(s)} = \emptyset$. Assume that $f(0) = 1$. Define* $\log f$ *in* $B(s)$ *by* $\log f(0) = 0$. *Then the following equation holds for* $|\mathfrak{z}| \leqq s$:

$$\begin{aligned}\log f(\mathfrak{z}) = {} & \frac{2}{\Phi(r)} \int_{S(r)} \log|f(\mathfrak{y})| \left[\frac{r^{2m}}{[r^2 - (\mathfrak{z}|\mathfrak{y})]^m} - 1\right] \sigma_r(\mathfrak{y}) - \\ & - \frac{1}{W_{m-1}} \int_{\mathfrak{M}(\nu_f) \cap B(r)} \nu_f(\mathfrak{y})\, l\left(\frac{(\mathfrak{z}|\mathfrak{y})}{r^2}\right) \omega_{m-1}(\mathfrak{y}) + \\ & + \frac{1}{W_{m-1}} \int_{\mathfrak{M}(\nu_f) \cap B(r)} \nu_f(\mathfrak{y})\, l\left(\frac{(\mathfrak{z}|\mathfrak{y})}{(\mathfrak{y}|\mathfrak{y})}\right) \omega_{m-1}(\mathfrak{y}) .\end{aligned}$$

The first integral a and the second integral b exist and are holomorphic in $B(r)$. The third integral c exists and is holomorphic on $B(s)$ only. Therefore

$$h = f e^{b-a}$$

is meromorphic in $B(r)$ with $\nu_h = \nu_f | B(r)$. Because $h = e^c$ on $B(s)$, the function h depends on the divisor of f alone. Therefore, suppose ν is a given non-negative divisor on G. A holomorphic function $f \not\equiv 0$ exists on

---

[2] This formula was proved in Stoll [*11*]. See also Stoll [*13*].

an open connected neighborhood H of $\overline{B(r)}$ with $H \subseteqq G$ such that $\nu_f = \nu | H$. Construct h as above. Then h depends on ν alone. This leads to the following theorem[3].

Theorem 1. *Let ν be a non-negative divisor on the open neighborhood G of $\overline{B(\mathfrak{a}, r)}$. Suppose that $\mathfrak{a} \notin \mathfrak{M}(\nu)$. Then one and only one holomorphic function h_ν exists such that:*

1. The function h_ν is holomorphic on $B(\mathfrak{a}, r)$.

2. $h_\nu(\mathfrak{a}) = 1$ holds.

3. Let s be any number in $0 < s < r$ such that $B(\mathfrak{a}, s) \cap \mathfrak{M}(\nu) = \emptyset$. Define $\log h_\nu$ *on $B(\mathfrak{a}, s)$ by* $\log h_\nu(\mathfrak{a}) = 0$. *For $\mathfrak{z} \in B(\mathfrak{a}, s)$*

$$\log h_\nu(\mathfrak{z}) = \frac{1}{W_{m-1}} \int\limits_{\mathfrak{M}(\nu) \cap B(\mathfrak{a}, r)} \nu(\mathfrak{y})\, l\left(\frac{(\mathfrak{z} - \mathfrak{a} | \mathfrak{y} - \mathfrak{a})}{(\mathfrak{y} - \mathfrak{a} | \mathfrak{y} - \mathfrak{a})}\right) \omega_{m-1}(\mathfrak{y} - \mathfrak{a}).$$

(If $\mathfrak{M}(\nu) \cap B(\mathfrak{a}, r) = \emptyset$, then $h_\nu \equiv 1$.) The function h_ν is called the associated function of ν on $B(\mathfrak{a}, r)$. In addition, it has the following properties:

4. $\nu_{h_\nu} = \nu | B(\mathfrak{a}, r)$.

5. For $0 < \vartheta < 1$ and $|\mathfrak{z} - \mathfrak{a}| \leqq \vartheta^4 r$

$$\log |h_\nu(\mathfrak{z})| \leqq \frac{8\,m}{(1 - \vartheta)^{3m}} [N_\nu(\mathfrak{a}, r) + n_\nu(\mathfrak{a}, r)].$$

6. Suppose that $L > 0$ and $0 < s < r$. Let $\mathfrak{N}$ be the set of all non-negative divisors ν on G such that $n_\nu(\mathfrak{a}, r) \leqq L$ and $\mathfrak{M}(\nu) \cap \overline{B(\mathfrak{a}, s)} = \emptyset$. Then, the family $\{h_\nu\}_{\nu \in \mathfrak{N}}$ is normal on $B(\mathfrak{a}, r)$ and avoids zero[4].

7. If ν_1 and ν_2 are non-negative divisors on G with

$$\mathfrak{a} \notin \mathfrak{M}(\nu_1) \cup \mathfrak{M}(\nu_2) \quad \textit{then} \quad h_{\nu_1 + \nu_2} = h_{\nu_1} \cdot h_{\nu_2}.$$

8. Let $\mathfrak{D}_\mathfrak{a}^+(G)$ be the set of all non-negative divisors ν on G with $\nu(\mathfrak{a}) = 0$. Let $h_\nu = \gamma(\nu)$ be the associated function of ν on $B(\mathfrak{a}, r)$. Let $\mathfrak{H}(B(\mathfrak{a}, r)$ be the Fréchet space of all holomorphic functions on $B(\mathfrak{a}, r)$.

Then the map

$$\gamma : \mathfrak{D}_\mathfrak{a}^+(G) \to \mathfrak{H}(B(\mathfrak{a}, r))$$

is continuous, that is, if $\{\nu_\lambda\}_{\lambda \in \Lambda}$ is a convergent net in $\mathfrak{D}_\mathfrak{a}^+(G)$ with $\nu = \lim\limits_{\lambda \to \Lambda} \nu_\lambda \in \mathfrak{D}_\mathfrak{a}^+(G)$, then

$$\gamma(\nu_\lambda) = h_{\nu_\lambda} \to h_\nu \quad \text{for} \quad \lambda \to \Lambda$$

where the convergence is uniform on every compact subset of $B(\mathfrak{a}, r)$.

---

[3] See Stoll [*13*].

[4] A family $\mathfrak{F} = \{f_\lambda\}_{\lambda \in \Lambda}$ of holomorphic functions on G *avoids zero* if and only if every convergent subset of $\mathfrak{F}$, whose convergence is uniform on each compact subset of G, has a limit which is not identically zero if restricted to a connectivity component of G. If G is connected and $\mathfrak{H}(G)$ is the Fréchet space of holomorphic functions on G, then this means that the closure of $\{f_\lambda | \lambda \in \Lambda\}$ in $\mathfrak{H}(G)$ does not contain 0.

Using the estimate $n_\nu(\mathfrak{a}, r) \geqq \nu(\mathfrak{a})$ the condition $\mathfrak{a} \notin \mathfrak{M}(\nu)$ can be eliminated to some extent.

Theorem 2.[5] *Let G be an open connected neighborhood of $\overline{B(\mathfrak{a}, r)}$. Let $L > 0$. Let $\mathfrak{N}$ be the set of all non-negative divisors on G such that $n_\nu(\mathfrak{a}, r) \leqq L$. Suppose that $0 < \tau_0 < \frac{1}{2} < \tau_1 < 1$. Then there exist a constant T, a point $\mathfrak{i}_\mu \in B(\mathfrak{a}, \tau_0 r)$ for each $\mu \in \mathfrak{N}$ and a holomorphic function h_μ on $B(\mathfrak{a}, \tau_1 r)$ for each $\mu \in \mathfrak{N}$ such that*

1. *For each $\mu \in \mathfrak{N}$: $\nu_h = \mu \mid B(\mathfrak{a}, \tau_1 r)$.*
2. *For each $\mu \in \mathfrak{N}$: $h_\mu(\mathfrak{i}_\mu) = 1$.*
3. *For each $\mu \in \mathfrak{N}$ and $\mathfrak{z} \in B(\mathfrak{a}, \tau_1 r)$:*

$$|h_\mu(\mathfrak{z})| \leqq T.$$

Hence, $\{h_\mu\}_{\mu \in \mathfrak{N}}$ is a normal family of holomorphic functions on $B(\mathfrak{a}, \tau_1 r)$ which avoids zero. It is not known to me if the functions h_μ can be chosen such that h_μ depends continuously on $\mu \in \mathfrak{N}$.

These two theorems show that the concept of a normal family of non-negative divisors as it was introduced here coincides with the concept of a normal family of non-negative divisors as introduced by OKA [7], OKA [8] and NISHINO [6]. In the case $m = 2$, NISHINO [6] and OKA [8] obtain results which are similar to the results mentioned in Theorems 1 and 2.

Let G be a complex manifold of dimension $m > 1$. An open subset B of G is said to be a *singular ball* on G if and only if a biholomorphic map α of an open neighborhood U of B onto an open subset U' of $\mathbf{C}^m$ exists such that $\alpha(B) = B(\mathfrak{a}, r)$ with $0 < r < \infty$. Let $\mathfrak{B}(G)$ be the set of all singular balls on G. If G is an open subset of $\mathbf{C}^m$, then let $\mathfrak{B}_g(G)$ be the set of all balls $B(\mathfrak{a}, r)$ with $\overline{B(\mathfrak{a}, r)} \subseteqq G$ and with $0 < r < \infty$. Obviously, $\mathfrak{B}_g(G) \subseteqq \mathfrak{B}(G)$.

Theorem 3.[6] *Let G be a complex manifold of dimension $m > 1$. Let χ be the exterior form of a continuous Hermitian metric on G. Let $\{\nu_\lambda\}_{\lambda \in \Lambda}$ be a net of non-negative divisors on G. Let ν be a non-negative divisor on G. Then*

$$\nu_\lambda \to \nu \quad \textit{for} \quad \lambda \to \Lambda$$

if and only if

$$V_{\nu_\lambda}(\bar{B}, \chi) \to V_\nu(\bar{B}, \chi) \quad \textit{for} \quad \lambda \to \Lambda$$

for every singular ball $B \in \mathfrak{B}(G)$. (If G is an open subset of $\mathbf{C}^m$, $m > 1$, it is sufficient to require this condition for the balls $B \in \mathfrak{B}_g(G)$ only.)

A consequence of these results is that the convergence of non-negative divisors on $\mathfrak{D}^+(G)$ satisfies the four conditions in KELLEY [4]

[5] See STOLL [13].
[6] See STOLL [13].

and therefore introduces a HAUSDORFF topology on $\mathfrak{D}^+(G)$ whose convergence is the given convergence. Moreover, for $B \in \mathfrak{B}(G)$, define the translation invariant pseudometric d_B in G by

$$d_B(\nu, \mu) = | V_\nu(B, \chi) - V_\mu(B, \chi)|$$

if $\nu \in \mathfrak{D}^+(G)$ and $\mu \in \mathfrak{D}^+(G)$. The family $\{d_B\}_{B \in \mathfrak{B}(G)}$ (respectively $\{d_B\}_{B \in \mathfrak{B}_g(G)}$ if G is an open subset of $\boldsymbol{C}^m$) defines a uniformity on $\mathfrak{D}^+(G)$ whose topology is the given topology on $\mathfrak{D}^+(G)$.

Another consequence of these results is[7]

MONTEL's theorem. *A family of non-negative divisors on G is normal if and only if the family is bounded on every compact subset of G.*

In the case $m = 2$, this theorem was formulated by OKA [7] in 1934. It was proved by NISHINO [6], 1962 and by RUTISHAUSER [10], 1950. However, I have difficulties to follow RUTISHAUSER's arguments.

Theorem 3 is wrong for $m = 1$. If $m > 1$ and K is compact in G, the volume $V_\nu(K)$ is not a continuous function of ν as the following example shows.

Define

$$f(z, w) = z^2, \quad \nu = \nu_f \qquad (\text{in } \boldsymbol{C}^2),$$

$$f_\lambda(z, w) = w - (\lambda z)^2, \quad \nu_\lambda = \nu_{f_\lambda} \qquad (\lambda \in \boldsymbol{N});$$

then

$$\nu_\lambda \to \nu \quad \text{for} \quad \lambda \to \infty.$$

Define

$$M = \{(z, w) | -1 < \operatorname{Re} z < 0; \ \ |\operatorname{Im} z| < 1; \ \ |w| < 1\},$$

$$K = \bar{M};$$

then

$$V_{\nu_\lambda}(M) = V_{\nu_\lambda}(K) = \pi(1 + \left(\frac{1}{2\lambda^2}\right) \to \pi \quad \text{for} \quad \lambda \to \infty$$

but

$$V_\nu(M) = 0 \neq \pi \neq 2\pi = V_\nu(K).$$

(Compare NISHINO [6].)

Instead of the convergence of divisors, the convergence of analytic sets can be considered. Let G be a complex manifold of dimension m. Let $\mathfrak{C}(G)$ be the set of all closed subsets of G. Then a point x of G is called a *cluster point* of a net $\mathfrak{F} = \{F_\lambda\}_{\lambda \in \Lambda}$ of closed subsets of G if and only if the set $\{\lambda | F_\lambda \cap U \neq \emptyset\}$ is cofinal in Λ for every open neighborhood U of x. A point x of G is said to be a *limit point* if for every neighborhood U of x an index $\lambda(x, U) \in \Lambda$ exists such that $U \cap F_\lambda \neq \emptyset$ for all $\lambda \geqq \lambda(x, U)$. The set $C(\mathfrak{F})$ of all cluster points and the set $L(\mathfrak{F})$ of all limit points of $\mathfrak{F}$ are closed. Obviously $C(\mathfrak{F}) \supseteqq L(\mathfrak{F})$. The net $\mathfrak{F}$ is said to be

[7] See STOLL [13].

convergent if and only if $L(\mathfrak{F}) = C(\mathfrak{F})$, and this set is defined to be the *limit of* $\mathfrak{F}$. This convergence satisfies the conditions of Kelly [4] and introduces on $\mathfrak{C}(G)$ a compact Hausdorff topology which has a countable base of open sets[8]. Hence a subset $\mathfrak{N}$ of $\mathfrak{C}(G)$ is compact if and only if the limit of every convergent subsequence of $\mathfrak{N}$ belongs to $\mathfrak{N}$.

Let $\mathfrak{A}_p(G)$ be the set of all analytic subsets of G which are either empty or are pure p-dimensional. Let χ be the exterior form of bidegree $(1, 1)$ which is associated to a continuous Hermitian metric on G. The subset $\mathfrak{A}_p(G)$ of $\mathfrak{C}(G)$ is not closed. However E. Bishop proved the following result.

Theorem of E. Bishop [1]. *Let* $\mathfrak{N} = \{N_\lambda\}_{\lambda \in N}$ *be a convergent sequence of elements* $N_\lambda \in \mathfrak{A}_p(G)$. *Suppose that the family* $\mathfrak{N}$ *is bounded on every compact subset of* G. *Then the limit of* $\mathfrak{N}$ *belongs to* $\mathfrak{A}_p(G)$, *i. e. the limit is either empty or a pure* p-*dimensional analytic subset of* G.

This implies immediately

Montel's Theorem for analytic sets. *Let* $\mathfrak{N} = \{N_\lambda\}_{\lambda \in \Lambda}$ *be a family of elements of* $\mathfrak{A}_p(G)$. *Let* $\mathfrak{N}$ *be bounded on every compact subset of* G. *Then* $\mathfrak{N}$ *is normal, that is, every subset* $\mathfrak{N}$ *has a convergent subset whose limit belongs to* $\mathfrak{A}_p(G)$.

Either theorem cannot be inverted as the following example shows:

$$N_\lambda = \{(z, (\lambda z)^\lambda) \mid z \in \boldsymbol{C}\} \quad \text{for} \quad \lambda \in \boldsymbol{N}.$$

$$N_\lambda \to N = \{0\} \times \boldsymbol{C} \quad \text{for} \quad \lambda \to \infty.$$

Take $K = \{(z, w) \mid |z| \leqq 1, |w| \leqq 1\}$. Then

$$V_{N_\lambda}(K) = \pi\left(\lambda + \frac{1}{\lambda^2}\right) \to \infty \quad \text{for} \quad \lambda \to \infty.$$

Let

$$\mathfrak{M} : \mathfrak{D}^+(G) \to \mathfrak{A}_{m-1}(G)$$

be the map which assigns to each non-negative divisor ν its support $\mathfrak{M}(\nu)$. Then $\mathfrak{M}$ is continuous. This remark can be applied to obtain Bishop's theorem and Montel's theorem for analytic sets in the case $p = m - 1$ from Montel's theorem for families of non-negative divisors.

The question arises if it is possible to represent families of non-negative divisors by families of holomorphic functions in a canonical way.

Problem I. *Let* G *be a connected complex manifold which is a* Cousin-*II-domain. Let* $\mathfrak{N} = \{\nu_\lambda\}_{\lambda \in \Lambda}$ *be a normal family of non-negative divisors on* G.

[8] This topology was at first considered by Hausdorff [3]. See also Busemann [2] and Rinow [9]. There, primarily the set $\mathfrak{C}_0(G)$ of all nonempty subsets is considered. On G, a finitely compact metric exists. According to Busemann, $\mathfrak{C}_0(G)$ is a finitely compact metric space. Hence, $\mathfrak{C}_0(G)$ is locally compact, has a countable base of open sets and is countable at infinite. Hence the Alexandroff compactification $\mathfrak{C}(G) = \mathfrak{C}_0(G) \cup \{\emptyset\}$ is compact and has a countable base of open sets.

Does there exist a normal family $\mathfrak{H} = \{h_\lambda\}_{\lambda \in \Lambda}$ *of holomorphic functions* h_λ *on* G *such that* $\mathfrak{H}$ *avoids zero and such that* $\nu_{h_\lambda} = \nu_\lambda$ *for all* $\lambda \in \Lambda$?

Problem II. *Let* G *be a connected complex manifold which is a* COUSIN-*II-domain. Let* $\mathfrak{N}$ *be a set of non-negative divisors on* G. *Let* $\mathfrak{H}(G)$ *be the Fréchet space of all holomorphic functions on* G. *Does there exist a continuous map*

$$h : \mathfrak{N} \to \mathfrak{H}(G) - \{0\}$$

such that $\nu_{h(\mu)} = \mu$ *for all* $\mu \in \mathfrak{N}$? *In other words, does there exist a continuous section over* $\mathfrak{N}$ *of the map* $\mathfrak{H}(G) - \{0\} \to \mathfrak{D}^+(G)$ *which is defined by* $f \to \nu_f$?

Problem I can be solved for $G = \boldsymbol{C}^m$ with $m > 1$. Problem II can be solved for $G = \boldsymbol{C}^m$ and for any normal set of non-negative divisors ν on $\boldsymbol{C}^m$ with $\nu(0) = 0$. This can be achieved by the following methods:

Let s be a non-negative, monotonically increasing function on the set $\boldsymbol{R}^+$ of all non-negative real numbers. Then q is said to be a *ballast function for* s if and only if the following properties hold.

1. *The function* q *is non-negative, monotonically increasing and of class* C^∞ *on* $\boldsymbol{R}^+$.

2. *The integral*

$$K_3(r, s, q) = \frac{2}{r} \int_{2r}^{\infty} s(t) \left(\frac{r}{t}\right)^{q(t)+2} \left[q(t) + 1 + t q'(t) \ln \frac{t}{r}\right] dt$$

exists for every positive number r.

In addition define:

$$u(t) = 3\,e(2 + \log(1 + q(t))).$$

$$K_2(r, s, q) = \frac{s(2r)\,u(2r)}{2^{q(r)+1}} + $$

$$+ \frac{1}{r} \int_{r}^{2r} s(t)\, u(t) \left(\frac{r}{t}\right)^{q(t)+2} \left[[q(t) + 1 + t q'(t) \ln \frac{t}{r}\right] dt.$$

If $s(t) = 0$ *for* $0 \leqq t \leqq a$ *with* $a > 0$, *define*

$$K_1(r, s, q) = \frac{1}{r} \int_{0}^{r} s(t)\, u(t)\, (q(t) + 1) \left(\frac{r}{t}\right)^{q(t)+2} dt,$$

$$K(r, s, q) = K_1(r, s, q) + K_2(r, s, q) + K_3(r, s, q).$$

Every non-negative, monotonically increasing function on $\boldsymbol{R}^+$ *admits a ballast function.*

If $x \in \boldsymbol{R}$, define the integer $[x]$ by $[x] - 1 < x \leqq [x]$.

Define

$$E(z,q) = (1-z)\,e^{\sum_{\mu=1}^{[p]} \frac{z^\mu}{\mu}}$$

$$e(z,q) = \frac{1}{(m-1)!}\frac{d^{m-1}}{dz^{m-1}}\left[z^{m-1}\log E(z,q)\right] \quad \text{for} \quad |z|<1,$$

where $\log E(0, q) = 0$.

Now, the KNESER integral[9] can be used to obtain the following result[10]:

Theorem 4. *Let ν be a positive divisor on $\mathbf{C}^m$ with $m > 1$ and with $\nu(0) = 0$. Let q be a ballast function $n_\nu(r)$. Then one and only one entire function h_ν called the canonical function for (ν, q) exists such that*

1. $\nu_{h_\nu} = \nu$.
2. $h_\nu(0) = 1$. *If r_0 is the largest positive number such that $B(r_0) \cap \cap\, \mathfrak{M}(\nu) = \emptyset$, define* $\log h_\nu$ *on* $B(r_0)$ *by* $\log h_\nu(0) = 0$.
3. *For* $|\mathfrak{z}| < r_0$

$$\log h_\nu(\mathfrak{z}) = \frac{1}{W_{m-1}} \int_{\mathfrak{M}(\nu)} \nu(\mathfrak{y})\, e\left(\frac{(\mathfrak{z}\,|\,\mathfrak{y})}{(\mathfrak{y}\,|\,\mathfrak{y})}, q(|\mathfrak{y}|)\right) \omega_{m-1}(\mathfrak{y}).$$

Moreover, if $|\mathfrak{z}| \leqq r$ *then*

$$\log |h_\nu(\mathfrak{z})| \leqq 8^{3m}[K(4r, n_\nu, q) + n_\nu(4r) + N_\nu(4r)].$$

Theorem 4 extends the results of STOLL [*11*] to functions of infinite order. Moreover it solves problem II. For, let $\mathfrak{N}$ be any normal set of non-negative divisors on $\mathbf{C}^m$ with $\nu(0) = 0$. Then

$$s(r) = \sup\{n_\nu(r) \mid \nu \in \mathfrak{N}\}$$

is finite and defines a non-negative, monotonically increasing function on $\mathbf{R}^+$. Let q be a ballast function for s. Then q is a ballast function for each n_ν. Let h_ν be the canonical function for (ν, q). Then *the map*

$$\gamma : \mathfrak{N} \to \mathfrak{H}(\mathbf{C}^m) - \{0\}$$

which is defined by $\gamma(\nu) = h_\nu$ is continuous and $\gamma(\nu) = h_\nu$ is an entire function whose divisor is ν. Also, Theorem 4 implies a solution to Problem I:

Theorem 5.[11] *Let $\mathfrak{N} = \{\nu_\lambda\}_{\lambda\in\Lambda}$ be a normal family of non-negative divisors on $\mathbf{C}^m$ with $m > 1$. Then a normal family $\{f_\lambda\}_{\lambda\in\Lambda}$ of entire functions f_λ exists such that*

1. *For each $\lambda \in \Lambda$, a point $\mathfrak{i}_\lambda \in B(1)$ exists with*

$$f_\lambda(\mathfrak{i}_\lambda) = 1.$$

2. *The function f_λ has the divisor $\nu_{f_\lambda} = \nu_\lambda$ for each $\lambda \in \Lambda$. Obviously, the family $\{f_\lambda\}_{\lambda\in\Lambda}$ avoids zero.*

[9] See KNESER [*5*] and STOLL [*11*].
[10] See STOLL [*13*].
[11] See STOLL [*13*].

References

[1] BISHOP, E.: Conditions for the analyticity of certain sets. (To appear.)
[2] BUSEMANN, H.: The geometry of geodesics. New York: Academic Press 1955, pp. 422.
[3] HAUSDORFF, F.: Mengenlehre. 3 ed. Berlin-Leipzig: W. de Gruyter, 1935, pp. 307.
[4] KELLEY, J. L.: General topology. Princeton: Van Nostrand., 1955, pp. 298.
[5] KNESER, H.: Zur Theorie der gebrochenen Funktionen mehrerer Veränderlichen. Jber. Dtsch. Math. Verein. **48**, 1—28 (1938).
[6] NISHINO, T.: Sur les familles de surfaces analytiques. J. Math. Kyoto Univ. **1**, 357—377 (1962).
[7] OKA, K.: Note sur les familles de fonctions analytiques multiformes etc. J. Sci. Hiroshima Univ. **A 4**, 94—98 (1934).
[8] — Sur les fonctions analytiques de plusieurs variables. X. Une mode nouvelle engendrant les domaines pseudoconvexes. Jap. J. Math. **32**, 1—12 (1962).
[9] RINOW, W.: Die innere Geometrie der metrischen Räume. Die Grundl. d. Math. Wiss. in Einzeldarst. **105**. Berlin-Göttingen-Heidelberg: Springer, 1961, pp. 520.
[10] RUTISHAUSER, H.: Über die Folgen und Scharen von analytischen und meromorphen Funktionen mehrerer Variabeln, sowie von analytischen Abbildungen. Acta Math. **83**, 287—304 (1954).
[11] STOLL, W.: Über die neuere Theorie der ganzen and meromorphen Funktionen bei mehreren komplexen Veränderlichen. Zulassungsarbeit z. Wiss. Prüfung. Tübingen 1949, pp. 287 (unpublished).
[12] — Ganze Funktionen endlicher Ordnung mit gegebenen Nullstellenflächen. Math. Zschr. **57**, 211—237 (1953).
[13] — Normal families of non-negative divisors. Math. Zschr. **84**, 154 — 218 (1964)

Department of Mathematics
University of Notre Dame
Notre Dame, Indiana

Boundaries of Complex Manifolds

By

J. J. KOHN *, **

1. Introduction

If M is a component of the boundary of a complex n-dimensional manifold X, then M has real dimension $2n - 1$ and at each point $x \in M$ the complexified tangent space T_x has a distinguished $(n - 1)$-dimensional subspace S_x which is the intersection of T_x with the holomorphic vectors at x. Thus, vector fields with values in $\bar{S}_x$ are the "tangential" Cauchy-Riemann operators.

* During the preparation of this paper, the author was partially supported by the National Science Foundation through a project at Brandeis University.

** Received June 8, 1964.

In this paper we study the potential theory associated with these operators. The cohomology associated with them which we call the "$\bar{\partial}_b$-cohomology" has been studied in [*10*] by means of the $\bar{\partial}$-Neumann problem on X. Here we work intrinsically on M and we need not assume that M is a boundary — we assume that it has subspaces S_x with certain properties, we call these manifolds "almost-complex", thus generalizing this terminology to odd dimensional manifolds. Structures such as those given by the S_x have been studied in [*7*] and [*13*].

From the point of view of partial differential equations, we are concerned here with a determined second order system of partial differential equations ($\Box_b \varphi = \alpha$) on a compact manifold M, which is not elliptic and yet possesses all the "qualitative" properties of an elliptic system (i. e. complete continuity and differentiability of weak solutions). The crucial steps in establishing these properties are the estimates in 5.3 and 6.8. The derivation of these estimates proceeds mostly along the lines of the methods developed in [*1*], [*6*], [*8*] and [*12*], although there are some new difficulties which are resolved here. Once these estimates are established we obtain the main results by applying a theorem proved in [*9*]. We give some indication of the proof of this theorem in the present case.

From the point of view of complex variables, the results here generalize the theory of harmonic integrals and we obtain the existence of the analogues of the operators in [*14*]. We also obtain finiteness of the dimension of certain cohomology groups which, in the case when M is a boundary, also follows from the results in [*2*], [*4*] and [*10*]. The work here is closely related to the extension problem for holomorphic functions, see [*3*], [*10*] and [*11*].

2. Definitions and notation

Let M be a compact C^∞ manifold of dimension $2n - 1$, with $n > 1$. Let T be the complexified tangent bundle of M whose fibre T_x is the complexified tangent space at x.

2.1. Definition. An *almost-complex structure* on M is given by a subbundle S of T with the following properties:

(a) For each $x \in M$ the fiber S_x of S over x is a subspace of T_x of complex dimension $n - 1$, such that $S_x \cap \bar{S}_x = \{0\}$. $\bar{S}_x$ denotes the set of conjugate vectors of S_x and we set $\bar{S} = \cup \bar{S}_x$.

(b) If η is a differential form of degree one which annihilates $S \oplus \bar{S}$ then $d\eta(s_1, s_2) = d\eta(\bar{s}_1, \bar{s}_2) = 0$ whenever $s_1, s_2 \in S$.

From now on we will assume that M is provided with an almost-complex structure and that T_x has a hermitian inner product under which S_x is orthogonal to $\bar{S}_x$ and which induces a Riemannian metric on M. Let K_x denote the orthogonal complement of $S_x \oplus \bar{S}_x$ so that:

$$T_x = S_x \oplus \bar{S}_x \oplus K_x \,. \tag{2.2}$$

We denote by $\mathscr{A}^m$ the space of m-forms; that is, these are maps of m-tuples of vectors in T_x which are multilinear *over the reals*, skew-symmetric and they depend differentiably (C^∞) on x. Denote by $\mathscr{D}^{p,q}$ the subset of $\mathscr{A}^{p+q}$ defined by:

$$\mathscr{D}^{p,q} = \{\varphi \in \mathscr{A}^{p+q} \mid \varphi(t_\cdot, \ldots, t_{p+q}) = 0 \quad \text{if} \quad \pi(\bar{t}_k) = 0 \quad \text{for } p \text{ of the } t_k \text{ or if } \pi(t_k) = 0 \quad \text{for } q \text{ of the } t_k\}, \tag{2.3}$$

where $\pi: T_x \to S_x$ is the orthogonal projection and $t_i \in T_x$ and φ is linear over $\boldsymbol{C}$ in the first p-variables and conjugate-linear in the remaining variables.

Thus we have

$$\mathscr{D}^{p,q} \approx \underbrace{S^* \wedge \ldots \wedge S^*}_{p\text{-times}} \wedge \underbrace{\bar{S}^* \wedge \ldots \wedge \bar{S}^*}_{q\text{-times}} \tag{2.4}$$

where S^* and $\bar{S}^*$ are the duals of S and $\bar{S}$ respectively. Furthermore we have the following decomposition:

$$\mathscr{A}^m = \bigoplus_{p+q=m} \mathscr{D}^{p,q} \oplus \mathscr{C}^m, \tag{2.5}$$

where $\mathscr{A}^m$ is the space of forms that annihilate $S \oplus \bar{S}$.

2.6. Definition. The map $\bar{\partial}_b: \mathscr{D}^{p,q} \to \mathscr{D}^{p,q+1}$ is defined by $\bar{\partial}_b \varphi = \prod_{p,q+1} d\varphi$, where $\varphi \in \mathscr{D}^{p,q}$ and $\prod_{p,q+1}: \mathscr{A}^{p+q+1} \to \mathscr{D}^{p,q+1}$ is the orthogonal projection on $\mathscr{D}^{p,q+1}$.

2.7. Definition. The almost-complex structure on M is called *integrable* if $\bar{\partial}_b^2 = 0$.

Finally, we denote by $<,>_x$ the product induced on $\mathscr{A}^m$ by the inner product on T_x; so that, if $\varphi, \psi \in \mathscr{A}^m$ then $<\varphi, \psi>$ is a C^∞ function on M. We define the inner product (φ, ψ) by:

$$(\varphi, \psi) = \int_M <\varphi, \psi> dV, \tag{2.8}$$

where dV denotes the volume element on M; we set:

$$|\varphi|^2 = <\varphi, \varphi> \quad \text{and} \quad \|\varphi\|^2 = (\varphi, \varphi). \tag{2.9}$$

3. The $\bar{\partial}_b$-cohomology

When the almost-complex structure on M is integrable, as we will assume from now on, we define the *$\bar{\partial}_b$-cohomology of type (p, q) space*, denoted by $H_b^{p,q}$ as follows:

$$H_b^{p,q} = \frac{\{\varphi \in \mathscr{D}^{p,q} \mid \bar{\partial}_b \varphi = 0\}}{\bar{\partial}_b \mathscr{D}^{p,q-1}}. \tag{3.1}$$

One of the aims of this paper is to show that, under certain circumstances, each cohomology class in $H_b^{p,q}$ has a unique representative which minimizes the norm. Given $\varphi \in \mathscr{D}^{p,q}$ with $\bar\partial_b\varphi = 0$ consider the cohomology class $\{\varphi + \bar\partial_b\psi\}$ and suppose that:

$$\|\varphi\| \leqq \|\varphi + \bar\partial_b\psi\| \quad \text{for all} \quad \psi \in \mathscr{D}^{p,q-1}. \tag{3.2}$$

Then, in particular, replacing ψ by $a\psi$ with any $a \in \boldsymbol{C}$, we obtain:

$$\|\varphi\|^2 \leqq \|\varphi + a\bar\partial_b\psi\|^2 = \|\varphi\|^2 + 2\operatorname{Re}\bar a(\varphi, \bar\partial_b\psi) + |a|^2\|\bar\partial_b\psi\|^2;$$

if $(\varphi, \bar\partial_b\psi) \neq 0$ we set

$$a = -\varepsilon\frac{(\varphi, \bar\partial_b\psi)}{|(\varphi, \bar\partial_b\psi)|\,\|\bar\partial_b\psi\|},$$

then

$$|(\varphi, \bar\partial_b\psi)| \leqq \varepsilon,$$

hence $(\varphi, \bar\partial_b\psi) = 0$ for all $\psi \in \mathscr{D}^{p,q-1}$.

3.3. Definition. The map $\vartheta_b : \mathscr{D}^{p,q} \to \mathscr{D}^{p,q-1}$ called the *formal adjoint of* $\bar\partial_b$ is defined by: if $\varphi \in \mathscr{D}^{p,q}$ then $\vartheta_b\varphi \in \mathscr{D}^{p,q-1}$ is that form such that $(\vartheta_b\varphi, \psi) = (\varphi, \bar\partial_b\psi)$ for all $\psi \in \mathscr{D}^{p,q-1}$.

The above shows that a representative φ of a cohomology class minimizes the norm if and only if $\vartheta_b\varphi = 0$.

3.4. Definition. A form $\varphi \in \mathscr{D}^{p,q}$ is called $\bar\partial_b$-*harmonic* if $\bar\partial_b\varphi = 0$ and $\vartheta_b\varphi = 0$. The set of $\bar\partial_b$-harmonic forms of type (p, q) is denoted by $\mathscr{H}_b^{p,q}$.

3.5. Definition. The map $\Box_b : \mathscr{D}^{p,q} \to \mathscr{D}^{p,q}$ is defined by

$$\Box_b\varphi = \bar\partial_b\vartheta_b\varphi + \vartheta_b\bar\partial_b\varphi.$$

We remark that:

$$\mathscr{H}_b^{p,q} = \{\varphi \in \mathscr{D}^{p,q} \mid \Box_b\varphi = 0\}, \tag{3.6}$$

since

$$(\Box_b\varphi, \varphi) = \|\bar\partial_b\varphi\|^2 + \|\vartheta_b\varphi\|^2.$$

Now we define the inner product $\boldsymbol{D}_b(\varphi, \psi)$ for $\varphi, \psi \in \mathscr{D}^{p,q}$ by:

$$\boldsymbol{D}_b(\varphi, \psi) = (\bar\partial_b\varphi, \bar\partial_b\psi) + (\vartheta_b\varphi, \vartheta_b\psi) + (\varphi, \psi), \tag{3.7}$$

and let

$$\boldsymbol{D}_b(\varphi)^2 = \boldsymbol{D}_b(\varphi, \varphi).$$

We say that $\boldsymbol{D}_b$ is *completely continuous* if whenever $\{\varphi_\nu\}$ is a sequence such that $\boldsymbol{D}_b(\varphi_\nu)$ is bounded then $\{\varphi_\nu\}$ has a convergent subsequence in the norm $\| \ \|$. It then follows that if $\boldsymbol{D}_b$ is completely continuous then $\mathscr{H}_b^{p,q}$ is finite dimensional.

Observe that given $\alpha \in \mathscr{D}^{p,q}$ then φ in $\mathscr{D}^{p,q}$ satisfies the equation $\Box_b \varphi + \varphi = \alpha$ if and only if $\boldsymbol{D}_b(\varphi, \psi) = (\alpha, \psi)$ for all $\psi \in \mathscr{D}^{p,q}$. To obtain existence theorems for this equation and to study the behaviour of $\boldsymbol{D}_b$ we will derive *a*-priori estimates, i.e. bounds for norms on φ in terms of norms on α.

4. The Levi form

Given $x \in M$ there exists a neighborhood U of x such that there exists one-forms $\zeta^1, \ldots, \zeta^{n-1}$ and η defined on U with the properties that $\zeta^j \in \mathscr{D}^{1,0}$, $\eta \in \mathscr{C}^1$ and that at each point of U they are linearly independent (over $\boldsymbol{C}$). Then any one-form on U can be expressed as a combination of $\zeta^1, \ldots, \zeta^{n-1}, \bar{\zeta}^1, \ldots, \bar{\zeta}^{n-1}, \eta$.

If ω is a one-form on U and u is a differentiable function on U we define the *derivative of u along* ω, denoted by u_ω, by:

$$u_\omega = \langle du, \omega \rangle . \tag{4.1}$$

Let $x^1, \ldots, x^{2n-1}$ be a system of local coordinates on U, then we have

$$u_{\zeta^i} = \sum a_i^k \frac{\partial u}{\partial x^k}, u_{\bar{\zeta}^j} = \sum \bar{a}_j^m \frac{\partial u}{\partial x^m} \quad \text{and} \quad u_\eta = \sum b^j \frac{\partial u}{\partial x^j} . \tag{4.2}$$

Thus every linear first order differential operator can be expressed in terms of these. Observe that:

$$u_{\bar{\zeta}^j \zeta^i} = \sum a_i^k \bar{a}_j^m \frac{\partial^2 u}{\partial x^k \partial x^m} + \text{first order terms}$$

and

$$u_{\zeta^i \bar{\zeta}^j} = \sum a_i^k \bar{a}_j^m \frac{\partial^2 u}{\partial x^k \partial x^m} + \text{first order terms.}$$

Hence the difference of these is a first order operator and we have:

$$u_{\bar{\zeta}^j \zeta^i} - u_{\zeta^i \bar{\zeta}^j} = \sum a_{ij}^k u_{\zeta^k} + \sum b_{ij}^k u_{\bar{\zeta}^k} + c_{ij} u_\eta . \tag{4.3}$$

Conjugating the above gives the same result as applying the negative of the above to u with i and j interchanged, hence we obtain

$$\bar{c}_{ij} \bar{u}_{\bar{\eta}} = - c_{ji} \bar{u}_\eta . \tag{4.4}$$

Now observe that $\mathscr{C}^1 = \bar{\mathscr{C}}^1$ and thus we may choose η to be purely imaginary (i.e. $\bar{\eta} = -\eta$ and the b_j in (4.2) are purely imaginary); then we have

$$\bar{c}_{ij} = c_{ji} . \tag{4.5}$$

The hermitian form defined by the c_{ij} is called the *Levi form.*

4.6. Proposition. *The number of non-zero eigen-values and the absolute value of the signature of the Levi form are indepedent of the choice of* ζ^j, η *and of the hermitian metric.*

The above can be proved by verifying that if η is any non-vanishing purely imaginary one-form which annihilates $S \oplus \bar{S}$ (see 2.1 (b)) then the hermitian form $d\eta(s_1, \bar{s}_2)$, defined on S, has the same number of non-zero eigen values and its signature has the same absolute value as the Levi form.

5. The basic estimate

If $\varphi \in \mathscr{D}^{p,q}$ then in a domain U of the basis $\zeta^i, \bar{\zeta}^i, \eta$ we have:

$$\varphi = \sum \varphi_{I\bar{J}} \zeta^{I\bar{J}}, \tag{5.1}$$

where $I = (i_1, \ldots, i_p)$ and $J = (j_1, \ldots, j_q)$ with $1 \leqq i_1 < \cdots < i_p \leqq \leqq n-1$, $1 \leqq j_1 < \cdots < j_q \leqq n-1$, the components $\varphi_{I\bar{J}}$ are C^∞ functions on U and $\zeta^{I\bar{J}} = \zeta^{i_1} \wedge \ldots \wedge \zeta^{i_p} \wedge \bar{\zeta}^{j_1} \wedge \ldots \wedge \bar{\zeta}^{j_q}$. If the support of φ is contained in U, we define the following semi-norms:

$$\|\varphi\|_\zeta^2 = \sum \int |\varphi_{I\bar{J}\zeta^k}|^2 dV \tag{5.2}$$

and

$$\|\varphi\|_{\bar{\zeta}}^2 = \sum \int |\varphi_{I\bar{J}\bar{\zeta}^k}|^2 dV .$$

The following theorem gives the basic estimate.

5.3. Theorem. *If $x_0 \in M$ and if the Levi form at x_0 has $\max(n-q, q+1)$ non-zero eigen-values of the same sign then there exists a neighborhood U of x_0 and a constant C such that:*

$$\|\varphi\|_\zeta^2 + \|\varphi\|_{\bar{\zeta}}^2 + \sum |\mathrm{Re} \int \varphi_{I\bar{J}\eta} \bar{\varphi}_{I\bar{J}} dV| \leqq C \boldsymbol{D}_b(\varphi)^2$$

for all $\varphi \in \mathscr{D}^{p,q}$ whose support lies in U.

Before proving the theorem we observe that if u and v are functions whose support lies in U then we have:

$$\int u_{\bar{\zeta}^i} \bar{v} dV = 0(\|u\|_{\bar{\zeta}} \|v\|) \tag{5.4}$$

$$\int u_{\zeta^i} \bar{v} dV = 0(\|u\| \|v\|_{\bar{\zeta}} + \|u\| \|v\|) \tag{5.5}$$

$$\begin{aligned}\int u_{\bar{\zeta}^j} \bar{v}_{\zeta^i} dV = \int u_{\zeta^i} \bar{v}_{\bar{\zeta}^j} dV - \int c_{ij} u_\eta \bar{v} dV + 0(\|u\|_{\bar{\zeta}} \|v\| + \\ + \|u\| \|v\|_{\bar{\zeta}} + \|u\| \|v\|) .\end{aligned} \tag{5.6}$$

(5.4) and (5.5) are obtained immediately from integration by parts and the Schwarz inequality and (5.6) follows by integrating by parts twice applying (4.3), (5.4) and (5.5).

5.7. Lemma. *If any eigen-value of the Levi form at x_0 is different from zero then there exists a neigborhood U of x and constants C and C' such that*

$$|\mathrm{Re} \int f u_\eta \bar{u} dV| \leqq C(\max_U |f| (\|u\|_\zeta^2 + \|u\|_{\bar{\zeta}}^2) + \max_U (|f_{\zeta^i}|, |f_{\bar{\zeta}^i}|) \|u\|^2)$$

and

$$\|u\|_\zeta^2 \leqq C'(\|u\|_{\bar\zeta}^2 + |\mathrm{Re}\int u_\eta \bar u\, dV| + \|u\|^2),$$

for all functions u with support in U and $f \in C^\infty(\overline{U})$.

Proof. First choose the ζ^i so that the Levi form is diagonal, let $\lambda_i = c_{ii}$ and number the ζ^i so that $0 < |\lambda_1(x_0)| \leqq |\lambda_2(x_0)| \leqq \cdots \leqq |\lambda_m(x_0)|$ and $\lambda_j(x_0) = 0$ for $j = m+1, \ldots, n-1$. Choose U so small that $|\lambda_i(x)| \geqq 1/2\,|\lambda_1(x_0)|$ and $|\lambda_j(x)| < \alpha$ for $x \in U$, $1 \leqq i \leqq m$ and $m+1 \leqq j \leqq n-1$, where α is a positive number which will be fixed later. Now by substituting $|\lambda_i|^{-1/2}\zeta^i$ for ζ^i, $1 \leqq i \leqq m$, we obtain a Levi form whose first m eigen-values are constant on U and have absolute value identically equal to one. The first inequality is then obtained by setting $i = j = 1$ and $v = \bar f u$ in (5.6). Now substituting $j = i$ and $v = u$ in (5.6) and summing on i we obtain:

$$\|u\|_\zeta^2 \leqq \text{const.}\,(\|u\|_{\bar\zeta}^2 + |\mathrm{Re}\int u_\eta \bar u\, dV| + \|u\|^2) + |\mathrm{Re}\int f u_\eta \bar u\, dV|,$$

where $f = \sum_{j=m+1}^{n-1} \lambda_j$. Now applying the first inequality to the second term on the right and choosing $\alpha = 1/nC$ we note that the coefficient of $\|u\|_\zeta^2$ on the right is less than one so that it can be absorbed in the left, thus proving the desired inequality.

5.8. Lemma. *If the ζ^i are orthonormal and further are chosen so that the Levi form is diagonal, i. e. $<\zeta^i, \zeta^j> = \delta^{ij}$ and $c_{ij} = \lambda_i \delta_{ij}$ on U, then for any φ whose support lies in U we have:*

$$\boldsymbol{D}_b(\varphi)^2 = \|\varphi\|_\zeta^2 + \sum_{I,J} \mathrm{Re}\int (\sum_{j\in J} \lambda_j)\varphi_{I\bar J\eta}\, \varphi_{I\bar J}\, dV + 0(\|\varphi\|_{\bar\zeta}\|\varphi\| + \|\varphi\|^2).$$

Proof. We have

$$\partial_b \varphi = \sum \varphi_{I\bar J \bar\zeta^j} \zeta \wedge \zeta + 0(\|\varphi\|) \tag{5.9}$$

and

$$\vartheta_b \varphi = -\sum \varepsilon^J_{jH} \varphi_{I\bar J\zeta^i} \zeta^H + 0(\|\varphi\|), \tag{5.10}$$

where

$$\varepsilon^J_{jH} = \begin{cases} 0 \quad \text{if} \quad jH \neq J \\ \text{sign of permutation} \quad jH \to J \quad \text{if} \quad jH = J. \end{cases}$$

From (5.9) we obtain:

$$\begin{aligned} \|\bar\partial_b \varphi\|^2 &= \|\varphi\|_{\bar\zeta}^2 - \sum_{j\in J} \int |\varphi_{I\bar J\bar\zeta}|^2 dV \\ &+ \sum_{i\neq j} \varepsilon^J_{jH}\, \varepsilon^K_{iH} \int \varphi_{I\bar J\bar\zeta^j} \varphi\, dV \\ &+ 0(\|\varphi\|_{\bar\zeta}\|\varphi\| + \|\varphi\|^2). \end{aligned}$$

The desired result is obtained by integrating by parts twice in the second and third term on the right and using (5.6).

Proof of 5.3. Choose the ζ^i so that they are orthonormal, i.e.

$$< \zeta^i, \zeta^j > = \delta^{ij},$$

and so that the Levi form is diagonal, we set $c_{ij} = \delta_{ij}\lambda_i$. By renumbering the ζ^i and, if necessary by replacing η with $-\eta$, we can suppose that $\lambda_i(x_0) > 0$ when $1 \leqq i \leqq r$, $\lambda_i(x_0) < 0$ if $r < i \leqq m$ and $\lambda_i(x_0) = 0$ if $m < i \leqq n-1$; where $1 < r \leqq m \leqq n-1$ and $r \geqq \max(n-q, q+1)$. We choose U so small that for all $x \in U$ we have $|\lambda_i(x) - \lambda_i(x_0)| < \alpha$, $1 \leqq i \leqq n-1$, where α is a positive number which will be fixed later.

If $0 < \varepsilon_0 < 1$ and if $\mathscr{V}$ is any subset of the set of triples $\{(I, J, k)\}$, we have, by applying (5.6):

$$\begin{aligned} \|\varphi\|^2_{\zeta} &\geqq \varepsilon_0 \|\varphi\|^2_{\zeta} + (1-\varepsilon_0) \sum_{(I,J,k)\in\mathscr{V}} \|\varphi_{IJ\bar{\zeta}}\|^2 \geqq \\ &\geqq \varepsilon_0 \|\varphi\|^2_{\zeta} - (1-\varepsilon_0) \sum_{(I,J,k)\in\mathscr{V}} \operatorname{Re} \int \lambda_k \varphi_{IJ_\eta} \overline{\varphi}_{IJ} dV + \\ &+ 0(\|\varphi\|_{\bar{\zeta}} \|\varphi\|^2 + \|\varphi\|^2). \end{aligned} \tag{5.11}$$

For any two disjoint subsets $\mathscr{P}$ and $\mathscr{N}$ of the set $\{(I, J)\}$ we define the subset $\mathscr{D}^{p,q}[\mathscr{P}, \mathscr{N}]$ of $\mathscr{D}^{p,q}$ by:

$$\begin{aligned} \mathscr{D}^{p,q}[\mathscr{P},\mathscr{N}] = \{\varphi \,\varepsilon\, \mathscr{D}^{p,q} \,|\, &\operatorname{Re}\int \varphi_{IJ_\eta} \overline{\varphi}_{IJ} dV > 0 \text{ when } (I,J) \in \mathscr{P}, \\ \operatorname{Re}\int \varphi_{IJ_\eta} \overline{\varphi}_{IJ} dV &< 0 \text{ when } (I,J) \in \mathscr{N} \\ \text{and} \quad \operatorname{Re}\int \varphi_{IJ_\eta} \overline{\varphi}_{IJ} dV &= 0 \text{ when } (I,J) \notin \mathscr{P} \cup \mathscr{N}\}. \end{aligned} \tag{5.12}$$

Since there are finitely many of these subsets and since $\mathscr{D}^{p,q}$ is their union it suffices to prove the result for $\varphi \in \mathscr{D}^{p,q}[\mathscr{P}, \mathscr{N}]$ with fixed $\mathscr{P}$ and $\mathscr{N}$. We define $\mathscr{V}$ by:

$$\begin{aligned} \mathscr{V} = \{(I, J, k) \,|\, 1 \leqq k \leqq r &\text{ when } (I, J) \in \mathscr{N} \\ \text{and } r < k \leqq m &\text{ when } (I, J) \in \mathscr{P}\}. \end{aligned} \tag{5.13}$$

Observe that when $\operatorname{Re}\int \varphi_{IJ_\eta} \overline{\varphi}_{IJ}\, dV$ is positive we choose k so that $\lambda_k < 0$ and when it is negative k is chosen so that $\lambda_k > 0$.

Now we apply (5.11) to the first term on the right in the equation of 5.8 and after collecting terms and replacing λ_k by $\lambda_k(x_0)$ we obtain:

$$\begin{aligned} \boldsymbol{D}_b(\varphi)^2 \geqq \varepsilon_0 \|\varphi\|^2_{\zeta} &+ a_{IJ} \operatorname{Re}\int \varphi_{IJ_\eta} \overline{\varphi}_{IJ} dV + \\ &+ \operatorname{Re}\int f \varphi_{IJ_\eta} \overline{\varphi}_{IJ} dV + \\ &+ 0(\|\varphi\|_{\bar{\zeta}} \|\varphi\| + \|\varphi\|^2), \end{aligned} \tag{5.14}$$

where

$$a_{IJ} = \begin{cases} \sum\limits_{j\in J, j\leqq r} \lambda_j(x_0) - \varepsilon_0 \sum\limits_{k\in J, k>r} \lambda_k(x_0) & \text{when} \quad (I,J) \in \mathscr{P} \\ -(1-\varepsilon_0) \sum\limits_{k\notin J, k\leqq r} \lambda_k(x_0) + \varepsilon_0 \sum\limits_{k\in J, k\leqq r} \lambda_k(x_0) & \text{when} \quad (I,J) \in \mathscr{N} \end{cases}$$

and f is a sum of the functions $\pm(\lambda_k(x_0) - \lambda_k(x))$ and hence $|f|$ can be made small by taking α small. First, we choose ε_0 so small that $a_{IJ} > 0$ when $(I, J) \in \mathscr{P}$. This can be done since there are r positive eigen-values and $r > q + 1$. Thus for some $j \in J$ we have $\lambda_j > 0$. We should also choose ε_0 small enough so that $a_{IJ} < 0$ when $(I, J) \in \mathscr{N}$. Again this can be done since $r \geqq n - q$ so that for some $k \notin J$ we have $\lambda_k > 0$. Setting $A = \min |a_{IJ}|$ we have:

$$\begin{aligned} \boldsymbol{D}_b(\varphi)^2 \geqq \varepsilon_0 \|\varphi\|_{\bar{\xi}}^2 &+ A\,|\operatorname{Re}\textstyle\int \varphi_{IJ_\eta}\overline{\varphi}_{IJ}\,dV| \\ &+ \operatorname{Re}\textstyle\int f\varphi_{IJ_\eta}\overline{\varphi}_{IJ}\,dV \\ &+ 0(\|\varphi\|_{\bar{\xi}}\|\varphi\| + \|\varphi\|^2). \end{aligned}$$

Now the desired inequality is obtained by setting α so small that the third term on the right is absorbed in the remaining terms by 5.7 and finally we use the inequality

$$\|\varphi\|_{\bar{\xi}}\|\varphi\| \leqq \text{small const.}\ \|\varphi\|_{\bar{\xi}}^2 + \text{large const.}\ \|\varphi\|^2$$

and the theorem is proved.

5.15. Corollary. *Under the same hypotheses as in Theorem 5.3 there exist a neigborhood U of x_0 and a constant $C > 0$ such that:*

$$|\operatorname{Re}\textstyle\int \varphi_{IJ_\eta}\overline{\psi}_{IJ}\,dV| \leqq C(\boldsymbol{D}_b(\varphi)^2 + \boldsymbol{D}_b(\psi)^2)$$

fvr all $\varphi, \psi \in \mathscr{D}^{p,q}$ whose support lies in U.

Proof. Observe that

$$\begin{aligned} 4\operatorname{Re}\textstyle\int \varphi_{IJ_\eta}\overline{\psi}_{IJ}\,dV &= \operatorname{Re}\textstyle\int(\varphi_{IJ} + \psi_{IJ})(\overline{\varphi}_{IJ} + \overline{\psi}_{IJ})\,dV \\ &\quad - \operatorname{Re}\textstyle\int(\varphi_{IJ} - \psi_{IJ})_\eta(\overline{\varphi}_{IJ} - \overline{\psi}_{IJ})\,dV, \end{aligned}$$

so the desired result is obtained by applying Theorem 5.3 to the forms $\varphi + \psi$ and $\varphi - \psi$ and the fact that $\boldsymbol{D}_b(\varphi + \psi)^2 \leqq 2(\boldsymbol{D}_b(\varphi)^2 + \boldsymbol{D}_b(\psi)^2)$.

6. The s-norms

If u is a compactly supported C^∞ function on $\boldsymbol{R}^{2n-1}$, we denote by $\hat{u}$ its Fourier transform, i. e.

$$\hat{u}(\xi) = \int e^{-ix\cdot\xi}u(x)\,dx, \tag{6.1}$$

where $\xi = (\xi^1, \ldots, \xi^m)$, $x = (x^1, \ldots, x^m)$, $x \cdot \xi = x^1\xi^1 + \cdots + x^m\xi^m$ and $dx = dx^1 \ldots dx^m$, $m = 2n - 1$.

For each real number s we define the norm $\|\ \|_s$ by:

$$\|u\|_s^2 = \int(1 + |\xi|^2)^s|\hat{u}(\xi)|^2\,d\xi. \tag{6.2}$$

Let $D_j = -\sqrt{-1}\,\frac{\partial}{\partial x^j}$ then there exist positive constants C_1 and C_2 such that:

$$C_1\|u\|_s^2 \leqq \sum_{j=1}^{2n-1}\|D_j u\|_{s-1}^2 \leqq C_2\|u\|_s^2. \tag{6.3}$$

The norm $\| \ \|_0$ is equivalent to the L_2-norm and if s is a positive integer $\| \ \|_s$ is equivalent to the sum of the L_2-norms of all derivatives of order less than or equal to s. We also have the generalized Schwarz inequality:

$$|(u, v)| \leqq \| u \|_s \| v \|_{-s} \leqq \text{small const.} \| u \|_s^2 + \text{large const.} \| v \|_s^2 . \tag{6.4}$$

We define the operator R_j by:

$$\widehat{R_j u}(\xi) = \frac{\xi_j}{|\xi_j|} \hat{u}(\xi) . \tag{6.5}$$

Then R_j is a singular integral operator and it is proven in [4] that if A is a first order differential operator with differentiable coefficients then:

$$\| A R_j u \| \leqq \text{const.} (\| R_j A u \| + \| u \|) . \tag{6.6}$$

If u is a function on M whose support lies in a coordinate neighborhood U, we define $\| u \|_s$ by considering U as a subset of $\boldsymbol{R}^{2n-1}$ and using (6.2). If $\varphi \in \mathscr{D}^{p,q}$ and if the support of φ lies in U then we define

$$\| \varphi \|_s^2 = \sum \| \varphi_{I\bar{J}} \|_s^2 , \tag{6.7}$$

where the components $\varphi_{I\bar{J}}$ are taken with respect to a system ζ^i as in (5.1).

6.8. Proposition. *If U is a coordinate neighborhood in M and if there exists a constant $C_0 > 0$ such that:*

$$\| \varphi \|_\zeta^2 + \| \varphi \|_{\bar{\zeta}}^2 + \sum | \text{Re} \int \varphi_{I\bar{J}\eta} \bar{\varphi}_{I\bar{J}} \, dV | \leqq C_0 \boldsymbol{D}(\varphi)^2$$

for all $\varphi \in \mathscr{D}^{p,q}$ whose support lies in U, then for any open set V with $\bar{V} \subset U$ there exists a constant C such that: $\| \varphi \|_{1/2} \leqq C \boldsymbol{D}_b(\varphi)$ for all $\varphi \in \mathscr{D}^{p,q}$ whose support lies in V.

Proof. First we show that if $\varphi, \psi \in \mathscr{D}^{p,q}$, with support in U then

$$|\text{Re}(D_j \varphi_{I\bar{J}}, \psi_{I\bar{J}})| \leqq \text{const.} (\boldsymbol{D}_b(\varphi)^2 + \boldsymbol{D}_b(\psi)^2) , \tag{6.9}$$

the constant is independent of φ and ψ. To see this, we write:

$$D_j u = \sum a_j^k u_{\zeta^k} + \sum b_j^k u_{\bar{\zeta}^k} + b_j u_\eta , \tag{6.10}$$

so that:

$$|\text{Re}(D_j \varphi_{I\bar{J}}, \psi_{I\bar{J}})| \leqq \sum |\text{Re}(a_j^k \varphi_{I\bar{J}\zeta^k}, \psi_{I\bar{J}})| + \sum |\text{Re}(b_j^k \varphi_{I\bar{J}\bar{\zeta}}, \psi_{I\bar{J}})| + |\text{Re}(b_j \varphi_{I\bar{J}\eta}, \psi_{I\bar{J}})| .$$

The first and second term on the right can clearly be bounded as required. To bound the third term we write:

$$|\text{Re}(b_j \varphi_{I\bar{J}\eta}, \psi_{I\bar{J}})| = |\text{Re}(\varphi_{I\bar{J}\eta}, b_j \psi_{I\bar{J}})| ,$$

which, by 5.15, is bounded by $C(\boldsymbol{D}_b(\varphi)^2 + \boldsymbol{D}_b(b_j \psi)^2)$. Thus (6.9) is established by observing that $\boldsymbol{D}_b(b_j \psi) \leqq \text{const.} \boldsymbol{D}_b(\psi)$.

It follows immediately from the definition of R_j that:

$$(D_j u, R_j u) = \int |\xi_j| \, |\hat{u}(\xi)|^2 d\xi$$

and hence

$$\|u\|_{1/2}^2 \leqq \text{const.} \left(\sum (D_j u, R_j u) + \|u\|^2 \right). \tag{6.11}$$

To conclude the proof it will suffice to show that:

$$(D_j \varphi_{I\bar{J}}, R_j \varphi_{I\bar{J}}) \leqq \text{const.} \, \boldsymbol{D}(\varphi)^2,$$

for all φ with support in V. Let ϱ be a differentiable function whose support lies in U and which is equal to 1 on V, then we have:

$$(D_j \varphi_{I\bar{J}}, R_j \varphi_{I\bar{J}}) = |\operatorname{Re}(D_j \varphi_{I\bar{J}}, \varrho R_j \varphi_{I\bar{J}})|$$

and the desired result is obtained by using (6.9) with $\psi_{I\bar{J}} = \varrho R_j \varphi_{I\bar{J}}$ and applying (6.6) to conclude that $\boldsymbol{D}_b(\varrho R_j \varphi) \leqq \text{const.} \, \boldsymbol{D}_b(\varphi)$.

6.12. Definition. M is called *r-strongly pseudo-convex* if M is covered with neighborhoods in which the forms $\zeta^1, \ldots, \zeta^n$ and η have been chosen so that in the intersection of two neighborhoods the η are positive multiples of each other and if the Levi form has at least r positive eigen-values at each point.

Observe that if the Levi form has r eigen-values of the same sign at each point and if $r \geqq n/2$ then the η can be chosen as in the above definition and M is r-strongly pseudo-convex. Now if $\{U_k\}$ is a covering of M by coordinate neighborhoods and if $\{\varrho_k\}$ is the associate partition of unity (i. e. the support of ϱ_k is contained in U_k and $\sum \varrho_k = 1$) then for any $\varphi \in \boldsymbol{D}^{p,q}$ we define $\|\varphi\|_s$ by:

$$\|\varphi\|_s = \sum_k \|\varrho_k \varphi\|_s. \tag{6.13}$$

We remark that $\boldsymbol{D}_b(\varrho_k \varphi) \leqq \text{const.} \, \boldsymbol{D}_b(\varphi)$, hence we obtain the following theorem by combining 5.3 with the above.

6.14. Theorem. *If M is r-strongly pseudo-convex, when $r = \max(n - q, q + 1)$, then there exist $C > 0$ such that $\|\varphi\|_{1/2} \leqq C \boldsymbol{D}_b(\varphi)$ for all $\varphi \in \mathscr{D}^{p,q}$.*

7. The main results

In [9] it is proven that the estimate in 6.14 implies very strong existence and regularity theorems. Below we give an outline of the method of [9]. First it is shown that for each non-negative integer s there exists a constant C_s such that:

$$\|\varphi\|_{s+1/2} \leqq C_s \|\Box_b \varphi + \varphi\|_{s-1/2} \tag{7.1}$$

for all $\varphi \in \mathscr{D}^{p,q}$.

The case $s = 0$ follows by observing that

$$\boldsymbol{D}_b(\varphi)^2 = (\Box_b\varphi + \varphi, \varphi) \leqq \|\Box_b\varphi + \varphi\|_{-1/2}\|\varphi\|_{1/2} \leqq$$
$$\leqq \text{large const.}\ \|\Box_b\varphi + \varphi\|_{-1/2}^2 + \text{small const.}\ \|\varphi\|_{1/2}^2,$$

and applying the estimate in 6.14. For $s = 1$ we apply 6.14 to $\sum \boldsymbol{D}_b(D_j\varrho_k\varphi)^2$ and prove that

$$\begin{aligned}\boldsymbol{D}_b(D_j\varrho_k\varphi)^2 &= -(\Box_b\varphi + \varphi, \varrho_k D_j^2\varrho_k\varphi)\\ &\quad + 0(\|\varphi\|\,\boldsymbol{D}_b(D_j\varrho_k\varphi) + \|\varphi\|_1^2)\\ &\leqq \text{l. c.}\ (\|\Box_b\varphi + \varphi\|_{1/2}^2 + \|\varphi\|^2)\\ &\quad + \text{s. c.}\ (\boldsymbol{D}_b(D_j\varrho_k\varphi)^2 + \|\varphi\|_{3/2}^2).\end{aligned}$$

Continuing in this way we obtain (7.1) for any s. Let $\boldsymbol{K}(\varphi, \psi)$ be defined by:

$$\boldsymbol{K}(\varphi, \psi) = \sum (\varrho_k D_j\varphi_{I\bar{J}}, \varrho_k D_j\psi_{I\bar{J}}), \tag{7.2}$$

where the $\{\varrho_k\}$ are a partition of unity associated with a covering $\{U_k\}$ by coordinate neighborhoods. For $\varepsilon > 0$ we define:

$$\boldsymbol{D}_b^\varepsilon(\varphi, \psi) = \boldsymbol{D}_b(\varphi, \psi) + \varepsilon\boldsymbol{K}(\varphi, \psi). \tag{7.3}$$

Now let $A^\varepsilon(\varphi)$ be defined by requiring

$$\boldsymbol{D}_b^\varepsilon(\varphi, \psi) = (A^\varepsilon(\varphi), \psi) \tag{7.4}$$

for all $\psi \in \mathscr{D}^{p,q}$. Then it is easy to see that A^ε is a strongly elliptic operator so that given any $\alpha \in \mathscr{D}^{p,q}$ there exists a unique $\varphi_\varepsilon \in \mathscr{D}^{p,q}$ such that $\alpha = A^\varepsilon(\varphi_\varepsilon)$. Furthermore, for each integer $s \geqq 0$ there exists $C_s > 0$, C_s being *independent of* ε, such that

$$\|\varphi_\varepsilon\|_{s+1/2} \leqq C_s\|\alpha\|_{s-1/2}. \tag{7.5}$$

Now by using Rellich's lemma and the diagonal process we conclude that there exists a subsequence of the φ_ε which converges in $\|\ \|_s$ for every s; hence the limit φ is in $\mathscr{D}^{p,q}$ and satisfies the equation $\Box_b\varphi + \varphi = \alpha$. Furthermore, using standard arguments, we obtain the following proposition:

7.6. Proposition. *If the estimate* 6.14 *holds then* $\boldsymbol{D}_b$ *is completely continuous, so that the operator* $(\Box_b + I)^{-1}$ *is completely continuous.* $\mathscr{H}_b^{p,q}$, *the null space of* $\Box_b$, *is finite dimensional and we have the orthogonal decomposition:* $\mathscr{D}^{p,q} = \Box_b\mathscr{D}^{p,q} \oplus \mathscr{H}_b^{p,q}$. *The equation* $\Box_b\varphi = \beta$ *has a solution* $\varphi \in \mathscr{D}^{p,q}$ *whenever* $\beta \in \mathscr{D}^{p,q}$ *and* $\beta \perp \mathscr{H}_b^{p,q}$.

Let $H_b : \mathscr{D}^{p,q} \to \mathscr{H}_b^{q,p}$ denote the orthogonal projection. Under the hypothesis of the above theorem we define the operator $N_b : \mathscr{D}^{p,q} \to \mathscr{D}^{p,q}$ by setting $N_b\alpha = \varphi$, where φ is the unique solution of $\Box_b\varphi = \alpha - H_b\alpha$ such that $H_b\varphi = 0$. Then we obtain the following theorem which gives the main results.

7.7. Theorem. *If M is r-strongly pseudo-convex with $r = \max(n - q, q + 1)$ then there exists an operator $N_b : \mathcal{D}^{p,q} \to \mathcal{D}^{p,q}$ with the following properties:*

(a) If $\varphi \in \mathcal{D}^{p,q}$ *then* $\varphi = \Box_b N_b \varphi + H_b \varphi$,

(b) $\bar\partial_b N_b = N_b \bar\partial_b$, $\vartheta_b N_b = N_b \vartheta_b$, $\Box_b N_b = N_b \Box_b$, $H_b N_b = N_b H_b = 0$,

(c) *N_b is completely continuous and for each non-negative s we have* $\| N \varphi \|_{s+1/2} \leqq \text{const.} \| \varphi \|_{s-1/2}$.

7.8. Corollary. Under the same assumptions as above each $\bar\partial_b$-cohomology class has a unique representative in $\mathcal{H}_b^{p,q}$; in fact if $\bar\partial_b \varphi = 0$ then $\varphi = \bar\partial_b(\vartheta_b N_b \varphi) + H_b \varphi$ and $H_b \varphi$ is the representative in $\mathcal{H}_b^{p,q}$.

7.9. Corollary. Again under the same assumptions if $\alpha \in \mathcal{D}^{p,q}$ then there exists a $\varphi \in \mathcal{D}^{p,q-1}$ such that: $\bar\partial_b \varphi = \alpha$ if and only if $H_b \alpha = \bar\partial_b \alpha = 0$; in fact, under those circumstances, we may choose $\varphi = \vartheta_b N_b \alpha$.

7.10. Corollary. If $n \geqq 3$ and if M is strongly pseudo-convex (i. e. all of the eigen-values of the Levi form are positive) then the orthogonal projection $H_b : \mathcal{D}^{p,0} \to \mathcal{H}_b^{p,0}$ is given by $H_b \alpha = \alpha - \vartheta_b N_b \bar\partial_b \alpha$, where $N_b : \mathcal{D}^{p,1} \to \mathcal{D}^{p,1}$ is the operator whose existence is given in 7.7. Under these circumstances there exists an operator $N_b : \mathcal{D}^{p,0} \to \mathcal{D}^{p,0}$ satisfying the properties (a) and (b) but not (c) of 7.7 and, moreover, N_b is given by $N_b = \vartheta_b N_b^2 \bar\partial_b$.

All of the above corollaries are easy consequences of 7.7.

References

[1] Ash, M. E.: The Neumann problem on strongly pseudo-convex multifoliate manifolds. Thesis, Princeton University, 1962.

[2] Andreotti, A., and H. Grauert: Théorèmes de finitude pour la cohomologie des espaces complexes. Bull. Soc. Math. France **90**, 193—259 (1962).

[3] Bochner, S.: Analytic and meromorphic continuation by means of Green's formula. Ann. Math. **44**, 652—673 (1943).

[4] Calderón, A. P., and A. Zygmund: Singular integral operators and differential equations. Amer. J. Math. **79**, 901—921 (1957).

[5] Ehrenpreis, L.: Some applications of the theory of distributions to several complex variables. Conference on Analytic Functions, 1957, 65—79.

[6] Hörmander, L.: Existence theorems for the $\bar\partial$ operator by L^2 methods. To appear.

[7] Hsu, C. J.: On some properties of π-structures of differentiable manifolds. Tohoku Math. J. **12**, 349—360 (1960).

[8] Kohn, J. J.: Harmonic integrals on strongly pseudo-convex manifolds I, II. Ann. Math. **78**, 112—148 (1963). II to appear in Ann. Math.

[9] —, and L. Nirenberg: Non-coercive boundary value problems. To appear in J. Pure App. Math.

[10] —, and H. Rossi: On the extension of holomorphic functions from the boundary of a complex manifold. To appear in Ann. Math.

[11] Lewy, H.: On the local character of the solutions of an atypical linear differential equation in 3 variables and a related theorem for functions of 2 complex variables. Ann. Math. 64, 514—522 (1956).
[12] Morrey, Jr., C. G.: The analytic embeddings of abstract real analytic manifolds. Ann. Math. 68, 159—201 (1958).
[13] Sasaki, S.: On differentiable manifolds with certain structures which are closely related to almost contact structure I, II (with Hatekayama, Y.). Tohoku Math. J. 12 (3), 459—476 (1960), and 13 (2), 281—294 (1961).
[14] Spencer, D. C.: Potential theory on almost-complex manifolds. Lectures on functions of a complex variable, Univ. of Michigan Press, 1955, 15—43.

Department of Mathematics
Brandeis University
Waltham, Mass.

Local Properties of Holomorphic Mappings*,**

By

H. Holmann

Introduction

The main purpose of this paper is to study local properties of holomorphic mappings between complex spaces using tangent spaces[1] to complex spaces.

There are suitable notions of regularity for holomorphic mappings and embeddings which generalize those for complex manifolds.

Definition 1. *A holomorphic mapping $\tau: X \to Y$ between complex spaces X and Y is called regular at $x \in X$ if there exists an open neighborhood U of x, an analytic subset S of U with $x \in S$ and a holomorphic retraction $r: U \to S$, such that $\tau | U = \tau | S \circ r$, where τ maps S biholomorphically onto an analytic subset of an open neighborhood of $\tau(x)$ in Y.*

By a slight alteration of this definition we obtain the notions of strongly and weakly regular mappings.

τ is called *strongly regular* at $x \in X$ if r is not only a holomorphic retraction but a projection[2] and it is called *weakly regular* at x if the restriction of τ to S is only injective. The strongly regular holomorphic mappings can be characterized as follows.

---

* This work has been supported by the National Science Foundation under Grant G-25224.
** Received June 8, 1964.
[1] See [2], [5], [6], [7].
[2] That means U can be realized as a product space $S \times P$, where P is a polycylinder, and r is the canonical projection onto the first component.

Theorem. *Let* $\tau: X \to Y$ *be a holomorphic mapping between complex spaces* X *and* Y. τ *is strongly regular at* $x \in X$ *if the corank*[3] *of* τ *is constant in a neighborhood of* x.

In general regular and strongly regular holomorphic mappings are different but in the case of complex manifolds both notions coincide. This is true because a holomorphic retraction of a complex manifold X onto an analytic subset Y does not exist if Y has singularities[4]. This comes out as a corollary of the following more general result about holomorphic retractions on complex spaces.

Theorem. *Let* $r: X \to Y$ *be a holomorphic retraction of a complex space* X *onto an analytic subspace* Y. *Then every point* $y \in Y$ *has a neighborhood* U *in* X, *which can be realized as an analytic set in the product* $P^m \times P^s$ *of two unit poly-cylinders of dimension* m *and* s *respectively, such that the following holds:*

(1) $r = p|U$, *where* $p: P^m \times P^s \to P^m \times \{0\}$ *is the canonical projection,*

(2) m' *and* $m + s$ *are equal to the embedding dimensions of* Y *and* X *respectively at the point* y.

This leads to the following definition of a regularly embedded complex subspace Y of a complex space X, which in the case X is a complex manifold means that Y is a complex submanifold of X.

Definition 2. *Let* Y *be a complex subspace of a complex space* X. *We say that* Y *is regularly embedded in* X *at the point* $y \in Y$ *if there exists a neighborhood* U *of* y *in* X *and a holomorphic retraction* $r: U \to U \cap Y$.

Necessary and sufficient conditions for the weak regularity of holomorphic mappings and the regularity of holomorphic embeddings are proved in the last section of this paper. There are other holomorphic mappings which are not regular but can be covered locally by regular mappings.

Theorem. *Let* $\tau: X \to Y$ *be an open holomorphic mapping between pure dimensional complex spaces* X *and* Y. *Then every point* $x \in X$ *has an open neighborhood* U, *there exist analytic coverings* $(\tilde{U}, \sigma)$ *of* U *and* $(\tilde{V}, \varrho)$ *of* $V := \tau(U)$ *and a regular holomorphic mapping* $\tilde{\tau}: \tilde{U} \to \tilde{V}$, *such that the following diagram is commutative:*

$$\begin{array}{ccc} \tilde{U} & \xrightarrow{\tilde{\tau}} & \tilde{V} \\ \sigma\downarrow & & \downarrow\varrho \\ U & \xrightarrow{\tau} & V. \end{array}$$

[3] See Definition 5.

[4] See [5], Theorem 5.2.

1. Complex tangent spaces

Let X be a (reduced) complex space with structure sheaf $\mathcal{O}(X)$ or simply $\mathcal{O}$.

Definition 3. *A linear mapping $T:\mathcal{O}_x \to \mathbf{C}$ is a tangent vector to X at $x \in X$ iff*

$$T(fg) = T(f)\,g(x) + f(x)\,T(g) \tag{1}$$

for all $f, g \in \mathcal{O}_x$.

The set of all tangent vectors to X at a point $x \in X$ forms a complex vector space $\mathcal{T}_x(X)$. The union $\mathcal{T}(X) := \bigcup_{x \in X} \mathcal{T}_x(X)$ shall be called the tangent space to X.

Proposition 1. $\dim \mathcal{T}_x(X) = \operatorname{emdim}_x X$, $x \in X$.

Here $\operatorname{emdim}_x X$ is defined by $\dim(m_x/m_x^2)$, where m_x denotes the maximal ideal of the $\mathbf{C}$-algebra $\mathcal{O}_x$. $\operatorname{emdim}_x X$ is also the smallest of all integers n, such that a neighborhood V of x can be embedded as an analytic set in a domain D of $\mathbf{C}^n$.

Proof. Each $T \in \mathcal{T}_x(X)$ vanishes on m_x^2 and also on the germs of constant functions because of the product rule (1). This induces a linear functional $t = \alpha(T)$ on m_x/m_x^2 such that the following diagram commutes:

$$\begin{array}{ccc} \mathcal{O}_x & \xrightarrow{\varphi} & m_x/m_x^2 \\ & {\scriptstyle T}\searrow \quad \swarrow {\scriptstyle t=\alpha(T)} & \\ & \mathbf{C} & \end{array}$$

where φ denotes the canonical projection. It can be checked easily that $\alpha:\mathcal{T}_x(X) \to (m_x/m_x^2)^*$ is a surjective isomorphism, α^{-1} being defined by $\alpha^{-1}(t) = t \circ \varphi$, $t \in (m_x/m_x^2)^*$.

Definition 4. *If $\tau:X \to Y$ is a holomorphic map then the differential $d\tau_x:\mathcal{T}_x(X) \to \mathcal{T}_{\tau(x)}(Y)$ of τ is defined by*

$$d\tau_x(T)(f) := T(f \circ \tau) \tag{2}$$

where $T \in \mathcal{T}_x(X)$ and $f \in \mathcal{O}_{\tau(x)}(Y)$.

The differentials $d\tau_x$, $x \in X$, are linear mappings and induce a map $d\tau:\mathcal{T}(X) \to \mathcal{T}(Y)$ such that the restriction $d\tau \,|\, \mathcal{T}_x(X)$ of $d\tau$ to $\mathcal{T}_x(X)$ is equal to $d\tau_x$. $d\tau$ behaves functorial (in a covariant way), that means if $\tau:X \to Y$ and $\sigma:Y \to Z$ are holomorphic mappings, then $d(\sigma \circ \tau) = d\sigma \circ d\tau$.

Proposition 2. [5] *$\tau:X \to Y$ is biholomorphic*[6] *in a neighborhood U_x of $x \in X$ iff $d\tau_x$ is non singular.*

---

[5] See [5], Lemma 2.5.

[6] That means $\tau(U_x)$ is an analytic subset of an open neighborhood of $\tau(x)$ in Y and $\tau: U_x \to \tau(U_x)$ is biholomorphic.

Let Y be a complex subspace of X and $i: Y \to X$ the injection mapping, then $di_y(\mathcal{T}_y(Y))$, $y \in Y$, is the following subspace of $\mathcal{T}_y(X)$:

Proposition 3.[7] *If $\mathcal{I}_y(Y) := \{f \in \mathcal{O}_y(X),\ f \circ i = 0\}$ denotes the stalk at y of the sheaf $\mathcal{I}(Y)$ of germs of holomorphic functions on X which vanish on Y, then*

$$di_y(\mathcal{T}_y(Y)) = \{T \in \mathcal{T}_y(X); T(f) = 0 \quad \text{for all} \quad f \in \mathcal{I}_y(Y)\}. \tag{3}$$

We want to define a natural complex structure on $\mathcal{T}(X)$, such that the projection $\pi: \mathcal{T}(X) \to X$, defined by $\pi(\mathcal{T}_x(X)) = x$, $x \in X$, and the differential $d\tau: \mathcal{T}(X) \to \mathcal{T}(Y)$ of a holomorphic mapping $\tau: X \to Y$ are always holomorphic.

If D is a domain in $\mathbf{C}^n$ we take the usual complex structure on $\mathcal{T}(D)$ which is induced by the bijective mapping

$$\varphi: D \times \mathbf{C}^n \to \mathcal{T}(D) \quad \text{where} \quad \varphi(y, \lambda) := \sum_{\nu=1}^{n} \lambda_\nu \frac{\partial}{\partial z_\nu}\bigg|_{z=y}$$

for $y = (y_1, \ldots, y_n) \in D$ and $\lambda = (\lambda_1, \ldots, \lambda_n) \in \mathbf{C}^n$. Let A be an analytic subset of the domain $D \subset C^n$, such that every stalk $\mathcal{I}_y(A)$, $y \in A$, of the sheaf $\mathcal{I}(A)$ is generated as an $\mathcal{O}_y(D)$-modul by a finite number of holomorphic functions $f_1, \ldots, f_k$ on D. $di: \mathcal{T}(A) \to \mathcal{T}(D)$ is injective and $di(\mathcal{T}(A))$ forms an analytic subset of $\mathcal{T}(D)$. In order to prove this one has to show that $B := \varphi^{-1}(di(\mathcal{T}(A)))$ is an analytic subset of $D \times \mathbf{C}^n$. Now this is true since B is equal to

$$\left\{(y, \lambda) \in A \times C^n; \sum_{\nu=1}^{n} \lambda_\nu b_{\varkappa\nu}(y) = 0, \quad \varkappa = 1, \ldots, k\right\}, \tag{4}$$

where $b_{\varkappa\nu}(y)$ are the holomorphic functions $\frac{\partial f_\varkappa}{\partial z_\nu}\Big|_{z=y}$ on D.

Since locally a complex space X can be realized as such an analytic subset of a domain D, we have locally defined a complex structure on $\mathcal{T}(X)$. Since this local complex structure is independent of the special local realization of X as an analytic set it induces a global complex structure on $\mathcal{T}(X)$. This structure is canonical in the sense described above. Let $\tau: X \to Y$ be a holomorphic mapping. We shall describe $d\tau$ in coordinates given by formula (4). Let us assume that neighborhoods of $x_1 \in X$ and $x_2 = \tau(x_1) \in Y$ are realized as analytic subsets A_i $(i = 1, 2)$ of domains $D_i \subset \mathbf{C}^{n_i}$, such that $\mathcal{I}_y(A_i)$ for each $y \in A_i$ is generated by functions $f_1^{(i)}, \ldots, f_{k_i}^{(i)}$ holomorphic on D_i. In coordinates

$$\left\{(y^{(i)}, \lambda^{(i)}) \in A_i \times \mathbf{C}^{n_i}; \sum_{\nu=1}^{n_i} \lambda_\nu^{(i)} b_{\varkappa\nu}^{(i)}(y^{(i)}) = 0, \quad \varkappa = 1, \ldots, k_i\right\} \tag{5}$$

[7] See [5], Lemma 2.2.

of $\mathcal{T}(A_i)$ with $b_\nu^{(i)}(y^{(i)}) := \left.\frac{\partial f_\varkappa^{(i)}}{\partial z_\nu}\right|_{z=y^{(i)}}$ the differential $d\tau$ is described as follows:

$$y^{(2)} = \tau(y^{(1)}); \lambda_\mu^{(2)} = \sum_{\nu=1}^{n_1} a_{\mu\nu}(y^{(1)})\,\lambda_\nu^{(1)}, \qquad \mu = 1, \ldots, n_2. \tag{6}$$

Every point $y_0^{(1)} \in A_1$ has a neighborhood U in D_i, such that τ is described by functions $h_1, \ldots, h_{n_2}$ holomorphic in U. $a_{\mu\nu}(y^{(1)})$ is then defined by $\left.\frac{\partial h_\mu}{\partial z_\nu}\right|_{z=y^{(1)}}$. Although $a_{\mu\nu}(y^{(1)})$ is not independent of the choice of the functions $h_1, \ldots, h_{n_2}$ the sum $\sum_{\nu=1}^{n_1} a_{\mu\nu}(y^{(1)}) \cdot \lambda_\nu^{(1)}$ is because the difference d_μ between two such functions h_μ and h'_μ vanishes on A_1 and because of (5) we have

$$\sum_{\nu=1}^{n_1} \lambda_\nu^{(1)} \left.\frac{\partial d_\mu}{\partial z_\nu}\right|_{z=y^{(i)}} = 0.$$

From the formulas (4) and (6) we can easily derive the following two propositions.

Proposition 4. *Let X be a complex space, then $\{x \in X; \operatorname{emdim}_x(X) \geqq r\}$ is an analytic subset of X for every non negative integer r.*

Proof. Locally X can be realized as an analytic set A in a domain $D \subset \mathbf{C}^n$, such that $\mathcal{T}(A)$ has local coordinates given by (4). Therefore we have $\operatorname{emdim}_y X = \dim \mathcal{T}_y(A) = n - \operatorname{rank}(b_{\varkappa\nu}(y))$. Since $\operatorname{emdim}_y X > r$ is equivalent to $\operatorname{rank}(b_{\varkappa\nu}(y)) < n - r$, the set $\{y \in X;\ \operatorname{emdim}_y X > r\}$ is nothing else but the common zeros of all determinants of $(n-r) \times$ $\times\ (n-r)$ submatrices of $(b_{\varkappa\nu}(y))$. This proves the statement of the proposition.

2. Rank and corank of holomorphic mappings

Let $\tau: X \to Y$ be a holomorphic mapping.

Definition 5. $rk_x(\tau) := \dim d\tau_x(\mathcal{T}_x(X)) = \operatorname{emdim}_x X - \dim \operatorname{kern}(d\tau_x)$ *and* $\operatorname{cork}_x(\tau) := \dim \operatorname{kern}(d\tau_x)$ *are called rank and corank respectively of τ at $x \in X$.*

Proposition 5. *$E_r(\tau) := \{x \in X;\ \operatorname{cork}_x(\tau) \geqq r\}$ is an analytic set in X for every non negative integer r.*

Proof. Locally $d\tau_x$ can be described by the equations (6). Therefore kern $d\tau_x$ is isomorphic to the nullspace of the matrix

$$\mathcal{A}(x) := \begin{pmatrix} (b_{\varkappa\nu}^{(1)}(x)) \\ (a_{\mu\nu}(x)) \end{pmatrix} \begin{array}{l} \varkappa = 1, \ldots, h \\ \nu = 1, \ldots, n_1 \\ \mu = 1, \ldots, n_2 \end{array}$$

where $b_{\varkappa\nu}^{(1)}$ and $a_{\mu\nu}$ are defined as in the equations (5) and (6). We have $\operatorname{cork}_x(\tau) = n - \operatorname{rank} \mathscr{A}(x)$. For the same reason as in the proof of Proposition 4 the set $\{x \in X; \operatorname{cork}_x(\tau) > r\}$ is equal to the set of common zeros of determinants of $(n-r) \times (n-r)$ submatrices of $\mathscr{A}(x)$. This proves the proposition.

Proposition 6. *The corank* $\operatorname{cork}_x(\tau)$ *of a holomorphic mapping* $\tau: X \to Y$ *at a point* $x \in X$ *is always greater than or equal to the embedding dimension* $\operatorname{emdim}_x \tau^{-1}(\tau(x))$ *of the fibre* $\tau^{-1}(\tau(x))$ *of* τ.

Proof. We have to prove that $\operatorname{kern} d\tau_x \supset \mathscr{T}_x(\tau^{-1}(\tau(x)))$. Let $T \in \mathscr{T}_x(\tau^{-1}(\tau(x)))$. For every $f \in \mathscr{O}_{\tau(x)}(Y)$ the germ $f \circ \tau \in \mathscr{O}_x(X)$ is equal to a constant germ plus a germ that vanishes on the fibre $\tau^{-1}(\tau(x))$ and therefore $d\tau_x(T)(f) = T(f \circ \tau) = 0$. This means $T \in \operatorname{kern}(d\tau_x)$.

Using Proposition 5 one can give another proof of the following theorem of R. Remmert[8].

Proposition 7. *Let* $\tau: X \to Y$ *be a holomorphic mapping, then* $F_r(\tau) := \{x \in X;\ d_x(\tau) \geqq r\}$ *with* $d_x(\tau) := \dim_x \tau^{-1}(\tau(x))$ *is an analytic subset of* X *for every non negative integer* r.

Proof. 1. First we show that $k := \min\{\operatorname{cork}_x(\tau); x \in X\}$ is equal to $\min\{d_x(\tau); x \in X\}$. Assume there is a point $x_0 \in X$ such that $d_{x_0}(\tau) < k$, then there exists a neighborhood U of x_0 such that $d_x(\tau) \leqq d_{x_0}(\tau) < k$ for all $x \in U$. This was proved by R. Remmert[9], using the Remmert-Stein-Thullen continuation theorem for analytic sets. On the other hand U contains an open subset U' without singular points such that $\operatorname{cork}_{x'}(\tau) = \text{constant} \geqq k$ for all $x' \in U'$. This is possible because of Proposition 5. Because of the implicit function theorem we have $d_{x'}(\tau) = \operatorname{cork}_{x'}(\tau) \geqq k$ for all $x' \in U'$, which contradicts the statement above that $d_{x'}(\tau) < k$ for all $x' \in U' \subset U$.

2. $X_1 := E_{k+1}(\tau)$ is an analytic subset of X. For the restriction $\tau_1 := \tau | X_1$ of τ to X_1 the following holds: $F_{k+1}(\tau) = F_{k+1}(\tau_1)$. This is true because $d_x(\tau) = k$ for all $x \in X - X_1$. By induction we define a decreasing sequence

$$X_1 \supset X_2 \supset \ldots \supset X_\sigma \supset X_{\sigma+1} \supset \ldots$$

of analytic subsets of X, setting $X_{\sigma+1} := E_{k+1}(\tau_\sigma)$ where $\tau_\sigma := \tau | X_\sigma$. Again $F_{k+1}(\tau) = F_{k+1}(\tau_\sigma)$ holds. For a fixed point $x_0 \in X$ there exists a neighborhood U such that

$$X_1 \cap U \supset X_2 \cap U \supset \ldots \supset X_\sigma \cap U \supset X_{\sigma+1} \cap U \supset \ldots$$

becomes stationary, i.e. there exists an index s such that $X_s \cap U = X_{s+1} \cap U$. This means that $\operatorname{cork}_x(\tau_s) \geqq k+1$ for all $x \in X_s \cap U$.

[8] See [4], Satz 17.

[9] See [4], Satz 15 and [3], Satz 16.

Because of what we proved under 1 also $d_x(\tau_s) \geqq k+1$ for all $x \in X_s \cap U$. Therefore $X_s \cap U = F_{k+1}(\tau_\sigma) \cap U = F_{k+1}(\tau) \cap U$, which shows that $F_{k+1}(\tau)$ is analytic in U.

In the same way one proves that $F_{k+2}(\tau) = F_{k+2}(\tau \mid F_{k+1}(\tau))$ is an analytic subset of $F_{k+1}(\tau)$ and therefore also an analytic subset of X. The proof of the analyticity of $F_r(\tau)$ for an arbitrary integer r goes by induction with respect to r.

It has to be mentioned that for a holomorphic mapping $\tau: X \to Y$ the set $\{x \in X; D_x(\tau) \geqq r\}$ with $D_x(\tau) := \operatorname{emdim}_x \tau^{-1}\tau(x)$ in general is not an analytic subset of X. For example let $X := \{(z_1, z_2, w) \in \mathbf{C}^3; z_1^2 - w z_2^3 = 0\}$ with the complex structure induced by $\mathbf{C}^3$. The holomorphic mapping $\tau: X \to \mathbf{C}^1$, given by $\tau(z_1, z_2, w) := w$, has not the property that $\{x \in X; D_x(\tau) \geqq 2\} = \{(0, 0, w) \in \mathbf{C}^3; w \neq 0\}$ is analytic in X, although it is locally analytic and its closure is analytic in X. The last statement may hold in general.

A holomorphic mapping of a complex space into another with constant corank has similar properties as a holomorphic map of a complex manifold into another with constant rank. In the sequel, P^k shall denote the polycylinder $\{z = (z_1, \ldots, z_k) \in \mathbf{C}^k; |z_\varkappa| < 1\}$.

Theorem 1. *Let X and Y be complex spaces, $\tau: X \to Y$ a holomorphic mapping with* $\operatorname{cork}_{x'}(\tau) = m$ *for all x' in a neighborhood U of a point $x \in X$.*

Then there exists a neighborhood U_x of x which can be realized as an analytic set $S \times P^m$ in $P^n \times P^m$ with $n + m = \operatorname{emdim}_x X$ such that the following holds:

1. *$\tau(U_x)$ is an analytic subset of a domain $G \subset Y$.*
2. *$\tau = j \circ p$, where $p: S \times P^m \to S$ is the natural projection and j is a biholomorphic mapping of S onto $\tau(U_x)$.*

Proof. 1. First we shall deal with the special case that $rk_x(\tau) = \operatorname{emdim}_{\tau(x)} Y$. If we set $n := rk_x(\tau)$ we have $n + m = \operatorname{emdim}_x X$. If U is a sufficiently small neighborhood of x, we can assume that U is realized as an analytic subset of a domain $D \subset \mathbf{C}^{n+m}$ and that $\tau(U)$ is contained in a neighborhood V of $\tau(x)$ which can be realized as an analytic subset of a domain $G \subset \mathbf{C}^n$. $\tau \mid U$ can be described by a system of functions $(g_1, \ldots, g_n)$ holomorphic on U. We may assume that there exist holomorphic functions $\hat{g}_\nu$ on D such that $\hat{g}_\nu \mid U = g_\nu$, $\nu = 1, \ldots, n$. Let $\hat{\tau}$ denote the mapping defined by $(\hat{g}_1, \ldots, \hat{g}_n)$. One checks easily that $rk_x(\hat{\tau}) = rk_x(\tau) = n$. If U and D are chosen small enough then $rk_{x'}(\hat{\tau}) = n$ for all $x' \in D$. We can assume that $D = P^n \times P^m$ and that the following holds: (1) $\hat{\tau}(D)$ is a domain in $\mathbf{C}^n$ (contained in G), (2) $\hat{\tau} = \hat{j} \circ \hat{p}$, where $\hat{p}: P^n \times P^m \to P^n$ is the natural projection and $\hat{j}$ is a biholomorphic mapping of P^n onto $\hat{\tau}(D)$. We have to show now that the subset U of $D = P^n \times P^m$ has the form $U = S \times P^m$, where S is an analytic subset of P^n. The fibre $\tau^{-1}(\tau(u))$ is contained in $\hat{\tau}^{-1}(\hat{\tau}(u)) = \hat{p}^{-1}(\hat{p}(u))$ for every $u \in U$. Since $\mathrm{cork}_u(\tau) = m$ for all $u \in U$ the fibres $\tau^{-1}(\tau(u'))$ are m-dimensional manifolds around every point $u' \in U' := U - N$, where N is the set of singular points of U. Therefore $\tau^{-1}(\tau(u'))$ is equal to $\hat{p}^{-1}(\hat{p}(u'))$ for every point $u' \in U'$. Since U is the closure of $\bigcup_{u' \in U'} \tau^{-1}(\tau(u'))$, it must have the form $S \times P^m$, where S is an analytic subset of P^n. Our theorem is proved for the special case that $rk_x(\tau) = \mathrm{emdim}_{\tau(x)} Y$ if we set $p := \hat{p} | S \times P^m$ and $j := \hat{j} | S$.

2. In general $r := \mathrm{emdim}_{\tau(x)} Y \geqq n := rk_x(\tau)$. If U is a sufficiently small neighborhood of x, we can again assume that U is realized as an analytic subset of a domain $D \subset C^{n+m}$ and that $\tau(U)$ is contained in a neighborhood V of $\tau(x)$ which can be realized as an analytic subset of a domain $G \subset C^r$. We furthermore can assume that $d\tau_x(\mathscr{T}_x(X)) = \mathscr{T}_{\tau(x)}(C^n)$, where C^n is regarded as a linear subspace of C^r spanned by the first n components of C^r. $\tau | U$ is described by a set $(g_1, \ldots, g_r)$ of holomorphic functions on U. $(g_1, \ldots, g_n)$ then defines a holomorphic mapping $\sigma: U \to C^n$ with the property that $\mathrm{cork}_u(\sigma) \geqq \mathrm{cork}_u(\tau) = m$ for all $u \in U$. Since $\mathrm{cork}_x(\sigma) = m$ consequently $\mathrm{cork}_u(\sigma) = m$ for all $u \in U$ if U is chosen small enough.

Part 1 of our proof can be applied because $rk_x(\sigma) = n$. We may therefore assume that U is realized as an analytic subset $S \times P^m$ of $P^n \times P^m$ such that the following holds:

(1) $\sigma(U)$ is an analytic subset of a domain $B \subset C^n$,

(2) $\sigma = i \circ p$, where $p: S \times P^m \to S$ is the natural projection and i is a biholomorphic mapping of S onto $\sigma(U)$.

Now $\sigma^{-1}(\sigma(u)) \supset \tau^{-1}(\tau(u))$ for all $u \in U$ and for the same reason as in part one of our proof we obtain $\sigma^{-1}(\sigma(u)) = \tau^{-1}(\tau(u))$. Therefore $\tau = j \circ p$, where j is an injective holomorphic mapping of S into $V \subset Y$. As a consequence of $\mathrm{cork}_u(i) = 0$, $u \in S$, also $\mathrm{cork}_u(j) = 0$ for all $u \in S$. Therefore $\tau(U) = j(S)$ is an analytic subset of G (if it was suitably chosen) and $j: S \to \tau(U)$ is biholomorphic.

3. Applications to complex transformation groups

Theorem 1 can also be expressed as follows:

Let X and Y be complex spaces, $\tau: X \to Y$ a holomorphic mapping with $\mathrm{cork}_{x'}(\tau) = m$ *for all x' in a neighborhood U of x. Then there exists a neighborhood U_x of x, an analytic subset S of U_x with $x \in S$ and a holomorphic retraction $r: U_x \to S$, such that $\tau = \tau | S \circ r$, where $\tau | S$ is a biholomorphic mapping of S onto an analytic subset of a neighborhood $V_{\tau(x)} \subset Y$ of $\tau(x)$.*

For applications of this theorem it is often necessary to know what choices of U_x and S are possible. Going through the proof of the theorem it is easily seen that U_x and $S \subset U_x$ can be chosen as follows. One may always assume that U is realized as an analytic subset of a k-dimensional polycylinder $P^k = P^n \times P^{k-n}$, where $k := \operatorname{emdim}_x X$ and $n = rk_x(\tau)$. If $\operatorname{kern}(d\tau_x) \cap \mathscr{T}_x(P^n \times \{0\}) = \{0\}$ then there exist arbitrarily small neighborhoods U_x of x which together with the analytic sets $S := U_x \cap (P^n \times \{0\})$ satisfy the theorem above.

An example of a holomorphic mapping with constant corank is the following: If L is an n-dimensional complex Lie transformation group of a complex space X, then the holomorphic mapping $\varphi: L \times X \to X$, which is defined by $\varphi(g, x) := g(x)$ for $g \in L$ and $x \in X$, has constant corank n. Since $rk_{(g,x)}\varphi = \operatorname{emdim}_{g(x)} X = \operatorname{emdim}_x X$ and $\operatorname{emdim}_{(g,x)}(L \times X) = n + \operatorname{emdim}_x X$, we obtain $\operatorname{cork}_{(g,x)}\varphi = n$. A consequence of this is the following[10]:

Theorem 2. *Let L be an n-dimensional complex Lie transformation group of a complex space X. If all the L-orbits $L(x) := \{g(x); g \in L\}$, $x \in X$, have the same dimension m, then each point $x \in X$ has a neighborhood of the form $U_x = A \times P^m$ where A is an analytic set in a polycylinder P^r, such that the sets $\{a\} \times P^m$, $a \in A$, are contained in an L-orbit.*

Proof. One sees easily that the mapping $\varphi_x: L \to X$, $\varphi_x(g) := g(x)$, has constant rank m for each $x \in X$. There is a neighborhood V of the neutral element $e \in L$ which can be realized as a product $Q^m \times Q^{n-m}$ of two unit polycylinders of dimension m and $n - m$ respectively with $e = 0$, such that $\varphi_x | V = j_x \circ p_x$ where $p_x: Q^m \times Q^{n-m} \to Q^m$ is the natural projection and $j_x: Q^m \to X$ is biholomorphic onto the part $\varphi_x(V)$ of the orbit $L(x)$.

There is a neighborhood W of x which can be realized as an analytic subset of a polycylinder $P^m \times P^r$ with $x = 0$ and $r + m = \operatorname{emdim}_x X$ such that $\mathscr{T}_0(P^m \times \{0\}) = \mathscr{T}_0(\varphi_x(V))$. From this we obtain by an easy calculation that

$$\operatorname{kern}(d\varphi_{(e,x)}) \cap \mathscr{T}_{(e,x)}[(Q^m \times \{0\}) \times (\{0\} \times P^r)] = \{0\}.$$

Consequently there exist a neighborhood $U_{(e,x)} \subset V \times W$ of (e, x) and a biholomorphic retraction $r: U_{(e,x)} \to S$, where $S := U_{(e,x)} \cap [(Q^m \times \{0\}) \times (\{0\} \times P^r)]$, such that $\varphi | U_{(e,x)} = \varphi | S \circ r$ and φ maps S biholomorphically onto an open neighborhood of x[11]. S again contains (relative to S) an open neighborhood of (e, x) of the form $(\hat{Q}^m \times \{0\}) \times A$, where $\hat{Q}^m$ is a polycylinder contained in Q^m and A is an analytic subset of $W \cap (\{0\} \times \hat{P}^r)$, $\hat{P}^r$ being a small polycylinder contained in P^r. If we identify $\hat{Q}^m \times \{0\}$ with $\hat{Q}^m$, then φ maps $\hat{Q}^m \times A$ biholomorphically onto

---

[10] See [*1*], Hilfssatz 2.

[11] $\varphi(S)$ is open since $\varphi(S) = \varphi(U_{(e,x)})$ and φ is an open mapping.

an open neighborhood U_x of x. The sets $\hat{Q}^m \times \{a\}$, $a \in A$, correspond to parts of an L-orbit.

4. Holomorphic retractions

Definition 6. *Let Y be a complex subspace of a complex space X. A holomorphic mapping $r: X \to Y$ is called a holomorphic retraction if $r \mid Y$ is the identity on Y.*

Theorem 3. *Let $r: X \to Y$ be a holomorphic retraction. Then every point $y \in Y$ has a neighborhood U in X which can be realized as an analytic set in the product $P^m \times P^s$ of two unit polycylinders of dimension m and s respectively such that the following holds:*

1. $r = p \mid U$, where $p: P^m \times P^s \to P^m \times \{0\}$ is the canonical projection,

2. $m + s = \text{emdim}_y X, \quad m = \text{emdim}_y Y$.

Before we prove Theorem 3 let us derive the following corollary from it.

Corollary. *Let D be a domain in $\mathbf{C}^n$, A an analytic subset of D and $r: D \to A$ a holomorphic retraction, then A is a submanifold of D and the fibres of r are non singular in a neighborhood of A.*

Proof. According to Theorem 3 every point $y \in A$ has a neighborhood U in D of the form $P^m \times P^s$, where $n = m + s$ and r is nothing else but the natural projection $p: P^m \times P^s \to P^m \times \{0\}$. $U \cap A$ is then equal to $P^m \times \{0\}$ and consequently y a nonsingular point of A. In U the fibres of r are manifolds biholomorphically equivalent to sets $\{a\} \times P^s$, $a \in P^m$.

Proof of Theorem 3. Let y be a point on Y with $\text{emdim}_y Y = m$. It is easily verified that also $rk_y(r) = m$. There exist neighborhoods $U_y \subset X$ and $V_y \subset Y$ of y, which can be realized as analytic sets in domains $D \subset \mathbf{C}^n$ and $B \subset \mathbf{C}^m$ respectively, such that $n = \text{emdim}_y X$ and $r(U_y) \subset V_y$. U_y and D can be suitably chosen, so that r can be regarded as the restriction of a holomorphic mapping $\hat{r}: D \to B$ with $rk_x(\hat{r}) = m$ for all $x \in D$. One may also assume that D has the form $Q^m \times Q^s$ with $m + s = n$ and $y = 0$ such that $\hat{r} = \sigma \circ q$ where $q: Q^m \times Q^s \to Q^m \times \{0\}$ is the canonical projection and σ is a biholomorphic map of $Q^m \times \{0\}$ onto a subdomain B' of B. $Y \cap U_y$ is mapped biholomorphically by $\hat{r}$ onto an analytic subset of a subdomain B'' of B'.

Consequently $q = \sigma^{-1} \circ \hat{r}$ maps $Y \cap U_y$ biholomorphically onto an analytic subset A of a subdomain G of $Q^m \times \{0\}$. $q^{-1}: A \to Y \cap U_y$ is described by a system of functions $(h_1, \ldots, h_n)$ holomorphic on A. One can assume that the functions h_ν, $\nu = 1, \ldots, n$, are also holomorphic in G and that $h_\mu = z_\mu$ for $\mu = 1, \ldots, m$, where $z_1, \ldots, z_m$ are coordinates of

Q^m and $z_{m+1}, \ldots, z_n$ are coordinates of Q^s. If we define a holomorphic mapping $\varphi: D \to \boldsymbol{C}^n$ by

$$\varphi:(z_1, \ldots, z_n) \to (z_1', \ldots, z_n'):$$
$$= (z_1, \ldots, z_m, z_{m+1} - h_{m+1}(z_1, \ldots, z_m), \ldots, z_n - h_n(z_1, \ldots, z_m))$$

we obtain a coordinate transformation which is biholomorphic in a neighborhood of the origin and keeps 0 fixed. In the new coordinates we can find a sufficiently small polycylinder $P^m \times P^s$ about the origin which is mapped biholomorphically by φ^{-1} onto a subdomain of $Q^m \times Q^s$. $U := \varphi(U_y) \cap (P^m \times P^s)$ is a realization of a neighborhood of y in X as an analytic subset of $P^m \times P^s$. If $P^m \times P^s$ is sufficiently small, $Y \cap U$ is an analytic subset of $P^m \times \{0\}$. One checks easily that the canonical projection $p: P^m \times P^s \to P^m \times \{0\}$ if restricted to U is nothing else but r.

Theorem 4. *Let X, Y be pure dimensional complex spaces, $\tau: X \to Y$ an open holomorphic mapping. Then every point $x \in X$ has an open neighborhood U and there exist open analytic coverings*[12] *$(\hat{U}, \sigma)$ of U and $(\hat{V}, \varrho)$ of $V := \tau(U)$ and an open holomorphic mapping $\tilde{\tau}$ such that the following holds:*

(1) *The following diagram is commutative*

$$\begin{array}{ccc} \tilde{U} & \xrightarrow{\tilde{\tau}} & \tilde{V} \\ \sigma\downarrow & & \downarrow\varrho \\ U & \xrightarrow{\tau} & V \end{array}$$

(2) *There exists a holomorphic retraction $r: \tilde{U} \to \tilde{S}$ of $\tilde{U}$ onto an analytic subset $\tilde{S}$ such that $\tilde{\tau} = \tilde{\tau}|\tilde{S} \circ r$, where $\tilde{\tau}|\tilde{S}$ is a biholomorphic mapping of $\tilde{S}$ onto $\tilde{V}$.*

(3) *$\tilde{U}$ and $\tilde{V}$ can be chosen as normal complex spaces if we drop the condition that σ, ϱ and $\tilde{\tau}$ are open mappings.*

We first prove the following two lemmas, which will be used in the proof of part (3) of Theorem 4.

Lemma 1. *Let $\tau: Y \to X$ be a holomorphic mapping, such that the image of an irreducible component of Y never lies completely in the set of singular points of X, then there exists a (unique) holomorphic mapping $\tau^*: Y^* \to X^*$, where (Y^*, μ) and (X^*, ν) are the normalizations of Y and X respectively, such that the following diagram is commutative:*

$$\begin{array}{ccc} Y^* & \xrightarrow{\tau^*} & X^* \\ \mu\downarrow & & \downarrow\nu \\ Y & \xrightarrow{\tau} & X \end{array}$$

[12] If $\tilde{X}$ and X are complex spaces, $\sigma: \tilde{X} \to X$ a surjective, proper, discrete holomorphic mapping, then $(\tilde{X}, \sigma)$ shall be called an analytic covering of X.

Proof. $\tilde{Y} := \{(y, x^*) \in Y \times X^*;\ \tau(y) = \nu(x^*)\}$ is an analytic subset of $Y \times X^*$. Let $\varphi: \tilde{Y} \to Y$ and $\psi: \tilde{Y} \to X^*$ be the following holomorphic mappings: $\varphi(y, x^*) := y$, $\psi(y, x^*) := x^*$ then $\tau \circ \varphi = \nu \circ \psi$. φ is a proper mapping since for any compact subset K of Y the set $\varphi^{-1}(K)$ is a closed subset of the compact set $K \times \nu^{-1}(\tau(K))$ and therefore also a compact set. φ is by definition a discrete map. $(\tilde{Y}^*, \varrho)$ may denote the normalization of $\tilde{Y}$. With $\tilde{\mu} := \varphi \circ \varrho$ and $\tilde{\tau}^* : \psi \circ \varrho$ the following diagram becomes commutative:

$$\begin{array}{ccc} \tilde{Y}^* & \xrightarrow{\tilde{\tau}^*} & X^* \\ {\scriptstyle\tilde{\mu}}\downarrow & & \downarrow{\scriptstyle\nu} \\ Y & \xrightarrow{\tau} & X. \end{array}$$

$\tilde{\mu}$ is a proper and discrete holomorphic mapping since φ and ϱ are proper and discrete and it maps $\tilde{Y}^* - \tilde{\mu}^{-1}(\tau^{-1}(M) \cup N)$ onto $Y - (\tau^{-1}(M) \cup N)$ biholomorphically, where M and N are the sets of singular points of X and Y respectively. $\tilde{\mu}^{-1}(\tau^{-1}(M) \cup N)$ may contain some irreducible components of $\tilde{Y}^*$, but it is nowhere dense in the union Y^* of the other irreducible components of $\tilde{Y}^*$. (Y^*, μ) with $\mu := \tilde{\mu} \mid Y^*$ is the normalization of Y which proves the lemma.

Lemma 2. *Let Y be a normal complex subspace of the complex space X such that no irreducible component of Y lies completely in the singular set of X. If $r: X \to Y$ is a holomorphic retraction then Y can also be embedded holomorphically in the normalization (X^*, ν) of X and there exists a holomorphic retraction $r^* : X^* \to Y$ such that the following diagram becomes commutative:*

$$\begin{array}{ccccc} & & X^* & & \\ & {\scriptstyle j^*}\nearrow & & \searrow{\scriptstyle r^*} & \\ Y & & \downarrow{\scriptstyle\nu} & & Y \\ & {\scriptstyle j}\searrow & & \nearrow{\scriptstyle r} & \\ & & X & & \end{array}$$

Here j and j^* denote the holomorphic embeddings of Y in X and X^* respectively.

Proof. r^* is defined by r and ν and because of Lemma 1 there exists a uniquely determined holomorphic mapping $j^* : Y \to X^*$, such that the above diagram becomes commutative. Since $r \circ j$ is the identity on Y, the same holds for $r^* \circ j^*$. This means that r^* is a holomorphic retraction of X^* onto Y.

Proof of Theorem 4. Since τ is open and X and Y are pure-dimensional there exists a neighborhood U for every point x such that the fibres of

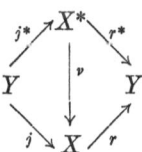

$\tau | U$ are pure dimensional and have all the same dimension[13] and such that U has an analytic subset $\tilde{V}$ with the following properties:

(1) $\tau(\tilde{V}) = \tau(U)$, (2) the restriction $\tau | \tilde{V}$ of τ to $\tilde{V}$ is a proper, discrete, open, holomorphic map. Now we define $V := \tau(U)$, $\tilde{U} := \{(u, \tilde{v}) \in U \times \tilde{V};\ \tau(u) = \tau(\tilde{v})\}$, $\varrho := \tau | \tilde{V}$. Then with $\sigma(u, \tilde{v}) := u$ and $\tilde{\tau}(u, \tilde{v}) := \tilde{v}$ for $(u, \tilde{v}) \in U$ the diagram in Theorem 4 becomes commutative. Since ϱ is proper also σ becomes proper because for every compact set K in U the set $\sigma^{-1}(K)$ is a closed subset of the compact set $K \times \times \varrho^{-1}(\tau(K))$ in $U \times \tilde{V}$ and therefore also compact. By definition σ is discrete and surjective. Since τ and ϱ are open also σ and $\tilde{\tau}$ are open. For reasons of symmetry it is sufficient to prove the openness of $\tilde{\tau}$. Let A and B be open subsets of U and $\tilde{V}$ respectively. We have to show that $\tilde{\tau}(\tilde{U} \cap (A \times B))$ is open in $\tilde{V}$. This follows immediately from the equation $\tilde{\tau}(\tilde{U} \cap (A \times B)) = B \cap \varrho^{-1}(\tau(A))$.

$\tilde{S} := \{(\tilde{v}, \tilde{v});\ \tilde{v} \in \tilde{V}\}$ is an analytic subset of $\tilde{U}$ which together with the holomorphic retraction $\tilde{r} : \tilde{U} \to \tilde{S}$, $\tilde{r}(u, \tilde{v}) := (\tilde{v}, \tilde{v})$, satisfies part (2) of Theorem 4.

To prove (3) we replace $\tilde{V}$ by its normalization $(\hat{V}, \mu)$, $\tilde{U}$ by $\hat{U} := \{(u, \hat{v}) \in U \times \hat{V};\ \tau(u) = \tau(\mu(\hat{v}))\}$ and the mappings ϱ, σ, $\tilde{\tau}$ by $\hat{\varrho} := \tau \circ \mu : \hat{V} \to V$, $\hat{\sigma} : \hat{U} \to U$, $\hat{\tau} : \hat{U} \to \hat{V}$, where $\hat{\sigma}(u, \hat{v}) := u$ and $\hat{\tau}(u, \hat{v}) := \hat{v}$ for $(u, \hat{v}) \in \hat{U}$. Again the parts (1) and (2) of Theorem 4 are satisfied, if we define $\hat{S} := \{(\tilde{v}, \hat{v}) \in \tilde{V} \times \hat{V};\ \tilde{v} = \mu(\hat{v})\}$ and $\hat{r} : \hat{U} \to \hat{S}$ by $\hat{r}(u, \hat{v}) := (\mu(\hat{v}), \hat{v})$ for $(u, \hat{v}) \in \hat{U}$. $\tilde{V}$ can be chosen such that no irreducible component of $\hat{S}$ lies completely in the singular set of $\hat{U}$. Because of Lemma 2 $\hat{S}$ can also be regarded as an analytic subset of the normalization $(\hat{U}^*, \nu)$ of $\hat{U}$ and there exists a holomorphic retraction $r : \hat{U}^* \to \hat{S}$ such that $\hat{r} \circ \nu = r$. If we define $\hat{\tau}^* : \hat{U}^* \to \hat{V}$ by $\hat{\tau}^* := \hat{\tau} \circ \nu$ and $\hat{\sigma}^* : \hat{U}^* \to U$ by $\hat{\sigma}^* := \hat{\sigma} \circ \nu$ then again $\tau \circ \hat{\sigma}^* = \hat{\varrho} \circ \hat{\tau}^*$ and $\hat{\tau}^* = \hat{\tau}^* | \hat{S} \circ r$ hold, where $\hat{\tau}^*$ maps $\hat{S}$ biholomorphically onto $\hat{V}$.

Example: Let $U := \boldsymbol{C}^2$, $V := \boldsymbol{C}$ and $\tau : U \to V$ the holomorphic mapping defined by $\tau(z_1, z_2) := z_1^2 - z_2^3$ for $(z_1, z_2) \in \boldsymbol{C}^2$. In this case we can choose $\tilde{V} = \boldsymbol{C}$, $\tilde{U} := \{(z_1, z_2, w) \in \boldsymbol{C}^3;\ z_1^2 - z_2^3 = w^2\}$ and define $\varrho : \tilde{V} \to V$ by $\varrho(w) := w^2$, $\tilde{\tau} : \tilde{U} \to \tilde{V}$ by $\tilde{\tau}(z_1, z_2, w) := w$; $\sigma : \tilde{U} \to U$ by $\sigma(z_1, z_2, w) := (z_1, z_2)$. $\tilde{\tau}$ can be decomposed in the following way $\tilde{\tau} = \tilde{\tau} | \tilde{S} \circ r$, where $\tilde{S} := \{(z_1, z_2, w) \in \tilde{U};\ z_2 = 0,\ z_1 = w\}$ and $r : \tilde{U} \to \tilde{S}$ is the holomorphic retraction $r(z_1, z_2, w) := (w, 0, w)$. $\tilde{\tau}$ maps S biholomorphically onto $\tilde{V}$.

[13] See [4], Satz 29.

The numbers of sheets of the coverings $(\tilde{V}, \varrho)$ and $(\tilde{U}, \sigma)$ are minimal because of the corollary to Theorem 3.

5. Regularity of holomorphic mappings and embeddings

If $\tau : X \to Y$ is a holomorphic mapping, then the sheaf $\mathscr{A}(\tau)$ of germs of τ-invariant holomorphic functions on X is defined by the following presheaf $\{A_U, r_V^U\}$, where A_U is the $\mathbf{C}$-algebra of τ-invariant holomorphic functions on the open set $U \subset X$ and $r_V^U : A_U \to A_V (U \supset V)$ the restriction mapping.

Theorem 5. *Let X, Y be complex spaces and $\tau : X \to Y$ a holomorphic mapping. Then the following statements are equivalent:*

(1) *τ is weakly regular at $x \in X$.*

(2) *The stalk $\mathcal{O}_x$ of the structure sheaf $\mathcal{O}$ of X is equal to the direct sum $\mathscr{I}_x \oplus \mathscr{A}_x(\tau)$ of an ideal $\mathscr{I}_x$ in $\mathcal{O}_x$ and the subalgebra $\mathscr{A}_x(\tau)$ of $\mathcal{O}_x$.*

(3) *There exists an analytic*[14] *algebra endomorphism φ_x of $\mathcal{O}_x$ such that $\varphi_x = \varphi_x^2$ and $\varphi_x(\mathcal{O}_x) = \mathscr{A}_x(\tau)$.*

Proof. (1) ⇒ (2): If τ is weakly regular at $x \in X$, then there exists a neighborhood U of x, an analytic subset S of U with $x \in S$ and a holomorphic retraction $r : U \to S$ such that $\tau | U = \tau | S \circ r$. Let $\mathscr{I}$ be the sheaf of germs of holomorphic functions on U which vanish on S. We have to prove that the stalk $\mathcal{O}_x$ is equal to the direct sum $\mathscr{I}_x \oplus \mathscr{A}_x(\tau)$. r induces an algebra endomorphism r_x of $\mathcal{O}_x$ which is defined by $r_x(f) = f \circ r$ for $f \in \mathcal{O}_x$. One checks easily that kern $(r_x) = \mathscr{I}_x$, $r_x | \mathscr{A}_x(\tau) =$ identity on $\mathscr{A}_x(\tau)$ and $r_x(\mathcal{O}_x) = \mathscr{A}_x(\tau)$. From this it follows that $\mathcal{O}_x = \mathscr{I}_x \oplus \mathscr{A}_x(\tau)$.

(2) ⇒ (3): Let $\mathcal{O}_x = \mathscr{I}_x \oplus \mathscr{A}_x(\tau)$, where $\mathscr{I}_x$ is an ideal in $\mathcal{O}_x$. The mapping $\varphi_x : \mathcal{O}_x \to \mathcal{O}_x$, defined by $\varphi_x(h + f) := f$ for $h \in \mathscr{I}_x$ and $f \in \mathscr{A}_x(\tau)$, is an analytic algebra endomorphism. We check the analyticity of φ_x. Let $f_\nu \in \mathscr{A}_x(\tau)$ and $h_\nu \in \mathscr{I}_x (\nu = 1, \ldots, n)$ with $f_\nu(x) + h_\nu(x) = 0$ and let F be a holomorphic function in a neighborhood of 0 in C^n. Then $F(f_1 + h_1, \ldots, f_n + h_n) = F(h_1, \ldots, h_n) + G$ with $G \in \mathscr{I}_x$, that means $\varphi_x(F(f_1 + h_1, \ldots, f_n + h_n)) = F(\varphi_x(f_1 + h_1), \ldots, \varphi_x(f_n + h_n))$. By definition $\varphi_x^2 = \varphi_x$ and $\varphi_x(\mathcal{O}_x) = \mathscr{A}_x(\tau)$.

(3) ⇒ (1): One checks easily that the ideal kern (φ_x) defines an analytic set S in an open neighborhood U of x with $x \in S$, such that the ideal $\mathscr{I}_x$ of germs $f \in \mathcal{O}_x$ which vanish on S is equal to kern (φ_x). The stalk $\mathcal{O}_x(S)$ of the structure sheaf $\mathcal{O}(S)$ of S can be identified with $\mathcal{O}_x / \mathscr{I}_x$. The

---

[14] Let A and B be stalks of germs of holomorphic functions, then an algebra-homomorphism $\varphi : A \to B$ is called analytic if $\varphi(F(f_1, \ldots, f_n)) = F(\varphi(f_1), \ldots, \varphi(f_n))$ for every function F holomorphic in a neighborhood of 0 in $\mathbf{C}^n$ and every set $f_1, \ldots, f_n$ of elements from the maximal ideal of A.

analytic endomorphism φ_x of $\mathcal{O}_x$ induces an analytic algebra-isomorphism $\varphi_x^*: \mathcal{O}_x(S) = \mathcal{O}_x | \mathcal{I}_x \to \mathcal{A}_x(\tau)$ such that $\varphi_x = \varphi_x^* \circ \pi$, where $\pi: \mathcal{O}_x \to \to \mathcal{O}_x | \mathcal{I}_x$ is the canonical analytical algebra-homomorphism. We can assume that U can be realized as an analytic subset of a domain $D \subset \mathbf{C}^n$ with $x = 0$ and that S is equal to the set of common zeros of functions $k_1, \ldots, k_s$ holomorphic on D. Let $z_1, \ldots, z_n$ be coordinates of C^n. The restriction f_ν of z_ν to S $(\nu = 1, \ldots, n)$ represents an element of the maximal ideal of $\mathcal{O}_x(S)$, while $g_\nu := \varphi_x^*(f_\nu)$ can be regarded as a holomorphic function on a neighborhood $\tilde{U} \subset U$ of x. Since $k_\sigma(f_1, \ldots, f_n) = 0$ for $\sigma = 1, \ldots, s$, we also have $k_\sigma(g_1, \ldots, g_n) = \varphi_x^*(k_\sigma(f_1, \ldots, f_n)) = 0$. Therefore $r := (g_1, \ldots, g_n)$ maps $\tilde{U}$ into S. If $i_x: \mathcal{O}_x \to \mathcal{O}_x(S)$ is defined by $i_x(f) := f | S$ for all $f \in \mathcal{O}_x$, by a simple calculation we obtain $i_x \circ \varphi_x^* =$ identity on $\mathcal{O}_x(S)$. This means that $r(z) = z$ for all $z \in S' := S \cap \tilde{\tilde{U}}$ for a sufficiently small open neighborhood $\tilde{\tilde{U}} \subset \tilde{U}$ of x. Now $U' := r^{-1}(S') \cap \tilde{\tilde{U}}$ is mapped onto S' by r and $r | S' =$ identity on S'. Hence $r: U' \to S'$ is a holomorphic retraction. By definition r is τ-invariant and therefore $\tau | S'$ injective. If U' is small enough we can assume that $\tau(U')$ is contained in an open neighborhood V of $\tau(x)$, which can be realized as an analytic set in a domain $D \subset \mathbf{C}^m$. The mapping $\tau: U' \to V$ can be described by a system of functions $(h_1, \ldots, h_m)$ holomorphic on U'. Since $\varphi_x^*: \mathcal{O}_x(S) \to \mathcal{A}_x(\tau)$ is bijective, we can assume that $h_\mu = \varphi_x^*(\hat{h}_\mu)$ for $\mu = 1, \ldots, m$, where $\hat{h}_\mu := h_\mu | S'$. Hence $\tau | S' \circ r$ is described by $(\hat{h}_1 \circ r, \ldots, \hat{h}_m \circ r) = (\varphi_x^*(\hat{h}_1), \ldots, \varphi_x^*(\hat{h}_m)) = (h_1, \ldots, h_m)$ and consequently $\tau | S' \circ r = \tau$ on U'.

There exist conditions for the regularity of holomorphic embeddings which are quite similar to those given in Theorem 5 for the regularity of holomorphic mappings and which can be proved almost the same way.

Theorem 6. *Let Y be a complex subspace of the complex space X, then the following statements are equivalent:*

(1) *Y is regularly embedded in X at $y \in Y$.*

(2) *The stalk $\mathcal{O}_y$ of the structure sheaf $\mathcal{O}$ of X is equal to the direct sum $\mathcal{I}_y(Y) \oplus \mathcal{A}_y$ of the ideal $\mathcal{I}_y(Y) \subset \mathcal{O}_y$ of germs of holomorphic functions at y which vanish on Y and an analytic*[15] *subalgebra $\mathcal{A}_y$ of $\mathcal{O}_y$.*

(3) *There exists an analytic algebra-isomorphism $\psi_y: \mathcal{O}_y(Y) \to \mathcal{O}_y$ of the stalk $\mathcal{O}_y(Y)$ of the structure sheaf $\mathcal{O}(Y)$ of Y into $\mathcal{O}_y$ such that $i_y \circ \psi_y$ is equal to the identity on $\mathcal{O}_y(Y)$. Here $i_y: \mathcal{O}_y \to \mathcal{O}_y(Y)$ associates with each $f \in \mathcal{O}_y$ its restriction to Y.*

---

[15] A subalgebra $\mathcal{A}_y$ of $\mathcal{O}_y$ is called analytic iff for every set $f_1, \ldots, f_n$ of elements of $\mathcal{A}_y$ with $f_\nu(y) = 0$ $(\nu = 1, \ldots, n)$ and every function F holomorphic on an open neighborhood of the origin in C^n the function $F(f_1, \ldots, f_n)$ belongs to $\mathcal{A}_y$.

References

[1] Holmann, H.: Komplexe Räume mit komplexen Transformationsgruppen. Math. Ann. **150**, 327–360 (1963).
[2] Kaup, W.: Holomorphe Vektorfelder und Transformationsgruppen komplexer Räume. Schriftenreihe des Math. Inst. der Universität Münster, Heft 24 (1963).
[3] Remmert, R.: Projektionen analytischer Mengen. Math. Ann. **130**, 410–441 (1956).
[4] — Holomorphe und meromorphe Abbildungen komplexer Räume. Math. Ann. **133**, 328–370 (1957).
[5] Rossi, H.: Local properties of holomorphic mappings. Ann. Math. **78**, 455 bis 467 (1963).
[6] Whitney, H.: Local properties of analytic varieties. Differential and combinatorial topology. Princeton Univ. Press 1964.
[7] — Tangents to an analytic variety. Princeton Notes (1964).

Department of Mathematics
University of California
Berkeley

Automorphic Forms and General Teichmüller Spaces*,**

By

L. Bers

This is a summary of results extending the theory of Teichmüller spaces [1, 2, 6, 7] to arbitrary Fuchsian groups and to open Riemann surfaces. The new point of view also sheds some light on the case of closed surfaces. Proofs will appear elsewhere.

1. Let D be a simply connected domain in $\boldsymbol{C} \cup \{\infty\}$ with more than one boundary point, $\lambda_D(z)|dz|$ the Poincaré line element in D, $q \geqq 2$ an integer, and G a discrete group of conformal self-mappings of D (which may be the trivial group $G = 1$). The Banach space $\mathscr{A}_q(D, G)$ of integrable automorphic forms ($\mathscr{B}_q(D, G)$ of bounded automorphic forms) consists of holomorphic functions $\varphi(z)$, $z \in D$, such that φ has a zero of order at least $2q$ at $z = \infty$ if $\infty \in D$, $(\varphi \circ A)(A')^q = \varphi$ for $A \in G$, and $\lambda_D^{2-q}\varphi \in L_1(D/G)$ ($\lambda_D^{-q}\varphi \in L_\infty(D)$). If $\Phi \in \mathscr{A}_q(D, 1)$, we denote by $\Theta_{G,q}\Phi$ the function $\sum_G (\Phi \circ A)(A')^q$. For $\varphi \in \mathscr{A}_q(D, G)$, $\psi \in \mathscr{B}_q(D, G)$ we denote by $(\varphi, \psi)_{D/G,q}$ the integral of $\lambda_D^{2-2q}\varphi\overline{\psi}$ over a fundamental domain D/G. Using the Bergman kernel and the methods of [8] one proves

* This paper represents results obtainted at the Courant Institute of Mathematical Sciences, New York University, with the United States Army Research Office, Contract No. DA-ARO-(D)-31-124-G156. Reproduction in whole or in part is permitted for any purpose of the United States Government.
** Received March 18, 1964.

Theorem 1. *$\Theta_{G,q}$ is a continuous mapping of $\mathscr{A}_q(D, 1)$ onto $\mathscr{A}_q(D, G)$.*

Theorem 2. *Every continuous linear functional on $\mathscr{A}_q(D, G)$ is uniquely representable as $(\varphi, \psi)_{D/G,q}$, $\psi \in \mathscr{B}_q(D, G)$.*

2. Now let D_1 and D_2 be disjoint Jordan domains with common boundary C, G a discrete group of Möbius transformations such that $A(D_1) = D_1$ for $A \in G$, and $q \geqq 2$ an integer. We assume that C admits a quasi-reflection h, that is an orientation-reversing automorphism $z \to h(z)$ of $\mathbf{C} \cup \{\infty\}$, of bounded excentricity, such that $h \circ h(z) = z$ for all z and $h(z) = z$ for $z \in C$. We also assume, for the sake of convenience, that $\{0, 1, \infty\} \subset C$. An important theorem of AHLFORS [*3*] states that h may be assumed to satisfy a uniform Lipschitz condition.

Theorem 3. *There exist canonical topological surjective isomorphisms a: $\mathscr{A}_q(D_2, G) \to \mathscr{A}_q(D_1, G)$ and b: $\mathscr{B}_q(D_1, G) \to \mathscr{B}_q(D_2, G)$ such that $(a\varphi, \psi)_{D_1/G,q} = (\varphi, b\psi)_{D_2/G,q}$.*

Sketch of proof. The definition of b is explicit and independent of G; it reads

$$(b\psi)(\zeta) = c_q \iint_{D_1} (\zeta - z)^{-2q} \lambda_{D_1}(z)^{-q} \overline{\psi(z)}\, dx\, dy, \quad \zeta \in D_2$$

where c_q is a constant. To show that b is surjective (for $G = 1$) one establishes, using Ahlfors' result, the reproducing formula

$$\hat{\psi}(\zeta) = c_q' \iint_{D_1} (\zeta - z)^{-2q} \hat{\psi}(h(z))(h(z) - z)^q h_{\bar{z}}(z)\, dx\, dy$$

for $\zeta \in D_2$, $\hat{\psi} \in \mathscr{B}_q(D_2)$, and appeals to Theorem 2. Let b': $\mathscr{B}_q(D_2, G) \to$ $\to \mathscr{B}_q(D_1, G)$ be defined similarly. One defines a by setting $a \circ \Theta_{G,q}\Phi$ $= \Theta_{G,q} \circ b'\Phi$ for $\Phi \in \mathscr{A}_q(D_2, 1) \subset \mathscr{B}_q(D_2, 1)$.

3. Let $M(D_1)$ denote the open unit ball in $L_\infty(D_1)$. For $\mu \in M(D_1)$ let $z \to w^\mu(z)$ be the quasi-conformal self-mapping of $\mathbf{C}$ satisfying $w^\mu(0) = 0$, $w^\mu(1) = 1$, $w_{\bar{z}} = \mu w_z$ in D_1, $w_{\bar{z}} = 0$ in D_2. Let $\varphi^\mu(z)$, $z \in D_2$, be the Schwarzian derivative of w^μ. By a theorem of NEHARI [*10*], $\varphi^\mu \in \mathscr{B}_2(D_2, 1)$.

If ψ belongs to the open unit ball in $\mathscr{B}_2(D_1, 1)$, we set $c\psi = \varphi^\mu$ where $\mu = \lambda_{D_1}^{-2} \bar{\psi}$. The following result is an extension of the theorems of NEHARI [*10*], AHLFORS-WEILL [*5*] and AHLFORS [*3*].

Theorem 4. *c is a real analytic homeomorphism of a neighborhood N_1 of 0 in $\mathscr{B}_2(D_1, 1)$ onto a neighborhood N_2 of 0 in $\mathscr{B}_2(D_2, 1)$. Also, c maps $N_1 \cap \mathscr{B}_2(D_1, G)$ onto $N_2 \cap \mathscr{B}_2(D_2, G)$.*

Sketch of proof. Using [*4*] one shows that $\mu \to \varphi^\mu$ is a real analytic mapping. So is the mapping c. The derivative of c at the origin is the mapping b of Theorem 3 (for $q = 2$). The second assertion follows by computing that $c((\psi \circ A)(A')^2) = (c\psi \circ A)(A')^2$ for $A \in G$.

4. Now let D_1 and D_2 be the upper and lower half-planes, U and L, respectively, so that G is a Fuchsian group. For $\mu \in M(U)$ let $z \to w_\mu(z)$ be the self-mapping of U satisfying

$$(w_\mu)_{\bar z} = \mu(w_\mu)_z, \quad w_\mu(0) = 0, \quad w_\mu(1) = 1, \quad w_\mu(\infty) = \infty.$$

$M(U, G) \subset M(U)$ denotes the subspace of those μ which satisfy $(\mu \circ A)\bar{A}'/A' = \mu$ for $A \in G$. If $\mu \in M(U, G)$, then $G^\mu = w^\mu G (w^\mu)^{-1}$ and $G_\mu = w_\mu G w_\mu^{-1}$ are groups of Möbius transformations.

The group Σ^* of all w_μ contains the normal subgroup Σ_* of those w for which $w(x) = x, x \in \mathbf{R}$. The coset of w_μ modulo Σ_* will be denoted by $[w_\mu]$. The image of $M(U, G)$ under the mapping $\mu \to [w_\mu]$ is the *Teichmüller space* $T(G)$. The distance in $T(G)$ is defined by

$$\delta_G([w_\mu], [w_\nu]) = \inf \log((1+s)/(1-s)), \qquad s = \operatorname{ess\,sup} |\sigma(z)|$$

where $\sigma \in M(U, G_\nu)$ and $w_\sigma \circ w_\nu = w_\mu$. (This definition of the Teichmüller space differs from the usual one if G is of the second kind.)

Theorem 5. *$T(G)$ is a complete metric space under δ_G. Also, $T(G)$ is a δ_1-closed subset of $T(1)$, and the metrics δ_1 and δ_G are equivalent on $T(G)$.*

The first statement is easily proved; the second follows from Theorem 6 below.

Using the method of [*7*a, p. 96] one verifies that for $\mu \in M(U)$ the element $\varphi^\mu \in \mathscr{B}_2(L, 1)$ depends only on $[w_\mu]$ and determines $[w_\mu]$. Also, $\varphi^\mu \in \mathscr{B}_2(L, G)$ if $\mu \in M(U, G)$. By Nehari's theorem [*10*] the norm of φ^μ is at most 6. Hence $\varrho: [w_\mu] \to \varphi^\mu$ is an injection of $T(G)$ into a bounded set in $\mathscr{B}_2(L, G)$.

Theorem 6. *ϱ is a homeomorphism of $T(G)$ onto a bounded domain in $\mathscr{B}_2(L, G)$, the component of $\varrho T(1) \cap \mathscr{B}_2(L, G)$ containing the origin.*

The proof follows easily from Theorem 4.

Remark. The openness of $\varrho T(1)$ was proved by Ahlfors in [*3*]. Recently Earle [*9*] obtained a direct derivation of Theorem 6 from Ahlfors' result.

From now on we identify $T(G)$ with $\varrho T(G)$. Using the classical theory of Fuchsian groups, or directly, one proves

Theorem 7. dim $T(G) < \infty$ *if and only if G is finitely generated and of the first kind.*

5. Let Γ_t denote the group of right translations in $T(1) = \Sigma^*/\Sigma_*$. Let Γ_r denote the group of 'rotations' in $T(1)$, i.e. of mappings $[w_\mu] \to$ $\to [w_\nu]$ where $\nu = (\mu \circ A)\bar{A}'/A'$, A being some fixed conformal self-mapping of U (so that $\varphi^\nu = (\varphi^\mu \circ A)(A')^2$). Let $\Gamma = \Gamma(1)$ be the group generated by Γ_t and Γ_r, and let $\Gamma(G)$ be the maximal subgroup of Γ mapping $T(G)$ into itself.

Theorem 8. *Every element of $\Gamma(1)$ is (uniquely) a rotation followed by a translation. Every element of $\Gamma(G)$ is a holomorphic self-mapping of $T(G)$ and a δ_G-isometry. If* $\dim T(G) < \infty$, *$\Gamma(G)$ is discrete.*

The proof is quite simple. We note two corollaries. $T(1)$ is a homogeneous domain. If $\dim T(G) < \infty$, $T(G)/\Gamma(G)$ is a normal complex space.

6. If G is a Fuchsian group, let $\wedge(G)$ denote the set of fixed points of elliptic elements of G, and set $U_G = U - (U \cap \wedge(G))$. A rather delicate argument which cannot be sketched here leads to

Theorem 9. *A conformal mapping of U_G/G onto $U_{G'}/G'$ induces a holomorphic homeomorphism of $T(G)$ onto $T(G')$.*

We may identify $T(G)$ with the Teichmüller space of the Riemann surface U_G/G.

7. If $\mu \in M(U)$, the domain $w^\mu(U)$ depends only on $\tau = [w_\mu] \in T(1)$. We write $w^\mu(U) = D(\tau)$. We define the fiber space $\widetilde{T}(G)$ over $T(G)$ as the space of pairs (τ, z) where $\tau \in T(G)$, $z \in D(\tau)$. There is a group $\underline{G}$ of holomorphic self-mappings of $\widetilde{T}(G)$ of the form $\tau \to \tau$,

$$z \to w^\mu \circ \overline{A} \circ (w^\mu)^{-1}(z),$$

where $A \in G$. Clearly $T(\widetilde{G})/\underline{G}$ is a fiber space over $T(G)$ with fibers $w^\mu(U)/G^\mu$. Using Theorem 3 one proves

Theorem 10. *Every $\varphi \in \mathscr{A}_q(U, G)$ (every $\varphi \in \mathscr{B}_q(U, G)$) is the restriction to $\tau = 0$ of a canonically defined holomorphic function $\Phi(\tau, z)$ in $\widetilde{T}(G)$, such that for every $\tau = [w^\mu]$ the mapping $\varphi(z) \to \Phi(\tau, z)$ is a topological isomorphism of $\mathscr{A}_q(U, G)$ (of $\mathscr{B}_q(U, G)$) onto $\mathscr{A}_q(D(\tau), G^\mu)$ (onto $\mathscr{B}_q(D(\tau), G^\mu)$), $q = 2, 3, \ldots$.*

For the case $\dim T(G) < \infty$ this theorem gives a new proof and refinement of the results in [7d].

If $\mu \in M(U, G)$, the isomorphism $A \to w_\mu \circ A \circ w_\mu^{-1}$ of G onto G_μ is called a quasi-conformal mapping. Theorem 10 has the

Corollary. *To every quasi-conformal mapping of a Fuchsian group G onto a Fuchsian group G' there belong topological isomorphisms of $\mathscr{A}_q(U, G)$ onto $\mathscr{A}_q(U, G')$ and of $\mathscr{B}_q(U, G)$ onto $\mathscr{B}_q(U, G')$, $q = 2, 3, \ldots$.*

For $\dim T(G) < \infty$ this contains and refines a classical result on cusp-forms.

References

[1] AHLFORS, L. V.: Some remarks on Teichmüller spaces of closed Riemann surfaces. Ann. Math. **74**, 171—191 (1961).

[2] — Teichmüller spaces. Proc. Intern. Congr. Math. Stockholm (1962), pp. 3—9.

[3] — Quasiconformal reflections. Acta Math. **109**, 291—301 (1963).

[4] —, and L. BERS: Riemann's mapping theorem for variable metrics. Ann. Math. **72**, 385—404 (1960).

[5] Ahlfors, L. V. and G. Weill: A uniqueness theorem for Beltrami equations. Proc. Amer. Math. Soc. 13, 975—978 (1962).
[6] Bers, L.: Spaces of Riemann surfaces. Proc. Intern. Congr. Math., Edinburgh (1958), pp. 349—361.
[7] — (a) Simultaneous uniformization. Bull. Amer. Math. Soc. 66, 94—97 (1960); (b) Spaces of Riemann surfaces as bounded domains, ibid. 98—103; (c) Correction, ibid. 67, 465—466 (1961); (d) Holomorphic differentials as functions of moduli, ibid., 206—210.
[8] — Completeness theorems for Poincaré series in one variable. Proc. Intern. Symp. Lin. Spaces, Jerusalem (1960), pp. 88—100.
[9] Earle, C. J.: The Teichmüller spaces for an arbitrary Fuchsian group. Bull. Amer. Math. Soc. (to appear).
[10] Nehari, Z.: The Schwarzian derivative and schlicht functions. Bull. Amer. Math. Soc. 55, 545—551 (1949).

Department of Mathematics
Columbia University
New York

The Extension Problem for Compact Submanifolds of Complex Manifolds I

(The Case of a Trivial Normal Bundle)*

By

Ph. A. Griffiths

Let X be a compact, complex submanifold of a V. We wish to consider over X certain *analytic objects*, such as: (i) a holomorphic vector bundle $E \to X$ (the notations are explained in § 1 below); (ii) a subspace $S \subset H^q(X, \mathscr{E})$; or (iii) a holomorphic mapping $f: X \to Y$ for some complex manifold Y. The extension problem we consider is, given an analytic object α over X, to find a corresponding analytic object β over V such that β restricted to X gives α.

We shall be primarily interested in the extension problem when V is a *germ of a neighborhood* of X, a concept which we now make precise. Let $\mathcal{O}$ be the sheaf of local rings of holomorphic functions on V and $\mathscr{I} \subset \mathcal{O}$ the ideal sheaf of X. Denote by $\mathscr{I}^\mu$ the μth power of $\mathscr{I}$ and set $\mathcal{O}^\mu = \mathcal{O}/\mathscr{I}^{\mu+1}$ (sheaf of jets of order μ in the normal parameter along X). The pair $(X, \mathcal{O}^\mu)$ then forms a ringed space X^μ, $X^\circ = X$. Also we set $\mathcal{O}^* = \mathcal{O}|X$ and denote by X^* the generalized complex space $(X, \mathcal{O}^*)$. Then X^* is a germ of a neighborhood of X and X^μ is a neighborhood of order μ of X in X^*.

* Received June 1, 1964.

Suppose now that we have an extension α^{μ} to X^{μ} of the analytic object α on X. Then the obstruction to extending α^{μ} to $X^{\mu+1}$ is given by a cohomology class $\omega(\alpha^{\mu})$. In case (i), $\omega(\alpha^{\mu}) \in H^2(X, \text{Hom}(\mathscr{E}, \mathscr{E}) \otimes \otimes \mathscr{I}^{\mu+1}/\mathscr{I}^{\mu+2})$; in case (ii), $\omega(\alpha^{\mu}) \in H^{q+1}(X, \mathscr{E} \otimes \mathscr{I}^{\mu+1}/\mathscr{I}^{\mu+2})$; and in (iii) $\omega(\alpha^{\mu}) \in H^1(X, \text{Hom}(\mathscr{T}(V), \mathscr{T}(Y)) \otimes \mathscr{I}^{\mu+1}/\mathscr{I}^{\mu+2})$.

There are two general statements which may hold for analytic objects of types (i), (ii), or (iii) and a germ of embedding $X \subset V$; these are

(I) there are only finitely many obstructions to the extension problem;

(II) a local extension exists if, and only if, a formal extension exists.

Of course, (I) and (II) are not always true; there are easy counter-examples to (I) and HIRONAKA has a counter-example to (II). Our program is to investigate the extension problem and its applications after making assumptions on the normal bundle N of X in V.

In this paper we shall essentially assume that N is trivial; this means, at least when $H^1(X, \mathscr{O}) = 0$, that V may be considered as an analytic fibre space over an analytic set D with one fibre being X. Thus the techniques in the theory of deformations of complex structure ([7] and [9]) are available to treat the extension problem. We are then able to prove I and II for analytic objects of types (i) or (ii) and to also derive several other results peculiar to the case of a trivial normal bundle.

For example, suppose that D is non-singular and of dimension 1 with parameter t. Write $\mathscr{V} = \bigcup_{t \in D} X_t$ where $X_0 = X$ and the X_t are the fibres of the projection of V onto D. Let $\mathscr{S}$ be a locally free analytic sheaf on V and $\mathscr{S}_0 = \mathscr{S}/I \cdot \mathscr{S}$. Denote by E^q the subspace of $H^q(X, \mathscr{S}_0)$ composed of extendible classes, and denote by J^q the subspace of E^q composed of extendible classes whose restrictions to X_t vanish for $t \neq 0$. (These may be called the jump classes.) Then there is a natural isomorphism

$$H^q(X, \mathscr{S}_0)/E^q \cong J^{q+1}.$$

If a class belongs to E^q, it is represented by a q-cocycle Z_t and, if it belongs to J^q, then Z_t is the coboundary of a $(q-1)$-cochain which has a pole at the point $t = 0$. If $\dim D = m > 1$, then Z_t is the coboundary of a $(q-1)$-cochain which has as polar locus an analytic set of dimension $(m-1)$ through 0.

As another illustration of our results, we are able to construct, for an analytic object α of type (i) or (ii), a maximal analytic subset V_α of V with $X \subset V_\alpha$ and such that α may be extended to V_α.

1. Notations and terminology

The basic object on which we shall work will be a compact, complex manifold. Let J be the almost-complex structure tensor of X acting on the complex tangent bundle $\boldsymbol{T}^{\#}(X)$; then $\boldsymbol{T}^{\#}(X) = \boldsymbol{T} + \boldsymbol{T}^*$ where $\boldsymbol{T}$ is the complex tangent bundle, which is the $\sqrt{-1}$ eigenbundle of J, and $\boldsymbol{T}^*$ is the conjugate bundle. A general holomorphic vector bundle over X will be written $E \to \boldsymbol{E} \to X$ where E, a complex vector space, is a typical fibre and $\boldsymbol{E}$ is the total space. The dual bundle is denoted by $E' \to \boldsymbol{E}' \to X$ and $\mathscr{E}$ is the sheaf of germs of holomorphic cross-sections of $E \to \boldsymbol{E} \to X$. We use the standard conventions: $\mathscr{T} = \Theta$, $\Lambda^P \mathscr{T}' = \Omega^P$, and $\mathcal{O}_X =$ sheaf of germs of holomorphic functions. For a holomorphic vector bundle $E \to \boldsymbol{E} \to X$, $\mathscr{A}^q(\boldsymbol{E})$ is the sheaf of germs of C^∞ $\boldsymbol{E}$-valued (o, q) forms over X.

We denote holomorphic principal bundles by $G \to P \overset{\pi}{\to} X$ where the complex Lie group G acts holomorphically on the right on the total space P. Over X, we have the holomorphic vector bundle $\boldsymbol{Q} = \boldsymbol{T}(P)/G$, and there is an onto bundle homomorphism $\pi: \boldsymbol{Q} \to \boldsymbol{T}$ with kernel $\boldsymbol{L} = P \times_G \mathfrak{g}$ where $\mathfrak{g}$ is the complex Lie algebra of G. Thus, we have the *fundamental bundle sequence* [*1*]

$$0 \to \boldsymbol{L} \to \boldsymbol{Q} \to \boldsymbol{T} \to 0 .$$

The sheaf $\mathscr{A}^\circ(\boldsymbol{Q})$ ($= C^\infty$ germs of sections of $\boldsymbol{Q}$) acts on $\mathscr{A}^\circ(\boldsymbol{E})$ as follows: a germ σ in $\mathscr{A}^\circ(\boldsymbol{E})$ is given by an E-valued C^∞ function $\hat{\sigma}$ on P satisfying $\hat{\sigma}(p \cdot g) = \varrho(g)\hat{\sigma}(p) (p \in P, g \in G)$. Let $\hat{\xi}$ be a germ of a right-invariant vector field on $P (=$germ ξ in $\mathscr{A}^\circ(\boldsymbol{Q}))$; then $\hat{\xi} \cdot \hat{\sigma}$ is again an E-valued C^∞ function on P satisfying the equivariance condition, and $\widehat{\xi \cdot \sigma} = \hat{\xi} \cdot \hat{\sigma}$. This action may be extended to a pairing $[\ ,\ ]: \mathscr{A}^p(\boldsymbol{Q}) \otimes \mathscr{A}^q(\boldsymbol{E}) \to \mathscr{A}^{p+q}(\boldsymbol{E})$ ([*11*]). In particular, we get a pairing $[\ ,\ ]: A^p(\boldsymbol{T}) \otimes A^q(\boldsymbol{T}) \to A^{p+q}(\boldsymbol{T})$.

A *deformation* $\{\mathscr{V} \overset{\tilde{\omega}}{\to} D\}$ of X is given by the following data: (i) An analytic subset D of an open neighborhood U of the origin in $\boldsymbol{C}^m$; (ii) An analytic space $\mathscr{V}$ and a proper holomorphic mapping $\tilde{\omega}: \mathscr{V} \to D$ such that $\tilde{\omega}$ has maximal rank and connected fibres $X_t = \tilde{\omega}^{-1}(t) (t \in D)$; and (iii) A holomorphic embedding $\iota: X \to \mathscr{V}$ such that $\tilde{\omega} \circ \iota = 0 \in D$. A mapping

$$\begin{array}{ccc} F: \mathscr{V} & \to & \mathscr{V}' \\ \downarrow \tilde{\omega} & & \downarrow \tilde{\omega}' \\ G: D & \to & D' \end{array}$$

between deformation spaces $\{\mathscr{V} \overset{\tilde{\omega}}{\to} D\}$ and $\{\mathscr{V}' \overset{\tilde{\omega}'}{\to} D'\}$ is given by a pair of holomorphic mappings $F: \mathscr{V} \to \mathscr{V}'$ and $G: D \to D'$ such that $\tilde{\omega}' \circ F = G \circ \tilde{\omega}$, $F \circ \iota = \iota'$, and such that F is biholomorphic on fibres.

For technical reasons, we introduce the notion of an *almost-complex deformation* $\{\mathscr{W} \overset{\tilde{\omega}}{\to} U\}$. This is given by:

(i) An open neighborhood U of 0 in $\boldsymbol{C}^m$;

(ii) An almost-complex manifold $\mathscr{W}$ and an almost-complex mapping $\tilde{\omega}: \mathscr{W} \to U$ such that $\tilde{\omega}$ has maximal rank, connected fibres, and such that each fibre $X_t = \tilde{\omega}^{-1}(t)\,(t \in U)$ is an almost-complex submanifold; and (iii) An almost-complex injection $\tau: X \to \mathscr{W}$ such that $\tilde{\omega} \circ \tau = 0 \in U$. If $\{\mathscr{V} \xrightarrow{\tilde{\omega}} D\}$ is a deformation, we say that $F: \mathscr{V} \to \mathscr{W}$ *embeds* $\{\mathscr{V} \xrightarrow{\tilde{\omega}} D\}$

$$\begin{array}{ccc} F:\mathscr{V} & \to & \mathscr{W} \\ \downarrow \tilde{\omega} & & \downarrow \tilde{\omega} \\ G: D & \to & U \end{array}$$

into $\{\mathscr{W} \xrightarrow{\tilde{\omega}} U\}$ if G is the injection of an analytic subset D of U into U, if F is differentiable embedding of $\mathscr{V}$ into $\mathscr{W}$ which induces an almost-complex injection on each fibre, and if the above diagram commutes. For any deformation $\{\mathscr{V} \xrightarrow{\tilde{\omega}} D\}$, we shall always *assume* that there exists an almost-complex deformation $\{\mathscr{W} \xrightarrow{\tilde{\omega}} U\}$ into which $\{\mathscr{V} \xrightarrow{\tilde{\omega}} D\}$ can be embedded.

Given a holomorphic principal bundle $G \to P \to X$, we define a *deformation* $\{G \to \mathscr{P} \xrightarrow{\pi} \mathscr{V} \xrightarrow{\tilde{\omega}} D\}$ (or just $\{G \to \mathscr{P} \to D\}$) to consist of a deformation $\{\mathscr{V} \xrightarrow{\tilde{\omega}} D\}$ of X together with a holomorphic principal bundle $G \to \mathscr{P} \xrightarrow{\pi} \mathscr{V}$ such that $\iota^{-1}(G \to \mathscr{P} \to \mathscr{V}) = G \to P \to X$. The auxiliary discussion above about deformations of X may now be carried over to deformations of $G \to P \to X$. In particular, the analogue of the assumption about the existence of ambient almost-complex deformations of X will be assumed to hold for deformations of $G \to P \to X$.

1.1. Graded complexes and Lie algebras

We recall that a *graded Lie algebra* is given by a graded vector space $A = \sum_{p \geqq 0} A^p$ together with a bracket operation $[\,,\,]: A^p \otimes A^q \to A^{p+q}$ such that:

$$[\varphi, \psi] = (-1)^{pq+1}[\psi, \varphi] \qquad (\varphi \in A^p,\ \psi \in A^q) \tag{1.1}$$

$$\begin{gathered}(-1)^{pr}[\varphi,[\psi,\eta]] + (-1)^{qr}[\eta,[\varphi,\psi]] + (-1)^{pq}[\psi,[\eta,\varphi]] = 0 \\ (\varphi \in A^p,\ \psi \in A^q,\ \eta \in A^r)\,.\end{gathered} \tag{1.2}$$

The notion of a homomorphism between graded Lie algebras is clear.

If $K = \sum_{q \geqq 0} K^q$ is a graded vector space, we say that K is over the graded Lie algebra A if there is a pairing $[\,,\,]: A^p \otimes K^q \to K^{p+q}$ such that

$$\begin{gathered}[\varphi,[\psi,\gamma]] + (-1)^{pq+1}[\psi,[\varphi,\gamma]] = [[\varphi,\psi],\gamma] \\ (\varphi \in A^p,\ \psi \in A^q,\ \gamma \in K)\,.\end{gathered} \tag{1.3}$$

We define $K = \sum_{q \geqq 0} K^q$ to be a *graded complex* (K, δ) if there are linear mappings $\delta: K^q \to K^{q+1}$ satisfying $\delta \circ \delta = 0$. We set $Z(K^p) =$ kernel $\delta: K^p \to K^{p+1}$, $H^p(K) = Z(K^p)/\delta K^{p-1}$, and $H(K) = \sum_{p \geqq 0} H^p(K)$.

A graded Lie algebra A which is also a graded complex (A, d) is called a *graded Lie algebra complex* if the following rule holds:

$$d[\varphi, \psi] = [d\varphi, \psi] + (-1)^p [\varphi, d\psi] \qquad (\varphi \in A^p,\ \psi \in A). \tag{1.4}$$

Finally, if (K, δ) is a graded complex and (A, d) is a graded Lie algebra complex, then (K, δ) is over (A, d) if K is over A and if

$$\delta[\varphi, \gamma] = [d\varphi, \gamma] + (-1)^p [\varphi, \delta\gamma] \qquad (\varphi \in A^p,\ \gamma \in K). \tag{1.5}$$

Suppose now that (K, δ) is over (A, d) as above, and let $t = (t^1, \ldots, t^m)$ be a variable point in $\boldsymbol{C}^m$. We write $K[t]$ for the graded vector space of formal power series in t with coefficients in K; $K\{t\}$ is the subspace of $K[t]$ consisting of the series with constant term equal to zero. We define $A[t]$ and $A\{t\}$ similarly, so that we have, e.g., $[A\{t\}, K[t]] \subset K\{t\}$. Let $\Phi(t) \in A^1\{t\}$.

Definition: We let

$$\Delta\Phi(t) = d\Phi(t) - [\Phi(t), \Phi(t)] \tag{1.6}$$

and say that $\Phi(t)$ is *integrable* if $\Delta\Phi(t) = 0$.

Let now $\Phi(t) \in A^1\{t\}$ and define $\delta_\Phi : K^q[t] \to K^{q+1}[t]$ by

$$\delta_\Phi(\Gamma(t)) = \delta\Gamma(t) - 2[\Phi(t), \Gamma(t)] \quad (\Gamma(t) \in K^q[t]). \qquad (1.6)'$$

Lemma 1.1. $\delta_\Phi \circ \delta_\Phi \Gamma(t) = -2[\Delta\Phi(t), \Gamma(t)]$.

Proof: $\delta_\Phi(\delta_\Phi\Gamma(t)) = \delta_\Phi(\delta\Gamma(t) - 2[\Phi(t), \Gamma(t)])$
$= \delta^2\Gamma(t) - 2\delta[\Phi(t), \Gamma(t)] - 2[\Phi(t), \delta\Gamma(t)]$
$+ 4[\Phi(t), [\Phi(t), \Gamma(t)]] = -2[\delta\Phi(t), \Gamma(t)]$
$+ 2[[\Phi(t), \Phi(t)], \Gamma(t)]$ by (1.4) and (1.2). Q.E.D.

If we now assume that $\Delta\Phi(t) = 0$, then $(K[t], \delta_\Phi)$ becomes again a graded complex; we set $Z^q{}_\Phi(K) = Z(K^q[t])$, $H^q{}_\Phi(K) = H^q(K[t])$, where the differential operator δ_Φ is given by (1.6). Our immediate goal is to study the relationship between $H^q(K)$ and $H^q{}_\Phi(K)$.

1.2. The formal extension problem in cohomology

Let (K, δ) be a graded complex over a graded Lie algebra complex (A, d), and let $\Phi(t) \in A^1\{t\}$ be an integrable element. Given $\gamma \in Z(K^q)$, we say that γ is *extendible* if there exists $\Gamma(t) \in K^q[t]$ such that $\Gamma(0) = \gamma$ and $\delta_\Phi\Gamma(t) = 0$. If $\sigma \in K^{q-1}$, then $\delta_\Phi(\sigma)$ is an extension of $\delta(\sigma)$, so that the following is justified:

Definition. A class $[\gamma] \in H^q(K)$ is *extendible* if there exists $\gamma \in Z(K^q)$ which represents $[\gamma]$ and is itself extendible.

Suppose now that $s \in \boldsymbol{C}$ is a single complex parameter, and let $K(s)$ be the formal power series in s and with coefficients in K which have finitely many terms with negative s-exponents; a $\sum(s) \in K(s)$ is thus written as

$$\sum(s) = \sum_{\mu=-N}^{\infty} \gamma_\mu s^\mu \qquad (\gamma_\mu \in K).$$

Definition. Let $\Gamma(s) \in Z^q_\Phi(K)$ be an extension of $\gamma \in Z(K^q)$. Then Γ is said to be a *jump extension* if there exists $\sum \in K^{q-1}(s)$ with $\delta_\Phi \sum = \Gamma$. We say that $\gamma \in Z(K^q)$ is a *jump cocycle* if there exists a jump extension Γ of γ.

If $\sigma \in K^{q-1}$, then $\delta_\Phi \sigma$ is a jump extension of $\delta\sigma$, so that we may define what is meant by a *jump class* in $H^q(K)$.

We let $E^q_\Phi \subset H^q(K)$ be the subspace of extendible classes, and $J^q_\Phi \subset E^q_\Phi$ the space of jump classes. Our first main result is the following

Theorem 1.1. *Assume that $H(K)$ is finite dimensional.*
(i) *There are only finitely many obstructions to extending a class in $H^q(K)$;*
(ii) *If $G^q_\Phi = H^q(K)/E^q_\Phi$, then there are on G^q_Φ and J^{q+1}_Φ canonical filtrations whose associated graded modules are naturally isomorphic.*

(1.2) Proof of Theorem 1.1. Let $\gamma \in Z(K^q)$. We wish to find $\Gamma(s) = \sum_{\mu=0}^{\infty} \gamma_\mu s^\mu$ with $\gamma_0 = \gamma$ and $\delta_\Phi \Gamma = 0$. Thus we must recursively solve the equations

$$\delta\gamma_N = 2 \sum_{\substack{\sigma+i=N \\ \sigma>0}}^{\infty} [\varphi \ , \gamma_i] \qquad (1.7)^N$$

with $\gamma_0 = \gamma$, and where we have written $\Phi(s) = \sum_{\nu=1}^{\infty} \varphi_\nu s^\nu$. Suppose that we have $\gamma_0, \ldots, \gamma_{N-1}$ such that $(1.7)^\mu$ is satisfied for $1 \leqq \mu \leqq N-1$; we say then that γ is *extendible to order* $N-1$. Let

$$\omega_N = 2 \sum_{\sigma+\iota=N} [\varphi_\sigma, \gamma_\iota]; \qquad (1.8)$$

then ω_N is the Nth *obstruction to extending* γ, or the *obstruction to extending γ to order N.*

Lemma 1.2. $\omega_N \in K^{q+1}$ and $\delta\omega_N = 0$.

Proof. We use (1.3) and (1.5) to calculate:

$$\begin{aligned}\delta\omega_N &= 2\sum_{\sigma+\iota=N} [d\varphi_\sigma, \gamma_\iota] - \sum_{\sigma+\iota=N} [\varphi_\sigma, \delta\gamma_\iota] \\ &= 2\sum_{\varrho+\sigma+\iota=N} [[\varphi_\varrho, \varphi_\sigma], \gamma_\iota] - 4\sum_{\varrho+\sigma+\iota=N} [\varphi_\varrho, [\varphi_\sigma, \gamma_\iota]] = 0.\end{aligned}$$

In the middle step we have used $(1.7)^\mu$ for $1 \leqq \mu \leqq N-1$. Q. E. D.

We come now to the main points in the formal theory.

Lemma 1.3. $\omega \in Z(K^{q+1})$ is a jump cocycle if, and only if, ω is an obstruction to extending some $\gamma \in Z(K^q)$.

Proof. Suppose first that ω is an obstruction. Then there exists $\gamma = \gamma_0, \gamma_1, \ldots, \gamma_{N-1} \in K^q$ with $\delta\gamma = 0$, $\delta\gamma_\mu = 2\sum_{\sigma+\iota=\mu}[\varphi_\sigma, \gamma_\iota]$ $(1 \leqq \mu \leqq N-1)$, and $\omega = 2\sum_{\sigma+\iota=N}[\varphi_\sigma, \gamma_\iota)$. Let $\gamma^{N-1}(s) = \sum_{\mu=0}^{N-1}\gamma_\mu s^\mu$, and define $\sum(s) \in K^q[D]$ by $\sum(s) = \underset{s}{\perp}_N\, \gamma^{N-1}(s)$. Then $\delta_\Phi \sum(s) = \delta\sum(s) - 2[\Phi(s), \sum(s)] = \underset{s}{\perp}_N(\delta\gamma^{N-1}(s) - 2[\Phi(s), \gamma^{N-1}(s)]) = \underset{s}{\perp}_N(\omega s^N + 0(s^{N+1}))$ $(0(s^{N+1})$ $=$ terms divisible by $s^{N+1})$ $= \Omega(s) \in K^{q+1}(D)$. Since $\Omega(0) = \omega$, the pair $(\Omega(s), \sum(s))$ makes ω into a jump cocycle.

Now suppose conversely that ω is a jump cocycle. Then there exists $\Omega(s) \in K^{q+1}(D)$, $\sum(s) \in K^q[D]$ such that:

(i) $\Omega(0) = \omega$;

(ii) $\delta_\Phi\Omega(s) = 0$; and

(iii) $\delta_\Phi\sum(s) = \Omega(s)$ $(s \neq 0)$.

Let N be the order of the pole of $\sum(s)$ at $s = 0$, and define $\gamma \in Z(K^q)$ by $\gamma = \lim_{s\to 0} s^N \sum(s)$. We may then reverse the above argument to find that ω is an Nth obstruction to extending $\gamma \in Z(K^q)$. Q. E. D.

Now the primary obstruction to extending $\gamma \in Z(K^q)$ is a well-defined class in $H^{q+1}(K)$, depending only on $[\gamma] \in H^q(K)$; this obstruction is given by $\omega_1 = [\varphi_1, \gamma]$. However, the higher obstructions do not depend upon $[\gamma] \in H^q(K)$ alone; if e.g., $\omega_1 = \delta\gamma_1$, then $\omega_2 = [\varphi_1, \gamma_1] + [\varphi_2, \gamma]$ is *a* secondary obstruction. But so is $\omega_2' = [\varphi_1, \gamma_1 + \sigma] + [\varphi_2, \gamma]$ for any $\sigma \in Z(K^q)$. We now show how to deal with this situation. First we observe:

Lemma 1.4. If the primary obstruction to extending $\gamma \in Z(K^q)$ is a non-zero class in $H^{q+1}(K)$, then no solution of the extension problem exists for γ.

Lemma 1.5. If ω is an Nth obstruction to extending $\gamma \in Z(K^q)$, and if, by different choices, γ is extendible to order N, then ω is an $N-1^{\text{st}}$ obstruction to extending some $\varrho \in Z(K^q)$.

Proof. We are given: (i) $\gamma = \gamma_0, \gamma_1, \ldots, \gamma_{N-1} \in K^q$ such that $(1.7)^\mu$ is satisfied for $1 \leqq \mu \leqq N-1$ and such that $\omega = 2\sum_{\sigma+\tau=N}[\varphi_\sigma, \gamma_\tau]$; and (ii) $\gamma = \gamma_0', \gamma_1', \ldots, \gamma_N' \in K^q$ such that $(1.7)^\mu$ is satisfied for $1 \leqq \mu \leqq N$.

Let $\varrho = \varrho_0 = \gamma_1 - \gamma_1'$, $\varrho_\mu = \gamma_{\mu+1}'$ $(1 \leqq \mu \leqq N-2)$. Then $\delta\varrho = \delta\gamma_1 - \delta\gamma_1' = [\varphi_1, \gamma] - [\varphi_1, \gamma] = 0$. Also, for $1 \leqq \mu \leqq N-2$, $2\sum_{\sigma+\tau=\mu}[\varphi_\sigma, \varrho_\tau]$ $= 2(\sum_{\sigma+\tau=\mu+1}[\varphi_\sigma, \gamma_\tau] - [\varphi_\sigma, \gamma_\tau']) = \delta\gamma_{\mu+1}' - \delta\gamma_{\mu+1}' = \delta\varrho_\mu$. Thus ϱ is extended to order $N-2$. But $2\sum_{\sigma+\tau=N-1}[\varphi_\sigma, \varrho_\tau] = 2\sum_{\sigma+\tau=N}[\varphi_\sigma, \gamma_\tau] - [\varphi_\sigma, \gamma'_\tau] = \omega - \delta\gamma'_N$ and thus ω is an $N-1^{st}$ obstruction to extending $\varrho \in Z(K^q)$. Q. E. D.

Corollary. If $\omega \in Z(K^{q+1})$ is a jump cocycle then there exists $\gamma^{\#} \in \in Z(K^q)$ such that the extension problem cannot be solved for $\gamma^{\#}$ and such that ω is an Nth obstruction to extending $\gamma^{\#}$ where N is maximal, i. e., there exists no extension of order N of $\gamma^{\#}$.

For $\omega \in Z(K^{q+1})$, $\gamma \in Z(K^q)$, we set: $N(\omega) = \inf\{N \in Z^+ \mid \omega$ is an Nth obstruction to extending some $\gamma' \in Z(K^q)\}$, $N^{\#}(\gamma) = \sup\{N \in Z^+ \mid \gamma$ is extendible to order $N-1\}$.

Lemma 1.6. (i) If ω is an obstruction of order $N(\omega)$ to extending $\gamma \in Z(K^q)$, then $N^{\#}(\gamma) = N(\omega)$. (ii) If ω is an obstruction of order $N^{\#}(\gamma)$ to extending $\gamma \in Z(K^q)$, then $N(\omega) = N^{\#}(\gamma)$.

Proof. (i) is just a restatement of Lemma 1.5. To see (ii), we assume that $N(\omega) < N^{\#}(\gamma)$ (since, in any case, $N(\omega) \leqq N^{\#}(\gamma)$); and we let $r = N^{\#}(\gamma) - N(\omega) > 0$. We know the following: (a) there exists $\gamma = \gamma_0, \gamma_1, \ldots, \gamma_{N-1}$ $(N = N^{\#}(\gamma))$ such that $(1.7)^\mu$ is satisfied for $1 \leqq \mu \leqq N-1$ and $\omega = 2\sum_{\sigma+\tau=N}[\varphi_\sigma, \gamma_\tau]$; (b) there exists $\varrho = \varrho_0, \varrho_1, \ldots, \varrho_{M-1}$ $(M = N(\omega))$ such that $(1.7)^\mu$ is satisfied for $1 \leqq \mu \leqq M-1$ and $\omega = 2\sum_{\sigma+\tau=M}[\varphi_\sigma, \varrho_\tau]$. Define now $\gamma = \gamma'_0, \gamma'_1, \ldots, \gamma'_{N-1} \in K^q$ as follows: $\gamma'_\mu = \gamma_\mu$ for $0 \leqq \mu \leqq r-1$; and $\gamma'_\nu = \nu_\nu - \varrho_{\nu-r}$ $(r \leqq \nu \leqq N-1)$. Then, for $1 \leqq \mu \leqq r-1$, $(1.7)^\mu$ is satisfied. For $r \leqq \nu \leqq N-1$, $\sum_{\sigma+\tau=\nu}[\varphi_\sigma, \gamma'_\tau] = \sum_{\sigma+\tau=\nu}[\varrho_\sigma, \gamma'_\tau] - \sum_{\sigma+\tau=\nu-r}[\varphi_\sigma, \varrho_\tau] = \delta\gamma_\nu - \delta\varrho_{\nu-r} = \delta(\gamma'_\nu)$. But, also, we have: $\sum_{\sigma+\tau=N}[\varphi_\sigma, \gamma'_\tau] = \sum_{\sigma+\tau=N}[\varphi_\sigma, \gamma_\tau] - \sum_{\sigma+\tau=N-r-M}[\varphi_\sigma, \varrho_\tau] = \omega - \omega = 0$. Thus γ is extendible to order N, which is a contradiction. Q. E. D.

Now J^{q+1}_Φ is a finite dimensional vector space; we choose a basis $\omega_1, \ldots, \omega_r$ and set $N_i = N(\omega_i)$ for $1 \leqq i \leqq R$. Then, to each ω_i, we may associate $\Omega_i(s) \in K^{q+1}[s]$, $\Sigma_i(s) \in K^q(s)$ such that $\Omega_i(0) = \omega_i$, $\delta_\Phi \Sigma_i = \Omega_i$, and $\Sigma_i(s)$ has a pole of order N_i in s. Since $N(\omega + \omega') \leqq \leqq \max(N(\omega), N(\omega'))$, if we set $N(q+1) = \max_{1 \leqq i \leqq R} N_i$, then, for each $\omega \in J^{q+1}_\Phi$, we may associate $\Omega(s) \in K^{q+1}[s]$, $\Sigma(s) \in K^q(s)$ such that $\Omega(0) = \omega$, $\delta_\Phi \Sigma = \Omega$, and such that $\Sigma(s)$ has a pole of order less than or equal to $N(q+1)$ in s.

Lemma 1.7. Let $\gamma \in Z(K^q)$. If, for some $N \geqq N(q+1)$, γ has an extension of order N, then γ may be extended.

Proof. If not, then $N(\gamma) < \infty$ and $N(\gamma) > N(q+1)$. Thus there exists $\omega \in J^{q+1}_\Phi$ which is an obstruction of order $N(\gamma)$ to extending γ. By Lemma 1.6, $N(\omega) = N(\gamma) > N(q+1)$, which is a contradiction. Q.E.D.

It is in the sense of Lemma 1.7 that there are only finitely many obstructions to the extension of cohomology.

Set now $M(q) = \sup_{\gamma \in G^q_\Phi} N(\gamma)$; by Lemma 1.7, $M(q) < \infty$. We define a filtration $\{G^q_M\}_{1 \leqq M \leqq M(q)}$ of G^q_Φ as follows: $G^q_M = \{\gamma \in G^q_\Phi : N^{\#}(\gamma) \geqq M\}$. Since $N^{\#}(\gamma + \gamma') \geqq \min(N^{\#}(\gamma), N^{\#}(\gamma'))$, the G^q_M are subspaces of G^q_Φ. Clearly $G_{M+1}^q \subset G^q_M$, and we set $G^q(M) = G^q_M / G_{M+1}^q$.

We also define a filtration $\{J^{q+1}_N\}_{1 \leqq N \leqq N(q+1)}$ of J^{q+1}_Φ: For $1 \leqq N \leqq$ $\leqq N(q+1)$, $J^{q+1}_N = \{\omega \in J^{q+1}_\Phi : N(\omega) \leqq N\}$. Clearly J^{q+1}_N is a vector subspace of J^{q+1}_Φ, and $J^{q+1}_{N+1} \supset J^{q+1}_N$. Set $J^{q+1}(N) = J^{q+1}_N / J^{q+1}_{N-1}$.

Lemma 1.8. Let Θ^N be the mapping which associates to each $\gamma \in G^q(N)$ its obstruction of order N in $J^{q+1}(N)$. Then Θ_N is well-defined, linear, and is an isomorphism for each N.

This result follows from Lemmas 1.2—1.7, and completes the proof of Theorem 1.1.

2.1. Norms and a harmonic theory

On our graded complexes (K, δ) we shall now assume the existence of a graded Banach space with norm $\| \ \|$ such that $\delta: K^q \to K^{q+1}$ is a bounded operator. In the case of a graded Lie algebra complex (A, d), we also assume that $[,]: A^p \times A^q \to A^{p+q}$ is bounded in each factor for $p + q \geqq 1$.

We shall also assume on (K, δ) a harmonic theory, given by linear transformations $\delta^*: K^{q+1} \to K^q$, $\boldsymbol{G}: K^q \to K^q$, and a projection $\pi_{\boldsymbol{H}}: K^q \to \boldsymbol{H}^q$ onto a finite dimensional subspace $\boldsymbol{H}^q \subset K^q$ such that the following hold:

(i) $\delta^* \circ \delta^* = 0$,

(ii) $\delta \circ \boldsymbol{G} = \boldsymbol{G} \circ \delta$ and $\delta^* \circ \boldsymbol{G} = \boldsymbol{G} \circ \delta^*$,

(iii) $\pi_{\boldsymbol{H}} \circ \boldsymbol{G} = \boldsymbol{G} \circ \pi_{\boldsymbol{H}} = \pi_{\boldsymbol{H}} \circ \delta = \delta \circ \pi_{\boldsymbol{H}} = \pi_{\boldsymbol{H}} \circ \delta^* = \delta^* \circ \pi_{\boldsymbol{H}} = 0$,

and (iv) every $\gamma \in K^q$ has a unique representation (Hodge decomposition)

$$\gamma = \pi_{\boldsymbol{H}}(\gamma) + \delta^* \delta \boldsymbol{G}(\gamma) + \delta\delta^* \boldsymbol{G}(\gamma) . \tag{2.1}$$

We also assume that $\delta^* \boldsymbol{G}: K^q \to K^q$ is bounded so that we have a Banach space direct sum decomposition $K^q = \boldsymbol{H}^q \oplus \delta K^{q-1} \oplus \delta^* K^{q+1}$. We denote by $\pi_{\boldsymbol{H}}$, π_δ, and $\pi_{\delta*}$ the respective projection operators.

The concept of homomorphism between graded complexes or graded Lie algebra complexes will now refer to bounded transformations which commute with the harmonic theories.

Let now (K, δ) be over (A, d), and let $\Phi(t) \in A^1\{t\}$ be convergent for small t. Then, in addition to the cohomology groups $H^q(K)$ and $H^q_\Phi(K)$, we may obviously form a whole family $H^q_t(K)$ of cohomology groups for each fixed t ($H^q_0(K) = H^q(K)$). Our object is now to establish the relationship between these three.

2.2. Existence theorems

Following the suggestion of Nijenhuis and Richardson, we shall now use the implicit function theorem in Banach spaces [2] (rather than doing successive approximations directly) to derive certain existence theorems. For notation, we denote by $N(*)$ a generic neighborhood of the origin in a Banach space $*$.

Let (A, d) be a normed graded Lie algebra complex. Define $\varphi \in A^1$ to be *semi-integrable* if $d\varphi - \pi_d[\varphi, \varphi] = 0$.

Lemma 2.1 ([12]). There exists $N(\pi_H(A^1))$, $N(\pi_{d^*}(A^1))$, and a differentiable mapping $p: N(\pi_H(A^1)) \to N(\pi_{d^*}(A^1))$ such that, if $\varphi \in N(\pi_H(A^1))$, $\psi \in N(\pi_{d^*}(A^1))$, then $\varphi + \psi$ is semi-integrable if, and only if, $\psi = p(\varphi)$.

Proof. Define a differentiable mapping $q: \pi_H(A^1) \times \pi_{d^*}(A^1) \to \pi_d(A^2)$ by $q(\varphi, \psi) = d(\varphi + \psi) - \pi_d[\varphi + \psi, \varphi + \psi]$. Then $D_2 q(0, 0) = d: \pi_{d^*}(A^1) \to \pi_d(A^2)$ is an isomorphism, and, by the implicit function theorem, there exist $N(\pi_H(A^1))$, $N(\pi_{d^*}(A^1))$, and $p: N(\pi_H(A^1)) \to N(\pi_{d^*}(A^1))$ such that, for $\varphi \in N(\pi_H(A^1))$, $\psi \in N(\pi_{d^*}(A^1))$, $q(\varphi, \psi) = 0$ if and only if $\psi = p(\varphi)$. But $q(\varphi, \psi) = 0$ if, and only if, $\varphi + \psi$ is semi-integrable.

Lemma 2.2. $p(\varphi)$ is defined by the equation

$$p(\varphi) = d^* G[\varphi + p(\varphi), \varphi + p(\varphi)]. \tag{2.2}$$

Proof. $d(\varphi + p(\varphi)) - \pi_d[\varphi + p(\varphi), \varphi + p(\varphi)] = 0$. Q.E.D.

For $\varphi \in N(\pi_H(A^1))$, we set $P(\varphi) = \varphi + p(\varphi)$, and we define a vector-valued holomorphic function h on $N(\pi_H(A^1))$ by

$$h(\varphi) = \pi_H[P(\varphi), P(\varphi)]. \tag{2.3}$$

Lemma 2.3. [9]. There exists $N(\pi_H(A^1))$ such that, for $\varphi \in N(\pi_H(A^1))$, $h(\varphi) = 0$ if, and only if, $P(\varphi)$ is integrable.

Proof. If φ is integrable, then, since $P(\varphi)$ is semi-integrable, we get that $\pi_{d^*}[P(\varphi), P(\varphi)] = -\pi_H[P(\varphi), P(\varphi)] = -h(\varphi)$. Thus $h(\varphi) = 0$.

Now we have that $d^* dG[P(\varphi), P(\varphi)] = 2d^* G[dP(\varphi), P(\varphi)] = 2d^* G[dd^* G[P(\varphi), P(\varphi)], P(\varphi)] = 2d^* G\{[h(\varphi), P(\varphi)] + [d^* dG[P(\varphi), P(\varphi)], P(\varphi)]\}$ by (1.1), (1.2), and (1.4). Setting $F(\varphi) = d^* dG[P(\varphi), P(\varphi)]$, if $h(\varphi) = 0$ we get $F(\varphi) = 2d^* G[F(\varphi), P(\varphi)]$ and thus $\|F(\varphi)\| \leq c\|P(\varphi)\| \cdot \|F(\varphi)\|$. However, this implies that $F(\varphi) = 0$ if $h(\varphi) = 0$ for $\varphi \in N(\pi_H(A^1))$. By (2.2), $\Delta P(\varphi) = h(\varphi) + F(\varphi)$. Q.E.D.

We let $V \subset N(\pi_H(A^1))$ be the analytic set through the origin defined by the zeroes of $h(\varphi)$; to each $\varphi \in V$ we have associated the integrable element $P(\varphi) \in A^1$.

Let now the graded complex (K, δ) be over the graded Lie algebra complex (A, d). Let $\varphi \in A^1$. We say that $\gamma \in K^q$ is *semi-closed relative to* φ if $\delta\gamma - \pi_\delta[\varphi, \gamma] = 0$.

For Banach spaces $S_1, S_2, L(S_1, S_2)$ is the Banach space of bounded linear transformations $T: S_1 \to S_2$.

Lemma 2.4. There exist $N(A^1)$, $N(L(Z^q, \pi_{\delta^*}(K^q))$ and a differentiable mapping $r: N(A^1) \to N(L(Z^q, \pi_{\delta^*}(K^q)))$ such that the following holds: If $\varphi \in N(A^1)$, $\gamma \in Z^q$, and $\sigma \in \pi_{\delta^*}(K^q)$, then $\gamma + \sigma$ is semi-closed relative to φ if, and only if, $\sigma = r(\varphi)\gamma$.

Proof. Define $s: A^1 \times Z^q \times \pi_{\delta^*}(K^q) \to \pi_\delta(K^{q+1})$ by $s(\varphi, \gamma, \sigma) = \delta(\gamma + \sigma) - 2\pi_\delta[\varphi, \gamma + \sigma]$. Then s is differentiable and $D_3 s(0, 0, 0) = \delta: \pi_{\delta^*}(K^q) \to \pi_\delta(K^{q+1})$ is an isomorphism. The existence of r now follows again from the implicit function theorem.

Lemma 2.5. $r(\varphi)\gamma$ is defined by

$$r(\varphi)\gamma = 2\delta^* G[\varphi, \gamma + r(\varphi)\gamma]. \tag{2.4}$$

Proof. $\delta(\gamma + r(\varphi)\gamma) - 2\pi_\delta[\varphi, \gamma + r(\varphi)\gamma] = 0$. Q.E.D.

For $\varphi \in N(A^1), \gamma \in Z^q$, we set $R_\Phi(\gamma) = \gamma + r(\varphi)\gamma$.

Now, for each $\gamma \in \boldsymbol{H}^q \subset Z^q$, we define a vector valued holomorphic function $h_\gamma(\varphi)$ on $N(\pi_{\boldsymbol{H}}(A^1))$ by

$$h_\gamma(\varphi) = \pi_{\boldsymbol{H}}[P(\varphi), R_\varphi(\gamma)], \tag{2.5}$$

where, by way of notation, we set $R_\varphi(\gamma) = R_{P(\varphi)}(\gamma)$.

Lemma 2.6. There exists a neighborhood $N(\pi_{\boldsymbol{H}}(A^1))$ such that, for $\varphi \in N(\pi_{\boldsymbol{H}}(A^1)) \cap V$, $\gamma \in \boldsymbol{H}^q$, $\delta R_\varphi(\gamma) - 2[P(\varphi), R_\varphi(\gamma)] = 0$ if, and only if, $h_\gamma(\varphi) = 0$.

Proof. If $\delta R_\varphi(\gamma) - 2[P(\varphi), R_\varphi(\gamma)] = 0$, then $\pi_{\boldsymbol{H}}[P(\varphi), R_\varphi(\gamma)] = -\pi_\delta[P(\varphi), R_\varphi(\gamma)] = 0$ since $R_\varphi(\gamma)$ is semi-closed relative to $P(\varphi)$.

Now assume that $\delta P(\varphi) = [P(\varphi), P(\varphi)]$. Then $\delta^*\delta G[P(\varphi), R_\varphi(\gamma)] = \delta^* G[[P(\varphi), P(\varphi)], R_\varphi(\gamma)] - 2\delta^* G\{[P(\varphi), [P(\varphi), R_\varphi(\gamma)]] + [P(\varphi), h_\gamma(\varphi)] + [P(\varphi), \delta^*\delta G[P(\varphi), R_\varphi(\gamma)]]\}$. Setting $E_\gamma(\varphi) = \delta^*\delta G[P(\varphi), R_\varphi(\gamma)]$, we conclude that, if $h_\gamma(\varphi) = 0$, $E_\gamma(\varphi) = 2\delta^* G[P(\varphi), E_\gamma(\varphi)]$, from which it follows that $E_\gamma(\varphi) = 0$ for $\varphi \in N(\pi_{\boldsymbol{H}}(A^1)) \cap V$. But $\delta R_\varphi(\gamma) - 2[P(\varphi), R_\varphi(\gamma)] = 2\{h_\gamma(\varphi) + E_\gamma(\varphi)\}$. Q.E.D.

For a subspace $S \subset \boldsymbol{H}^q$, we let $V(S) \subset V$ be the analytic set defined by $V(S) = \{\varphi \in V: h_\gamma(\varphi) = 0 \text{ for all } \gamma \in S\}$. Then, to each $\varphi \in V(S)$, $\gamma \in S$, we have associated an element $R(\gamma)$ which satisfies

$$\delta_{P(\varphi)} R_\varphi(\gamma) = \delta R_\varphi(\gamma) - 2[P(\varphi), R_\varphi(\gamma)] = 0 \tag{2.6}$$

We close this section with the following remark. Suppose that we are given a neighborhood U of the origin in any $\boldsymbol{C}^m$ and a holomorphic mapping $t \to \Phi(t)$ of U into A^1 such that the locus $Z = \{t \in U: \Delta\Phi(t) = 0\}$

is an analytic set Z. Then, for any subspace $S \subset \boldsymbol{H}^q$, we may construct the analytic set $Z(S) \subset Z$ just as $V(S) \subset V$ was constructed above.

2.3. A continuity property of cohomology

We shall now combine the results in 1.2 and 2.2. Before doing this, we must first establish a certain continuity property for cohomology.

Assume that $\varphi \in N(A^1)$ is integrable, and recall that any $\xi \in K^q$ which is semi-closed relative to φ may be uniquely written as $\xi = R_\varphi(\gamma) = \gamma + r(\varphi)\gamma$ where $\gamma \in Z(K^q)$ and $r(\varphi)\gamma \in \pi_{\delta^*}(K^q)$. Since φ is integrable, if $\tau \in K^{q-1}$, then $\delta_\varphi(\tau) = \delta\tau - 2[\varphi, \tau]$ is closed relative to φ, and we may write $\delta_\varphi(\tau) = R_\varphi(\gamma)$ where $\gamma \in Z(K^q)$, and, in fact,

$$\gamma = \delta\tau - 2\,\delta\delta^* G[\varphi, \tau] - 2\,\pi_H[\varphi, \tau]\,. \tag{2.7}$$

Define a linear mapping $\Lambda_\varphi : \pi_{\delta^*}(K^{q-1}) \subset Z(K^q)$ by $\Lambda_\varphi(\tau) = \delta\tau - 2\,\delta\delta^* G[\varphi, \tau] - 2\pi_H[\varphi, \tau]$, so that $R_\varphi \Lambda_\varphi(\tau) = \delta_\varphi(\tau)$. Finally, define a bounded linear mapping

$$\begin{aligned} &\equiv_\varphi : \pi_{\delta^*}(K^{q-1}) \to \pi_\delta(K^q) \quad \text{by} \\ &\equiv_\varphi(\tau) = \delta\tau - 2\,\delta\delta^* G[\varphi, \tau]\,. \end{aligned} \tag{2.8}$$

For $\varphi \in N(A^1)$, $\equiv_\varphi$ is an isomorphism.

Now let $Z_\varphi(K^q) \to K^q$ be the kernel of δ_φ. Then every $\xi \in Z_\varphi(K^q)$ may be written as $\xi = R_\varphi \gamma$ for some $\gamma \in Z(K^q)$. Clearly Λ_φ is an injection of $\pi_{\delta^*}(K^{q-1})$ into $Z_\varphi(K^q)$, and, since $R_\varphi(\Lambda_\varphi(\pi_{\delta^*}(K^{q-1}))) \subset \delta_\varphi(K^{q-1})$, we have $\dim H^q_\varphi(K) \leqq \dim \{Z(K^q)/\Lambda_\varphi(\pi_{\varphi^*}(K^{q-1}))\}$.

Lemma 2.7. $\dim\{Z(K^q)/\Lambda_\varphi(\pi_{\delta^*}(K^{q-1}))\} = \dim H^q(K)$.

Proof. Write $Z^q(K) = \pi_H(K^q) \times \pi_\delta(K^q) = \pi_H(K^q) \times \delta(\pi_{\delta^*}(K^{q-1}))$. We define $\Theta_\varphi : \pi_{\delta^*}(K^{q-1}) \to \pi_H(K^q)$ by $\Theta_\varphi(\tau) = \pi_H[\varphi, \tau]$. We then write $\pi_{\delta^*}(K^{q-1}) = W_\varphi \times V_\varphi$ where W_φ is the kernel of Θ_φ and where V_φ is a complementary finite dimensional subspace. On W_φ, we have then that $\Lambda_\varphi = \equiv_\varphi$. If we let $X_\varphi = \Lambda_\varphi(W_\varphi)$; then X_φ is a closed subspace with codim (X_φ) in $\pi_{\delta^*}(K^{q-1}) = \dim V_\varphi$, and

$$\dim\{Z(K^q)/\Lambda_\varphi(\pi_{\delta^*}(K^{q-1}))\} \leqq \dim H^q(K) + \dim V_\varphi.$$

In fact, it is easily seen that
$\dim\{Z(K^q)/\Lambda_\varphi(\pi_{\delta^*}(K^{q-1}))\} = \dim H^q(K) + \dim V_\varphi - \dim \Lambda_\varphi(V_\varphi) = \dim H^q(K)$, provided that φ is sufficiently near to 0. This proves the Lemma.

Thus we have a diagram

$$\begin{array}{ccc} Z^q(K) & \xrightarrow{R_\varphi} & Z^q_\varphi(K) \\ \uparrow{\scriptstyle\delta} & & \uparrow \\ \pi_{\delta^*}(K^{q-1}) & \xrightarrow{R_\varphi \circ \Lambda_\varphi} & \delta_\varphi(K^{q-1}) \end{array}$$

and a mapping

$\sigma: Z^q(K)/\pi_\delta^*(K^{q-1}) \cong Z^q_\varphi(K)/R_\varphi \circ \Lambda_\varphi(\pi_\delta^*(K^{q-1})) \xrightarrow{\text{onto}} Z^q_\varphi(K)/\delta_\varphi(K^{q-1})$; i.e. $\sigma: H^q(K) \xrightarrow{\text{onto}} H^q_\varphi(K)$. This is the continuity property of cohomology which we were seeking.

2.4. The extension problem in cohomology for normed complexes

Let U be an open neighborhood of the origin in C^m and $t \to \Phi(t)$ holomorphic mapping of U into A^1. Suppose furthermore that the locus $D = \{t \in U : \Delta\Phi(t) = 0\}$ is an analytic set. The *extension problem in cohomology* is the following: Given $\gamma \in H^q(K)$, to find a holomorphic mapping $\Gamma: U \to K^q$ such that $\Gamma(0) = \gamma$ and $\delta\Gamma(t) - 2[\Phi(t), \Gamma(t)] = 0$ for $t \in D$.

For fixed $t \in D$, denote by $H^q_t(K)$ the cohomology computed with the differential operator $\delta - 2[\Phi(t), \;]$. We summarize our results in the following theorems:

Theorem 2.1. *There are finitely many obstructions to the extension problem for $\gamma \in H^q(K)$. Furthermore, if a formal solution exists, then an actual solution exists.*

Theorem 2.2. *The extension problem for $\gamma \in H^q(K)$ can be solved if dim $H^{q+1}_t(K)$ is independent of t. If dim $H^{q+1}_t(K)$ is independent of t, then dim $H^q_t(K)$ is independent of t if, and only if, dim $H^{q-1}_t(K)$ is independent of t.*

Finally, suppose that $m = 1$ and $D = U$.

Theorem 2.3. *In the notations of Theorem* 1.1, $H^q_t(K) = E^q_\Phi/J^q_\Phi$ *for* $t \neq 0$.

The proofs of Theorems 2.1 — 3 are immediate from what we have done; the continuity property of cohomology, together with the finitely many obstructions, are sufficient to assure convergence in the formal statement of Theorem 1.1. It is perhaps worth noting that Lemma 2.7 includes, in particular, the usual statements about upper-semi-continuity of cohomology, while Theorem 2.3 implies the invariance of the Euler characteristic $\chi_t(K) = \sum_{q=0} (-1)^q \dim H^q_t(K)$. Indeed, for $t \neq 0$, $\chi_t(K)$

$$= \sum_q (-1)^q \{\dim E^q_\Phi - \dim J^q_\Phi\} = \sum_q (-1)^q \{\dim H^q(K) - \dim G^q_\Phi(K) - \dim J^q_\Phi\} = \sum_q (-1)^q \{\dim H^q(K) - \dim J^{q+1}_\Phi(K) - \dim J^q_\Phi(K)\}$$

$$= \sum_q (-1)^q \dim H^q(K).$$

Finally, we have the following

Theorem 2.4. *Let $S \subset H^q(K)$ be a subspace. Then the extension problem can be solved over $D(S) \subset D$, and $D(S)$ is a maximal such analytic set.*

2.5. The extension problem for exact sequences of graded Lie algebra complexes

Let (A, d), (B, d), and (C, d) be graded Lie algebra complexes which form an exact sequence $0 \to A \overset{i}{\to} B \overset{\pi}{\to} C \to 0$. Thus π and i are bounded mappings which commute with the harmonic theories. Clearly, π maps integrable elements of B^1 into integrable elements in C^1; we wish to know when π maps the integrable elements of B^1 *onto* those in C^1.

Theorem 2.5. *Assume $H^2(A) = 0$. Then there exist $N(C^1)$, $N(B^1)$, and a differentiable mapping $T: N(C^1) \to N(B^1)$ such that $\pi \circ T$ = Indentity and such that, for $\varphi \in N(C^1)$ $T(\varphi)$ is integrable if, and only if, φ is integrable.*

Proof. We first make an assumption concerning the exact sequence $0 \to A \to B \to C \to 0$ which will be satisfied in our applications.

Namely, we suppose that there exist bounded linear maps $\omega \in \mathrm{Hom}(C^i, B^i)$ $(i = 1, 2)$ and $\Omega \in \mathrm{Hom}(C^1, A^2)$ which satisfy the following: $\pi \circ \omega = \text{Identity}$ and $d(\omega\varphi) = \Omega(\varphi) + \omega(d\varphi)$ for $\varphi \in C^1$. Thus the general element ψ in B^1 such that $\pi(\psi) = \varphi \in C^1$ is of the form $\psi = \omega(\varphi) + \gamma$ where $\gamma \in A^1$. We recall that $\psi \in B^1$ is semi-integrable if $d(\psi) - \pi_d[\psi, \psi] = 0$.

Lemma 2.8. There exist $N(C^1)$ and $N(\pi_{d^*}(A^1))$ and a differentiable mapping $t: N(C^1) \to N(\pi_{d^*}(A^1))$ such that, for $\varphi \in N(C^1)$, $\gamma \in N(\pi_{d^*}(A^1))$, $\omega(\varphi) + \gamma \in B^1$ is semi-integrable if, and only if, $\gamma = t(\varphi)$.

Proof. If we define a differentiable mapping $u: C^1 \times \pi_{d^*}(A^1) \to \pi_d(B^1)$ by $u(\varphi, \gamma) = d(\omega(\varphi) + \gamma) - \pi_d[\omega(\varphi) + \gamma, \omega(\varphi) + \gamma]$, then $D_2 u(0,0) = d$. The result then follows from the implicit function theorem.

Lemma 2.9. If φ is integrable, then $\gamma = t(\varphi)$ is given by $\gamma = d^*G([\gamma + \omega(\varphi), \gamma + \omega(\varphi)] - \omega[\varphi, \varphi] - \Omega(\varphi))$.

Proof. $\pi([\gamma + \omega(\varphi), \gamma + \omega(\varphi)] - \omega[\varphi, \varphi] - \Omega(\varphi)) = [\varphi, \varphi] - [\varphi, \varphi] = 0$; thus $\gamma \in \pi_{d^*}(A^1)$. We must show that $\gamma + \omega(\varphi)$ is semi-integrable if $d\varphi = [\varphi, \varphi]$. We have $d(\omega(\varphi)) = \omega([\varphi, \varphi]) + \Omega(\varphi)$, and thus $d(\gamma + \omega(\varphi)) - \pi_d[\gamma + \omega(\varphi), \gamma + \omega(\varphi)] = d(\omega(\varphi)) + dd^*G([\gamma + \omega(\varphi), \gamma + \omega(\varphi)]) - dd^*G(d(\omega(\varphi)) - \pi_d[\gamma + \omega(\varphi), \gamma + \omega(\varphi)] = 0$. Q.E.D.

For $\varphi \in N(C^1)$, we set $T(\varphi) = \varphi + t(\varphi) \in B^1$.

Lemma 2.10. Assume $H^2(A) = 0$. Then there exists $N'(C^1) \subset N(C^1)$ such that, for $\varphi \in N'(C^1)$, $T(\varphi)$ is integrable if, and only if, φ is.

Proof. Assume that φ is integrable; then $dT(\varphi) - dd^*G[T(\varphi), T(\varphi)] = 0$ by Lemma 2.9. Thus $\Delta T(\varphi) = \pi_H[T(\varphi), T(\varphi)] + d^*dG[T(\varphi), T(\varphi)]$. But, since $H^2(A) = 0$, $\pi_H[T(\varphi), T(\varphi)] = \pi_H(\pi[T(\varphi), T(\varphi)]) = \pi_H[\pi T(\varphi), \pi T(\varphi)] = \pi_H[\varphi, \varphi] = 0$. Set $\Lambda = d^*dG[T(\varphi), T(\varphi)]$. Then $\Delta T(\varphi) = \Lambda$. But $\Lambda = 2d^*G[dT(\varphi), T(\varphi)] = 2d^*G[[T(\varphi), T(\varphi)],$

$T(\varphi)] + 2d^*G[A, T(\varphi)]$. Thus $\| A\| \leqq c\| T(\varphi)\| \| A\|$; since $T(0) = 0$, the Lemma follows. Q.E.D.

This completes the proof of Theorem 2.5.

3. Some results on complex manifolds, fibre bundles, and deformations of complex structures

3.1. Deformations of complex structures

Let X be a compact, complex manifold. Denote by $\mathscr{A}^q(\boldsymbol{T})$ the sheaf of germs of vector-valued (0, q) forms of class $C^{k-q+\alpha}$ (in the sense of [8]), and set $C^q = H^\circ(X, \mathscr{A}^q(\boldsymbol{T}))$. Then the graded vector space $C = \sum_{p \geqq 0} C^p$ has a differential operator $\bar{\partial}: C^p \to C^{p+1}$ and a bracket $[\,,\,]: C^p \otimes C^q \to \to C^{p+q}$ such that (1.1), (1.2), and (1.4) are satisfied.

We may define $\|\ \|$ on C^q by setting $\|\ \| = \|\ \|_{k-q+\alpha}$ $(k \gg 0)$ where the latter norm was defined in [8], § 4. We set $d = \bar{\partial}$ on C and, by taking a C^∞ Hermitian metric on X, we may define d^* as the adjoint $\mathfrak{d}$ of $\bar{\partial}$. The harmonic theory for (C, d) is then taken as the harmonic theory relative to the Hermitian metric on X; e. g., $\boldsymbol{G}$ is the usual Green's operator. In fact, using the potential-theoretic lemma in [8], § 4, it is easily seen that (C, d) becomes a normed graded Lie algebra complex as prescribed in § 1.

Let J_0 be the almost complex structure underlying the complex structure on X.

Lemma 3.1. There is a one-to-one correspondence between almost-complex structures J on X, which are sufficiently close to J_0, and elements $\Phi \in C^1$ which are near to 0. The integrability condition is

$$\bar{\partial}\Phi - [\Phi, \Phi] = 0\,. \tag{3.1}$$

Proof. An almost-complex structure J is given by a family of "admissible frames" $e^{\#} = (e_1, \ldots, e_n;\ e_1^*, \ldots, e_n^*)$ where the e_α are complex tangent vectors and e_α^* is the complex conjugate of e_α. We write $e^{\#} = (e, e^*)$; given $e^{\#}$, the admissible frames are of the form $(A\,e, \bar{A}\,e^*)$ where $A \in GL(n, \boldsymbol{C})$.

We let P_0 and Q_0 be the projections, associated to J_0, onto the vectors of type (1,0) and (0,1) respectively; P and Q fulfill similar functions for J. Let $z = (z^1, \ldots, z^n)$ be local holomorphic coordinates on X. If J is close to J_0, then Q_0 will be non-singular on Image (Q), and we may uniquely choose a J-admissible frame $e^{\#} = (e(z), e^*(z))$ such that $Q_0(e^*) = \left(\frac{\partial}{\partial \bar{z}^1}, \ldots, \frac{\partial}{\partial \bar{z}^n}\right)$. Then

$$e_\alpha^* = \frac{\partial}{\partial \bar{z}^\alpha} - \sum_{\beta=1}^{n} \Phi_{\bar{\alpha}}^{\beta} \frac{\partial}{\partial z^\beta}\,, \quad \text{and} \quad e_\alpha = \frac{\partial}{\partial z^\alpha} - \sum_{\beta=1}^{n} \overline{\Phi_{\bar{\alpha}}^{\beta}} \frac{\partial}{\partial \bar{z}^\beta}\,. \tag{3.2}$$

From (3.2), it follows that $\Phi = \sum_{\alpha,\beta} \Phi^{\alpha}_{\bar\beta} \frac{\partial}{\partial z^{\alpha}} \otimes \overline{dz^{\beta}}$ is a tensor and defines an element of C^1. (An intrinsic representation of Φ is given in Lemma 3.3 below.)

Now the co-frame $\omega^{\#} = (\omega, \omega^*)$ where $\omega = (\omega^1, \ldots, \omega^n)$ which is dual to $e^{\#}$ is defined by $\langle \omega^{\alpha}, e_{\beta} \rangle = \delta^{\alpha}_{\beta}$ $(\alpha, \beta = 1, \ldots, n)$. It then follows that $\omega^{\alpha} = dz^{\alpha} + \sum \Phi^{\alpha}_{\bar\beta} d\bar z^{\beta}$. The *Frobenius integrability condition* is written symbolically as $d\omega^{\alpha} \equiv 0 \pmod{\omega}$, which means that $d\omega^{\alpha}$ should be in the exterior ideal generated by $\omega^1, \ldots, \omega^n$. We have that

$$d\omega^{\alpha} = \sum \frac{\partial \Phi^{\alpha}_{\bar\beta}}{\partial z^{\gamma}} dz^{\gamma} \wedge d\bar z^{\beta} + \sum \frac{\partial \Phi^{\alpha}_{\bar\beta}}{\partial \bar z^{\tau}} d\bar z^{\tau} \wedge d\bar z^{\beta}.$$

Since $-dz^{\alpha} \equiv \sum \Phi^{\alpha}_{\bar\beta} d\bar z^{\beta} \pmod{\omega}$ and $d\bar z^{\gamma} \equiv \bar\omega^{\gamma} \pmod{\omega}$, it follows that

$$d\omega^{\alpha} \equiv \sum \left(\frac{\partial \Phi^{\alpha}_{\bar\beta}}{\partial \bar z^{\tau}} d\bar z^{\tau} \right) \wedge d\bar z^{\beta} - \sum \Phi^{\gamma}_{\bar\tau} \frac{\partial \Phi^{\alpha}_{\bar\beta}}{\partial z^{\gamma}} d\bar z^{\tau} \wedge d\bar z^{\beta} \pmod{\omega};$$

this equation says, *by definition*, that $d\omega \equiv \bar\partial \Phi - [\Phi, \Phi] \pmod{\omega}$. Since this argument is reversible, the lemma follows. Q. E. D.

Remarks. (i) If J is integrable and, if $\xi^1, \ldots, \xi^n$ are local holomorphic coordinates for J, then $\Phi = \{\Phi^{\alpha}_{\bar\beta}\}$ is defined locally by

$$\frac{\partial \xi^{\alpha}}{\partial \bar z^{\tau}} = \sum_{\gamma=1}^{n} \frac{\partial \xi^{\alpha}}{\partial z^{\gamma}} \Phi^{\gamma}_{\bar\tau}. \tag{3.3}$$

Thus, in an intrinsic form, introducing holomorphic coordinates for J is *equivalent* to solving locally the linear equation $\bar\partial \xi = 2[\Phi, \xi]$, for a local vector-valued function ξ which gives a differentiable coordinate for the C^{∞} structure on X.

(ii) If J is an almost-complex structure near to J_0, and if $f: X \to X$ is a diffeomorphism near the identity, then f transforms J into a new almost-complex structure $J \circ f_*$ near to J_0. If $\Phi \in C^1$ corresponds to J, we denote by $f_*(\Phi) \in C^1$ the element corresponding to $J \circ f_*$.

Let now $V \subset \pi_H(C^1)$ be the germ of an analytic set defined in § 2.2.

Proposition A. There exists a deformation $\{\mathscr{V} \xrightarrow{\tilde\omega} V\}$ of X such that $\tilde\omega^{-1}(\varphi)$ has the integrable almost-complex structure $P(\varphi)$ given in Lemma 2.2.

Following KURANISHI, we call an element $\varphi \in C^1$ which satisfies $d^*\varphi = 0$ *extremal*. If $\varphi \in \pi_H(C^1)$, then $P(\varphi)$ is clearly extremal.

Proposition B. (KURANISHI). Let $\varphi \in N(C^1)$. Then there exists a diffeomorphism $f: X \to X$ such that $f_*(\varphi)$ is extremal. If furthermore $\varphi = \varphi(s)$ depends differentiably on s, then we may assume that $f(s)$ depends differentiably on s.

From this, we have

Proposition 3.1. The family $\{\mathscr{V} \xrightarrow{\tilde{\omega}} V\}$ is universal for differentiable families of complex structures.

Proof. Let $\{\mathscr{V}' \xrightarrow{\tilde{\omega}} D\}$ be a differentiable family (cf. [7]) such that $X_d = \tilde{\omega}^{-1}(d)$ $(d \in D)$ has the almost complex structure represented by $\Phi(d) \in C^1$. Then $\Phi(d)$ is differentiable in d, and we may find diffeomorphisms $f(d)$ such that $f(d)_* \Phi(d)$ is extremal. But $f(d)_* \Phi(d)$ is integrable, hence semi-integrable, and thus, by Lemma 2.1, $f(d)_* \Phi(d) = P(\varphi(d))$ where $\varphi(d) \in \pi_H(C^1)$ and is differentiable in $d \in D$. If we define $G: D \to V$ by $G(d) = \varphi(d)$, then we may define

$$\begin{array}{ccc} F: \mathscr{V}' & \to & \mathscr{V} \\ \downarrow & & \downarrow \\ G: D & \to & V \end{array} \qquad \text{by } F(d, x) = (G(d), f(d)x) \qquad (d \in D, x \in X),$$

and this proves the universality of $\{\mathscr{V} \xrightarrow{\tilde{\omega}} V\}$.

Finally, we also need

Proposition C. If $\{\mathscr{V}' \xrightarrow{\tilde{\omega}} D\}$ is a complex analytic deformation, then we may choose

$$\begin{array}{ccc} F: \mathscr{V}' & \to & \mathscr{V} \\ \downarrow & & \downarrow \\ G: D & \to & V \end{array}$$

to be a complex analytic mapping.

Propositions A, B, and C are consequences of the results of KURANISHI given in his paper in this volume.

3.2. Deformations of holomorphic fibre bundles

Let X be a compact complex manifold, G a complex Lie group, and $G \to P \to X$ a holomorphic principal fibre bundle. We consider the fundamental bundle sequence (see [1] and [11])

$$0 \to \boldsymbol{L} \to \boldsymbol{Q} \to \boldsymbol{T} \to 0. \tag{3.4}$$

The sheaves $\mathscr{L}, \mathscr{Q}, \mathscr{T}$ are sheaves of Lie algebras and $0 \to \mathscr{L} \to \mathscr{Q} \to \mathscr{T} \to 0$ is an exact sequence of sheaves of Lie algebras. We set $A^q = H^0(X, \mathscr{A}^q(\boldsymbol{L}))$, $B^q = H^0(X, \mathscr{A}^q(\boldsymbol{Q}))$. Then, if $A = \sum_{q \geq 0} A^q$ and $B = \sum_{q \geq 0} B^q$, and if $d = \bar{\partial}: A^q \to A^{q+1}$ and $d = \bar{\partial}: B^q \to B^{q+1}$, (A, d) and (B, d) become graded Lie algebra complexes. In fact, there are natural homomorphisms of graded Lie algebras $\iota: A \to B$ and $\pi: B \to C$ such that

$$0 \to A \xrightarrow{\iota} B \xrightarrow{\pi} C \to 0 \tag{3.5}$$

is an exact sequence of complexes.

Now let J, J' be almost-complex structures on the differentiable manifolds P, X respectively.

Definition 3.1. The pair (J, J') makes $G \to P \to X$ an *almost-complex fibre bundle* if the following three conditions are satisfied: (i) J is G-invariant where G acts on P on the right; (ii) the almost-complex structure on P/G induced by J is J' (i. e. $\pi_* J = J' \pi_*$); and (iii) J restricted to a fibre gives the integrable almost-complex structure on G (this makes sense by (i) and since G is a complex Lie group).

Let (J_0, J_0') be the given integrable almost-complex fibre bundle structure on $G \to P \to X$.

Lemma 3.2. (i) There is a one-to-one correspondence between almost-complex fibre bundle structures (J, J') on P, which are sufficiently close to (J_0, J_0'), and elements $\psi \in B^1$ which are sufficiently near 0 and which satisfy $\pi(\psi) = \Phi$ where Φ corresponds to J' using Lemma 3.1.

(ii) J is integrable if, and only if,

$$\bar{\partial}\psi - [\psi, \psi] = 0. \tag{3.6}$$

Proof. Using the above notation and Lemma 3.1, J is given by an element $\psi \in D^1 = \Gamma_\infty(P, \mathrm{Hom}(\boldsymbol{T}(P)^*, \boldsymbol{T}(P)))$. By (i) in Definition 3.1, $\psi \in \Gamma_\infty(P, \mathrm{Hom}(\boldsymbol{T}(P)^*/G, \boldsymbol{T}(P)^*/G)) \cong \Gamma_\infty(X, \mathrm{Hom}(\boldsymbol{Q}^*, \boldsymbol{Q}))$. By (iii) in the Definition, ψ will annihilate vertical vectors, and thus $\psi \in$ $\in \Gamma_\infty(X, \mathrm{Hom}(\boldsymbol{Q}^*/\boldsymbol{L}^*, \boldsymbol{Q})) \cong B^1$. Finally, from the proof of Lemma 3.1, it is clear that (ii) implies that $\pi(\psi) = \Phi$. By reversing this argument, we get (i) in Lemma 3.2.

Also, (ii) in the Lemma follows from (ii) in Lemma 3.1 by using the embedding $B^1 \subset D^1$ and the fact (mentioned above) that $\bar{\partial}$ and $[\,,\,]$ on B are induced from these operations on D. Q. E. D.

By speaking of $(0, q)$ forms with values in $\boldsymbol{Q}$ of class $C^{k-q+\alpha}$, we may, just as above, make (B, d) (where $B = \sum_{p \geqq 0} B^p$ and $d = \bar{\partial}$) into a graded Lie algebra complex in the sense of § 1. Furthermore, the same remarks which were made about (C, d) and deformations of complex structure now make sense and are true for (B, d) and deformations of bundle structure. For example, where we spoke above of diffeomorphisms of X, we must now speak of bundle diffeomorphisms; one such is given by a diffeomorphism $f: P \to P$ such that $f(pg) = f(p)g$ $(p \in P, g \in G)$. In particular, we have now analogues of Propositions A, B, C, and 3.1; we shall assume that the reader has translated these into the language of complex fibre bundles, and we shall refer to them as Propositions A', B', C' and 3.1′ respectively. The proofs of Propositions A', B', and C' are similar to those of Propositions A, B, and C; as above, Proposition 3.1′ follows from the other three.

3.3. Perturbation of differential operators

Let X_0 be a fixed C^∞ manifold, G a complex Lie group, and let $G \to P \to X$, $G' \to P' \to X'$ be two "close" (cf. § 3.2) complex fibre bundle structures on a fixed C^∞ principal bundle $G \to P_0 \to X_0$. If $\varrho: G \to GL(E)$ is a finite-dimensional holomorphic linear representation of G, then we may construct two holomorphic vector bundles: $E \to \boldsymbol{E} = P \times_G E \to X$, and $E \to \boldsymbol{E}' = P' \times_G E \to X'$. These are both the same C^∞ bundle $E \to \boldsymbol{E}_0 \to X_0$. If $\mathscr{A}^0(\boldsymbol{E})$ is the sheaf of C^∞ cross-sections of $E \to \boldsymbol{E}_0 \to X_0$, then there are differential operators $\bar\partial$ and $\bar\partial'$ on $\mathscr{A}^0(\boldsymbol{E})$ corresponding to the two complex structures involved. From Lemma 3.2, we know that $G \to P' \to X'$ is uniquely prescribed from $G \to P \to K$ by an element $\psi \in B^1$ which satisfies (3.6). We want to find an expression for $\bar\partial'$ in terms of $\bar\partial$ and ψ.

Let $U \subset\subset \boldsymbol{R}^{2n}$ be a contractible open set, and let J, J' be two integrable almost-complex structures on U. Let $z = (z_1, \ldots, z_n)$ and $\xi = (\xi_1, \ldots, \xi_n)$ be holomorphic coordinates for J, J' respectively. Define a section $\partial\xi$ of $\operatorname{Hom}(\boldsymbol{T}, \boldsymbol{T}')$ by $\partial\xi = \sum_{\alpha=1} \frac{\partial}{\partial\xi_\alpha} \oplus \partial\xi_\alpha$ where ∂ is taken with respect to J. Assume that $\partial\xi$ is non-singular, i. e. J' is close to J. Define now a section $\bar\partial\xi$ of $\operatorname{Hom}(\boldsymbol{T}^*, \boldsymbol{T}')$ by $\bar\partial\xi = \sum_{\alpha=1} \frac{\partial}{\partial\xi_\alpha} \oplus \bar\partial\xi_\alpha$. Then the element $\Phi \in C^1$ (relative to J) which defines J' from J is given by

$$\bar\partial\xi = \partial\xi \circ \Phi \qquad \text{(see (3.3))}. \tag{3.7}$$

We now write $J' = J_\Phi$, $\boldsymbol{T}' = \boldsymbol{T}_\Phi$, $\boldsymbol{T}'^* = \boldsymbol{T}_\Phi^*$, and $\bar\partial' = \bar\partial_\Phi$. Let P, Q be the projections of $\boldsymbol{T}^\#$ on $\boldsymbol{T}$, $\boldsymbol{T}^*$ respectively; and let P_Φ, Q_Φ be the projections of $\boldsymbol{T}^\#$ on $\boldsymbol{T}_\Phi$, $\boldsymbol{T}_\Phi^*$. Now, for a function f and a vector v, $\langle \bar\partial_\Phi f, v\rangle = \langle df, Q_\Phi(v)\rangle$; thus, we seek to find Q_Φ in terms of Φ, P, and Q.

Lemma 3.3. $\boldsymbol{T}_\Phi^* = (I - \Phi)\boldsymbol{T}^*$.

Proof. A vector v lies in $\boldsymbol{T}_\Phi^*$ if, and only if, $\langle d\xi_\alpha, v\rangle = 0$ $(\alpha = 1, \ldots, n)$. But $\langle d\xi_\alpha, v\rangle = \langle \bar\partial\xi_\alpha + \partial\xi_\alpha, v\rangle = \langle \partial\xi_\alpha, (I + \Phi)v\rangle$ (by (3.7)). Since $\Phi \circ \Phi = 0$, $\boldsymbol{T}_\Phi^* = (I - \Phi)\boldsymbol{T}^*$. Q. E. D.

We now determine P_Φ and Q_Φ in a purely algebraic fashion. On a $2n$-dimensional real vector space V, let J_1 and J_2 be complex structures $(J_1^2 = -I = J_2^2)$ where J_2 is close to J_1. Write:

$$\mathfrak{V} = V \otimes_{\boldsymbol{R}} \boldsymbol{C} = \begin{cases} U_1 \oplus W_1; & J_1\text{-decomposition where } W_1^* = U_1 \\ U_2 \oplus W_2; & J_2\text{-decomposition where } W_2^* = U_2 . \end{cases}$$

Let P_1, Q_1 and P_2, Q_2 be the projection operators associated to J_1 and J_2 respectively, and suppose that we have $T \in \operatorname{Hom}(W_1, U_1)$ such that $W_2 = (I - T)W_1$. (Lemma 3.3). Letting now * be conjugation with respect to J_1, we then have: $U_2 = (I - T^*)U_1$, $U_1 = (I + T^*)U_2$, and $W_1 = (I + T)W_2$. Since J_2 is close to J_1, $(I - TT^*)$ is invertible, and we set $S_1 = (I - TT^*)^{-1}$. Also, we define $S_2 = T^* \circ S_1 = S_1^* \circ T^*$ (using $(I - A)^{-1} = I + A^2 + A^2 + \ldots$ for a linear transformation A). Then, since $I = (I - TT^*)\, S_1 = (I - T^* + T^* - TT^*)\, S_1 = (I - T^*)\, S_1 + (I - T) S_2$, we get:

$$\begin{aligned} P_1 &= (I - T^*) S_1 P_1 + (I - T) S_2 P_1 \\ Q_1 &= (I - T) S_1^* W_1 + (I - T^*) S_2^* Q_1 . \end{aligned} \tag{3.8}$$

Lemma 3.4. $P_2 = (I - T^*) S_1 P_1 + (I - T^*) S_2^* Q_1$ and $Q_2 = (I - T) S_2 P_1 + (I - T) S_1^* Q_1$.

Proof. Let $V \in U_1$. Then, by (3.8),

$$v = (I - T^*) A_1 P_1(v) + (I - T) S_2 P_1(v)$$

and, since

$$(I - T^*) S_1 P_1(v) \in U_2 \quad \text{and} \quad (I - T) S_2 P_1(v) \in W_2 ,$$

we have verified the formula for P_2 on U_1. The other verifications are similar. Q. E. D.

Now, setting $S' = (I + T)\, Q_2$, $S' \in \mathrm{Hom}(W_2, W_1)$ and establishes a vector space isomorphism. We have that

$$S' = S_2 \circ P_1 + S_1^* \circ Q_1 \tag{3.9}$$

by Lemma 4.4 and since $T^2 = 0$. Similarly, $S'^* \in \mathrm{Hom}(U_2, U_1)$ and is an isomorphism. Combining, we get an isomorphism

$$S : \Lambda^p U_1' \oplus \Lambda^q W_1' \to \Lambda^p U_2' \oplus \Lambda^q W_2' .$$

Applying now this result to $U \subset\subset \boldsymbol{R}^{2n}$, we get an isomorphism $S_\Phi : \mathscr{A}^{p,q} \to \mathscr{A}^{p,q}_\Phi$ where $\mathscr{A}^{p,q}_\Phi$ is the sheaf of C^∞ (p, q) forms relative to J_Φ.

Lemma 3.5. For $\omega \in \mathscr{A}^{p,q}$, $\bar\partial_\Phi (S_\Phi \omega) = S_\Phi(\bar\partial \omega - 2[\Phi, \omega])$.

Remark. As in [3], $[\Phi, \omega] = d\omega \wedge \Phi + (-1)^{p+q} d(\omega \wedge \Phi)$ where $\wedge$ is the contraction operation given by equation (2.7) in [3].

Proof. Since $\bar\partial^2 = 0 = \bar\partial^2_\Phi$, $\bar\partial_\Phi d = - d\bar\partial_\Phi$, and, by Proposition 4.5 in [3], we may prove Lemma 3.5 when $\omega = f$ is a function. In this case, as operators on $\boldsymbol{T}^{\#}$, $S_\Phi(\bar\partial f - 2[\Phi, f]) = (\bar\partial f - df \wedge \Phi) \circ S_\Phi$. On the other hand, $\bar\partial_\Phi (S_\Phi f) = \bar\partial_\Phi f = df \circ Q_\Phi =$ (by Lemma 3.4) $df \circ (I - \Phi) \circ \circ S_2 \circ P_1 + (I - \Phi) \circ S_1^* \circ Q_1) = df \circ (I - \Phi) \circ (S_2 \circ P_1 + S_1^* \circ Q_1) = (\partial f + \bar\partial f) \circ (I - \Phi) \circ S_\Phi$(by (4.9)) $= (\bar\partial f - \partial f \circ \Phi) \circ S_\Phi$ (since $\bar\partial f \circ \Phi = 0 = \partial f \circ S_\Phi) = (\bar\partial f - df \wedge \Phi) \circ S_\Phi = S_\Phi(\bar\partial f - 2[\Phi, f])$. Q. E. D.

Return now to the situation at the beginning of this section. We let $\mathscr{A}^q(\boldsymbol{E})$ be the sheaf of germs of $C^{k-q+\alpha}(0, q)$ forms with values in $E \to \to \boldsymbol{E} \to X$ and $\mathscr{A}^q(\boldsymbol{E}')$ the sheaf of germs of $C^{k-q+\alpha}(0, q)'$ forms (i. e. $(0, q)$ forms relative to the J' structure) with values in $E \to \boldsymbol{E}' \to X'$.

Theorem 3.1. *There exists a linear isomorphism of sheaves* $S_\varphi : \mathscr{A}^q(\boldsymbol{E}) \to \to \mathscr{A}^q(\boldsymbol{E}')$ *which is bounded in* $\| \;\|$ *and which has the property: For a germ* η *in* $\mathscr{A}^q(\boldsymbol{E})$,

$$\bar{\partial}' S_\Psi(\eta) = S_\Psi(\bar{\partial}\eta - 2[\Psi, \eta])\,. \tag{3.10}$$

Proof. The proof is an easy consequence of Lemma 3.5 together with the standard remarks to the effect that: (i) bundle-valued forms on X' are given by ordinary vector-valued forms on P' which satisfy an equivariance condition, and (ii) the operator $\bar{\partial}'$ on bundle-valued forms on X' corresponds to $\bar{\partial}'$ on ordinary forms on P'. Q. E. D.

3.4. Complexes over analytic sets and the cohomology of fibered analytic spaces

Let $G \to P \to X$ be a holomorphic principal bundle, let $\Psi(t) \in B^1\{t\}$, and suppose that the equation $\Delta\Psi(t) = 0$ defines a germ $D_\Psi = D$ of an analytic set in $\boldsymbol{C}^m$; by § 3.2, we then get a deformation $\{\mathscr{P} \to \mathscr{V} \to D\}$ of $G \to P \to X$. Let $\varrho : G \to GL(E)$ be a holomorphic representation and consider the holomorphic vector bundle $E \to \boldsymbol{E}^{\#} = \mathscr{P} \times_G E \to \mathscr{V}$. We wish to describe the groups $H^q(\mathscr{V}, \mathscr{E}^{\#})$ in terms of differential forms.

With no loss of generality we may assume that D is Stein. Furthermore, we may choose a differentiable isomorphism $h : \mathscr{V} \tilde{\underset{C^\infty}{=}} X \times D$ such that: (i) $h|\,\tilde{\omega}^{-1}(0)$ is holomorphic; and (ii) if $E \to \boldsymbol{E} \to X \times D$ is the trivial extension of $E \to \boldsymbol{E} = P \times_G E \to X$, then $h^{-1}(E \to \boldsymbol{E} \to X) \tilde{\underset{C^\infty}{=}} E \to \to \boldsymbol{E}^{\#} \to \mathscr{V}$.

Consider now the sheaf $\mathscr{A}^q(\boldsymbol{E})$ of germs of $C^\infty(0, q)$ forms with values in $\boldsymbol{E}$. Over an open set $U_1 \subset X$, the sections of this sheaf form a Fréchèt space $\boldsymbol{A}^q(\boldsymbol{E})(U_1)$ by taking the family of norms to be $\{\| \;\|_{k+\alpha}^{V_\gamma}\}$ where $k = 0, 1, \ldots$ and V_γ runs over a countable family of compact subsets of U_1 which generate the topology of U_1. We now define the sheaves $\mathscr{A}^q(\boldsymbol{E}) \hat{\otimes} \mathscr{O}_D$ over V; this definition will be done via h. If $U_1 \subset X$ and $U_2 \subset D$, then the sections of $\mathscr{A}^q(\boldsymbol{E}) \hat{\otimes} \mathscr{O}_D$ over $h^{-1}(U_1 \times U_2)$ are given by the holomorphic functions $\varphi : U_2 \to \boldsymbol{A}^q(\boldsymbol{E})(U_1)$. Thus we may write $\varphi(x, t) = \sum_{\mu=0}^{\infty} f_\mu(t)\, \varphi_\mu(x)$ where $f_\mu(t)$ is holomorphic in U_2, $\varphi_\mu(x) \in \in \boldsymbol{A}^q(\boldsymbol{E})(U_1)$, and, for compact sets $K_1 \subset U_1$, $K_2 \subset U_2$ and for an integer k, $\sum_\mu |f_\mu(t)|\, \|\varphi_\mu(x)\|_{k+\alpha}^{K_1}$ converges uniformly for $t \in K_2$.

We define $\boldsymbol{D} : \mathscr{A}^q(\boldsymbol{E}) \hat{\otimes} \mathscr{O}_D \to \mathscr{A}^{q+1}(\boldsymbol{E}) \hat{\otimes} \mathscr{O}_D$ by

$$\boldsymbol{D}\,\varphi(x, t) = \bar{\partial}\varphi(x, t) - 2[\Psi(t), \varphi(x, t)]\,. \tag{3.11}$$

From our definition, it is clear that $\boldsymbol{D}\varphi(x, t)$ is a germ of section of $\mathscr{A}^{q+1}(\boldsymbol{E}) \hat{\otimes} \mathscr{O}_D$; by the remark at the end of § 3.3, $\boldsymbol{D}^2 = 0$.

Lemma 3.6. D satisfies a Poincaré lemma.

Proof. Let $\varphi(x, t)$ be a section of $\mathscr{A}^q(E) \hat{\otimes} \mathcal{O}_D$ over $U_1 \times U_2$ which satisfies $D\varphi(x, t) = 0$.

Now we shall use the definition of deformation of complex structure given in [8]. Namely, there exists open neighborhoods $U_1' \subset U_1$, $U_2' \subset U_2$ and a bi-C^∞ mapping $h_{U'}: U' \to P' \times U_2' (U' = U_1' \times U_2')$ where $P' \subset C^n$ is a polycylinder, and such that $h_{U'}$ locally trivializes the deformation in the following sense: The transform of D by $h_{U'}$ is the operator $\bar{\partial}'$ in the holomorphic coordinates z' in P'. This follows from the Newlander-Nirenberg theorem [10] together with Theorem 3.1. Also $\varphi(x, t)$ is transformed into $\varphi(x', t)$ $(x' \in P', t \in U_2')$ where $\varphi(x', t)$ is still holomorphic in t. From this point of view, the D-Poincaré lemma is essentially the $\bar{\partial}$-Poincaré lemma with holomorphic dependence on t. As is well-known, this is permissible. Q.E.D.

It is perhaps worth remarking that, conversely, the D-Poincaré lemma implies the Newlander-Nirenberg theorem.

Theorem 3.2. (i) *There exists an injection* $j: \mathscr{E}^\# \to \mathscr{A}^0(E) \hat{\otimes} \mathcal{O}_D$ *such that:*

$$0 \to \mathscr{E}^\# \to \mathscr{A}^0(E) \hat{\otimes} \mathcal{O}_D \xrightarrow{D} \cdots \xrightarrow{D} A^q(E) \hat{\otimes} \mathcal{O}_D \xrightarrow{D} \cdots \tag{3.12}$$

is an exact sequence of sheaves over V.

(ii) $H^q(V, \mathscr{A}^q(E) \hat{\otimes} \mathcal{O}_D) = 0 \qquad (q \geqq 1)$; *and*

(iii) *The* D*-cohomology of the complex* $\ldots \to \Gamma(\mathscr{V}, \mathscr{A}^q(E) \hat{\otimes} \mathcal{O}_D) \xrightarrow{D} \Gamma(\mathscr{V}, \mathscr{A}^{q+1}(E) \hat{\otimes} \mathcal{O}_D) \to \cdots$ *represents the sheaf cohomology* $H^*(\mathscr{V}, \mathscr{E}^\#)$.

Proof. (iii) follows from (i) and (ii) by the standard sheaf argument ([6]). By Lemma 3.6, (3.12) is exact except perhaps at $\mathscr{E}^\# \xrightarrow{j} A^0(E) \hat{\otimes} \mathcal{O}_D \xrightarrow{D} \mathscr{A}^1(E) \hat{\otimes} \mathcal{O}_D$; exactness here follows from (3.10). Finally, it has been pointed out to me by L. BUNGART that (ii) follows from a suitable generalization, given in BUNGART's thesis (Princeton University, 1962), of the Künneth formula of GROTHENDIECK ([5]). We use here that $H^p(X, \mathscr{A}^q(E)) = 0 = H^p(D, \mathcal{O}_D) \quad (p \geqq 1)$. Q.E.D.

4. The extension problem for fibered complex-analytic varieties

4.1. The extension problem in cohomology

Let $X \subset \mathscr{V}$ be a germ of an embedding such that $\mathscr{V}$ may be considered as a deformation $\{\mathscr{V} \xrightarrow{\tilde{\omega}} D\}$ of the compact, complex manifold X. Let $E^\# \to \mathscr{V}$ be a holomorphic vector bundle and $E = E^\# | X$. By Theorems 3.3 and 3.2, the extension problem for $H(X, \mathscr{E})$ fits into the formal framework built in §§ 1 and 2. Our main results are then the following:

Theorem 4.1. *There are only finitely many obstructions to extending a class* $\gamma \in H^q(X, \mathscr{E})$ *to* $H^q(\mathscr{V}, \mathscr{E}^{\#})$. *Furthermore, if a formal extension exists, then an actual one does also.*

Theorem 4.2. *Let* $S \subset H^q(X, \mathscr{E})$ *be a subspace. Then the extension problem for* S *can be solved over* $D(S) \subset D$ *and* $D(S)$ *is a maximal such analytic set.*

Set $X_t = \tilde{\omega}^{-1}(t)$ $(t \in D)$ and $\boldsymbol{E}_t = \boldsymbol{E}^{\#} | X_t$.

Theorem 4.3. *The extension problem for* $H^q(X, \mathscr{E})$ *can be solved if dim* $H^{q+1}(X_t, \mathscr{E}_t)$ *is independent of* t. *If dim* $H^{q+1}(X_t, \mathscr{E}_t)$ *is independent of* t, *then dim* $H^q(X_t, \mathscr{E}_t)$ *is locally constant if, and only if, dim* $H^{q-1}(X_t, \mathscr{E}_t)$ *is locally constant.*

Suppose that dim $D = 1$.

Theorem 4.4. *For* $t \neq 0$, $H^q(X_t, \mathscr{E}_t)$ *is isomorphic to* $\{$*extendable classes in* $H^q(X, \mathscr{E})\}/\{$*Jump classes in* $H^q(X, \mathscr{E})\}$.

4.2. The extension problem for analytic fibre bundles

Let $X \subset \mathscr{V}$ be as above in 4.1, and suppose that $G \to P \to X$ is a holomorphic principal bundle. We wish to find a principal bundle $G \to \mathscr{P} \to \mathscr{V}$ such that $\mathscr{P} | X = P$.

From the fundamental bundle sequence $0 \to \boldsymbol{L} \to \boldsymbol{Q} \to \boldsymbol{T} \to 0$ (§ 1) we have seen (§ 3.2) that we get an exact sequence of graded Lie algebra complexes.

$$0 \to A \xrightarrow{\iota} B \xrightarrow{\pi} C \to 0\,. \tag{4.1}$$

The deformation $\{\mathscr{V} \xrightarrow{\tilde{\omega}} D\}$ may, by § 3.1, be given by an element $\Phi(t) \in C^1\{t\}$. By Lemma 3.2, the extension problem for $G \to P \to X$ is the same as the extension problem for exact sequences of graded Lie algebra complexes considered in § 2.5. In order to apply the result there and in § 3.2, we must first settle two technical points: (i) We must assure that the mappings in (4.1) may be made compatible with harmonic theories on A, B, and C; and (ii) We must produce the maps $\omega \in \operatorname{Hom}(C^i, B^i)$ $(i = 1, 2)$ and $\Omega \in \operatorname{Hom}(C^1, A^2)$ which satisfy $\pi \circ \omega =$ Identity and $d(\omega\varphi) = \Omega(\varphi) + \omega(d\varphi)$ $(\varphi \in C^1)$.

Now (i) is easily arranged by choosing an Hermitian metric in B which induces a C^∞ splitting of (4.1) so that, for each $x \in X$, $\boldsymbol{Q}_x = \boldsymbol{L}_x \oplus \boldsymbol{T}_x$. In this case, ι and π clearly commute with d^* (= adjoint of $\bar{\partial}$), hence with $\square$, and finally with the Green's operators $\boldsymbol{G}$.

We may thus deal with (ii).

For a holomorphic vector bundle $E \to \boldsymbol{E} \to X$, we set $A^q(\boldsymbol{E}) = H^0(X, \mathscr{A}^q(\boldsymbol{E}))$; $H^q(\boldsymbol{A}(\boldsymbol{E}))$ is the qth Dolbeault cohomology group. Let $0 \to \boldsymbol{E}' \xrightarrow{\iota} \boldsymbol{E} \xrightarrow{\pi} \boldsymbol{E}'' \to 0$ be an exact sequence of analytic vector bundles over X. From the exact sequences $0 \to \boldsymbol{A}^q(\boldsymbol{E}') \to \boldsymbol{A}^q(\boldsymbol{E}) \to \boldsymbol{A}^q(\boldsymbol{E}'') \to 0$, we get the exact cohomology sequence $\to H^q(\boldsymbol{A}(\boldsymbol{E})) \to H^q(\boldsymbol{A}(\boldsymbol{E}'')) \xrightarrow{\delta^q}$ $\xrightarrow{\delta^q} H^{q+1}(\boldsymbol{A}(\boldsymbol{E}')) \to$. Consider the bundle $\mathrm{Hom}(\boldsymbol{E}'', \boldsymbol{E}')$. There is a natural pairing $\circ$: $\boldsymbol{A}^p(\mathrm{Hom}(\boldsymbol{E}'', \boldsymbol{E}')) \otimes \boldsymbol{A}^q(\boldsymbol{E}'') \to \boldsymbol{A}^{p+q}(\boldsymbol{E}')$ which satisfies

$$\bar{\partial}(\xi \circ \eta) = \bar{\partial}\xi \circ \eta + (-1)^p \xi \circ \bar{\partial}\eta \quad (\xi \in \boldsymbol{A}^p(\mathrm{Hom}(\boldsymbol{E}'', \boldsymbol{E}')), \eta \in \boldsymbol{A}^q(\boldsymbol{E}'')). \tag{4.2}$$

Lemma 4.1. There exists an element $\Omega \in H^1(\boldsymbol{A}(\mathrm{Hom}(\boldsymbol{E}'', \boldsymbol{E}'))$ such that: (i) For any $\eta \in H^q(\boldsymbol{A}(\boldsymbol{E}''))$, $\Omega \circ \eta \in H^{q+1}(\boldsymbol{A}(\boldsymbol{E}'))$;
(ii) $\delta^q(\eta) = \Omega \circ \eta$.

Proof. Let I be the identity in $H^0(A(\boldsymbol{E}'', \boldsymbol{E}''))$ and choose $\omega \in A^0(\mathrm{Hom}(\boldsymbol{E}'', \boldsymbol{E}))$ such that $\pi \circ \omega = I$. Then $\Omega = \bar{\partial}\omega \in \boldsymbol{A}^1(\mathrm{Hom}(\boldsymbol{E}'', \boldsymbol{E}'))$ and it is easily checked that it satisfies the required conditions. Q.E.D.

In the case of the fundamental sequence $0 \to \boldsymbol{L} \to \boldsymbol{Q} \to \boldsymbol{T} \to 0$, the element $\omega \in \Gamma_\infty(X, \mathrm{Hom}(\boldsymbol{T}, \boldsymbol{T}(P)/G))$ geometrically gives a C^∞ connection of type (1, 0) in $G \to P \to X$. The tensor Ω is a (1,1) form with values in $\boldsymbol{L}$, and gives the curvature of ω. Clearly ω and Ω satisfy the requirements of (ii) above.

Theorem 4.5. *The extension problem for $P \to X$ can be solved if $H^2(X, \mathscr{L}) = 0$. If $H^1(X, \mathscr{L}) = 0$, then any solution to the extension problem is unique.*

Proof. We need only prove uniqueness. Suppose that we have two analytic principal bundles $G \to \mathscr{P} \to \mathscr{V}$, $G \to \tilde{\mathscr{P}} \to \mathscr{V}$ which are both extensions of $G \to P \to X$. Let $\varrho: G \to GL(E)$ be any holomorphic linear representation, and form the associated bundles $E \to \boldsymbol{E}^\# \to V$, $E \to \tilde{\boldsymbol{E}}^\# \to V$. The vector bundle $\mathrm{Hom}(\boldsymbol{E}^\#, \tilde{\boldsymbol{E}}^\#)$ is an extension to $\mathscr{V}$ of $\mathrm{Hom}(\boldsymbol{E}, \boldsymbol{E}) \cong \boldsymbol{L}$ over $X \cong \tilde{\omega}^{-1}(0)$. Since $H^1(X, \mathscr{L}) = 0$, there exists, by Theorem 4.1, an extension $\Gamma \in H^0(\mathscr{V}, \mathrm{Hom}(\mathscr{E}^\#, \tilde{\mathscr{E}}^\#))$ of any class $\gamma \in H^0(X, \mathrm{Hom}(\mathscr{E}, \mathscr{E}))$. Taking γ = identity, it follows that Γ establishes a bundle equivalence between $\boldsymbol{E}^\#$ and $\tilde{\boldsymbol{E}}^\#$. Q.E.D.

5. Some examples and applications

5.1. The extension problem for the groups $\mathrm{H}^p(\mathrm{X}, \Omega^q)$

Let $\{\mathscr{V} \xrightarrow{\tilde{\omega}} D\}$ be a deformation of X and let Ω^q be the sheaf of germs of holomorphic $(q, 0)$ forms on X. We set $h^{p,q} = \dim H^p(X, \Omega^q)$ and also $h_t^{p,q} = \dim H^p(X_t, \Omega_t^q)$ $(t \in D)$. Finally, we let $b_r = \dim H^r(X, \boldsymbol{C})$.

Proposition 5.1. If, for some r, $\sum_{p+q=r} h^{p,q} \leq b_r$, then the extension problem can be solved for all the groups $H^p(X, \Omega^q)$ $(p + q = r)$. In particular, if X is Kähler, then the extension problem can be solved for the groups $H^p(X, \Omega^q)$.

Proof. We recall the inequality of FRÖLICHER: $\sum_{p+q=r} h^{p,q} \geqq b_r$. From this, using upper semi-continuity, it follows that $h_t^{p,q}$ is locally a constant function of t. We complete the proof by giving a proof of the inequality $\sum_{p+q=r} h^{p,q} \geqq b_r$.

If $A^{p,q}$ is the vector space of global $C^\infty(p, q)$ forms on X, then $A = \sum_{p,q} A^{p,q}$ forms a double complex with differential operators $d, \partial, \bar\partial$ satisfying $d = \partial + \bar\partial$ and $\partial\bar\partial + \bar\partial\partial = 0$. By [4], § 4.8, there is a spectral sequence $\{E_r^{p,q}\}$ such that E_∞ is associated to $H^*(A) \cong H^*(X, C)$ (de Rham), and such that $E_1^{p,q} = H_{\bar\partial}(A^{p,q}) \cong H^p(X, \Omega^q)$ (Theorem of DOLBEAULT). But then clearly $\sum_{p+q=r} \dim E_1^{p,q} \geqq \dim H^r(X, C)$. Q.E.D.

5.2. An example of a non-extendible abelian differential

Let F be the complex Lie group of complex matrices

$$f = \begin{pmatrix} 1 & z_1 & z_3 \\ 0 & 1 & z_2 \\ 0 & 0 & 1 \end{pmatrix}$$; let $\Gamma \subset F$ be the discrete subgroup

of matrices $\gamma = \begin{pmatrix} 1 & \gamma_1 & \gamma_3 \\ 0 & 1 & \gamma_2 \\ 0 & 0 & 1 \end{pmatrix}$ where the γ_i are

Gaussian integers. The manifold $X = F/\Gamma$ is a compact, complex manifold which was first discussed by Iwasawa. A basis for the abelian differentials on X is given by the right-invariant holomorphic Maurer-Cartan forms on X. These are: $\omega_1 = dz_1$, $\omega_2 = dz_2$, $\omega_3 = -z_2 dz_1 + dz_3$. The dual holomorphic vector fields are:

$$\Theta_1 = \frac{\partial}{\partial z_1} + z_2 \frac{\partial}{\partial z_3}, \Theta_2 = \frac{\partial}{\partial z_2}, \Theta_3 = \frac{\partial}{\partial z_3}.$$

The element

$$\varphi = \Theta_2 \otimes \bar\omega_2 \left(= \frac{\partial}{\partial z_2} \otimes \overline{dz_2}\right)$$

gives a non-zero element of $H^1(X, \Theta)$, and $[\varphi, \varphi] \equiv 0$. Thus $\Phi(t) = t\varphi$ gives a 1-parameter family $\{\mathscr{V} \to D\}$ of deformations of X.

Consider the abelian differential $\omega = \omega_3 = -z_2 dz_1 + dz_3 \in H^0(X, \Omega^1)$. We have: $[\varphi, \omega] = d\omega \,\bar\wedge\, \varphi - d(\omega \,\bar\wedge\, \varphi) = \partial\omega \,\bar\wedge\, \varphi = (\omega_1 \wedge \omega_2) \,\bar\wedge\, \Theta_2 \otimes \bar\omega_2 = \omega_1 \otimes \bar\omega_2 \neq 0$ in $H^1(X, \Omega^1)$. Thus the extension problem for ω cannot be solved over $\{\mathscr{V} \to D\}$.

5.3. Stability of automorphisms under deformations

Let X be as above and let Γ be the identity component of the complex Lie group of analytic automorphisms of X. Then Γ acts analytically in the bundle $C^n \to T \to X$, and there are induced holomorphic representations $\varrho^q: \Gamma \to GL(H^q(X, \Theta))$.

Suppose that we have a deformation of $X, \{\mathscr{V} \to D\}$, given by $\Phi(t) = \sum_{\mu=1}^{\infty} \varphi_\mu(t) \in C^1\{t\}$, and assume that the classes φ_1^α span $H^1(X, \Theta)$ $(\varphi_1(t) = \sum_{\alpha=1}^{m} \varphi_1^\alpha t_\alpha)$. Let Γ_t be the identity component of the analytic automorphism group of $X_t (t \in D)$.

Proposition 5.2. For $t \neq 0$, $\Gamma_t \subseteqq \mathrm{Ker}(\varrho^1)$.

Remark. This Theorem shows how we may think of automorphisms as being "exceptional phenomena".

Proof. Let $\mathfrak{g}_t$ be the complex Lie algebra of $\Gamma_t: \mathfrak{g}_0 = \mathfrak{g}$. The infinitesimal representation of Γ on $H^1(X, \Theta)$ is given by the bracket [,]; i.e. for $\gamma \in \mathfrak{g}$, $\varphi \in H^1(X, \Theta)$, we have that $d\varrho^1(\gamma)(\varphi) = [\gamma, \varphi]$. Thus the subspace of $H^0(X, \Theta)$ for which the extension problem can be solved is a subspace of $\{\mathrm{Ker}\, d\varrho^1\} \subseteqq \mathfrak{g}$. The theorem now follows. Q.E.D.

Corollary. The "general" deformation of a simply-connected compact homogeneous complex manifold is non-homogeneous.

5.4. Extension of holomorphic mappings

Let X be a compact, complex manifold, let $\boldsymbol{G}$ be a Grassmann variety, and let $f: X \to \boldsymbol{G}$ be a holomorphic mapping. Let $G \to B \to \boldsymbol{G}$ be the universal principal bundle over $\boldsymbol{G}$ and let $E \to \boldsymbol{F} \to \boldsymbol{G}$ be the universal vector bundle. Set $P = f^{-1}(B)$, $\boldsymbol{E} = f^{-1}(\boldsymbol{F})$. Let $0 \to \boldsymbol{L} \to \boldsymbol{Q} \to \boldsymbol{T} \to 0$ be the fundamental bundle sequence of $G \to P \to X$.

Theorem 5.1. (i) *There exists a maximal germ of deformation* $\{\mathscr{V}_f \xrightarrow{\tilde{\omega}} D_f\}$ *of* X *for which there is a holomorphic mapping* $F: \mathscr{V}_f \to \boldsymbol{G}$ *such that* $F \mid \tilde{\omega}^{-1}(0) = f$.

(ii) *If* f *is an embedding, then* F *is an embedding on fibres.*

(ii) $\{\mathscr{V}_f \to D_f\}$ *coincides with Kuranishi's family if*

$$H^2(X, \mathscr{L}) = 0 = H^1(X, \mathscr{E}).$$

Proof. Giving $f: X \to \boldsymbol{G}$ is equivalent to giving an analytic vector bundle $E \to \boldsymbol{E} \to X$ such that the sections $H^0(X, \mathscr{E})$ generate each fibre $\boldsymbol{E}_x (x \in X)$. Thus, in order to extend f to a deformation $\{\mathscr{V} \to D\}$ of X, we must solve the extension problems for $E \to \boldsymbol{E} \to X$ and $H^0(X, \boldsymbol{E})$ over $\mathscr{V}$. We first solve the extension problem for $E \to \boldsymbol{E} \to X$ in a maximal way, and then we solve the extension problem for $H^0(X, \mathscr{E})$ in a maximal way. The rest is clear. Q.E.D.

5.5. The direct image theorem for fibered analytic spaces

Let $\mathscr{V}$ be a complex space and let $\tilde{\omega}: \mathscr{V} \to D$ be a proper holomorphic mapping of maximal rank whose fibres are connected and non-singular. Let $\boldsymbol{E}^{\#} \to V$ be a holomorphic vector bundle; set $X_t = \omega^{-1}(t)$ and $\boldsymbol{E}_t = \boldsymbol{E}^{\#} | X_t$ for $t \in D$.

Proposition 5.3. The sets $D^{q,k} = \{t \in D | \dim H^q(X_t, \mathscr{E}_t) \geqq k\}$ are analytic subsets of D.

In fact, from §§2 and 3 we have the following more general result, which is still a special case of GRAUERT's Theorem [*Ein Theorem der analytischen Garbentheorie und die Modulräume komplexer Strukturen.* Inst. Hautes Études Sci., Publ. Math 5, 1—64 (1960)].

Proposition 5.4. The direct image sheaves $\mathscr{R}^q(\tilde{\omega}, \mathscr{E}^{\#})$ are coherent analytic sheaves over D.

Remark. In §1.3 we have exhibited explicitly a finite set of generators and relations for each stalk

$$\mathscr{R}^q(\tilde{\omega}, \mathscr{E}^{\#})_t = \lim_{U \supset \{t\}} H^q(\tilde{\omega}^{-1}(U), \mathscr{E}^{\#} | \tilde{\omega}^{-1}(U)).$$

5.6. On the local triviality of certain analytic fibre spaces

We give another application which generalizes a theorem of KODAIRA-SPENCER. Let $\mathscr{V}, D$ be compact, complex manifolds and $\tilde{\omega}: \mathscr{V} \to D$ a proper holomorphic mapping of maximal rank. Set $X_t = \tilde{\omega}^{-1}(t)$ $(t \in D)$ and suppose that $H^1(X_t, \Theta_t) = 0$ for all $t \in D$. Suppose furthermore that, for some t_0, X_{t_0} is bi-holomorphically equivalent to a rational homogeneous manifold G/U where G and U are suitable complex algebraic Lie groups. (Then automatically $H^1(X_{t_0}, \Theta_{t_0}) = 0$.)

Theorem 5.2. *$\mathscr{V}$ is a locally trivial fibre bundle over D with typical fibre G/U. In fact, G acts on $\mathscr{V}$ as a complex Lie group of bi-holomorphic transformations and $D = \mathscr{V}/G$.*

Remark. The assumption $H^1(X_t, \Theta_t) = 0$ is necessary, as the family of Hirzebruch surfaces shows (cf. § 5.8 below).

Proof. Since $\dim H^1(X_t, \Theta_t) = 0$, we may locally solve the extension problem for $H^0(X_{t_0}, \Theta_{t_0}) \cong \mathfrak{g}$ where $\mathfrak{g}$ is the complex Lie algebra of G. But then it is easy to see that there exists on $\mathscr{V}$ a complex Lie algebra, isomorphic to $\mathfrak{g}$, of vertical holomorphic vector fields. From this, it follows that G acts on $\mathscr{V}$ effectively as a group of bi-regular transformations. Then we may form the holomorphic vector bundle $\boldsymbol{T}(\mathscr{V})/G$ over D, and there is an onto bundle mapping $\boldsymbol{T}(\mathscr{V})/G \xrightarrow{\pi} \boldsymbol{T}(D) \to 0$ $(\pi = \tilde{\omega}_*)$. Thus we get over D an exact sequence of holomorphic vector bundles $0 \to \boldsymbol{L} \to \boldsymbol{T}(\mathscr{V})/G \to \boldsymbol{T}(D) \to 0$, and we let $0 \to \mathscr{L} \to \Xi \to \Sigma \to 0$ be the corresponding exact sheaf sequence. It is easy to see that, since

$H^1(X_t, \Theta_t) = 0$, this exact sheaf sequence is locally split. Thus, given locally n independent holomorphic vector fields $\sigma_1, \ldots, \sigma_n$ on D, there exist n independent G-invariant holomorphic vector fields $\gamma_1, \ldots, \gamma_n$ on $\mathscr{V}$ such that $\tilde{\omega}_*(\gamma_j) = \sigma_j$. But then these holomorphic vector fields may be used to give a local holomorphic cross-section $\sigma: D \to \mathscr{V}$ passing through any point $v \in \mathscr{V}$. The Theorem now follows.

5.7. An interpretation of the integrability equation 1.6

Let (A, d) be a graded Lie algebra complex and let $\varphi(t) = \sum_{\mu=1}^{\infty} \varphi_\mu t^\mu \in$ $\in A^1\{t\}$ satisfy

$$d\varphi(t) - [\varphi(t), \varphi(t)] = 0. \tag{5.1}$$

If we let $\psi(t) = \frac{\partial \varphi(t)}{\partial t} \in A^1[t]$, then by differentiating (5.1) we get

$$d\psi(t) - 2[\varphi(t), \psi(t)] = 0. \tag{5.2}$$

Since $\psi(0) = \varphi_1 = \left.\frac{\partial \varphi(t)}{\partial t}\right]_{t=0} \in H^1(A)$, we have

Proposition 5.5. If $\varphi_1 \in H^1(A)$ is tangent to an integrable family given by $\varphi(t) \in A^1\{t\}$, then φ_1 is extendible along this family.

Remark. This Proposition is rather obvious geometrically. Moreover, it is clear that (in case $\varphi(t)$ converges), for each fixed t_0, $\psi(t_0) \in H^1{}_{\varphi(t_0)}(A)$ is tangent to an integrable family based at $\varphi(t_0)$. Observe also that the converse to Proposition 5.5 is true.

5.8. Automorphisms and jumping of structure

Our next application concerns the following remark of Mumford: Let $\{\mathscr{V} \xrightarrow{\tilde{\omega}} D\}$ be an algebraic family of algebraic varieties and X a variety such that $X_t \cong X$ for $t \neq 0$ but $X_0 \neq X$ (this is a so-called *jumping of structure*). Then $\dim H^0(X_0, \Theta_0) \geqq 1$.

By the theory of § 1 we can show explicitly where the holomorphic vector field on X_0 comes from:

Proposition 5.6. Let $\{\mathscr{V} \xrightarrow{\tilde{\omega}} D\}$ be an analytic family of compact, complex manifolds where there is a jumping of structure. Then $\dim H^i(X_0, \Theta_0) \geqq \dim H^1(X_t, \Theta_t) + 1$ $(i = 0, 1; t \neq 0)$. More precisely, there exists a jump class $\psi \in H^1(X_0, \Theta_0)$ which obstructs an element $\theta \in H^0(X_0, \Theta_0)$.

Proof. For simplicity, assume $\dim D = 1$. We record two obvious remarks: (i) if a germ of deformation $\mathscr{V} \xrightarrow{\tilde{\omega}} D$ is trivial, then the tangent $\tau_0 \in H^1(X_0, \Theta_0)$ is zero; (ii) if a germ of deformation $\mathscr{V} \xrightarrow{\tilde{\omega}} D$ is not trivial, then, for any neighborhood U of $0 \in D$, there exists a $t \in U$ such that $\tau_t \neq 0$ in $H^1(X_t, \Theta_t)$.

The family $\{\mathscr{V} \xrightarrow{\tilde{\omega}} D\}$ is given by a holomorphic function $\varphi(t) \in A^1\{t\}$ satisfying $d\varphi(t) - [\varphi(t), \varphi(t)] = 0$. By § 5.7 above, the elements $\psi_t = \frac{\partial \varphi(t)}{\partial t}$ give a family of classes in $H^1(X_t, \Theta_t)$ which are tangent at X_t to the deformation $\{\mathscr{V} \xrightarrow{\tilde{\omega}} D\}$. The germ of deformation which $\{\mathscr{V} \xrightarrow{\tilde{\omega}} D\}$ defines is not trivial at 0 but is trivial at any $t \neq 0$. Thus $\psi = \psi_0$ is a non-zero class in $H^1(X_0, \Theta_0)$ but ψ_t defines the zero class in $H^1(X_t, \Theta_t)$ for $t \neq 0$. Thus the element $\psi \in H^1(X_0, \Theta_0)$ is a jump class (i. e. it drops off into a boundary for $t \neq 0$), and by § 1 it obstructs an element $\theta \in H^0(X_0, \Theta_0)$. Q.E.D.

Remark. It is perhaps interesting to compare this with Mumford's argument in the algebraic case. Let D be the affine line and $\mathscr{W} = X \times D$. Since $\{\mathscr{V} \xrightarrow{\tilde{\omega}} D\}$ ($\tilde{\omega}$ is now a regular map) is a jumping of structure, there is a meromorphic mapping $F: \mathscr{W} \to \mathscr{V}$ defined except on $X \times \{0\}$. Now $\partial/\partial t$ is a regular vector field on $\mathscr{W}$ and so $F_*(\partial/\partial t)$ has at worst a finite pole on $X_0 = \tilde{\omega}^{-1}(0)$. Thus, for some smallest integer m, $t^m F_*(\partial/\partial t)$ is a regular vector field θ on $\mathscr{V}$; we shall show that θ is tangent to X_0.

Now, since
$$\begin{array}{ccc} \mathscr{W} & \xrightarrow{F} & \mathscr{V} \\ \downarrow & & \downarrow \tilde{\omega} \\ D & \to & D \end{array}$$

commutes, θ restricted to each fibre X_t has constant projection on D. If $\tilde{\omega}_*(\theta) \neq 0$ on X_0, then the local 1-parameter group generated by θ would move X_0 biregularly onto X_{t_0} for some $t_0 \neq 0$. However, this is impossible by assumption. Q.E.D.

References

[1] Atiyah, M. F.: Complex analytic connexions in fibre bundles. Trans. Amer. Math. Soc. **85**, 181—207 (1957).

[2] Dieudonné, J.: Foundations of modern analysis. Academic Press 1962.

[3] Frölicher, A., and A. Nijenhuis: Theory of vector-valued differential forms, I. Derivations in the graded ring of differential forms. Nederl. Akad. Wetensch. Proc. Ser. A. vol. **59**, 338—359 (1956).

[4] Godement, R.: Théorie des faisceaux. Paris: Hermann 1958.

[5] Grothendieck, A.: Produits tensoriels topologiques et espaces nucléaires. Mem. Amer. Math. Soc. No. 16, 1955.

[6] Hirzebruch, F.: Neue topologische Methoden in der algebraischen Geometrie. Ergebn. Math. Grenzgeb., Heft 9, 1956.

[7] Kodaira, K., and D. C. Spencer: On deformations of complex analytic structures, I—II. Ann. Math. **67**, 328—466 (1958).

[8] —, L. Nirenberg, and D. C. Spencer: On the existence of deformations of complex analytic structures. Ann. Math. **68**, 450—459 (1958).

[9] Kuranishi, M.: On the locally complete families of complex analytic structures. Ann. Math. 75, 536—577 (1962).
[10] Newlander, A., and L. Nirenberg: Complex analytic coordinates in almost complex manifolds. Ann. Math. 65, 391—404 (1957).
[11] Nickerson, H. K.: On differential operators and connections. Trans. Amer. Soc. 99, 509—539 (1961).
[12] Nijenhuis, A., and R. Richardson: On the Kuranishi family of complex structures (to appear).

Department of Mathematics
University of California
Berkeley, California

New Proof for the Existence of Locally Complete Families of Complex Structures*, **

By

M. Kuranishi

Introduction

The purpose of the present paper is to give a simpler new proof and some improvement of the theory the writer developed in [5]. We start with an explanation of the problem. Take a compact C^∞ manifold $\boldsymbol{M}$ and a complex analytic structure M on $\boldsymbol{M}$. We ask to what extent we can deform the structure M. By "deform the structure M" we mean that we have a parameter space T with reference point t_0 and an assignment of a complex analytic structure M_t for each t in T with $M_{t_0} = M$ in such a way that M_t depends nicely on t. Now, assume that we have such a family $\{M_t : t \in T\}$, a space S with reference point s_0, and a nice mapping $\tau : S \to T$ with $\tau(s_0) = t_0$. Then the assignment $s \to M_{\tau(s)}$ is a family of deformations of M, which is called the family induced by τ from the family $\{M_t : t \in T\}$. To answer the question posed above, we would like to construct a universal family, i. e. a family $\{M_t : t \in T\}$ such that any family of deformations of M is homeomorphic to a family induced from $\{M_t : t \in T\}$. Among such universal families, we also like to have one which is, in a sense, the most economical one. We are here interested in the local aspect of the theory, i. e. in the germs of families of deformations of M at the reference points.

In studying this question, you can consider differentiable, real analytic, or complex analytic families of deformations. In the present

* This work has been partially supported by the National Science Foundation under Grant NSF GP-1904.
** Received May 27, 1964.

paper we study the theory for complex analytic families of deformations over parameter spaces which are analytic sets. By the nature of the theory it turns out that it is not wise to consider only non-singular manifolds as parameter spaces. If we are interested in C^∞ (or real analytic) families of deformations over non-singular manifolds, we see easily by our proof that the family we construct is universal. However, if we consider complex analytic families, we have to prove that the inducing mapping of analytic parameter sets is holomorphic. Thus we have an additional problem. This problem was solved affirmatively in [*5*]. Our new method also gives a simple new proof for this additional problem provided the 1-dimensional cohomology group with coefficients in the sheaf of germs of holomorphic vector fields does not change on a neighborhood of the reference point. Moreover, we can use the notion of complex analytic family which is more general than the one adopted in [*5*].

The main tool we use is that of strongly elliptic differential equations on compact manifolds, about which one may consult [*3*]. In Preliminaries, we recall some fundamental notions and theorems we are going to use. Especially the representation of a complex analytic structure on $\boldsymbol{M}$ by differential forms, about which we follow Kodaira, Nirenberg, and Spencer [*4*], is the key opening the way to apply the theory of partial differential equations. In § 1 we introduce the notion of a complex analytic family of complex structures. § 2 is devoted to the proof of the main theorem. About the implicit function theory on Banach spaces we use in § 2 one may consult [*6*].

Preliminaries

Let T_r be the tangent vector bundle of $\boldsymbol{M}$. The complex vector bundle $T = \boldsymbol{C} \otimes_{\boldsymbol{R}} T_r$ is called the complex tangent vector bundle of $\boldsymbol{M}$. (This T has nothing to do with parameter spaces T used in the introduction.) T has the conjugation mapping ι induced by the mapping: $\boldsymbol{C} \ni z \to \to \bar{z} \in \boldsymbol{C}$. For a subset S of T, $\iota(S)$ is called the conjugate of S. Fix a complex analytic structure M on $\boldsymbol{M}$. Denote by T', T'' the complex vector bundles of type (1, 0), (0, 1) with respect to M, respectively. We have the direct sum decomposition $T = T' \oplus T''$, and T' is the conjugate of T''. By an almost complex structure on $\boldsymbol{M}$ we mean a C^∞ direct sum decomposition $T = T_1' \oplus T_1''$ such that T_1' is the conjugate of T_1''. If we denote by M' the almost complex structure, we set $T_1' = T'(M')$ and $T_1'' = T''(M')$. A complex analytic structure M_1 on $\boldsymbol{M}$ naturally gives rise to an almost complex structure, which we still denote by M_1. Namely, $T'(M_1)$ and $T''(M_1)$ are the bundles of complex tangent vectors of type (1, 0) and (0, 1) with respect to M_1, respectively. In this way, the set of complex analytic structures on $\boldsymbol{M}$ can be considered as a subset of almost complex analytic structures. An almost complex structure is called integrable when it comes from a complex analytic structure.

Let M' be an almost complex structure on M such that $T''_x(M')$ is very close to T''_x for each $x \in M$, where the index x indicates the fiber over x. Then it is clear that there is a C^∞ fiber mapping $\omega_{M'}$ of T'' into T' (i. e. a mapping which induces the identity mapping on the base space and linear mappings among fibers) such that

$$T''(M') = \{-\omega_{M'}(X) + X : X \in T''\}. \tag{1}$$

Because $T'(M')$ is the conjugate of $T''(M')$, the formula shows that M' is completely determined by $\omega_{M'}$. We will say that M' has finite distance to M when we can find $\omega_{M'}$ as above. Conversely, if we have a C^∞ fiber mapping $\omega : T'' \to T'$ then the formula similar to (1) defines an almost complex structure having finite distance to M, which we denote by M_ω. Thus

$$T''(M_\omega) = \{-\omega(X) + X : X \in T''\}.$$

Hence we have a one-to-one correspondence between the set of almost complex structures on M having finite distance to M and the set of fiber mappings of T'' to T'.

Let A^p be the space of C^∞ differential forms of type $(0, p)$ of the complex analytic structure M with values in T'. In particular, A^1 is the space of fiber mappings of T'' into T'. Hence the set of almost complex structures on M having finite distance to M is indexed by elements in A^1 as indicated above. We have the exterior derivative $\bar\partial : A^p \to A^{p+1}$. Moreover, for $\varphi \in A^p$ and $\psi \in A^q$, the bracket $[\varphi, \psi]$ is defined as follows: Take a complex analytic local coordinate $(z^1, \ldots, z^n)$ of M, and express $\varphi = X_{\alpha_1 \ldots \alpha_p} d\bar z^{\alpha_1} \cdots d\bar z^{\alpha_p}$, $\psi = Y_{\beta_1 \ldots \beta_q} d\bar z^{\beta_1} \cdots d\bar z^{\beta_q}$ where $X_{\alpha_1 \ldots \alpha_p}$ and $Y_{\beta_1 \ldots \beta_q}$ are complex vector fields of type $(1, 0)$ and anti-symmetric with respect to the indices, then the local expression of $[\varphi, \psi]$ is given by

$$[X_{\alpha_1 \ldots \alpha_p}, Y_{\beta_1 \ldots \beta_q}]\, d\bar z^{\alpha_1} \cdots d\bar z^{\alpha_p} d\bar z^{\beta_1} \cdots d\bar z^{\beta_q}.$$

$[\varphi, \psi]$ is bilinear and satisfies the following formulae:

$$[\varphi, \psi] = (-1)^{pq+1}[\psi, \varphi] \tag{2}$$

$$\bar\partial[\varphi, \psi] = [\bar\partial\varphi, \psi] + (-1)^p[\varphi, \bar\partial\psi] \tag{3}$$

$$(-1)^{pr}[\varphi, [\psi, \Theta]] + (-1)^{pq}[\psi[\Theta, \varphi]] + (-1)^{qr}[\Theta, [\varphi, \psi]] = 0 \tag{4}$$

for $\Theta \in A^r$. One of the key observations in our discussion is the following: An almost complex structure M_ω is integrable if and only if ω satisfies the partial differential equation

$$\bar\partial\omega - [\omega, \omega] = 0$$

(cf. (3)′ [4]). Thus we have a way of applying the theory of partial differential equations to our deformation theory.

We are interested in families of deformations of M up to homeomorphism. So we are going to formulate the notion in a form suitable for our approach. Let f be a diffeomorphism of $\boldsymbol{M}$. Take $\omega \in A^1$. Then there is a unique almost complex structure M' on M such that f induces an isomorphism of M' to M_ω. If M' again has finite distance to M, there is a unique $\Theta \in A^1$ such that $M' = M_\Theta$. In this case, we set $\Theta = \omega \circ f$. For a given ω, it is clear that $\omega \circ f$ is defined for f sufficiently near (in the topology of 1-jets) the identity mapping. Let U be the domain of a local coordinate (z). Assume that $f(U_1) \subseteqq U$ for an open subset U_1 of U. Then one sees that the expression of $\Theta = \omega \circ f$ on U_1 in terms of (z) is given by the following formula:

$$(\partial f^\alpha/\partial \bar z^\beta) + w^\alpha_\gamma(f(z))\,(\partial \bar f^\gamma/\partial \bar z^\beta) = ((\partial f^\alpha/\partial z^\delta) + w^\alpha_\gamma(f(z))\,(\partial \bar f^\gamma/\partial z^\delta))\,\Theta^\delta_\beta(z) \quad (5)$$

where $\omega = X_\alpha d\bar z^\alpha$, $X_\alpha = w^\beta_\alpha(z)\,(\partial/\partial z^\beta)$, and similarly for Θ. In particular, $\omega \circ f$ is defined when the matrix $((\partial f^\alpha/\partial z^\delta) + w^\alpha_\gamma(f(z))\,(\partial \bar f^\gamma/\partial z^\delta))$ is non-singular. It is convenient for our purpose to index f near the identity mapping by elements in A^0, i. e. by complex vector fields of type (1, 0). Let us fix a hermitian metric on M. For a real vector field X on $\boldsymbol{M}$, let $(e'(X))\,(x)$ be the end point of the geodesic drawn from the point x by the initial velocity X_x after the time interval 1. Thus $e'(X)$ is a C^∞ mapping of M into M. If X is sufficiently small, $e'(X)$ is a diffeomorphism of $\boldsymbol{M}$. Take $\xi \in A^0$. Then $\xi + \iota(\xi)$, where ι is the conjugation mapping, is a real vector field on $\boldsymbol{M}$. We set $e(\xi) = e'(\xi + \iota(\xi))$. By the construction it is clear that there is a C^∞ mapping B of T to M such that

$$e(\xi)\,(x) = B(\xi_x) \quad (6)$$

and, in terms of a local coordinate (z), $\xi = \xi^\beta(\partial/\partial z^\beta)$

$$(\partial B^\alpha(\xi)/\partial \xi^\beta)_{\xi=0} = \delta^\alpha_\beta, \quad (\partial B^\alpha(\xi)/\partial \bar\xi^\beta)_{\xi=0} = 0\,. \quad (7)$$

This observation together with the formula (5) imply

$$\omega \circ e(\xi) = \omega + \bar\partial \xi + R(\omega, \xi) \quad (8)$$

where

$$R(t\omega, t\xi) = t^2 R(\omega, \xi, t) \quad (9)$$

for a real parameter t, with $R(\omega, \xi, t)$ depending differentiably on t for small t. Moreover, in terms of a local coordinate (z), $R(\omega, \xi)$ is a C^∞ function R_1 of

$$\omega^\alpha_\beta(z), \quad \omega^\alpha_\beta(e(\xi)\,(z)), \quad \xi^\alpha(z), \quad \partial\xi^\alpha/\partial z^\beta, \quad \partial\xi^\alpha/\partial \bar z^\beta\,. \quad (10)$$

$\omega \circ e(\xi)$ is defined when ω and the 1-jet of ξ are sufficiently small, i. e. the variables (10) are in the domain of R_1.

With respect to the hermitian metric we fixed, we have the formal adjoint operator $\mathfrak{d}$. The strongly elliptic differential operator $\mathfrak{d}\bar{\partial} + \bar{\partial}\mathfrak{d} = \Box$ on A^p plays the central role in our discussion. So in the remainder of this section we collect known facts about $\Box$ which we will use.

We denote by (φ, ψ) the L_2 inner product of $\varphi, \psi \in A^p$. By the definition $(\bar{\partial}\varphi, \psi) = (\varphi, \mathfrak{d}\psi)$. Let H^p be the space of harmonic forms in A^p, i. e. the space of all $\varphi \in A^p$ such that $\bar{\partial}\varphi = \mathfrak{d}\varphi = 0$, or equivalently $\Box\varphi = 0$. The dimension of H^p is finite. There is a linear mapping $\boldsymbol{H}$ of A^p onto H^p such that $\boldsymbol{H}$ on H^p is the identity mapping and such that the kernel is the space of all φ orthogonal to H^p. Also we have a linear mapping G of A^p into A^p, the range of which is equal to the kernel of $\boldsymbol{H}$ and such that $\boldsymbol{H} + G\Box$ is the identity mapping. Moreover

$$G\bar{\partial} = \bar{\partial}G, \quad G\mathfrak{d} = \mathfrak{d}G. \tag{11}$$

Hence, by setting $Q = \mathfrak{d}G$ we have

$$\varphi = \boldsymbol{H}\varphi + \bar{\partial}Q\varphi + Q\bar{\partial}\varphi \quad (\varphi \in A^p), \tag{12}$$

$$Q \circ Q = \mathfrak{d} \circ Q = Q \circ \mathfrak{d} = \boldsymbol{H} \circ Q = Q \circ \boldsymbol{H} = 0. \tag{13}$$

For an open set U of $R^{n'}$ and for complex valued C^∞ functions u, v on $\overline{U}$, we define the SOBOLEV inner product $(u, v)_k$ by

$$\Sigma_D \int_U D u(x) \cdot \overline{D v(x)}\, dx$$

where D runs through all partial derivatives of order up to k. We set $|u|_k = ((u, u)_k)^{1/2}$, or more precisely $|u|_k^U$. This is a norm on the space of C^∞ functions on $\overline{U}$. Moreover, if U' is relatively compact in U, we can find a constant c such that

$$|uv|_k^{U'} < c\, |u|_k^U\, |v|_k^U, \tag{14}$$

provided k is sufficiently large. This is a corollary of the fundamental lemma of SOBOLEV: For $k > n'/2$ and for $x \in U'$

$$|D^l u(x)| < c\, |u|_{k+l}^U \tag{15}$$

where D^l is an arbitrary partial derivative of order l and c is a constant depending only on k, l, U', U. Take an open covering U_j with C^∞ local coordinate $(x^{(j)})$ which can be extended to $\overline{U}_j$. For $\varphi \in A^p$, denote by φ_{jB} a component of the expression of φ in terms of the local coordinate $(x^{(j)})$. We define $(|\varphi|_k)^2 = \Sigma_{j,B} |\varphi_{jB}|_k^2$. The equivalent class of the norm $|\ |_k$, for a fixed k, is independent of the choice of the above covering. For a constant c, we have for arbitrary φ in A^p

$$|\boldsymbol{H}\varphi|_k < c|\varphi|_k, \tag{16}$$

$$|Q\varphi|_{k+1} < c|\varphi|_k. \tag{17}$$

§1. Families of complex structures

Let G be an open subset of the complex euclidean space $\boldsymbol{C}^m$. Take an analytic subset S in G. Denote by S_r the set of regular point of S. Let $\mathscr{U}$ be an open set of $\boldsymbol{M} \times S$. Let F be a mapping of $\mathscr{U}$ into a manifold N. We say that F is of class C^∞ when F is continuous and the following two conditions are satisfied: (1) F is of class C^∞ on $\mathscr{U} \cap (\boldsymbol{M} \times S_r)$, (2) for any open set V of $\boldsymbol{M}$ with C^∞ chart (x) and for any subset S_1 of S such that $V \times S_1 \subseteqq \mathscr{U}$ and $F(V \times S_1)$ in a domain of a C^∞ chart of N, $(\partial^k/\partial x_{i_1} \ldots \partial x_{i_k}) F_j$ is a continuous function on $V \times S_1$ for any partial derivative $\partial^k/\partial x_{i_1} \ldots \partial x_{i_k}$. Let u^* be a continuous mapping of $\mathscr{U}$ onto $U \times S$, where U is an open set of $\boldsymbol{C}^n$. Defining $\mathscr{U}_s$ by $\mathscr{U} \cap (U \times s) = \mathscr{U}_s \times s$ for a fixed s in S, we have a mapping u_s of $\mathscr{U}_s$ into U by the formula: $u_s(p) = \varrho_1(u^*(p, s))$, where ϱ_1 is the projection of $U \times S$ onto U. We say that u^* is a special chart of $\boldsymbol{M} \times S$ when it satisfies the following conditions: $\varrho_2 \circ u^* = \varrho$, where ϱ (resp. ϱ_2) is the canonical projection of $\boldsymbol{M} \times S$ (resp. $U \times S$) onto S, (2) for each fixed s, u_s is a C^∞ chart of $\boldsymbol{M}$, and (3) u^*, considered as a mapping into $U \times G$, is of class C^∞.

By a (LT) complex structure on $\boldsymbol{M} \times S$ we mean that we are given a collection of special charts of $\boldsymbol{M} \times S$ the members of which are called charts of the structure, satisfying the following two conditions: (1) the domains of the charts of the structure cover $\boldsymbol{M} \times S$, (2) a special chart is a chart of the structure if and only if the transition mapping from this special chart to any chart of the structure is a holomorphic mapping (between open subsets of $\boldsymbol{C}^n \times S$). When we have a (LT) complex structure on $\boldsymbol{M} \times S$, the charts of the structure induce a complex analytic structure on the submanifold $\boldsymbol{M} \times s$ for each fixed s, which we naturally identify with a complex analytic structure on $\boldsymbol{M}$. This structure is called the structure induced on the fiber over s by the (LT) complex analytic structure of $\boldsymbol{M} \times S$. In the following we might also mean by charts of the structure the restrictions to open subsets of domains of charts defined above.

Take an analytic set S' in an ambient manifold G'. Let τ be a holomorphic mapping of S into S'. Consider a continuous fiber mapping $\boldsymbol{f}$ of $(\boldsymbol{M} \times S, S)$ to $(\boldsymbol{M} \times S', S')$ which induces τ on the base spaces. Then, for any fixed s in S, we have a continuous mapping f^s of $\boldsymbol{M}$ into itself defined by the formula: $\boldsymbol{f}(p, s) = (f^s(p), \tau(s))$. Let $\mathscr{M}_S$ and $\mathscr{M}_{S'}$ be (LT) complex structures on $\boldsymbol{M} \times S$ and $\boldsymbol{M} \times S'$, respectively. By a holomorphic homomorphism of $\mathscr{M}_S$ to $\mathscr{M}_{S'}$ over τ we mean a fiber mapping $\boldsymbol{f}$ of $\boldsymbol{M} \times S$ to $\boldsymbol{M} \times S'$ over τ such that $\boldsymbol{f}$ is of class C^∞ and holomorphic with respect to charts of the structures and such that each f^s is a diffeomorphism of $\boldsymbol{M}$. When τ is a holomorphic homeomorphism of S to S' we say that $\boldsymbol{f}$ is a holomorphic homeomorphism of S to S' when $\boldsymbol{f}$ and $\boldsymbol{f}^{-1}$ are holomorphic homomorphism over τ and τ^{-1}, respectively.

Let $\{M_s; s \in S\}$ be a family of deformations of a complex analytic structure M on $\boldsymbol{M}$. We say that the family is a complex analytic family when we are given a (LT) complex analytic structure $\mathcal{M}_S$ on $\boldsymbol{M} \times S$ such that M_s is equal to the structure on the fiber over s induced by $\mathcal{M}_S$. So the complex analytic family is more adequately denoted by $\mathcal{M}_S$. Denote by s_0 the reference point of S. Let τ be a holomorphic mapping of S' to S sending the reference point s_0' to s_0. τ can be prolonged to a mapping i_τ of $\boldsymbol{M} \times S'$ to $\boldsymbol{M} \times S$ by $i_\tau(x, s') = (x, \tau(s'))$. Let u^* be a chart of $\mathcal{M}_S$. It may happen that $u^* \circ i_\tau$ is not of class C^∞. This is due to the fact that our notion of C^∞ mapping on analytic sets is too naive (cf. the definition in the beginning of this section). If the image of S' is in the singular locus of S, a C^∞ property of u^* may not imply the corresponding C^∞ property of $u^* \circ i_\tau$. When i_τ is of class C^∞ we say that τ is compatible with $\mathcal{M}_S$. In this case the induced family $\{M_{\tau(s')} : s' \in S'\}$ is a complex analytic family. Let $\{M_s' : s \in S\}$ be a complex analytic family associated with a (LT) complex analytic structure $\mathcal{M}_S'$ of $\boldsymbol{M} \times S$. We say that the two families are holomorphically homeomorphic (over the identity mapping of S) when there is a holomorphic homeomorphism $\boldsymbol{f}$ of $\mathcal{M}_S'$ to $\mathcal{M}_S$ over the identity mapping. The writer does not know at present whether a complex family $\mathcal{M}_S$ is uniquely, up to homeomorphism, determined by a collection $\{M_s : s \in S\}$. However, for convenience, we may write $\mathcal{M}_S = \{M_s : s \in S\}$ as far as no confusion can occur. Let $\mathcal{M}_{S'} = \{M_{s'}'; s' \in S'\}$ be a complex analytic family of deformations of M. Let τ be a holomorphic mapping of S' to S compatible with $\mathcal{M}_S$. Then it is easy to establish the following proposition: $\mathcal{M}_{S'}$ is holomorphically homeomorphic to the family induced by τ from $\mathcal{M}_S$ if and only if there is a continuous family of diffeomorphisms $f^{s'}$, with $f^{s_0'}$ = the identity mapping, which depend differentiably on s' and such that $f^{s'}$ is a holomorphic mapping of $M_{s'}'$ to $M_{\tau(s')}$ for all $s' \in S'$. In the above we mean, by saying that $f^{s'}$ depend differentiably on s', that the mapping $(p, s') \to (f^{s'}(p), s')$ of $\boldsymbol{M} \times S'$ onto itself is of class C^∞ in terms of charts of the structures.

Let $\mathcal{M}_S$ be a complex analytic family of deformations of M over S with a reference point s_0. By replacing S by a small open neighborhood of s_0, if necessary, we may assume that each structure of $\mathcal{M}_S$ has finite distance to M. Therefore $\mathcal{M}_S$ is also represented by a continuous mapping ω of S into A^1, the space of complex $(1, 0)$-type tangent vector valued differential forms of type $(0, 1)$, such that $\omega(s_0) = 0$ and

$$\bar{\partial}\omega(s) - [\omega(s), \omega(s)] = 0$$

for all s in S. The assumption that $\mathcal{M}_S$ is a complex analytic family may not imply that $\omega(s)$ depends holomorphically on s, since if ω represents a complex analytic family then $\omega'(s) = \omega(s) \circ f^s$ also represents a complex analytic family. However, if $\omega(s)$ is a holomorphic mapping of S

into A^1 satisfying the above conditions and if each $\omega(s)$ isa real analytic differential form then we can prove that $\omega(s)$ represents a complex analytic family.

Let T be an analytic set with a reference point t_0. Take a complex analytic family $\mathscr{M}_T$ of deformations of M over T with reference point t_0. As above, we may assume that $\mathscr{M}_T$ is represented by a continuous mapping $\varphi(t)$ of T into A^1. Then the following is clear: $\mathscr{M}_S$ is holomorphically homeomorphic to a family induced from $\mathscr{M}_T$ if and only if we can find (1) a holomorphic mapping τ of S into T sending s_0 to t_0 which is compatible with $\mathscr{M}_T$ and (2) a continuous family f^s of diffeomorphisms of $\boldsymbol{M}$ which depend differentiably on s and f^{s_0} = the identity mapping in such a way that, for all s in S,

$$\omega(s) = (\varphi(\tau)(s)) \circ f^s .$$

We say that a complex analytic family $\mathscr{M}_T$ of deformations of M is a complex analytic universal family (at the reference point) when, for any complex analytic family $\mathscr{M}_S$ of deformations of M, the restriction of $\mathscr{M}_S$ to an open neighborhood of the reference point s_0 in S is holomorphically homeomorphic to a family induced from $\mathscr{M}_T$. Now we state our main

Theorem. *For any complex analytic compact manifold M, there exists a universal family of deformations of M.*

§2. Proof of the main theorem

By the previous discussion, it is clear that our task is

(i) to find a family $\varphi(t)$ of solutions of the equation

$$\bar{\partial}\varphi - [\varphi, \varphi] = 0 \tag{18}$$

where φ is in A^1, which is parametrized by t in an analytic set T and $\varphi(t_0) = 0$ for a reference point t_0, and which can be considered as representing a complex analytic family $\mathscr{M}_T$, and

(ii) the family $\varphi(t)$ is such that any complex analytic family of deformations of M over S with reference point s_0 represented by a mapping $\omega(s)$ can be written in a neighborhood of s_0 as

$$\omega(s) \circ f^s = \varphi(\tau(s))$$

where τ is a holomorphic mapping (compatible with $\mathscr{M}_T$) of a neighborhood of s_0 into T such that $\tau(s_0) = t_0$ and where f^s is a continuous family of diffeomorphisms of $\boldsymbol{M}$ which depend differentiably on s and such that f^{s_0} = the identity mapping.

Set $f^s = e(\xi^s)$ where $\xi^s \in A^0$. Then the linear part of the above two equations are $\bar{\partial}\varphi = 0$ and $\omega + \bar{\partial}\xi = \varphi$ (cf. (8)). Thus the linearization of our problem is the following: To find a finite dimensional vector subspace of the space of cocycles of A^1 which forms a set of representatives of cohomology classes. By the classical theory of harmonic integrals, such a subspace is obtained by taking all cocycles φ such that $\mathfrak{d}\varphi = 0$; here we fix a hermitian metric on M. So we might conjecture that a small neighborhood of the set

$$\Phi = \{\varphi \in A^1; \bar{\partial}\varphi - [\varphi, \varphi] = \mathfrak{d}\varphi = 0\}$$

is a universal family. This is what we are going to prove. Thus our proof is divided into two parts: (I) To show that a small neighborhood of the set Φ in the $|\ |_k$-norm topology forms a complex analytic family, and (II) to show that the family is holomorphically universal.

(I) Set $\Box = \mathfrak{d}\bar{\partial} + \bar{\partial}\mathfrak{d}$. Then for any $\varphi \in \Phi$, $\Box\varphi - \mathfrak{d}[\varphi, \varphi] = 0$. Apply Green's operator G to this equality (cf. § 2). Since $\varphi = \boldsymbol{H}\varphi + G\Box\varphi$, it follows that $\varphi - Q[\varphi, \varphi] = -\boldsymbol{H}\varphi$, where $Q = G\mathfrak{d} = \mathfrak{d}G$. Hence Φ is a subset of

$$\Psi = \{\varphi \in A^1; \varphi - Q[\varphi, \varphi] \in H^1\},$$

where H^r is the space of harmonic forms in A^2. Consider the mapping $F: A^1 \ni \varphi \to \varphi - Q[\varphi, \varphi] \in A^1$. By (17) and by the fact that $[\varphi, \varphi]$ is induced by a quadratic mapping of the 1-jet of φ, we have

$$|F(\varphi)|_k < c|\varphi|_k$$

for a constant c. Hence F can be extended to a continuous mapping of the completion B^1 of A^1 into itself, which we still denote by F. Since Q is linear and $[\varphi, \psi]$ is bilinear, F is a holomorphic mapping of the Banach space B^1 into itself. The differential of F at the origin is clearly the identity mapping. Hence a neighborhood of the origin is mapped bijectively by F to a neighborhood of the origin, and its inverse F^{-1} is also holomorphic. Hence if we choose a sufficiently small positive number ε and if we set

$$W = \{t \in H^1; |t|_k < \varepsilon\},$$

then there is a holomorphic injective mapping $\varphi(t)$ of W into B^1 such that $F(\varphi(t)) = t$. $\varphi(t)$ is a differential form of class k_1 for an integer $k_1 < k - n/2$ by the lemma of SOBOLEV and we still have the equality: $\Box\varphi(t) - \mathfrak{d}[\varphi(t), \varphi(t)] = 0$, since $\Box Q = \Box\mathfrak{d}G = \mathfrak{d}\Box G = \mathfrak{d}$. It is a standard theorem in the theory of strongly elliptic partial differential equations that the solutions of this equation are of class C^∞. Hence $\varphi(t) \in A^1$ and depend holomorphically on t. By the construction it is clear that the image of the mapping $\varphi(t)$ covers a neighborhood of Ψ in the $|\ |_k$-norm topology, and a neighborhood of Φ in the same topology is contained in the image of the mapping $\varphi(t)$. Moreover, we have the equality

$$\varphi(t) - Q[\varphi(t), \varphi(t)] = t. \tag{19}$$

Now, we set out to find a necessary and sufficient condition for $\varphi(t)$ being in Φ. Since $\mathfrak{d}Q = \mathfrak{d}\mathfrak{d}G = 0$, clearly $\mathfrak{d}\varphi(t) = 0$. As for the other condition, since $\bar{\partial}\varphi(t) = \bar{\partial}Q[\varphi(t), \varphi(t)]$, we have by (12)

$$\begin{aligned}\bar{\partial}\varphi(t) - [\varphi(t), \varphi(t)] &= \bar{\partial}Q[\varphi(t), \varphi(t)] - [\varphi(t), \varphi(t)]\\ &= -Q\bar{\partial}[\varphi(t), \varphi(t)] - \boldsymbol{H}[\varphi(t), \varphi(t)].\end{aligned}$$

Since the images of Q and $\boldsymbol{H}$ are orthogonal in the L_2 inner product, $\bar{\partial}\varphi(t) - [\varphi(t), \varphi(t)] = 0$ if and only if $\boldsymbol{H}[\varphi(t), \varphi(t)] = Q\bar{\partial}[\varphi(t), \varphi(t)] = 0$. Now we claim that the condition $\boldsymbol{H}[\varphi(t), \varphi(t)] = 0$ implies $Q\bar{\partial}[\varphi(t), \varphi(t)] = 0$. Namely, we have

$$\begin{aligned}Q\bar{\partial}[\varphi(t), \varphi(t)] &= 2\,Q[\bar{\partial}\varphi(t), \varphi(t)] && \text{(by (3))}\\ &= 2\,Q[\bar{\partial}Q[\varphi(t), \varphi(t)], \varphi(t)] && \text{(by (19))}\\ &= 2\,Q[[\varphi(t), \varphi(t)], \varphi(t)] - 2\,Q[Q\bar{\partial}[\varphi(t), \varphi(t)], \varphi(t)] && \text{(by (12))}\\ &= -2\,Q[Q\bar{\partial}[\varphi(t), \varphi(t)], \varphi(t)] && \text{(by (4))}.\end{aligned}$$

Hence if we set $\xi(t) = Q\bar{\partial}[\varphi(t), \varphi(t)]$, we have by (17)

$$|\xi(t)|_k < c_1 |\xi(t)|_k |\varphi(t)|_k .$$

Therefore $\xi(t) = 0$, provided we choose ε so small that $|\varphi(t)|_k < c_1^{-1}$. Thus we found that $\varphi(t) \in \Phi$ if and only if $\boldsymbol{H}[\varphi(t), \varphi(t)] = 0$. $h(t) = \boldsymbol{H}[\varphi(t), \varphi(t)]$ is a holomorphic mapping of W into the finite dimensional complex euclidean space H^2. Hence $T = \{t \in W; h(t) = 0\}$ is an analytic set and $\{\varphi(t); t \in T\}$ covers a neighborhood of Φ in the $|\ |_k$-norm topology. In order to conclude the part (I) of our proof, it remains to show that $\{\varphi(t) : t \in T\}$ represents a complex analytic family of deformations of M over T (with reference point the origin). To see this, let U be an open subset of M with complex analytic chart $(z) = (z_1, \ldots, z_n)$ of M. Write down the expression

$$\varphi_\alpha^\beta(z, t)\, d\bar{z}_\beta (\partial/\partial z_\alpha)$$

of $\varphi(t)$ for $t \in W$. Consider the following differential equation for unknown functions $Z_\gamma(z, t)$ and $\lambda_\gamma(z, t)$ $(\gamma = 1, \ldots, n)$ on $U \times W$:

$$\partial Z_\gamma/\partial \bar{z}_\beta = \varphi_\alpha^\beta(z, t)\, \partial Z_\gamma/\partial z_\alpha + \lambda_\gamma(z, t) \tag{20}$$

with the conditions:

$$Z_\gamma(z, 0) = z_\gamma \quad \text{and} \quad \lambda_\gamma(z, t) = 0 \quad \text{for} \quad t \in T . \tag{20$'$}$$

Then, by the definition M_φ, a solution $Z_\gamma(z, t)$ of (20) and (20)$'$ gives a complex analytic chart of $M_{\varphi(t)}$ for $t \in T$. In order that the family $\{\varphi(t) : t \in T\}$ is complex analytic, it is sufficient to show that the above is solved with the following additional condition:

$$Z_\gamma(z, t) \text{ is holomorphic in } t \in W \text{ for any fixed } z \in U. \tag{20$''$}$$

Surprisingly it does not seem easy to solve (20) with the additional conditions (20)′ and (20)″. In our case our manifold M has a real analytic manifold structure, i. e. the underlying real analytic manifold of M. Hence we may take a real analytic hermitian metric on M, by means of the imbedding theorem of real analytic compact manifolds ([1] or [7]). Then $\varphi(t)$ is a real analytic differential form, since it is a solution of real analytic elliptic differential equation. Since $\varphi(t)$ can be expressed as a power series in t, $\varphi_\alpha^\beta(z, t)$ is real analytic in (z, t). Then we can solve (20) with the additional conditions (20)′ and (20)″ by using the classical theorem of involutive differential systems. Thus $\{\varphi(t); t \in T\}$ is a complex analytic family. In later discussions, we need the following additional condition on Z_γ: Let $Z_{\gamma|1}(z, t)$ be the linear part of $Z_\gamma(z, t)$ in t, i.e. $Z_\gamma(z, t) \equiv z_\gamma + Z_{\gamma|1}(z, t) \pmod{t^2}$. Then

$$(\partial Z_{\gamma|1}/\partial \bar{z}_\beta)\, d\bar{z}_\beta (\partial/\partial z_\gamma) \tag{20$'''$}$$

is the expression of the differential form for $t \in W \subseteqq A^1$.

We can impose this condition because the linear part of $\varphi(t)$ is the identity mapping t by the construction and because we can make $\lambda_\gamma(z, t) \equiv 0 \pmod{t^2}$. For the details of the construction of $Z_\gamma(z, t)$, we refer to Lemma 1.2, p. 544 [5]. For each U with complex analytic chart (z), we construct $Z_\gamma^U(z, t)$ as above. For U and U' let $h^{UU'}(\zeta, t)$ be the change of charts from $Z^{U'}$ to Z^U. Let $h_1^{UU'}(\zeta, t)$ be the linear part of $h^{UU'}(\zeta, t)$ in the power series expansion in t. Set $\Theta_1^{UU'}(t) = h_{1\gamma}^{UU'}(\xi, t)$ $(\partial/\partial z_\gamma)$. Then (20)‴ implies that $\Theta^{UU'}(t)$ is a holomorphic vector field on $U \cap U'$. Let Θ be the sheaf of germs of holomorphic vector fields on M. $\{\Theta_1^{UU'}(t)\} = \Theta_1(t)$ is a Čech cocycle with coefficients in Θ. Then it follows by (20)‴ that $\Theta_1(t)$ belongs to the cohomology class of $H^1(M, \Theta)$ corresponding to $t \in H$ by the Dolbeault isomorphism.

(II) By definition, A^0 is the vector space of C^∞ complex tangent vector fields of type (1, 0) on M, and H^0 is that of holomorphic vector fields on M. Let ${}^\perp A^0$ be the orthogonal complement of H^0 in A^0 with respect to the L_2 inner product. The topology considered will be the $|\ |_k$-norm topology, unless otherwise stated. In order to complete the part (II) of our proof, we prove the following

Proposition. *We can find a neighborhood* $\boldsymbol{U}$ *of the origin in* A^1 *and a neighborhood* $\boldsymbol{W}$ *of the origin in* ${}^\perp A^0$ *such that for any* ω *in* $\boldsymbol{U}$ *there is a unique* $\xi = \xi(\omega)$ *in* $\boldsymbol{W}$ *with the condition*

$$\mathfrak{d}(\omega \circ e(\xi)) = 0\,. \tag{21}$$

Moreover, if S *is a manifold and* $\omega(s)$ *is a continuous mapping of* S *into* $\boldsymbol{U}$ *which depends differentiably on* s, $\xi^s = \xi(\omega(s))$ *is also continuous in* s *and depends differentiably on* s.

Proof. By (8) we find that (21) is satisfied if and only if

$$\mathfrak{d}\omega + \square\,\xi + \mathfrak{d}R(\omega, \xi) = 0\,.$$

Applying G, we see that the above is equivalent with

$$\xi + G\,\mathfrak{d}\omega + G\,\mathfrak{d}R(\omega, \xi) = 0\,, \tag{22}$$

because ξ is in ${}^{\perp}A^0$. Take a neighborhood $\boldsymbol{U}_1$ (resp. $\boldsymbol{W}_1$) of the origin in A^1 (resp. ${}^{\perp}A^0$) such that R is defined on $\boldsymbol{U}_1 \times \boldsymbol{W}_1$. Let F be the mapping

$$\boldsymbol{U}_1 \times \boldsymbol{W}_1 \ni (w, \xi) \to \xi + G\,\mathfrak{d}\omega + Q\,R(\omega, \xi) \in {}^{\perp}A^0\,.$$

By (10) we find that $R(\omega, \xi)$ is a continuous function on $A^1 \times A^0$ with the $|\;|_k$-norm topology into A^0 with the $|\;|_{k-1}$-norm topology. Therefore by (17) we see that F is a continuous mapping of $\boldsymbol{U}_1 \times \boldsymbol{W}_1$ into ${}^{\perp}A^0$, all with the $|\;|_k$-norm topology, provided we choose sufficiently small $\boldsymbol{U}_1$ and $\boldsymbol{W}_1$. Hence F can be extended to a mapping of the completion of the domain to the completion of ${}^{\perp}A^0$. By (9) and by applying the implicit function theory in Banach spaces, we find that, for a sufficiently small $\boldsymbol{U}$, any $\omega \in \boldsymbol{U}$ admits a unique solution $\xi(\omega)$ of the equation (22) in the completion of ${}^{\perp}A^0$ which is sufficiently small. Since $\square\,\xi + \mathfrak{d}R(\omega, \xi) + \mathfrak{d}\omega = 0$, $\xi(\omega)$ is of class C^∞ when ω is of class C^∞. Since $\xi(\omega)$ is induced by a C^∞ mapping of the Banach spaces, differential dependence of ξ on ω is immediate.

Now we can prove that the complex analytic family $\mathscr{M}_T$ represented by $\{\varphi(t) : t \in T\}$ constructed in (I) is universal at the origin. Let $\mathscr{M}_S$ be a complex analytic family of deformations of M over an analytic set S with reference point s_0 represented by $\omega(s)$. By the definition, $\omega(s)$ is C^∞ in the usual sense for s in the set S_r of regular points of S. For our purpose we can assume without loss of generality that $\omega(s) \in \boldsymbol{U}$ for all $s \in S$. By the above proposition, $\mathfrak{d}(\omega(s) \circ f^s) = 0$ where $f^s = e(\xi(\omega(s)))$. Since $\omega(s) \circ f^s$ is integrable, $\omega(s) \circ f^s = \varphi(\tau(s))$, for a continuous mapping τ of S into T with $\tau(s_0) = 0$. Clearly τ restricted to S_r is a C^∞ mapping into W. Since $\mathscr{M}_T$ has charts $Z^U(z, t)$ which can be extended to a real analytic mapping of $U \times W$, it is clear that τ is compatible.

We will show that τ is holomorphic under the assumption that $\dim H^1(M_t, \Theta_t)$ is independent of t, where Θ_t is the sheaf of germs of holomorphic vector fields of $M_t = M_{\varphi(t)}$. Let U be a sufficiently small domain of a complex analytic chart (z) of M. Take a chart $k^U(z, s)$ of $\mathscr{M}_S$. By the construction,

$$Z^U_\alpha(f^s(z), \tau(s)) = g^U_\alpha(k^U(z, s), s)$$

where $g^U_\alpha(\zeta, s)$ is holomorphic in ζ for each fixed s, and is of class C^∞ on (ζ, s) provided $s \in S_r$. Let $k^{UU'}$ (resp. $h^{UU'}$) be the transition mapping from $k^{U'}$ (resp. $Z^{U'}$) to k^U (resp. $k^{U'}$). Denoting by ζ^U the general variable in the range of k^U, we see immediately that

$$g_\alpha^U(k^{UU'}(\zeta^{U'}, s), s) = h_\alpha^{UU'}(g^{U'}(\zeta^{U'}, s), \tau(s)). \tag{23}$$

Denote by ξ^{U_s} the chart $k^U(\cdot, s)$ of the complex analytic structure $M(s)$ associated with s in $\mathscr{M}_S$. u^{U_s} denotes the chart $g^U(\xi^{U_s}, s)$ of $M(s)$. Let $(t_1, \ldots, t_m)$ be a linear complex coordinate of H^1. For each partial derivative $\partial/\partial\bar{s}$ of S at a regular point s, we set

$$\Theta^{UU'}(s) = \sum_\lambda \frac{\partial h_\alpha^{UU'}}{\partial t_\lambda}(u^{U'_s}, \tau(s))\frac{\partial \tau^\lambda}{\partial \bar{s}}\left(\frac{\partial}{\partial u_\alpha^{U_s}}\right),$$

$$\xi^U(s) = \frac{\partial g_\alpha^U}{\partial \bar{s}}(\zeta^{U_s}, s)\left(\frac{\partial}{\partial u_\alpha^{U_s}}\right).$$

The first (resp. the latter) is a holomorphic vector field of $M(s)$ on $U \cap U'$ (resp. U). By applying $(\partial/\partial\bar{s})$ on (23), we find

$$\Theta^{UU'}(s) = \xi^U(s) - \xi^{U'}(s).$$

Thus $\{\Theta^{UU'}(s)\}$ represents the zero element of $H^1(M(s), \Theta_s)$. By the last statement in (I), $t \to \Theta_1(t)$ is an isomorphism of H^1 to $H^1(M, \Theta)$. On the other hand, since dim $H^1(M(s), \Theta_s)$ is independent of s, we have continuous dependence of $H^1(M(s), \Theta_s)$ (cf. [3]). Hence $\Theta^{UU'}(s)$ represents the zero of $H^1(M(s), \Theta_s)$ only if $\partial/\partial\bar{s} = 0$. Hence τ is holomorphic on the set of regular points of S. Since τ is continuous on S, τ is holomorphic.

Finally, we remark that our argument also proves the following: If M' is a complex analytic structure on M such that the element $\omega \in A^1$ representing M' is sufficiently close to zero in the $|\,|_k$-norm topology, at least one member M_t of $\mathscr{M}_T$ is homeomorphic to M'.

References

[1] GRAUERT, H.: On Levi's problem and the imbedding of real-analytic manifolds. Ann. Math. **68**, 460—472 (1958).

[2] KODAIRA, K., and D. C. SPENCER: On deformations of complex analytic structures I—II. Ann. Math. **67**, 328—466 (1958).

[3] — — On deformations of complex analytic structes III. Ann. Math. **71**, 43—76 (1960).

[4] —, L. NIRENBERG, and D. C. SPENCER: On the existence of deformations of complex analytic structures. Ann. Math. **68**, 450—459 (1958).

[5] KURANISHI, M.: On the locally complete families of complex analytic structures. Ann. Math **75**, 536—577 (1962).

[6] LANG, S.: Introduction to differential manifolds. Interscience Publishers (1962).

[7] MORREY, C. B.: The analytic embedding of abstract real-analytic manifolds. Ann. Math. **68**, 159—201 (1958).

Department of Mathematics
Columbia University
New York

Algebraic Function Fields on Complex Analytic Spaces*,**

By

N. Kuhlmann

It has turned out that the theory of holomorphic mappings of complex spaces as developed by R. Remmert and K. Stein (see [*13*], [*24*], [*25*], [*26*], [*27*], [*29*], [*34*], [*35*], [*36*], [*37*], for a different approach see E. Bishop [*6*], H. Rossi [*30*]) is an excellent tool in exploring the relationship between algebraic and analytic dependence of meromorphic functions (see for instance [*26*], [*27*], [*34*], [*37*]). The purpose of the present paper is to describe some applications of the mapping theory in this direction.

R. Remmert invented in [*26*] a strikingly simple method how to apply his well known result concerning proper holomorphic mappings.

In [*20*] a slightly improved mapping theorem was obtained: *let* $\tau: X \to Y$ *be a holomorphic map of the (reduced) complex analytic space* X *into the (reduced) complex analytic space* Y. *Assume that there exists for each* $Q \in \tau(X)$ *a neighborhood* V_Q *and in* X *a compact set* X'_Q *such that* $\tau(X'_Q) \cap V_Q = \tau(X) \cap V_Q$ *holds. Then* $\tau(X)$ *is locally analytic in* Y.

A special case is of interest: *let* $\tau: X \to Y$ *be a holomorphic map of the (reduced) irreducible complex-analytic space* X *into the reduced complex analytic space* Y. *Assume there is a compact set* X' *in* X *such that* $\tau(X') = \tau(X)$ *holds. Then* $\tau(X)$ *is analytic in* Y.

If one replaces in [*26*] Remmert's theorem for proper holomorphic mappings by the above one, one gets an improvement of known results (see for instance [*26*], [*31b*]):

Let X *be an irreducible reduced complex analytic space,* Γ *a group of holomorphic automorphisms of* X, X/Γ *compact. Then* Γ*-automorphic meromorphic functions are analytically dependent if and only if they are algebraically dependent. The field* $\boldsymbol{K}^{\Gamma}(X)$ *of* Γ*-automorphic meromorphic functions on* X *is an algebraic function field.*

The proof of the special mapping theorem needed here is much shorter and simpler than the proof of the general form in [*20*]. It seems therefore of some value to present this short proof (see §2).

I want to describe another method which is more complicated but which gives a stronger result.

Of great interest during the recent years was the Siegel modular group Γ which operates in the Siegel upper half plane X of a complex dimension >1. It turned out that the field $\boldsymbol{K}^{\Gamma}(X)$ of Γ-automorphic

* This is essentially an abstract of a talk which was given at the Conference on Complex Analysis at the University of Minnesota, March 16—March 22, 1964.

** Received June 8, 1964.

meromorphic functions is an algebraic function field, though the quotient is not compact. (See for instance SATAKE and BAILY [*31a*], [*4*], SIEGEL [*33*]).

ANDREOTTI and GRAUERT found out that one can treat the Siegel modular group within the frame work of a certain type of automorphism groups, which they called pseudoconcave automorphism groups (see [*1*], [*2*], [*3*]; in this connection ROSSI [*31*] has to be mentioned). Satisfactory results however were only obtainable in the case of *properly discontinuous*, pseudoconcave automorphism groups.

We consider now irreducible, reduced, complex analytic spaces X and automorphism groups Γ on X with the following property:

There exists a relatively compact subset $D \subset\subset X$ *of* X, *such that for every non constant* Γ*-automorphic function* f *the relation* $f(\bar{D}) = \boldsymbol{P}^1$ *holds, where* $\bar{D}$ *is the closed hull of* D *in* X.

One can prove *under this assumption that the field* $\boldsymbol{K}^{\Gamma}(X)$ *of* Γ*-automorphic meromorphic functions on* X *is an algebraic function field;* an outline of the proof is given in §3.

This assumption is for instance fulfilled in the case of pseudoconcave automorphism groups on normal complex spaces (whether they are properly discontinuous or not).

One of the main tools for the proof is the theory of analytic decomposition of K. STEIN [*34, 35, 36*].

Now, the generalized mapping theorem of REMMERT and the theory of analytic decomposition are both fundamentally dependent on the following mapping theorem of R. REMMERT [*24, 25*]: *let* $\tau: X \to Y$ *be a holomorphic mapping of the complex analytic space* X *into the complex analytic space* Y *which degenerates nowhere. Then every point* $P \in X$ *has a neighborhood* U_P *such that* $\tau(U_P)$ *is locally analytic in* Y.

Therefore an interest in this theorem seems to be justified. I present a proof of this theorem which also works in the case of analytic Hausdorff spaces over algebraically closed, complete non-discrete valued ground fields (see [*10*]), simply by substituting complex analytic concepts by the equivalent ones of the general case (see § 1. 2.).

This proof depends mainly on the following special case of GRAUERT's theorem [*12a*], which is not too difficult to get (see for instance HOUZEL [*10*], Exposé 19): *If* $\tau: X \to Y$ *is a proper, holomorphic map with discrete fibers then the direct image* $\tau_*(\mathfrak{O}_X)$ *of the structure sheaf* $\mathfrak{O}_X$ *of* X *is a coherent sheaf of* $\mathfrak{O}_Y$*-modules.*

I thank Prof. Dr. STOLL for fruitful discussions about some of the topics.

§ 1

In order to avoid a lengthy introduction of concepts let us assume that the spaces we consider are complex analytic spaces in the sense of

GRAUERT, GROTHENDIECK and SERRE (see for instance [*16*]), i. e., the stalks of the structure sheaf can have nilpotent elements. But all statements of this chapter (unless explicitely otherwise stated) are true for analytic Hausdorff spaces over algebraically closed, complete, non-discrete valued fields in the sense of GROTHENDIECK [*10*], Exposé 9. Proofs will be given in such a way that one gets the results for these more general ground fields simply by substituting complex analytic concepts by the equivalent ones of the general case.

1. Let me first state some results of HOUZEL, SERRE and GROTHENDIECK which can be found in [*10*], Exposés 13, 18—21.

a. Some notations are needed: let $\boldsymbol{C}\{z_1, \ldots, z_n\}$ be the ring of convergent power series in the n variables $z_1, \ldots, z_n$ over the field $\boldsymbol{C}$ of complex numbers. A $\boldsymbol{C}$-algebra B is called a holomorphic algebra if B is as a $\boldsymbol{C}$-algebra isomorphic to a quotient $\boldsymbol{C}\{z_1, \ldots, z_n\}/\mathfrak{a}$, where $\mathfrak{a}$ is an ideal in $\boldsymbol{C}\{z_1, \ldots, z_n\}$.

Let A and B be holomorphic algebras with a $\boldsymbol{C}$-homomorphism of algebras $\tilde{\varphi}: A \to B$. By virtue of $\tilde{\varphi}$ we consider B as an A-algebra by defining $a \cdot b = \varphi(a) \cdot b$ for all $a \in A$, $b \in B$. $\mathfrak{m}$ be the maximal ideal of A, $\mathfrak{n}$ that of B. B is called quasifinite over A if for a suitable natural integer n the relation $\mathfrak{n}^n \subset \mathfrak{m} B$ holds.

The following theorem is true: *B is quasifinite over A if and only if B is finite over A.*

b. We need further: *let X and Y be complex analytic spaces with the structure sheafs $\mathfrak{O}_X$, $\mathfrak{O}_Y$. $x \in X$, $y \in Y$ be points of X resp. Y and $\tilde{\beta}: \mathfrak{O}_{Y,y} \to \mathfrak{O}_{X,x}$ a $\boldsymbol{C}$-homomorphism of algebras* (such a map $\tilde{\beta}$ is always a local homomorphism of local rings, i. e. maps the maximal ideal into the maximal ideal). *Then there exists a neighborhood U of x with a holomorphic map $\beta: U \to Y$, $\beta(x) = y$, such that the homomorphism $\beta^*: \mathfrak{O}_{Y,y} \to \mathfrak{O}_{X,x}$, which is induced by β, is exactly $\tilde{\beta}$.*

c. *Let X be a complex analytic space with a structure sheaf $\mathfrak{O}_X$, $x \in X$ be a point, A a finite $\mathfrak{O}_{X,x}$-algebra. Then there exists a neighborhood U of x with a coherent $\mathfrak{O}_X | U$-algebra $\mathfrak{A}$* (i. e. $\mathfrak{A}$ is a sheaf of $\mathfrak{O}_X | U$-algebras which is coherent as a sheaf of $\mathfrak{O}_X | U$-modules) *such that $\mathfrak{A}_x$ is isomorphic to A as an $\mathfrak{O}_{X,x}$-algebra.*

d. X be — as before — a complex analytic space with a structure sheaf $\mathfrak{O}_X$. $\mathfrak{A}$ be a coherent $\mathfrak{O}_X$-algebra. For each point $P \in X$ exists a neighborhood $U = U(P)$ with finitely many sections $f_1, \ldots, f_r \in \Gamma(U, \mathfrak{A})$ which generate over $\mathfrak{O}_U = \mathfrak{O}_X | U$ the sheaf $\mathfrak{A} | U$. U can be chosen so small that there exist polynomials $p_i = p_i(T_i) = T_i^{n_i} + a_1^{(i)} T_i^{n_i - 1} + \cdots + a_{n_i}^{(i)}$ with indeterminates T_i and coefficients $a_j^{(i)} \in \Gamma(U, \mathfrak{O}_X)$, $1 \leqq i \leqq r$, $1 \leqq j \leqq n$, for which $p_i(f_i) = 0$. $\mathfrak{O}_U[T_1, \ldots, T_r]/(p_1, \ldots, p_r)$ is a coherent $\mathfrak{O}_U$-sheaf with a natural $\mathfrak{O}_U$-sheaf homomorphism

$$\mathfrak{O}_U[T_1, \ldots, T_r]/(p_1, \ldots, p_r) \to \mathfrak{A}_U := \mathfrak{A} \mid U$$

which maps T_i onto f_i. The kernel is coherent. Let us assume that U is chosen small such that the kernel is already generated over U by finitely many sections $g_1, \ldots, g_s \in \Gamma(U, \mathfrak{O}_U[T_1, \ldots, T_r]/(p_1, \ldots, p_r))$. Let $G_1, \ldots, G_s$ be sections in $\Gamma(U, \mathfrak{O}_U[T_1, \ldots, T_r])$ which are mapped onto $g_1, \ldots, g_s$ by the homomorphism

$$\mathfrak{O}_U[T_1, \ldots, T_r] \to \mathfrak{O}_U[T_1, \ldots, T_r]/(p_1, \ldots, p_r).$$

$T_1, \ldots, T_r$ can be considered as coordinates of the space $\boldsymbol{C}^r$ and $p_1, \ldots, p_r$, $G_1, \ldots, G_s$ as sections $\in \Gamma(U \times \boldsymbol{C}^r, \mathfrak{O}_{U \times \boldsymbol{C}^r})$. **SPECAN** $(\mathfrak{A}_U)$ be the support of $\mathfrak{O}_{U \times \boldsymbol{C}^r}/(p_1, \ldots, p_r, G_1, \ldots, G_s)$. **SPECAN** $(\mathfrak{A}_U)$, provided with the sheaf

$$\mathfrak{O}_{U \times \boldsymbol{C}^r}/(p_1, \ldots, p_r, G_1, \ldots, G_s) \mid \mathbf{SPECAN}\,(\mathfrak{A}_U) =: \mathfrak{O}_{\mathbf{SPECAN}(\mathfrak{A}_U)},$$

is a complex analytic space which is (up to a canonical isomorphism) uniquely defined by $\mathfrak{A}_U$. In a natural way one can glue together for all $P \in X$ the corresponding spaces $\mathbf{SPECAN}(\mathfrak{A}_{U(P)})$ and gets a complex analytic space $\mathbf{SPECAN}(\mathfrak{A})$ with a structure sheaf $\mathfrak{O}_{\mathbf{SPECAN}(\mathfrak{A})}$ and a proper holomorphic map $\pi_{\mathfrak{A}}: \mathbf{SPECAN}(\mathfrak{A}) \to X$ which has discrete fibers. $\pi_{\mathfrak{A}}$ $(\mathbf{SPECAN}(\mathfrak{A}))$ *coincides with the support of* $\mathfrak{A}$ *and the image sheaf* $\pi_{\mathfrak{A}*}$ $(\mathfrak{O}_{\mathbf{SPECAN}(\mathfrak{A})})$ *is isomorphic to* $\mathfrak{A}$. *For each* $P \in X$ *the stalk* $\mathfrak{A}_P$ *is isomorphic to the direct product*

$$\prod_{Q \in \pi_{\mathfrak{A}}^{-1}(P)} \mathfrak{O}_{\mathbf{SPECAN}(\mathfrak{A}),\, Q}.$$

Each $Q \in \pi_{\mathfrak{A}}^{-1}(P)$ *is uniquely associated to a maximal ideal* $\mathfrak{n}_Q$ *(and vice versa) in* $\mathfrak{A}_p$ *such that the quotient ring* $(\mathfrak{A}_p)_{\mathfrak{n}_Q}$ *is isomorphic to* $\mathfrak{O}_{\mathbf{SPECAN}(\mathfrak{A}),\, Q}$.

Let $\tilde{X}$ *be a complex analytic space with the structure sheaf* $\mathfrak{O}_{\tilde{X}}$ *and a holomorphic map* $\pi: \tilde{X} \to X$ *which is proper and has discrete fibers. Then the image* $\mathfrak{A} = \pi_*(\mathfrak{O}_{\tilde{X}})$ *is a coherent* $\mathfrak{O}_X$*-algebra and there exists a biholomorphic map* $\tau: \tilde{X} \to \mathbf{SPECAN}(\mathfrak{A})$ *such that the diagram*

$$\begin{array}{ccc} \tilde{X} & \xrightarrow{\tau} & \mathbf{SPECAN}\,(\mathfrak{A}) \\ {\scriptstyle \pi}\downarrow & \swarrow {\scriptstyle \pi_{\mathfrak{A}}} & \\ X & & \end{array}$$

is commutative. If A *is an analytic subset of* $\tilde{X}$, *then* $\pi(A)$ *is analytic in* X *and is of the same dimension as* A.

e. Let X be a complex analytic space with the structure sheaf $\mathfrak{O}_X$. For $P \in X$ the dimension $\mathbf{dim}_P X$ of X in P be defined by $\mathbf{dim}_P X :=$ **altitude** of $\mathfrak{O}_{X,P}$ (see [21], p. 24). *If X is in P irreducible*, i. e. if $\mathfrak{O}_{X,P}$ has as zero divisors at most nilpotent elements, *then there exists a neighborhood U of P such that for all $Q \in U$ the relation* $\mathbf{dim}_Q X = \mathbf{dim}_P X$ *holds.* $\mathbf{dim}\, X$ be defined by $\mathbf{dim}\, X := \sup\limits_{P \in X} \mathbf{dim}_P X$. M be a locally analytic subset of X, i. e., for each point $P \in X$ there exists on X a neighborhood U_P of P with finitely many sections $f_1, \ldots, f_r \in \Gamma(U_P, \mathfrak{O}_X)$, whose common zero set is exactly $M \cap U_P$ (in other words: $M \cap U_P$ is the support of $\mathfrak{O}_{U_P}/(f_1, \ldots, f_r)\, \mathfrak{O}_{U_P}$, where $\mathfrak{O}_{U_P} := \mathfrak{O}_X \,|\, U_P$).

If we provide $M \cap U_P$ with the structure sheaf

$$[\mathfrak{O}_{U_P}/(f_1, \ldots, f_r)\, \mathfrak{O}_{U_P}] \,|\, M \cap U_P\,,$$

then $\mathbf{dim}_P M$ is well defined. $\mathbf{dim}_P M$ *does not depend on the special choice of $f_1, \ldots, f_r$.* Let $\mathfrak{C}_{M \cap U_P}$ be the sheaf of all germs in $\mathfrak{O}_{U_P}$ which vanish on M. $\mathfrak{C}_{M \cap U_P}$ *is $\mathfrak{O}_{U_P}$-coherent* (Cartan-Oka [9], [22]). Let us furnish $M \cap U_P$ with the structure sheaf

$$[\mathfrak{O}_{U_P}/(\mathfrak{C}_{M \cap U_P})] \,|\, M \cap U_P\,.$$

As

$$\mathbf{altitude}\, (\mathfrak{O}_{U_P}/(\mathfrak{C}_{M \cap U_P}) \,|\, M \cap U_P)_P = \mathbf{altitude}\, (\mathfrak{O}_{U_P}/(f_1, \ldots, f_r)\mathfrak{O}_{U_P})_P,$$

it is clear, that $\mathbf{dim}_P M$ is independent of $f_1, \ldots, f_r$.

If M is an open subset of X, we will always assume that M be provided with the sheaf $\mathfrak{O}_M := \mathfrak{O}_X \,|\, M$.

f. Let X, Y_1, Y_2 be complex analytic spaces with the structure sheaves $\mathfrak{O}_X$, $\mathfrak{O}_{Y_1}$, $\mathfrak{O}_{Y_2}$. $\varphi_1 : X \to Y_1$, $\varphi_2 : X \to Y_2$ be holomorphic maps.

The complex analytic spaces $X \times X$, $Y_2 \times Y_2$ are well defined and $\varphi_1 \times \varphi_2 : X \times X \to Y_1 \times Y_2$ is a well defined holomorphic map.

The diagonal Δ of $X \times X$ is analytic in $X \times X$. $\mathfrak{J}_\Delta$ be the (coherent) sheaf of germs in $\mathfrak{O}_{X \times X}$ which vanish on Δ. Let us provide Δ with the structure sheaf $[\mathfrak{O}_{X \times X}/\mathfrak{J}_\Delta] \,|\, \Delta$. We have a canonical injective holomorphic map $i : \Delta \to X \times X$ and a natural biholomorphic map $j : X \to \Delta$ of X onto Δ. In the following I will always denote the holomorphic map $(\varphi_1 \times \varphi_2) \circ i \circ j$ by $\varphi_1 \dot{\times} \varphi_2$.

Let $q_1, \ldots, q_r$ be holomorphic sections in $\Gamma(X, \mathfrak{O}_X)$. These sections define a holomorphic map $\gamma : X \to \boldsymbol{C}^r$ in the following way: for $Q \in X$ let

$$\gamma(Q) = (q_1(Q), \ldots, q_r(Q)) \in \boldsymbol{C}^r\,.$$

The associated homomorphism

$$\mathfrak{O}_{\boldsymbol{C}^r, \gamma(Q)} \to \mathfrak{O}_{X,Q}$$

maps the i-th coordinate function onto the germ of q_i in $\mathfrak{O}_{X,Q}$.

2. Let X and Y be complex analytic spaces with the structure sheaves $\mathfrak{O}_X$ resp. $\mathfrak{O}_Y$. $\varphi : X \to Y$ be a holomorphic map. If A is an analytic subset of Y then $\varphi^{-1}(A)$ is analytic in X.

For $P \in X$ let the rank $\mathbf{rk}(\varphi, P, X)$ of φ in P be defined by

$$\mathbf{rk}(\varphi, P, X) := \mathbf{dim}_P X - \mathbf{dim}_P \varphi^{-1}(\varphi(P)).$$

Let us assume that P is a point on X and that X is in P irreducible. Then there exists a neighborhood V of P such that for all $Q \in V$ the relation $\mathbf{dim}_Q X = \mathbf{dim}_P X$ holds (see for instance [*10*], Exposé [*20*], in dealing with a general ground field k).

Let us assume *that for all* $Q \in V$ *the relation* $\mathbf{rk}(\varphi, Q, X) \leqq \mathbf{rk}(\varphi, P, X)$ *holds*. Let $\mathbf{dim}_P X = n$, $\mathbf{dim}_P \varphi^{-1}(\varphi(P)) = d$. The maximal ideal in $\mathfrak{O}_{\varphi^{-1}(\varphi(P)), P}$ has a parameter system of d elements $q_1, \ldots, q_d$ (see [*21*], section 24). These elements are induced by sections in $\mathfrak{O}_X$ over a certain neighborhood of P; for the sake of convenience let V be already such a neighborhood and let us denote these sections also by $q_1, \ldots, q_d$. $\mathfrak{O}_M$ be the sheaf $\mathfrak{O}_V/(q_1, \ldots, q_d)$. $\mathfrak{O}_V$, restricted to its support M. M furnished with $\mathfrak{O}_M$ is clearly a complex analytic space. V be chosen so small that $M \cap \varphi^{-1}(\varphi(P)) = P$.

We define a "restriction map" $\varphi | M : M \to Y$ in the following way: For $Q \in M$ be $(\varphi | M)(Q) = \varphi(Q)$, and the associated $\mathbf{C}$-homomorphism $\mathfrak{O}_{Y, \varphi(Q)} \to \mathfrak{O}_{M, Q}$ be given by the composition of the homomorphisms $\mathfrak{O}_{Y, \varphi(Q)} \to \mathfrak{O}_{X, Q}$ (which is well defined as φ is a holomorphic map) and

$$\mathfrak{O}_{X, Q} \to \mathfrak{O}_{X, Q}/(q_1, \ldots, q_d)\mathfrak{O}_{X, Q} = \mathfrak{O}_{M, Q}.$$

$\varphi | M$ will be abbreviated by $\bar{\varphi}$. We have $\bar{\varphi}^{-1}(\varphi(P)) = P$, i.e.

$$\mathfrak{m}_{Y, \varphi(P)} \mathfrak{O}_{M, P} \supset \mathfrak{m}_{M, P}^t$$

for a suitable natural number t ("Nullstellensatz" of Hilbert). Here $\mathfrak{m}_{Y, \varphi(P)}$, $\mathfrak{m}_{M, P}$ denote the maximal ideals in $\mathfrak{O}_{Y, \varphi(P)}$ resp. $\mathfrak{O}_{M, P}$. This implies by l. a. the finiteness of $\mathfrak{O}_{M, P}$ over $\mathfrak{O}_{Y, \varphi(P)}$. There exists a neighborhood W of $\varphi(P)$ with a coherent $\mathfrak{O}_W$-algebra $\mathfrak{A}$, $\mathfrak{O}_W := \mathfrak{O}_Y | W$, for which $\mathfrak{A}_{\varphi(P)}$ is isomorphic to $\mathfrak{O}_{M, P}$ as a $\mathfrak{O}_{W, \varphi(P)}$-algebra. Let us consider $\pi_{\mathfrak{A}} : \mathbf{SPECAN}(\mathfrak{A}) \to W$. $\mathfrak{A}_{\varphi(P)}$ has only one maximal ideal. By l. d. the fiber $\pi_{\mathfrak{A}}^{-1}(\varphi(P))$ of $\pi_{\mathfrak{A}} : \mathbf{SPECAN}(\mathfrak{A}) \to W$ contains therefore only one point P' and $\mathfrak{O}_{\mathbf{SPECAN}(\mathfrak{A}), P'}$ is $\mathfrak{O}_{Y, \varphi(P)}$-isomorphic to $\mathfrak{O}_{M, P}$. By l. b. there exist neighborhoods $\tilde{V} \subset V \cap M$ of P on M and $\tilde{V}'$ of P' on $\mathbf{SPECAN}(\mathfrak{A})$ with a surjective biholomorphic map $\tau : \tilde{V} \to \tilde{V}'$ such that $\varphi(\tilde{V}) = \pi_{\mathfrak{A}}(\tilde{V}')$ and that the diagram

$$\begin{array}{ccc} \tilde{V} & \xrightarrow{\tau} & \tilde{V}' \\ {\scriptstyle \varphi|\tilde{V}}\downarrow & & \downarrow{\scriptstyle \pi_{\mathfrak{A}}|\tilde{V}'} \\ & \pi_{\mathfrak{A}}(\tilde{V}') & \end{array}$$

is commutative. $\pi_{\mathfrak{A}}(\tilde{V}')$ is naturally a neighborhood of $\varphi(P)$ on the support S of $\mathfrak{A}$. As $\mathfrak{A}$ is coherent, S is analytic. It follows: P has on M a neighborhood $\tilde{V}$ such that $\varphi(\tilde{V})$ is analytic in $\varphi(P)$ and such that $\mathbf{dim}_P M = \mathbf{dim}_{\varphi(P)} \varphi(\tilde{V})$.

I want to show: The neighborhood V of P on X can be chosen such that $\varphi(V)$ coincides in a suitable neighborhood of $\varphi(P)$ with $\varphi(\tilde{V})$ and such that for all $Q \in V$ the relation $\mathbf{rk}(\varphi, Q, X) = \mathbf{rk}(\varphi, P, X)$ holds. We will further see that $\mathbf{dim}_P M = \mathbf{rk}(\varphi, P, X)$, i.e. that $\mathbf{dim}_{\varphi(P)} \varphi(\tilde{V})$ $(= \mathbf{dim}_{\varphi(P)} \varphi(V)) = \mathbf{rk}(\varphi, P, X)$ holds.

$q_1, \ldots, q_d$ define a holomorphic map $\gamma: V \to C^d$ (in the sense of l. f.). Let $\psi := (\varphi \mid V) \dot{\times} \gamma: V \to Y \times C^d$. By construction, P is an isolated point of $\psi^{-1}(\psi(P))$. If we denote by $\mathfrak{m}_{Y \times C^d, \psi(P)}$ the maximal ideal in $\mathfrak{O}_{Y \times C^d, \psi(P)}$ and by $\mathfrak{m}_{X,P}$ the maximal ideal in $\mathfrak{O}_{X,P}$, there exists a natural integer

$$t > 0 \quad \text{with} \quad \mathfrak{m}^t_{X,P} \subset \mathfrak{m}_{Y \times C^d, \psi(P)} \mathfrak{O}_{X,P} .$$

By l.a. and l.d. there exist a neighborhood of P on X (V be already this neighborhood), a neighborhood W^* of $\psi(P)$ on $Y \times C^d$, a coherent $\mathfrak{O}_{W^*}$-algebra $\mathfrak{A}^*$ with a map $\pi_{\mathfrak{A}^*}: \mathbf{SPECAN}(\mathfrak{A}^*) \to W^*$, which has only one point P^* in the fiber $\pi_{\mathfrak{A}^*}^{-1}(\psi(P))$, a neighborhood V^* of P^* on $\mathbf{SPECAN}(\mathfrak{A}^*)$ with a surjective biholomorphic map $\tau^*: V \to V^*$, $\tau^*(P) = P^*$, $\psi(V) \subset \pi_{\mathfrak{A}^*}(V^*)$, such that the diagram

$$\begin{array}{ccc} V & \xrightarrow{\tau^*} & V^* \\ \psi\downarrow & & \downarrow \pi_{\mathfrak{A}^*} | V^* \\ & \pi_{\mathfrak{A}^*}(V^*) & \end{array}$$

is defined, commutative and such that $\psi: V \to \pi_{\mathfrak{A}^*}(V^*)$ is proper. $\pi_Y: Y \times C^r \to Y$ be the projection map. We have $\varphi \mid V = \pi_Y \circ \psi$. The support S^* of $\mathfrak{A}^*$ is in W^* analytic. $\pi_{\mathfrak{A}^*}(V^*) = \psi(V)$ is on S^* a neighborhood of $\psi(P)$ and therefore in $\psi(P)$ analytic.

We have for $Q \subset V$ the relation $\mathbf{rk}(\varphi, Q, X) \leqq \mathbf{rk}(\varphi, P, X)$, i.e. $\mathbf{dim}_Q \varphi^{-1}(\varphi(Q)) \geqq \mathbf{dim}_P \varphi^{-1}(\varphi(P)) = d$. Because of $\varphi \mid V = \pi_Y \circ \psi$ clearly $\mathbf{dim}_Q \varphi^{-1}(\varphi(Q)) \leqq \mathbf{dim}_{\psi(Q)} \pi_Y^{-1}(\pi_Y(\psi(Q)) \leqq d$ holds; note that for every analytic set A in V the dimensions of A and of $\psi(A)$, which is analytic in $\psi(V)$, coincide (see l. d.). Therefore we have

$$\mathbf{dim}_Q \varphi^{-1}(\varphi(Q)) = \mathbf{dim}_P(\varphi^{-1}(\varphi(P))) , \quad \text{i.e.} \quad \mathbf{rk}(\varphi, Q, X) = \mathbf{rk}(\varphi, P, X) .$$

C_d be furnished with a $(z_1, \ldots, z_d)$-coordinate system. $\psi(M)$ is in $\psi(V) \subset Y \times C^d$ characterized by $z_1 = \cdots = z_d = 0$. As $\psi(M)$ intersects all fibers of π_Y through $\psi(V)$, the sets $\varphi(M) = \pi_Y(\psi(M))$ and $\varphi(V) = \pi_Y(\psi(V))$ coincide. Therefore $\varphi(V)$ is locally analytic Y.

It is easy to see that $\varphi(V)$ is in a neighborhood of $\varphi(P)$ the common zero set of finitely many sections in $\mathfrak{O}_Y$ which do not depend on $z_1, \ldots, z_d$. Therefore there exists a neighborhood V_1 of $\psi(P)$ on $\psi(V)$ which can be written as $V_1 = W_1 \times W_2$ where W_1 is a neighborhood of $\varphi(P)$ on $\varphi(V)$

and where W_2 is an open set in C^d. Therefore it is clear that $\mathbf{dim}_{\varphi(P)}\varphi(V) = \mathbf{dim}_{\psi(P)}\psi(V) - d = \mathbf{dim}_P V - d = n - d = \mathbf{rk}(\varphi, P, X)$.

Let us now drop the assumption that X is irreducible in P. There exists a neighborhood U of P with finitely many coherent ideal sheafs

$$\mathfrak{J}_1, \ldots, \mathfrak{J}_s \text{ in } \mathfrak{O}_U, \bigcap_{i=1}^{s} \mathfrak{J}_i = (0),$$

such that for $1 \leqq i \leqq s$ the support X_i of $\mathfrak{O}_X/\mathfrak{J}_i$ provided with the structure sheaf $\mathfrak{O}_{X_i} := \mathfrak{O}_X/\mathfrak{J}_i \,|\, X_i$ contains P and is irreducible in P. $X_1, \ldots, X_s$ are called the local components of X in P.

We say that φ does not degenerate in P if there exists a neighborhood $\tilde{U} \subset U$ of P on X such that for $1 \leqq i \leqq s$ $\mathbf{dim}_P[\varphi^{-1}(\varphi(P)) \cap X_i] \leqq \mathbf{dim}_Q[\varphi^{-1}(\varphi(Q)) \cap X_i]$ for $Q \in U$ holds.

We can easily extend our previous considerations to this case by considering the local components of X in P and we get the theorem:

Let X and Y be complex analytic spaces with a holomorphic map φ: $X \to Y$. *The set of points of X where φ does not degenerate is open and dense on X. If φ does not degenerate in $P \in X$, then there exists a neighborhood V of P such that $\varphi(V)$ is a local analytic subset of Y of dimension*

$$\mathbf{rk}(\varphi, P, X).$$ [1]

3. We use the notations of 2. The set E_φ of all points of X where φ degenerates is called the set of degeneration of φ.

We will prove: The set E_φ of degeneration of φ is analytic on X. Actually we prove the stronger theorem: *X and Y be complex analytic spaces with a holomorphic map $\varphi: X \to Y$. The set E_m of all points $Q \in X$ with* $\mathbf{dim}_Q \varphi^{-1}(\varphi(Q)) \geqq m$ *is analytic in X* ([*24*], [*25*]).

The proof will be given by induction on $\mathbf{dim}\, X$. [2] The trivial generalization of this proof for arbitrary algebraically closed, complete, non-discrete valued ground fields k holds only for **char** $(k) = 0$. By transition to a suitable map with the same fibers the generalization for **char** $(k) \neq 0$ is possible. But I will not discuss this. For $\mathbf{dim}\, X = 0$ nothing is to prove. Let us assume that the above-stated theorem is true for spaces of a dimension $\leqq n$.

Evidently it is enough to discuss the case where X is normal and where Y is the space C^r. Let us denote the coordinate functions in C^r by z_i, $1 \leqq i \leqq r$, and the lifted functions $\varphi^*(z_i)$ by f_i.

Let P be a point of X. There exists a neighborhood V of P, an open set U in C^n, $n = \mathbf{dim}_P X$, with a surjective proper holomorphic map $\pi: V \to U$

[1] After the submission of this paper I learned that a similar proof has been produced by Mr. Kiehl, Heidelberg.

[2] The present proof is essentially the same as in [*24*], [*25*]. The only difference is that (in order to be as general as possible) I avoid an application of the continuation theorem [*28*].

whose fibers are discrete and which is locally biholomorphic outside the pure 1-codimensional analytic set B of branch points of π; $\mathfrak{O}_{X,P}$ is a separable algebraic extension of $\mathfrak{O}_{C^n,\pi(P)}$. (This situation can also be arranged in the case of a general ground field k as described above; $\mathbf{char}(k) = 0$ is unnecessary; see [*10*], Exposé 21).

Let the coordinate functions in U be $x_1, \ldots, x_n$, the lifted functions $\pi^*(x_1), \ldots, \pi^*(x_n)$. These provide a manifold parameter system in the points of $X - B$. Here the partial derivatives $\partial/\partial(\pi^*(x_i))$, $1 \leqq i \leqq n$, are well defined. As $\mathfrak{O}_{X,P}$ is separably generated over $\pi^*(\mathfrak{O}_{C^n,\pi(P)})$, it is possible to extend these derivations to derivations in a whole neighborhood of P. Here $\pi^*(\mathfrak{O}_{C^n,\pi(P)})$ is the isomorphic image of $\mathfrak{O}_{C^n,\pi(P)}$ in $\mathfrak{O}_{X,P}$. Let the quotient field of $\mathfrak{O}_{X,P}$ be generated over the quotient field of $\pi^*(\mathfrak{O}_{C^n,\pi(P)})$ by the element $\vartheta \in \mathfrak{O}_{X,P}$ which fulfills an equation $\vartheta^s + a_1\vartheta^{s-1} + \cdots + a_s = 0$ with coefficients

$$a_i \in \pi^*(\mathfrak{O}_{C^n,\pi(P)}), \quad 1 \leqq i \leqq s.$$

$a_1, \ldots, a_s, \vartheta$ are induced by functions which are holomorphic in a neighborhood of $\pi(P)$. Let us use for these functions the same notations as for the germs they induce. Define

$$\frac{\partial\vartheta}{\partial(\pi^*(x_i))} = \left(\sum_{\sigma=1}^{s}\left(\frac{\partial}{\partial(\pi^*(x_i))}\, a_\sigma\right)\vartheta^{s-\sigma}\right) \Big/ \left(\sum_{\sigma=1}^{s}(s-\sigma)\,a_\sigma\,\vartheta^{s-\sigma-1}\right).$$

This enables us to define $\dfrac{\partial}{\partial(\pi^*(x_i))}$ in a whole neighborhood of P. Let $Q \in V - B$ be a point with a neighborhood $V_Q \subset V - B$ such that in all points of V_Q the rank of the jacobian $(\partial f_i/\partial(\pi^*(x_j))$ is equal or less than the rank in Q; we express this situation by saying that the jacobian has locally in Q maximal rank. If the jacobian has in $Q \in V - B$ locally maximal rank then there exists a neighborhood such that in all points of this neighborhood the jacobian has the same rank as in Q and φ does not degenerate in Q.

The set E^0 of points of $V - B$ where the jacobian does not have locally maximal rank is clearly analytic in $V - B$. As the functions $\dfrac{\partial}{\partial(\pi^*(x_i))} f_i$ are meromorphic in a neighborhood of P (let us assume, that V already is such a neighborhood), we can extend E^0 to an analytic set E of V (apply the first two theorems which are mentioned in 4) which is thin in V, i.e. whose complement $V - E$ is dense in V. Application of the preceding paragraph and the induction hypothesis to $\pi|E: E \to Y$ delivers the analyticity of $E_m \cap V$. Because if $\pi|V$ degenerates in a point then this point must necessarily lie in E.

4. Some more or less technical results are needed (see [*25*], especially § 6). Throughout this chapter X be a reduced complex analytic space with the structure sheaf $\mathfrak{O}_X$. Let $\mathfrak{M}_X$ be the sheaf of quotient rings of $\mathfrak{O}_X$.

A simple, but very useful result is: Let *M and G be analytic sets in X. Then the closed hull* $\overline{(X-M)\cap G}$ of $(X-M)\cap G$ *is analytic in X* (compare [25], p. 365 and my paper Projektive Modifikationen komplexer Räume, Math. Ann. **130**, 217—238 (1960), p. 224. section b).

The following two theorems are consequences: *Let N be a thin analytic set in X, f a meromorphic function on X, i. e. a section in* $\Gamma(X,\mathfrak{M}_X)$ *which is a holomorphic function in* $X-N$, Z_0 *be the set of zeros of f in* $X-N$. *Then the closure* $\bar{Z}_0$ *of* Z_0 *in X is analytic in X.* — $f_1,\ldots,f_r$ be meromorphic functions on X, which are holomorphic functions outside the thin analytic set N. These functions define a holomorphic map $\mu: X-N \to \overset{r}{\underset{1}{\times}} \boldsymbol{P}^1$ of $X-N$ into the r-fold projective space $\overset{r}{\underset{1}{\times}} \boldsymbol{P}^1$ which maps $P\in X-N$ onto $\{(1,f_1(P);\ldots;\{1,f_r(P))\}\in \overset{r}{\underset{1}{\times}}\boldsymbol{P}^1$.Let $G_\mu:=\{(P,\mu(P))\mid P\in X-N\}$ be the graph of μ, a set which is analytic in $(X-N)\times\left(\overset{r}{\underset{1}{\times}}\boldsymbol{P}^1\right)$. Then *the closed hull* $G_{f_1,\ldots,f_r}$ *of* G_μ *in* $X\times\left(\overset{r}{\underset{1}{\times}}\boldsymbol{P}^1\right)$ *is analytic in* $X\times\left(\overset{r}{\underset{1}{\times}}\boldsymbol{P}^1\right)$ and independent of the special thin analytic set N which we choose outside of which $f_1,\ldots,f_r$ are holomorphic. The projection map π_X: $G_{f_1,\ldots,f_r}\to X$ is biholomorphic over $X-N$. (See [25] and my paper mentioned at the end of the previous section.) We say that $f_1,\ldots,f_r$ define a meromorphic map of X into $\overset{r}{\underset{1}{\times}}\boldsymbol{P}^1$ and call $G_{f_1,\ldots,f_r}$ the graph of this meromorphic map. Every biholomorphic map $\gamma: X\to X$ of X onto itself induces a sheaf homomorphism $\gamma^*:\gamma^{-1}(\mathfrak{M}_X)\to\mathfrak{M}_X$; here $\gamma^{-1}(\mathfrak{M}_X)$ shall denote the topological inverse image of $\mathfrak{M}_X$. A meromorphic function $f\in\Gamma(X,\mathfrak{M}_X)$ is called γ-automorphic or γ-invariant if $\gamma^*(\gamma^{-1}(f))=f$. Let me note that $\gamma^*(\gamma^{-1}(f))=f\circ\gamma$ if f is holomorphic.

Let Γ be a group $\{\gamma\}$ of biholomorphic maps $\gamma: X\to X$. f is called Γ-automorphic or Γ-invariant if f is γ-invariant for all $\gamma\in\Gamma$.

The following is clear: *If* $f_1,\ldots,f_r$ *are* Γ*-automorphic then all* $\gamma\in\Gamma$ *can be lifted to biholomorphic maps* $\tilde{\gamma}: G_{f_1,\ldots,f_r}\to G_{f_1,\ldots,f_r}$.

If f is a meromorphic function on X and G_f the graph of the meromorphic map of X into $\boldsymbol{P}^1$ defined by f with the projections $\pi_X: G_f\to X$, $\pi: G_f\to\boldsymbol{P}^1$, we denote for $D\subset X$ by $f(D)$ the set

$$\pi(\pi_x^{-1}(D))\subset\boldsymbol{P}^1 .$$

We speak of G_f as the graph of f. A point $P\in X$ is called an indeterminate point of f if π_X is not biholomorphic over a neighborhood of P.

If P is a normal point of X then f is indeterminate in P if and only if $f(P)=\boldsymbol{P}^1$.

It is not difficult to see that *the set E of indeterminate points of f is analytic on X.*

Let Z be a complex analytic space with a holomorphic map $\tau: Z \to X$. $\tau^{-1}(E)$ *be thin on* Z, i.e., $Z - \tau^{-1}(E)$ dense on Z. *We can lift f in a natural way to a meromorphic function on Z, denoted by $f \circ \tau$.* As $\tau^{-1}(E)$ lies thin on Z, there exists for each $P \in \tau(Z)$ a neighborhood with a representation f_1/f_2 of f in U for which $f_2 \circ (\tau | \tau^{-1}(U_p))$ is a non-zero divisor in each point of $\tau^{-1}(U_p)$. Therefore $f_1 \circ (\tau | \tau^{-1}(U_p))/f_2 \circ (\tau | \tau^{-1}(U_p))$ is a well defined meromorphic function in $\tau^{-1}(U_p)$ which represents here $f \circ \tau$. If Z is a local-analytic subset of X and $\tau: Z \to X$ the canonical injective mapping we denote $f \circ \tau$ also by $f | Z$.

§ 2

All complex analytic spaces in this and the following paragraph are reduced complex analytic spaces.

a. A proof of the following special case of a theorem in [*19*] shall be presented: *Let $\tau: X \to Y$ be a holomorphic map of the irreducible complex analytic space X in the complex analytic space Y. Let there be a compact set $X^1 \subset X$ with $\tau(X^1) = \tau(X)$. Then $\tau(X)$ is analytic in Y and of dimension* $\mathbf{rk}(\tau, X)$.

Proof. Let E_τ be the (analytic) set of degeneration of τ, $F = \tau(E_\tau \cap X^1)$. F is closed in Y because $E_\tau \cap X^1$ is compact. First we prove that $\tau(X)$ is analytic in the points $P \in \tau(X) - F$. As $\tau^{-1}(P) \cap X^1$ is compact and as E_τ is closed in X there exists a neighborhood of $\tau^{-1}(P) \cap X^1$ where τ does not degenerate. $\tau^{-1}(P) \cap X^1$ can therefore be covered by finitely many open sets W_i, $i = 1, \ldots, s$, whose images $\tau(W_i)$ are analytic in P (by § 1.2.). $W := \bigcup_{i=1}^{s} \tau(W_i)$ coincides in a certain neighborhood of P with $\tau(X)$ and has in P the dimension $\mathbf{rk}(\varphi, X) := \sup_{Q \in X} \mathbf{rk}(\varphi, Q, X)$ by § 1.2. Therefore $\tau(X)$ is analytic in P. As P is an arbitrary point of $(X) - F$ and as $\tau(X) - F = \tau(X^1) \cap (Y - F)$ is closed in $Y - F$ the set $\tau(X) - F$ is analytic in $Y - F$ and pure-dimensional. Let us assume $E_\tau \cap X^1 \neq \emptyset$. E_{τ_2} be the set of degeneration of

$$\tau_2 := \tau | E_\tau, \quad F_2 := \tau(E_{\tau_2} \cap X^1).$$

F_2 is closed in Y.

Let T be a point of $F - F_2$. $\tau_2^{-1}(T) \cap X^1$ is compact and can therefore by § 1.2. be covered by a finite number of sets $W_1^1, \ldots, W_{t_1}^1$ which are open on E_τ, have no point in common with E_{τ_2} and whose images $\tau(W_i^1)$, $1 \leq i \leq t_1$, are analytic in T. I.e. there exists a neighborhood $U_1 \subset Y - F$ of T such that $W_1 := \bigcup_{i=1}^{t_1} \tau(W_i^1)$ is analytic in U. By § 1.2. the relation

$$\dim_T W_1 = \sup_{Q \in \tau_2^{-1}(T) \cap X^1} \mathbf{rk}(\tau_2, Q, E_\tau) < \mathbf{rk}(\tau, X) = \dim(\tau(X) \cap (Y - F))$$

holds. Application of the extension theorem for analytic sets of REMMERT and STEIN [*28*] delivers the analyticity of $\tau(X) \cap U_1$ in U_1. Therefore $\tau(X) \cap (Y - F_2)$ is analytic in $Y - F_2$. Now one considers the set of degeneration E_{τ_3} of $\tau_3 := \tau | E_{\tau_2}$ and so on. As $\dim E_\tau > \dim E_{\tau_2} > \dim E_{\tau_3}, \ldots$, this procedure stops after a finite number of steps.

b. Let X be as before an irreducible complex analytic space, Γ be a group $\{\gamma\}$ of biholomorphic maps $\gamma: X \to X$ of X onto itself, $f_1, \ldots, f_2$ Γ-automorphic meromorphic functions, let $G_{f_1, \ldots, f_r}$ be the graph of the meromorphic map of X into $\overset{r}{\underset{1}{\times}} \boldsymbol{P}^1$, $\pi_X : G_{f_1, \ldots, f_r} \to X_1$, and $\pi_r : G_{f_1 \ldots f_r} \to \overset{r}{\underset{1}{\times}} \boldsymbol{P}^1$ the projections.

Assume that on X there exists a compact set X^1 such that every point of X is Γ-equivalent to at least one point of X^1. Γ^1 be the group $\{\gamma^1\}$ of biholomorphic maps γ^1: $G_{f_1 \ldots f_r} \to G_{f_1 \ldots f_r}$, which we get by lifting the elements $\gamma \in \Gamma$. Then every point of $G_{f_1 \ldots f_r}$ is Γ^1-equivalent to at least one point of $G^1 = \pi_X^{-1}(X^1)$ and consequently $\pi_r(G^1)$ and $\pi_r(G_{f_1 \ldots f_r})$ coincide. π_X is proper; therefore G^1 is compact and a. delivers the analyticity of $\pi_r(G_{f_1 \ldots f_r})$ in $\overset{r}{\underset{1}{\times}} \boldsymbol{P}^1$. By CHOW's theorem [*11*] $\pi_r(G_{f_1 \ldots f_r})$ must be an algebraic subset of $\overset{r}{\underset{r}{\times}} \boldsymbol{P}^1$.

This implies: *If* $\mathbf{rk}(\pi_{r_1}, G_{f_1 \ldots f_r}) < r$, *then* $\pi_r(G_{f_1 \ldots f_r})$ *is a proper algebraic subset of the projective space* $\overset{r}{\underset{1}{\times}} \boldsymbol{P}^1$. Consequently there exists a polynomial $q(z_1, \ldots, z_r) \in \boldsymbol{C}[z_1, \ldots, z_r]$, $q \neq 0$, such that for all points $P \in X$, where $f_1, \ldots, f_r$ are simultaneously holomorphic functions, $q(f_1(P), \ldots, f_r(P)) = 0$ holds.

One calls $f_1, \ldots, f_r$ algebraically dependent if the latter is fulfilled. $f_1, \ldots, f_r$ are called analytically dependent if $\mathbf{rk}(\pi_r, G_{f_1 \ldots f_r}) < r$. Using these notations the result above can be expressed as: *Let X be an irreducible complex analytic space, Γ a group of biholomorphic maps of X onto itself. Let there be a compact set X^1 in X such that every point of X is Γ-equivalent to at least one point in X^1. Then Γ-automorphic meromorphic functions are analytically dependent if and only if they are algebraically dependent.* This method, invented by REMMERT in [*26*], allows to prove more: *The field $\mathbf{K}^\Gamma(X)$ of Γ-automorphic functions on X is an algebraic function field.* But I will not discuss this.

§ 3

The following considerations depend strongly on the theory of analytic decomposition of K. STEIN as developed in [*18, 34, 35, 36*].

1. Let X, Y, Z be complex analytic spaces with holomorphic mappings $\varphi: X \to Y$, $\varphi_0: X \to Z$.

One says φ_0 depends analytically on φ if for all $x \in X$ the rank $\mathbf{rk}(\varphi, x, X)$ of φ in x coincides in x with the rank $\mathbf{rk}(\varphi \dot{\times} \varphi_0, x, X)$ of the map $\varphi \dot{\times} \varphi_0: X \to X \times Z$ which maps $y \in X$ onto $(\varphi(y), \varphi_0(y)) \in X \times Z$. In other words: φ_0 depends analytically on φ if and only if φ_0 is constant on all connected fiber components of φ.

φ and φ_0 are called related if and only if φ depends analytically on φ_0 and φ_0 depends analytically on φ.

The pair (φ_0, Z) is called a complex base and Z a complex base space corresponding to φ if the following is fulfilled: φ_0 is surjective, φ_0 and φ are related; if $\psi: X \to V$ is a holomorphic map into a complex space V which depends analytically on φ then there exists a holomorphic map $\tilde{\psi}: Z \to V$ such that the diagram

$$\begin{array}{ccc} X & \xrightarrow{\psi} & V \\ & {\scriptstyle\varphi_0}\searrow \quad \nearrow{\scriptstyle\tilde{\psi}} & \\ & Z & \end{array}$$

is commutative.

We need the main theorem of [*36*]: *let X be irreducible and normal, $\varphi: X \to Y$ a holomorphic map which degenerates nowhere. Then there exists a complex base (φ_0, Z); Z is normal and pure r-dimensional where* $r = \mathbf{rk}(\varphi, X)$. This theorem has been extended to the non-normal case in [*37*]; but this sharper result will not be needed.

2. X, Y, Z, φ, φ_0 shall have the same meaning as in the theorem mentioned above. Let g be a meromorphic function on X. It is known that the set E of points (so called indeterminate points of g) where neither g nor $1/g$ are holomorphic functions is analytic on X. $g \mid X - E$ defines a holomorphic map of $X - E$ into $\mathbf{P}^1$. g is called analytically dependent on φ if $g \mid X - E$ depends analytically on $\varphi \mid X - E$.

We need: *if g depends analytically on φ then g can be taken down to a meromorphic function on Z, i.e., on Z there exists a meromorphic function $\tilde{g}$ with $\tilde{g} \circ \varphi_0 = g$.*

The proof uses a continuation theorem for meromorphic functions, the continuation theorem for proper analytic coverings in [*35*] and depends further on an analysis of the proof in [*36*].

3. Let X be an irreducible complex analytic space, Γ a group of biholomorphic maps of X onto itself, and $f, f_1, \ldots, f_r$ Γ-automorphic meromorphic functions on X, $f_1, \ldots, f_r$ being analytically independent, but $f, f_1, \ldots, f_r$ analytically dependent.

Let X_0 be the maximal (open) set of manifold points $P \in X$ with the following property: $f_1, \ldots, f_r$ are holomorphic functions in a neighborhood U_P of P and the map $\mu_P: U_P \to \mathbf{C}^r$, defined by $\mu_P(Q) = (f_1(Q),$

$\dots, f_r(Q)) \in C^r$ for $Q \in U_P$, has in P a jacobian of rank r. In other words: X_0 is the maximal set of manifold points $P \in X$, where $f_1, \dots, f_r$ form a subsystem of a manifold coordinate system. In these manifold coordinate systems the partial derivatives $\partial f/\partial f_i$, $1 \leqq i \leqq r$, are defined. As f depends analytically on $f_1, \dots, f_r$, these partial derivatives $\partial f/\partial f_i$ are independent of the special completion of $f_1, \dots, f_r$ to a local manifold coordinate system and therefore well defined meromorphic functions on X_0.

We need: $\partial f/\partial f_i$, $1 \leqq i \leqq r$, *can be extended to a meromorphic function on* X *which is* Γ*-automorphic.*

The proof employs a technique which is used in algebraic geometry to extend derivations of fields to field extensions.

4. Let X be an irreducible, complex analytic space with the structure sheaf 0_X, Γ a group of biholomorphic mappings of X onto itself, and $\boldsymbol{K}^\Gamma(X)$ the field of Γ-automorphic meromorphic functions on X.

Assumption: there exists an open relatively compact subset $D \subset\subset X$ such that for all nonconstant $f \in \boldsymbol{K}^\Gamma(X)$ the relation $f(\bar{D}) = \boldsymbol{P}^1$ holds; here $\bar{D}$ is the closure of D in X. — Such a set D exists for instance if there exists a compact subset D' of X such that all points of X are Γ-equivalent to at least one point of D'. We need the

Theorem: *This assumption is also fulfilled in the case of a pseudoconcave group* Γ *and a normal complex analytic space* X.

Let me remind the reader (see [1], [3]): Γ is called pseudoconcave if there exists an open, relatively compact set $D \subset\subset X$ such that every point $P \in \partial D$ is Γ-equivalent to an interior point $P' \in D$ or to a pseudoconcave boundary point $P' \in \partial D$ of D. A point $P' \in \partial D$ is called a pseudoconcave boundary point of D if there exists a fundamental system of neighborhoods $\mathfrak{A} = \{U\}$ of P' such that for all

$$U \in \mathfrak{A} \quad \text{the set} \quad \{P \in U \,|\, |f(P)| \leqq \sup_{Q \in U \cap D} |f(Q)| \quad \text{for every} \quad f \in \Gamma(U, \mathfrak{O}_X)\}$$

contains a neighborhood of P'.

The proof of the theorem is not difficult: let D be such an open, relatively compact set in X, which characterizes the pseudoconcavity of the group Γ. Let f be a non-constant meromorphic function in $\boldsymbol{K}^\Gamma(X)$. — Assume that f has an indeterminate point Q in the closure $\bar{D}$ of D. Then $f(Q) = \boldsymbol{P}^1$ and we have a fortiori $f(\bar{D}) = \boldsymbol{P}^1$.

Therefore it is enough to give a proof in the case where f has no indeterminate points in a neighborhood $\tilde{D}$ of D, i. e. where f defines a holomorphic mapping of $\tilde{D}$ into $\boldsymbol{P}^1$. Assume that $f(\bar{D}) \neq \boldsymbol{P}^1$. Without restriction we can assume $0 \in f(\bar{D})$, $f(\bar{D}) \subset C$. $f(\bar{D})$ is compact and therefore closed and bounded. There exists a circle ∂K_ϱ of largest radius ϱ with center in 0 and $\partial K_\varrho \cap f(\bar{D}) \neq \emptyset$. There exists a point $P \in D$ with $f(P) \in \partial K_\varrho \cap f(\bar{D})$. P must necessarily be a pseudoconcave boundary point of D. Because otherwise P would be Γ-equivalent to an interior point P_1 of D and a whole neighborhood of P_1 would be mapped onto a neighborhood of $f(P)$, as $f|\tilde{D}$ is by § 1.2. an open map. In this case ∂K_ϱ cannot be the circle with largest radius ϱ with center in 0 and $\partial K_\varrho \cap f(\bar{D}) \neq \emptyset$.

As P is a pseudoconcave boundary point of D there exists a neighborhood U of P, such that $\{P' \in U \,|\, |f(P')| \leqq \sup_{Q \in U \cap D} |f(Q)|\}$ contains a neighborhood U_1 of P. This means that the open set $f(U_1)$, a neighborhood of $f(P)$, lies in the interior of the circle ∂K_ϱ. But this cannot be as $f(P) \in \partial K_\varrho$. Therefore the assumption $f(\bar{D}) \neq \boldsymbol{P}^1$ is wrong.

5. X, Γ, $\boldsymbol{K}^\Gamma(X)$ have the same meaning as in 4. The assumption in 4. be fullfilled, i. e. there exists a compact subset D of X such that for all nonconstant $f \in \boldsymbol{K}^\Gamma(X)$ the relation $f(\bar{D}) = \boldsymbol{P}^1$ holds.

X^* be the normalisation of X with the normalisation map $\tilde{\pi}^*$: $X^* \to X$. — It can easily be seen that each biholomorphic map $\gamma: X \to X$ can be lifted to a biholomorphic map $\gamma^*: X^* \to X^*$. Γ^* be the group of biholomorphic maps one gets by lifting the biholomorphic maps of Γ to X^*. $\boldsymbol{K}^\Gamma(X)$ is in a natural way isomorphic to $\boldsymbol{K}^{\Gamma^*}(X^*)$, the field of all Γ^*-automorphic meromorphic functions on X^*. — For each nonconstant $f^* \in \boldsymbol{K}^{\Gamma^*}(X^*)$ the relation $f^*(\tilde{\pi}^{*-1}(\bar{D})) = \boldsymbol{P}^1$ holds. — *This justifies the restriction to normal spaces from now on.*

X be normal, $f_1, \ldots, f_r \in \boldsymbol{K}^\Gamma(X)$ be a maximal set of analytically independent meromorphic functions of $\boldsymbol{K}^\Gamma(X)$ ($r \leq \dim X$). Because of 1.4. we can assume without any restriction that these functions have no indeterminate points. $f_1, \ldots, f_r$ define a holomorphic map $\mu: X \to \mathop{\times}\limits_1^r \boldsymbol{P}^1$. E_μ be the (analytic) set of degeneration of μ. E_1 be the (analytic) set of points where E_μ is not singularity free in X. As X is normal, the set S of singular points of X is at least 2-codimensional. *Therefore there exists a neighborhood $\tilde{D}$ of $E_1 \cup S$ such that for all non-constant functions $f \in \boldsymbol{K}^\Gamma(X)$ the relation $f(\bar{D} - \tilde{D}) = \boldsymbol{P}^1$ holds.* — Here we used a theorem of Grauert and Remmert [*13*]: *M be a pure k-dimensional analytic set of the complex-analytic space Y with countable basis. Then there exists a neighborhood $V(M)$ of M, which does not contain a $(k+1)$-dimensional analytic set of Y.*

6. $\mu | X - E_\mu$ is non-degenerate. $(\mu_0, \tilde{X}_0)$ be the corresponding complex base which exists (see [*36*]). By 2. the restricted functions $f|X_0$, $X_0 := X - E_\mu$, $f \in \boldsymbol{K}^\Gamma(X)$, can be taken down to $\tilde{X}_0$. $\tilde{\boldsymbol{K}}_0$ be the field of meromorphic functions we get this way on $\tilde{X}_0$.

There exists a holomorphic map $\mu_0: X_0 \to \mathop{\times}\limits_1^r \boldsymbol{E}^1$ with discrete fibers such that the diagram

$$\begin{array}{ccc} X_0 & \xrightarrow{\mu|X_0} & \overset{r}{\underset{1}{\times}} \boldsymbol{P}^1 \\ & {\scriptstyle \mu_0} \searrow \quad \nearrow {\scriptstyle \tilde{\mu}_0} & \\ & \tilde{X}_0 & \end{array}$$

is commutative.

7. Before proceeding let us compile some further tools.

a. $X, \Gamma, \boldsymbol{K}^\Gamma(X), D$ shall have the same meaning as in 6, f be a function in $\boldsymbol{K}^\Gamma(X)$ with the graph G_f and the projection maps $\pi_X: G_f \to X$, $\pi: G_f \to \boldsymbol{P}^1$. Let $\pi^*: G_f^* \to G_f$ be the normalization map, (τ_f, Y_f) the complex base corresponding to $\pi \circ \pi^*$. Y_f is a Riemann surface. We need the Theorem: *Y_f is compact.*

Assume Y_f were not compact. Then by [*12*] (see also [*5*]) there exists a non-constant holomorphic function g on Y_f. $g \circ \tau$ is a non-constant holomorphic function on G_f^* which can be taken down to a non-constant meromorphic function $g_1 \in \boldsymbol{K}^\Gamma(X)$ on X. But g clearly violates the condition $g_1(\bar{D}) = \boldsymbol{P}^1$ because $g_1(\bar{D}) \subset \boldsymbol{C}$.

b. The following theorem of Hadamard [*18*] will be used: *let*

$$h(w) = \sum_0^\infty h_j (w-a)^j$$

be holomorphic at $w = a$, *and let*

$$D_j^\mu(h;a) := \begin{vmatrix} h_j & \dots & h_{j+\mu} \\ \cdot & \cdot\,\cdot\,\cdot & \cdot \\ h_{j+\mu} & \dots & h_{j+2\mu} \end{vmatrix}$$

Then $M^+ := (\lim_{\mu \to \infty} (\limsup_{j \to \infty} |D_j^\mu(h;a)|^{1/j})^{1/\mu})^{-1}$ *is the radius of meromorphy.*

8. Now we continue 6. f be a function of $\tilde{\boldsymbol{K}}_0$. $\overset{r}{\underset{1}{\times}} \boldsymbol{P}^1$ be furnished with an inhomogeneous $(z_1, \dots, z_r)$-coordinate system, $K_{a_1 \dots a_{i-1} a_{i+1} \dots a_r}$ be the line given by

$$z_1 = a_1, \dots, z_{i-1} = a_{i-1}, \quad z_{i+1} = a_{i+1}, \dots, z_r = a_r.$$

$\tilde{K}_{a_1 \dots a_{i-1} a_{i+1} \dots a_r}$ be the lifted curve

$$\tilde{\mu}_0^{-1}(K_{a_1 \dots a_{i-1} a_{i+1} \dots a_r}), \quad 1 \leqq i \leqq r.$$

We can arrange that for a given f_0 the restrictions $f_0 | K_{a_1 \dots a_{i-1} a_{i+1} \dots a_r}$ are always well defined.

With the help of 3, 7a, 7b and [*34*], [*36*] one can prove that

$$f_0 | K_{a_1 \dots a_{i-1} a_{i+1} \dots a_r}$$

is always algebraic over z_i (let us identify z_i and $z_i \circ \tilde{\mu}_0$). This implies that f is algebraic over $z_1, \dots, z_r$. It is not too difficult to see that the degree of f over the field $C(z_1, \ldots, z_r)$ in the r indeterminates $z_1, \ldots, z_r$ is bounded.

It follows: $\tilde{K}_0$ is an algebraic function field of transcendence degree r over C. But $\tilde{K}_0$ is isomorphic to $K^\Gamma(X)$, *therefore* $K^\Gamma(X)$ *is an algebraic function field with transcendence degree r over* C.

References

[1] Andreotti, A.: Complex pseudoconcave spaces and automorphic functions. Proc. Intern. Congr. Math., 306—308 (1962).

[2] — Théorèmes de dépendence algébrique sur les espaces complexes pseudoconcaves. Bull. Soc. Math. France **91**, 1—38 (1963).

[3] —, and H. Grauert: Algebraische Körper von automorphen Funktionen. Nachr. Akad. Wiss. Göttingen Math.-phys. Kl. II, 39—48 (1961).

[4] Baily, W. L.: Satake compactification of V^*. Amer. J. Math. **80**, 348—364 (1958).

[5] Behnke, H., and F. Sommer: Theorie der analytischen Funktionen einer komplexen Veränderlichen. Berlin-Göttingen-Heidelberg: Springer 1962.

[6] Bishop, E.: Mappings of partially analytic spaces. Amer. J. Math. **83**, 209.

[7] — Partially analytic spaces. Amer. J. Math. **83**, 669—692 (1961).

[8] Bochner, S., and W. T. Martin: Several complex variables. Princeton 1948.

[9] Cartan, H.: Idéaux et modules de fonctions analytiques de variables complexes. Bull. Soc. Math. France, **78**, 28—64 (1950).

[10] — Séminaire Henri Cartan, 13e année, 1960/61.

[11] Chow, W. L.: On compact analytic varieties. Amer. J. Math. **71**, 893—914 (1949).

[12] Florack, H.: Reguläre und meromorphe Funktionen auf nicht geschlossenen Riemannschen Flächen. Schriften a.d. Math. Inst. d. Univ. Münster, Heft 1 (1948).

[12a] Grauert, H.: Ein Theorem der analytischen Garbentheorie. Inst. Hautes Études Sci. Publications Mathématiques, 1—64, No. 5 (1960).

[13] —, and R. Remmert: Zur Theorie der Modifikationen I. Stetige und eigentliche Modifikationen komplexer Räume. Math. Ann. **128**, 274—296 (1955).

[14] — — Komplexe Räume. Math. Ann. **136**, 245—318 (1958).

[15] — — Bilder und Urbilder analytischer Garben. Ann. Math. **68**, 393—443 (1958).

[16] —, and H. Kerner: Deformationen von Singularitäten komplexer Räume. Math. Ann. **153**, 236—260 (1963).

[17] Hadamard, J.: Essai sur l'étude des fonctions données par leurs développements de Taylor. Journ. Math. Pures Appl. (4), 8, 101—186 (1892).

[18] Koch, K.: Die analytische Projektion. Schriftenreihe des Math. Inst. d. Univ. Münster, Heft 8 (1954).

[19] Kuhlmann, N.: Über holomorphe Abbildungen komplexer Räume. Arch. Math. **15**, 81—90 (1964).

[20] — Algebraische Funktionenkörper auf komplexen Räumen. (Will appear in the Weierstrass-Festband.)

[21] Nagata, M.: Local rings. Interscience Tracts in Pure and Applied Mathematics, Number 13 (1962).

[22] Oka, K.: Sur les fonctions analytiques de plusieurs variables VII. Bull. Soc. Math. France 78, 1—27 (1950).

[24] Remmert, R.: Projektionen analytischer Mengen. Math. Ann. **130**, 410—441 (1956).
[25] — Holomorphe und meromorphe Abbildungen komplexer Räume. Math. Ann. **132**, 328—370 (1957).
[26] — Meromorphe Funktionen in kompakten komplexen Räumen. Math. Ann. **132**, 277—288 (1956).
[27] — Analytic and algebraic dependence of meromorphic functions. Amer. J. Math. **82**, 891—899 (1960).
[28] —, and K. Stein: Über die wesentlichen Singularitäten analytischer Mengen. Math. Ann. **126**, 263—306 (1953).
[29] — — Eigentliche holomorphe Abbildungen. Math. Zschr. **73**, 159—189 (1960).
[30] Rossi, H.: Analytic spaces with compact subvarieties. Math. Ann. **146**, 129—145 (1962).
[31] — Attaching analytic spaces to analytic spaces along a pseudoconcave boundary. Mimeographed notes.
[31a] Satake, I.: On the compactification of the Siegel space. J. Ind. math. Soc. (new series) **20**, 259—281 (1956).
[31b] Serre, J. P.: Fonctions automorphes, Exp. II in Sém. E.N.S., 1953—1954.
[32] Siegel, C. L.: Meromorphe Funktionen auf kompakten analytischen Mannigfaltigkeiten. Nachr. Akad. Wiss.Göttingen Math.-phys. Kl., math.-phys.-chem. Abt., 71—77 (1955).
[33] — Über die algebraische Abhängigkeit von Modulfunktionen n-ten Grades. Nachr. Akad. Wiss. Göttingen Math.-phys. Kl. II, 257—272 (1960).
[34] Stein, K.: Analytische Projektion komplexer Mannigfaltigkeiten. Coll. sur les fonct. de plus. var., Brussels 1953, 97—107.
[35] — Analytische Zerlegungen komplexer Räume. Math. Ann. **132**, 63—93 (1956).
[36] — Die Existenz komplexer Basen zu holomorphen Abbildungen. Math. Ann. **136**, 1—8 (1958).
[37] — Maximale holomorphe und meromorphe Abbildungen I. Amer. J. Math. **85**, 298—315 (1963).
[38] Wolffhardt, K.: Existenzbedingungen für maximale holomorphe und meromorphe Abbildungen. Thesis, Munich, 1963.

Department of Mathematics
University of Notre Dame
Notre Dame, Ind.

Riemann Surfaces with the Absolute AB-maximum Principle*, **

By

H. L. Royden

Let W be an open Riemann surface. We say that W satisfies the absolute AB-maximum principle if every subregion V of W with compact relative boundary Γ has the property that each function which is

* This research sponsored by the Army Research Office (Durham) under Project 1323 M.
** Received June 3, 1964.

bounded and analytic on $V \cup \Gamma$ assumes its maximum on Γ. Henceforth, we shall restrict ourselves to subregions whose relative boundaries consist of a finite number of piecewise analytic curves.

If f is a function analytic on $V \cup \Gamma$, then $f[\Gamma]$ consists of a number of piecewise analytic curves in the plane. If $z \notin f[\Gamma]$ we define the valence $\nu(z)$ of f at z to be the number of points $p \in V$ at which $f(p) = z$, and define the index $i(z)$ to be $(2\pi)^{-1}\int\limits_{\Gamma} d \arg(f(p) - z)$. Thus, if $V \cup \Gamma$ is compact, we have $\nu(z) = i(z)$. For points $z \in f[\Gamma]$ which are not images of points where Γ is not differentiable (corners), we define $\nu(z)$ as the number of $p \in V$ with $f(p) = z$ plus one-half the number of $p \in \Gamma$ with $f(p) = z$. If we make a suitable definition of the fractional multiplicity with which we count points on Γ where Γ is not differentiable, we still have $\nu(z) = i(z)$ when $V \cup \Gamma$ is compact. Let us call $\delta(z) = i(z) - \nu(z)$ the deficiency of f at z with respect to the region V. If V and V' are two subregions of W with compact relative boundaries and if V differs from V' by a set with compact closure, then the deficiency of f at a point z is the same with respect to V as it is with respect to V'.

A closed set E in the plane is called a Painlevé null set if there are no non-constant bounded analytic functions defined on the complement of E. If E is a Painlevé null set and U an open set, then each bounded analytic function defined in $U \sim E$ can be extended to be an analytic function in all of U. Conversely, if $E = F \cup G$ with F a Painlevé null set and G having the property that each point of G has a neighborhood U such that every bounded analytic function in $U \sim G$ can be extended to be analytic in all of U, then E is a Painlevé null set (except for the trivial case in which E is the entire plane). It is well known that a Painlevé null set is totally disconnected and hence does not separate the plane, i. e. that its complement is connected.

With these definitions in mind, we state the following theorem:

Theorem 1. *Let W be a Riemann surface satisfying the absolute AB-maximum principle, and let V be a subregion of W whose relative boundary Γ consists of a finite number of piecewise analytic closed curves. Let f be a bounded analytic function on $V \cup \Gamma$. Then f has bounded valence, and in fact the deficiency δ of f is non-negative. Moreover, $\{z : \delta(z) > 0\}$ is a Painlevé null set.*

Proof. Since $f[\Gamma]$ is a finite collection of piecewise analytic curves in the plane, they have a finite maximum winding number N. Hence if we show that $\delta \geqq 0$, then the valence of f is at most N. Let $E_k = \{z : \delta(z) \geqq k\}$. Since f is an open mapping, each E_k is closed. Clearly E_{N+1} is empty, and therefore a Painlevé null set.

Assume E_{k+1} is a Painlevé null set, and suppose for the moment that E_k has no interior points. Then we shall show that E_k is a Painlevé null set. Let $D_k = E_k - E_{k+1}$, and let $z \in D_k$. Since δ is unaltered by the removal from V of a compact set with nice boundary, we may alter V so that z has a neighborhood U so that no point of $U \cap D_k$ is assumed (on $\bar{V}$) by f, while each point of $U \sim D_k$ is assumed (on V) by f. If $U \cap D_k$ were not a Painlevé null set, there would be a non-constant bounded analytic function g defined in the complement of a closed subset of $U \cap D_k$. By the maximum principle g must take a larger value at some point of U than its maximum in the complement of U. Hence $g \circ f$ must take a larger value in V than on Γ, contradicting the absolute AB-maximum principle for W. Thus $U \cap D_k$ is a Painlevé null set and so removable for bounded analytic functions. Since this is true for each $z \in D_k$ and E_{k+1} is a Painlevé null set, we conclude that E_k is also a Painlevé null set.

There remains to be considered only the case that D_k has interior points. Let z be a boundary point of the interior of D_k in the complement of E_{k+1}. Then by altering V suitably, we can find an open disc U containing z such that f assumes (on V) no values in $D_k \cap U$ and assumes (on V) some values in U. Since $D_k \cap U$ has interior points, we can find a rational function g whose only pole is in the interior of $D_k \cap U$ and which is larger in U than in the complement of U. Then $g \circ f$ is a bounded analytic function on $\bar{V}$ which is larger at those points of V which are mapped into U than it is on Γ. Since this contradicts the absolute AB-maximum principle for W, we see that the interior of D_k can have no boundary points in the complement of E_{k+1}. Thus the interior of D_k is both open and closed in the complement of E_{k+1}, and since the complement of E_{k+1} is connected, D_k must be the entire complement of E_{k+1}.

Since f is bounded, D_0 contains a neighborhood of infinity. Thus we see that D_k has interior points only if $k = 0$, and that D_0 is the complement of E_1. Thus $\delta = 0$ except on the Painlevé null set E_1 where $\delta \geqq 1$. This completes the proof of the theorem.

Let us assume the hypotheses of Theorem 1 and also that $W \sim V$ is compact. Then an argument similar to the preceding but involving compositions of f with bounded harmonic functions instead of bounded analytic functions establishes the following theorem:

Theorem 2. *Assume the hypotheses of Theorem 1 and that $W \sim V$ is compact. Then W is parabolic if $\{z : \delta(z) > 0\}$ has capacity zero, and W admits non-constant bounded harmonic functions if $\{z : \delta(z) > 0\}$ has positive capacity.*

As a corollary we have the following theorem of Kuramochi:

Corollary. Let $W \in 0_{HB} \sim 0_G$, and let V be a subregion of W whose closure is not compact, but which has compact boundary relative to W. Then $V \in 0_{AB}$.

HEINS [1] had shown that if W has a single parabolic end (and *a fortiori* satisfies the absolute AB-maximum principle) then each bounded analytic function f on an end of W has finite valence, and the set where f is deficient consists of a single point. OZAWA [2] has shown by methods from the theory of cluster sets that under the hypotheses of Theorem 1 the bounded analytic function f must have bounded valence.

References

[1] HEINS, M.: Riemann surfaces of infinite genus. Ann. Math. **55**, 296—317 (1952).
[2] OZAWA, M.: Meromorphic functions on some Riemann surfaces. To appear in Proc. Amer. Math. Soc.

Department of Mathematics
Stanford University
Stanford, California

A Remark on Non-compact Quotients of Bounded Symmetric Domains*,**

By

A. ANDREOTTI and E. VESENTINI

Let D be a bounded symmetric domain of $\boldsymbol{C}^n$, and let G be the group of all holomorphic automorphisms of D. It follows from a general result of H. CARTAN that G, endowed with the compact open topology, is a Lie group.

Let Γ be a properly discontinuous group of holomorphic automorphisms of D, i. e. a discrete subgroup of G. It is generally suspected that, if D has no irreducible components of complex dimension 1, then every discrete subgroup Γ of G such that vol (D/Γ) is finite, is commensurable with an arithmetic group. This would imply that there are no other families of discontinuous groups containing Γ except the trivial ones, i. e. those obtained by operating on Γ by a family of inner automorphisms of G. Under the additional assumption that D/Γ be compact, this has been proven to be the case in [5] and in [8] as a consequence of a general rigidity theorem. When D/Γ is not compact, it has been conjectured by A. SELBERG that at least the following should be true:

Let Γ_1 and Γ_2 be two properly discontinuous groups acting on D in such a way that: 1) Γ_1 is an arithmetic group; 2) there exist fundamental

* Supported by the National Science Foundation through research contracts at Brandeis University and Harvard University.
** Received June 2, 1964.

domains F_1 and F_2 for Γ_1 and Γ_2 and a compact set $K \subset D$ satisfying the following condition

$$F_1 - F_1 \cap K = F_2 - F_2 \cap K.$$

Then Γ_2 is commensurable with Γ_1.

This conjecture would imply that there are only trivial families of discontinuous groups Γ containing the arithmetic group Γ_1 and keeping the "part at infinity" of D/Γ rigid. This statement has been established in [4]. In that paper we have shown that any such family is locally trivial, provided that one of the fibers D/Γ behaves at infinity as if Γ is an arithmetic group. More precisely we assume that:

i) the bounded symmetric domain D has no irreducible components of complex dimension 1;

ii) Γ is finitely generated;

iii) D/Γ is a strongly pseudoconcave space.

Both of the latter two conditions are satisfied by arithmetic groups. The concavity assumption has been verified in particular cases in [1], [9] and by K. G. RAMANATHAN (unpublished); it has been established in general by A. BOREL (unpublished).

By a general theorem of A. SELBERG [8], any finitely generated properly discontinuous group Γ contains a subgroup of finite index acting freely on D. Since conditions ii) and iii) are stable by commensurability, we can assume that Γ acts on D without fixed points.

In this report we outline briefly some of the main results of [4].

§ 1. Pseudoconcave manifolds

a) Let X be a complex manifold of pure complex dimension n.

A C^∞ function $\Phi : X \to \mathbf{R}$ is said to be *strongly q-pseudoconvex* at the point $x_0 \in X$ if the Levi form

$$\mathbf{L}(\Phi) = \sum \left(\frac{\partial^2 \Phi}{\partial z^\alpha \partial \bar{z}^\beta} \right)_{x_0} u^\alpha \overline{u^\beta}$$

(where $z^1, \ldots, z^n$ are local complex coordinates at x_0) has at least $n - q$ positive eigenvalues.

One shows easily that, if the C^∞ function Φ is strongly q-pseudoconvex at the point x_0, then there exists a bi-holomorphic imbedding

$$\tau : D^{n-q} \to U$$

of the unit disk $D^{n-q} = \{t = (t_1, \ldots, t_{n-q}) \in \mathbf{C}^{n-q} | \sum t_i \bar{t}_i < 1\}$ in a neighborhood U of x_0 such that $\tau(0) = x_0$, and that the Levi form of the function $\Phi \circ \tau$ is positive definite at $t = 0$.

Using this property, we can generalize the above definition as follows:

A *continuous* function $\Phi : X \to \mathbf{R}$ is called strongly q-pseudoconvex at the point $x_0 \in X$ if

i) there exist a neighborhood U of x_0 in X and finitely many real valued C^∞ functions $\Phi_1, \ldots, \Phi_k$ such that

$$\Phi(x) = \operatorname{Sup}(\Phi_1(x), \ldots, \Phi_k(x)) \qquad \forall\, x \in U;$$

ii) there exists a bi-holomorphic imbedding

$$\tau : D^{n-q} \to U$$

such that $\tau(0) = x_0$, and that for each i, $1 \leqq i \leqq k$, the Levi form of $\Phi_i \circ \tau$ is positive definite at $t = 0$.

This more general class of functions seems to appear naturally in the study of discontinuous groups and does not involve, for the purpose we have in mind, any additional complication.

We say that the manifold X is *strongly q-pseudoconcave* if we can find a compact set $K \subset X$ and a continuous function $\Phi : X \to \mathbf{R}$ such that

α) Φ is strongly q-pseudoconvex at each point of $X - K$;

β) for any $c > \operatorname{Inf} \Phi$ the sets

$$B_c = \{x \in X \mid \Phi(x) > c\}$$

are relatively compact in X.

b) From now on we shall assume X to be connected.

Let X be strongly q-pseudoconcave, with $0 \leqq q \leqq n - 1$. The function Φ cannot have a relative maximum at any point where it is strongly q-pseudoconvex. It follows from this fact that, for any constant c, such that $\min_K \Phi > c > \operatorname{Inf}_X \Phi$, the closure of B_c is

$$\bar{B}_c = \{x \in X \mid \Phi(x) \geqq c\}.$$

Furthermore there exists a c_1, $\min_K \Phi > c_1 > \operatorname{Inf}_X \Phi$, such that, for any $c < c_1$, the sets B_c are connected.

Let $\pi : \tilde{X} \to X$ be the universal covering of X. One proves that, if the fundamental group $\pi_1(X)$ is finitely generated, then there exists a constant c_2, $\min_K \Phi > c_2 > \operatorname{Inf}_X \Phi$ such that for any $c < c_2$ the sets $\pi^{-1}(B_c)$ are connected.

c) Let $\mathcal{O}$ denote the sheaf of germs of holomorphic functions on X. Let A be an open set in X. We say that X is an *analytic completion* of A if the restriction map

$$H^0(X, \mathcal{O}) \to H^0(A, \mathcal{O})$$

is an isomorphism.

If X is an analytic completion of A, and if Y is a Stein manifold, then any holomorphic map

$$f : A \to Y$$

extends, in a unique way, to a holomorphic map

$$\tilde{f} : X \to Y .$$

It follows that, if A has an analytic completion X which is holomorphically complete, this is unique up to an isomorphism which is the identity on A. We say in this case that X is the *envelope of holomorphy* of A.

The following theorem holds.

Theorem I. *Let X be a connected complex manifold of complex dimension n. Let $\pi : \tilde{X} \to X$ be the universal covering of X. If the following two conditions are satisfied:*

i) *X is strongly q-pseudoconcave for some q, with $0 \leqq q \leqq n - 2$;*

ii) *the fundamental group $\pi_1(X)$ is finitely generated,*

then there exists a constant $c_3 > \operatorname{Inf}_X \Phi$ such that, for any $c < c_3$, $\pi^{-1}(B_c)$ has $\tilde{X}$ as an analytic completion.

In particular, if $\tilde{X}$ is holomorphically complete, then for any $c < c_3$, $\tilde{X}$ is the envelope of holomorphy of $\pi^{-1}(B_c)$.

§ 2. Families of uniformizable structures

a) A differentiable family of complex manifolds is the set of the following data:

a differentiable manifold $\mathscr{V}$,

a differentiable manifold M,

a differentiable, surjective map $\omega : \mathscr{V} \to M$,

satisfying the following conditions:

i) ω is of maximal rank at each point of $\mathscr{V}$,

ii) for every point $x \in \mathscr{V}$ we can find:

a neighborhood W of x in $\mathscr{V}$,

a neighborhood U of $\omega(x)$ in M,

an open set S in some numerical space $\boldsymbol{C}^n$, and

a diffeomorphism $\varphi : U \times S \to W$ such that

(a) $pr_U = \omega \circ \varphi$,

(b) if $\varphi_i : U_i \times S_i \to W_i$ $(i = 1, 2)$ are any two such diffeomorphisms, then $\varphi_2^{-1} \circ \varphi_1$ is an isomorphism of $\varphi_1^{-1}(W_1 \cap W_2)$ onto $\varphi_2^{-1}(W_1 \cap W_2)$, structure sheaves being the sheaves of germs of C^∞ functions holomorphic on the fibers of the projections $pr_{U_i} (i = 1, 2)$.

For any $t \in M$, $\omega^{-1}(t) = X_t$ has a natural structure of a complex manifold. We will take as structural sheaf on $\mathscr{V}$ the sheaf of germs of C^∞ functions, whose restrictions to the fibers of ω are holomorphic.

Analogously one defines a complex analytic (or holomorphic) family of complex manifolds (cf. [*2*]).

Let X_0 be a complex manifold. A (differentiable) deformation of X_0 is the set of the following data:

a differentiable family of complex manifolds $(\mathscr{V}, \omega, M)$,

a point $m_0 \in M$,

an isomorphism $i: X_0 \to \omega^{-1}(m_0)$.

In a similar way one defines a complex analytic (or holomorphic) deformation.

The above statements generalize Kodaira-Spencer's definitions of deformations of *compact* complex manifolds. The definitions of equivalent or locally equivalent deformations and classes of local deformations are as in [7].

Any deformation $(\mathscr{V}, \omega, M)$ of X_0, which is equivalent to the deformation $(X_0 \times M, pr_M, M)$, is called a *trivial deformation of* X_0.

Definition. A deformation $(\mathscr{V}, \omega, M)$ of X_0 is called *rigid at infinity* if there exists a compact set $K_0 \subset X_0$ and an isomorphism

$$g: (X_0 - K_0) \times M \to \mathscr{V}$$

onto an open subset of $\mathscr{V}$, such that

$$\omega \circ g = pr_M,$$

and that $\omega | \mathscr{V} - \mathrm{Im}\, g$ is a proper map.

b) Let $(\mathscr{V}, \omega, M)$ be a family of complex manifolds over a connected and simply connected manifold M. Let $\pi: \tilde{\mathscr{V}} \to \mathscr{V}$ be the universal covering manifold of $\mathscr{V}$. Then $(\tilde{\mathscr{V}}, \omega \circ \pi, M)$ is a new family of complex manifolds over M.

Let D be a complex manifold. We say [cf. *10*] that $(\mathscr{V}, \omega, M)$ is a family of complex manifolds uniformizable on the manifold D if there exists an isomorphism

$$\sigma: \tilde{\mathscr{V}} \to D \times M$$

(structure sheaves being the sheaves of germs of C^∞ functions holomorphic respectively on the fibers of $\omega \circ \pi$ and of pr_M) so that the following diagram is commutative

$$\begin{array}{ccc} \tilde{\mathscr{V}} & \xrightarrow{\sigma} & D \times M \\ \pi\downarrow & & \downarrow pr_M \\ \mathscr{V} & \xrightarrow{\omega} & M \end{array}$$

We shall always assume $\mathscr{V}$ to be connected. This implies that $\tilde{\mathscr{V}}$ is connected and simply connected. Hence D is connected and simply connected, and, for each $t \in M$,

$$D \times \{t\} \xrightarrow{\pi \circ \sigma^{-1}} X_t = \omega^{-1}(t)$$

is the universal covering of X_t.

The fundamental group $\Gamma = \pi_1(V)$ can be viewed as the group of automorphisms of the universal covering $\pi : \tilde{\mathscr{V}} \to \mathscr{V}$. It follows from the previous remark that, for each $t \in M$, $\Gamma = \pi_1(X_t)$.

Let G be the group of all holomorphic automorphisms of the complex manifold D. By means of σ we identify Γ with $\sigma \Gamma \sigma^{-1}$ as a group of automorphisms of $D \times M$. Every element $\gamma \in \Gamma$ represents then a map $D \times M \to D \times M$ given by equations of type

$$\gamma : \begin{cases} z \to \gamma(z, t) \\ t \to t \end{cases} \quad z \in D, \quad t \in M, \tag{1}$$

where, for every $t \in M$, $\gamma(z, t) \in G$.

We assume that G (with the compact open topology) has the structure of a Lie group. By a well known theorem of H. CARTAN, this is the case when D is a bounded domain of $\mathbf{C}^n$.

c) We remark that complex analytic families of uniformizable structures on bounded domains are locally trivial. One has in fact the following result.

Theorem II. *Let D be a bounded domain in $\mathbf{C}^n$. Any complex analytic family $(\mathscr{V}, \omega, M)$ over the unit ball $M = \{t = (t_1, \ldots, t_m) \in \mathbf{C}^m \mid \sum t_i \bar{t}_i < 1\}$ of complex manifolds uniformizable on D is trivial.*

The proof is straightforward: Using the notations of § 2b) we have a group Γ of complex analytic automorphisms of type (1), where now $\gamma(z, t)$ is a holomorphic function of $z = (z_1, \ldots, z_n) \in D$ *and* $t = (t_1, \ldots, t_m) \in M$, with values in D.

Any automorphism of $D \times M$ is an isometry of the Bergman metric of $D \times M$. If ds_D^2 and ds_M^2 are the Bergman metrics of the bounded domains D and M, then the Bergman metric of $D \times M$ is expressed by

$$ds^2_{D \times M} = ds_D^2 + ds_M^2 .$$

Let γ be any element of Γ written as in (1). Let $(z_0, t_0) \in D \times M$. Consider the automorphism γ_0 of $D \times M$ given by

$$\gamma_0 : \begin{cases} z \to \gamma(z, t_0) \\ t \to t . \end{cases}$$

The automorphism $\gamma \circ \gamma_0^{-1}$ has an expression of the following type

$$\gamma \circ \gamma_0^{-1} : \begin{cases} z \to A \cdot (z - z_0) + B \cdot (t - t_0) + 0(2) \\ t \to t , \end{cases}$$

where A and B are constant matrices. Since $\gamma \circ \gamma_0^{-1}$ is an isometry of the Bergman metric, then $B = 0$. On the other hand $\gamma \circ \gamma_0^{-1}$ is the identity on $D \times \{t_0\}$; then $A = I$. Thus the linear part of $\gamma \circ \gamma_0^{-1}$ is the identity. Since $D \times M$ is a bounded domain, this implies, by a theorem of H. CARTAN

[6], that $\gamma \circ \gamma_0^{-1}$ is the identity, i. e. $\gamma = \gamma_0$. This means that γ does not depend on t.

We remark that there are no restrictions on the dimension of D. Thus, for instance, if $(\mathscr{V}, \omega, M)$ is the family of curves of genus $g > 1$ over the Teichmüller space M, then the uniformizing parameter of $X_t = \omega^{-1}(t)$ $(t \in M)$ over the unit circle cannot depend analytically on t. This fact was first pointed out to us by L. Bers.

d) Let M be the unit ball of $\mathbf{R}^m$.

Theorem III. *Let $(\mathscr{V}, \omega, M)$ be a differentiable family of deformations of the manifold $X_0 = \omega^{-1}(0)$. We assume that:*

i) *the family $(\mathscr{V}, \omega, M)$ is a family of uniformizable structures on a bounded symmetric domain D, none of whose irreducible components has complex dimension one;*

ii) *the deformation is rigid at infinity;*

iii) *X_0 is strongly q-pseudoconcave, with $0 \leqq q \leqq \dim_{\mathbf{C}} X_0 - 2$;*

iv) *the fundamental group of X_0 is finitely generated.*

Then the whole deformation of X_0 is trivial.

To prove this theorem one first remarks that all fibers X_t play the same role, and that it is sufficient to show that $\mathscr{V}$ is locally trivial at $t = 0$. The main tools of the proof of this fact are the following:

1) Let B be a relatively compact subset of X_0. We can assume that the inverse image $\tilde{B}$ of B in D has D as its envelope of holomorphy, in view of Theorem I.

2) Some of the results of chapter 3 of [5] can be rephrased saying that the complex tangent bundle to X_t is $W^{(0,1)}$-elliptic with respect to the Bergman metric (see [3] for the definition of W-ellipticity). The same is also true if the Bergman metric is perturbed on X_t on a compact set, provided that $|t|$ is sufficiently small.

3) By 2) and an argument of [2] one can show that any compact set $B \subset X_0$ can be deformed trivially inside the family $\mathscr{V}$.

This partial local trivialization of B "extends" to a local trivialization of $\mathscr{V}$, in view of 1).

References

[1] Andreotti, A., und H. Grauert: Algebraische Körper von automorphen Funktionen. Nachr. Akad. Wissensch. Göttingen, 1961, 39—48.

[2] —, and E. Vesentini: On the pseudorigidity of Stein manifolds. Ann. Sc. Norm. Sup. (Pisa) (3) **16**, 213—223 (1962).

[3] — — Les théorèmes fondamentaux de la théorie des espaces holomorphiquement complets. Séminaire C. Ehresmann, Vol. IV, 1—31 (1962—1963).

[4] — — On deformations of discontinuous groups. Acta Mathematica 112, (1964).

[5] Calabi, E., and E. Vesentini: On compact, locally symmetric Kähler manifolds. Ann. Math. **71**, 472—507 (1960).

[6] CARTAN, H.: Les fonctions de deux variables complexes et le problème de la représentation conforme. J. Math. pur. appl. (9) **10**, 1–114 (1931).
[7] KODAIRA, K., and D. C. SPENCER: On deformations of complex analytic structures, I and II. Ann. Math. **67**, 328–466 (1958).
[8] SELBERG, A.: On discontinuous groups in higher dimensional symmetric spaces. Contrib. Function Theory (Bombay) 1960, 147–164.
[9] SPILKER, J.: Algebraische Körper von automorphen Funktionen. Math. Ann. **149**, 341–360 (1963).
[10] WEIL, A.: On discrete subgroups of Lie groups, II. Ann. Math. **75**, 578–602 (1962).

Istituto Matematico
Università
Pisa, Italia

Pseudo-convex Domains in Linear Topological Spaces*

By

H. J. BREMERMANN

1. Introduction

In [2] it was shown that the notion "pseudo-convex domain" may be extended to infinite dimensional Banach spaces. In this paper we will extend it to linear topological spaces. We show that many basic results that hold for finite dimension remain true.

2. Definition, basic properties

Let X be a complex linear topological space. A *region* is an open set in X. A *domain* a connected region. A region $D \subset X$ is called *pseudo-convex* if and only if every intersection of D with a finite dimensional translated linear subspace of X is pseudo-convex. (We recall: A finite dimensional region G is pseudo-convex if and only if there exists a function V plurisubharmonic in G that tends to infinity at every finite boundary point of G).

This very general notion implies some of the usual properties of pseudo-convex regions: *the intersection of a family of pseudoconvex regions, if open, is a pseudo-convex region. Every subset of X has a pseudo-convex envelope. The union of a monotone increasing sequence of pseudo-convex regions is pseudo-convex. The exterior of a bounded set cannot be a pseudo-convex region, except for dimension one.* These properties follow immediately from the definition and from corresponding theorems for

* Received May 21, 1964.

finite dimension by taking restrictions to finite dimensional translated linear subspaces.

Convex regions. A region in X is called convex if and only if every connected component is convex. *Any convex region D in X is pseudo-convex.* Indeed, the intersection with any finite dimensional translated linear subspace is convex. For finite dimension any convex region is pseudo-convex [*1*]. Hence D is pseudo-convex.

Tube domains. Given a real linear topological space X'. Define a complex linear topological space as follows: $X = X' \times X'$ with the product topology. Define $x + iy = (x, y) \in X' \times X'$ and define multiplication with complex numbers $\lambda = \mu + i\nu$ as follows: $\lambda(x + iy) = \mu x - \nu y + i(\nu x + \mu y)$. A tube domain T_B is the product of a domain $B \subset X'$ with X' with complex vector space structure as above. B is called the *base*. Let $a_0, a_1, \ldots, a_n \in X'$. Consider the finite dimensional translated linear subspace $\{x \mid x = a_0 + \lambda_1 a_1 + \cdots + \lambda_n a_n\}$, $\lambda_1, \ldots, \lambda_n$ complex parameters. Its intersection with T_B is a finite dimensional tube region with base $B' \cap \{x \mid x = a_0 + \mu_1 a_1 + \cdots + \mu_n a_n\}$, where $\mu_j = \mathrm{Re}\,\lambda_j$. If T_B is pseudo-convex, then the finite dimensional tube region $T_{B'}$ must be pseudo-convex, hence convex [*3*]. Thus: If T_B is a pseudo-convex tube domain, then the intersection of B with any finite dimensional translated linear subspace of X' is convex. And conversely, if the latter is true, then T_B is pseudo-convex.

If now X' is polygonally connected, then a polygon connection of any two points in B is contained in a finite dimensional translated linear subspace, hence they are in the same component of B'. B' is convex. Hence the line segment connecting the two points is in B'. Hence B' is a domain. Hence: *If X' is polygonally connected, then T_B is a pseudo-convex tube domain if and only if B is a convex domain.*

A real valued function is *plurisubharmonic* in a region $D \subset X$ if and only if its restriction to the intersection of D with any finite dimensional translated linear subspace is plurisubharmonic. Analogously one defines *convex* functions and (complex valued) *holomorphic* functions. If D is a region such that there exists a plurisubharmonic (convex) function that tends to infinity at every finite boundary point of D, then D is pseudo-convex (convex). Indeed, the restriction of the function to a finite dimensional translated linear subspace has the same property and hence by the corresponding theorem for finite dimension any such intersection is pseudoconvex (convex). Hence D is pseudo-convex (convex, respectively).

3. Distance function

A neighborhood V of zero is called *circled*, if with each point $z^{(0)} \in V$ all points $\lambda z^{(0)}$ belong to V for $|\lambda| \leqq 1$. V is called *radial* at zero if it contains a line segment through zero in each direction. We will write rV for $\{z = rz', z' \in V\}$ and $V(z^{(0)})$ for $\{z | z - z^{(0)} \in V\}$.

Let V be a fixed circled and radial neighborhood of zero. Note that $rV(z^{(0)}) \in V(z^{(0)})$ for $r \leqq 1$. Let V be such that for each point $z^{(0)} \in V$ there exists an r such that $rV(z^{(0)}) \subset V$. Note that we do not require V to be convex. If X is a normed linear space, then $V = \{z | \|z\| < < c\}$, where c is a constant, satisfies the requirements on V. We define a boundary distance function as follows:

$$\Delta(z) = \sup r \ni rV(z) \in D .$$

If X is a normed linear space and $V = \{z | \|z\| < c\}$, then $\Delta(z)$ is Lipschitz continuous with constant 1. In the following we will assume that X and V are such that $\Delta(z)$ is continuous.

We also introduce a distance function "in a direction a" as follows: Let $a \in X$. Let $\delta = \sup r \ni ra \in V$. In the following, let $\delta(a) = 1$. Then we define

$$\Delta_a(z) = \sup r \ni z + \lambda a \in D \quad \text{for all} \quad |\lambda| \leqq r .$$

Note that $V = \{z | z = \lambda a, |\lambda| \leqq 1, \delta(a) = 1\}$. Thus we have

$$\Delta(z) = \inf_{\delta(a)=1} \Delta_a(z) .$$

It is easy to see that $\Delta_a(z)$ is lower-semicontinuous.

Theorem. *Let D be open $\subset X$. Let X and V be such that $\Delta(z)$ is continuous in D. Then D is pseudo-convex if and only if $-\log \Delta(z)$ is plurisubharmonic in D.*

Proof. If $-\log \Delta(z)$ is plurisubharmonic, then its restriction to any finite dimensional translated linear subspace is plurisubharmonic and it tends to infinity at every boundary point. Hence every such intersection is pseudo-convex. Hence D is pseudo-convex.

To prove the converse, it is enough to show that $-\log \Delta_a(z)$ is plurisubharmonic for each a. Indeed, $-\log \Delta(z) = \sup_{\delta(a)=1} -\log \Delta_a(z)$ and $-\log \Delta(z)$ is continuous. Hence, if $-\log \Delta_a(z)$ is plurisubharmonic for each a, then $-\log \Delta(z)$ is plurisubharmonic.

Suppose that $-\log \Delta_a(z)$ would not be plurisubharmonic. Then there would exist a one-dimensional (complex) translated linear subspace E such that the restriction of $-\log \Delta_a(z)$ is not subharmonic. Then there would exist a disc and a harmonic function that majorizes $-\log \Delta_a(z)$ at the boundary but not everywhere in the interior. Then we construct as usual [2] a violation of the "Kontinuitätssatz" in the subspace spanned by a and E. This is a contradiction since the intersection with this subspace is pseudo-convex. Hence $-\log \Delta_a(z)$ is plurisubharmonic. That proves the theorem.

Pseudo-convexity is a local property in the following sense: Let X be a normed linear space. Let there be an $\varepsilon > 0$ such that for each finite boundary point $z^{(0)}$ the intersection $\{z \mid \|z - z^0\| < \varepsilon\} \cap D$ is pseudo-convex. Then D is pseudo-convex. Proof. In the case of a normed linear space $-\log \Delta(z)$ is continuous. The distance function of the intersection $\{z \mid \|z - z^{(0)}\| < \varepsilon\} \cap D$ is plurisubharmonic. For $\|z - z^{(0)}\| < \varepsilon/2$ it coincides with the distance function of D. Hence $\sup(-\log \Delta(z), -\log \varepsilon/4)$ is plurisubharmonic in D and tends to infinity at every finite boundary point. Hence D is pseudo-convex.

Analytic discs. Let Λ be the unit disc in the complex λ-plane. Let $\Lambda \to D$ be a C^2 map of the closed disc into a region $D \subset X$ such that every plurisubharmonic function in D, when restricted to Λ, is subharmonic in the interior of Λ. The image of Λ together with the map is called an *analytic disc*. We will denote the image of Λ again by Λ. For finite dimensional spaces this definition reduces to the following: After introducing a basis the component functions are either holomorphic or anti-holomorphic: Let $z_\nu = f_\nu(\lambda)$, $\nu = 1, \ldots, n$, then $f_\nu(\lambda)$ is holomorphic, or $\bar{f}_\nu(\lambda)$ is holomorphic.

Kontinuitätssatz. Let $\{\Lambda_i\}$ be a sequence of analytic discs in a region D. We say that *the Kontinuitätssatz holds for D* iff $\inf \Delta(\partial \Lambda_i) > 0$ implies $\inf \Delta(\Lambda_i) > 0$. (Here $\Delta(\partial \Lambda_i) = \inf_{z \in \partial \Lambda_i} \Delta(z)$).

Theorem. *Let X be such that $-\log \Delta(z)$ is continuous. Then the Kontinuitätssatz holds for a region $D \subset X$ if and only if D is pseudo-convex.*

Proof. If the Kontinuitätssatz holds for D, then it holds for the intersection of D with translated finite dimensional linear subspaces. For finite dimension the theorem is true [1]. Hence any such intersection is pseudo-convex, hence D is pseudo-convex. Conversely, if D is pseudo-convex and if $-\log \Delta(z)$ is continuous, then $-\log \Delta(z)$ is plurisubharmonic. Hence $\sup_{z \in \partial \Lambda_i} -\log \Delta(z) = \sup_{z \in \Lambda_i} -\log \Delta(z)$, which is equivalent to $\Delta(\partial \Lambda_i) = \Delta(\Lambda_i)$. Hence the Kontinuitätssatz holds.

4. Regions of holomorphy

We defined already the notion of holomorphic function. A holomorphic function that coincides with a given function in an open set is called a *holomorphic continuation*. A boundary point $z^{(0)}$ of a region D is a *singular* point of a function f, holomorphic in D, if and only if the following condition is satisfied: There exists a neighborhood $N(z^{(0)})$ such that for none of the components of $N(z^{(0)}) \cap D$ there is a domain N', $z^{(0)} \subset N' \subset N(z^{(0)})$, for which there is a holomorphic continuation of f from the component into N'. A boundary point of a region is called *essential* if and only if there exists a function that is holomorphic in the region and that is singular at the given point. A region is called a *region of holomorphy* if and only if there exists a function that is holomorphic in the region and for which every boundary point is singular.

Theorem. *Let X be such that* $-\log \Delta(z)$ *is continuous. Let D be a region such that every boundary point is essential, then D is pseudo-convex.*

Corollary. *If D is a region of holomorphy, then D is pseudoconvex.*

Proof. Suppose D would not be pseudo-convex. Then $-\log \Delta(z)$ would not be plurisubharmonic, hence not subharmonic on some analytic plane. Then one constructs a contradiction to the assumption that all boundary points are essential. This construction is analogous to [2].

The converse problem whether any pseudo-convex region is a region of holomorphy is an open question.

References

[1] Bremermann, H. J.: Complex convexity. Trans. Amer. Math. Soc. **82**, 17–51 (1956).
[2] — Holomorphic functionals and complex convexity in Banach spaces. Pac. J. Math. **7**, 811–831 (1957). (Errata in Vol. 7 at the end.)
[3] — The envelopes of holomorphy of tube domains in infinite dimensional Banach spaces. Pac. J. Math. **10**, 1149–1153 (1960).
[4] Hille, E., and R. S. Phillips: Functional analysis and semi-groups. Amer. Math. Soc. Publ., Revised Ed., New York, 1957.

Department of Mathematics
University of California
Berkeley, California

Connections for a Class of Pseudogroup Structures*

By

R. C. Gunning

1. Introduction

Affine connections have long been familiar in differential geometry, among other things for providing a measure of deviation from flatness for manifolds; here flatness means the existence of a system of local coordinates such that the transition functions between any two coordinate systems are affine transformations. The intention of the present note is to show how similar connections can be introduced, in real C^∞ or in complex analytic manifolds, associated to some general classes of pseudogroup structures, for the purpose of investigating the existence of systems of

* Received April 7, 1964.

local coordinates with transition functions lying in the pseudogroups. The point of view adopted here is analytical rather than geometrical, in the sense that connections are envisaged as arising from the formal properties of certain systems of partial differential equations rather than as deriving from the splittings of vector bundles into vertical and horizontal components. The classical affine and projective connections of course appear again, and the familiar relationship between them seems perhaps simpler in this context.

2. The general form of the connections

The set of germs of real analytic automorphisms at the origin in $\boldsymbol{R}^n$, modulo terms of order greater than r in the power series expansion, form a Lie group under composition; this group will be denoted by $G^r(n, \boldsymbol{R})$, and will be called the *real general r-fold group*. The complex general r-fold group $G^r(n, \boldsymbol{C})$ arises in the same manner from the germs of complex analytic automorphisms at the origin in $\boldsymbol{C}^n$. These two groups have much the same formal structure and will be treated concurrently, writing $G^r(n)$ to stand for either of them. The product of two elements ξ, η of $G^r(n)$ will be written $\xi \circ \eta$, and the inverses, ξ^{-1}, η^{-1}. Note that $G^1(n)$ is just the usual general linear group of rank n. If F is a non-singular C^∞ mapping defined in an open subset $U \subset \boldsymbol{R}^n$ (or a non-singular holomorphic mapping defined in an open subset $U \subset \boldsymbol{C}^n$), then to each point $p \in U$ there corresponds a unique element $(D^r F)(p) \in G^r(n)$; and the mapping $D^r F : U \to G^r(n)$ is also a C^∞ (or holomorphic) mapping. The value $(D^r F)(p)$ can be considered either as the Taylor expansion of F at p up to order r, or as the set of all partial derivatives of F at p of orders at most r. If F and G are two mappings such that their composition $F \circ G$ is defined in an open set U, then obviously $D^r(F \circ G)(p) = D^r F(Gp) \circ D^r G(p)$ for all $p \in U$.

Let M be an n-dimensional C^∞ manifold (or an n-dimensional complex analytic manifold). There is then a covering of M by open subsets $U_\alpha \subset M$, to each of which there corresponds a topological homeomorphism $F_\alpha : U_\alpha \to V_\alpha$ onto an open subset V_α of $\boldsymbol{R}^n$ (or $\boldsymbol{C}^n$); moreover, for each non-empty intersection $U_\alpha \cap U_\beta \subset M$ the coordinate transition functions $F_{\alpha\beta} = F_\alpha \circ F_\beta^{-1} : F_\beta(U_\alpha \cap U_\beta) \to F_\alpha(U_\alpha \cap U_\beta)$ are C^∞ (or holomorphic) mappings. To each point $p \in U_\alpha \cap U_\beta$ associate the element

$$\xi^r_{\alpha\beta}(p) = D^r F_{\alpha\beta}(F_\beta p) \in G^r(n). \tag{1}$$

If $p \in U_\alpha \cap U_\beta \cap U_\gamma$ then clearly $\xi^r_{\alpha\beta}(p) \circ \xi^r_{\beta\gamma}(p) = D^r F_{\alpha\beta}(F_\beta p) \circ D^r F_{\beta\gamma}(F_\gamma p) = D^r(F_{\alpha\beta} \circ F_{\beta\gamma})(F_\gamma p) = \xi^r_{\alpha\gamma}(p)$; therefore the mappings $\xi^r_{\alpha\beta} : U_\alpha \cap U_\beta \to G^r(n)$ define a principal $G^r(n)$-bundle over the manifold M. This bundle will be denoted by $\Xi^r(M)$, and is the principal bundle associated to the r-th order jet bundle of EHRESMANN [3]; in particular, for $r = 1$, it is just the tangent bundle of M.

There now arises the problem whether the manifold M admits a coordinate covering such that the transition elements $\xi^r_{\alpha\beta}$ lie in a subgroup H of $G^r(n)$; this is the "problem of the reduction of pseudogroup structures" of M, where the relevant pseudogroup consists of all mappings F such that $D^r F(x) \in H$ for all x in the domain of definition of F. The problem decomposes naturally into two components: (a) the problem of reducing the structural group of the bundle $\Xi^r(M)$ to the subgroup $H \subset G^r(n)$; (b) the problem of realizing this reduction of the bundle by an appropriate change of the coordinatization of the manifold M. For part (a) the problem is that of the existence of C^∞ (or holomorphic) mappings $\eta_\alpha : U_\alpha \to G^r(n)$ such that in each non-empty intersection $U_\alpha \cap U_\beta$

$$\eta_\alpha \circ \xi^r_{\alpha\beta} \circ \eta_\beta^{-1} \in H \subset G^r(n); \tag{2}$$

while for part (b) the problem is that of the existence of C^∞ (or holomorphic) mappings $G_\alpha : V_\alpha \to \boldsymbol{R}^n$ (or $\boldsymbol{C}^n$) such that

$$\eta_\alpha(p) = D^r G_\alpha(F_\alpha p) \quad \text{for all} \quad p \in U_\alpha . \tag{3}$$

For the following general class of subgroups $H \subset G^r(n)$ the reduction problem can be handled very conveniently. Suppose that $\mathscr{V}$ is a finite-dimensional real (or complex) vector space, and that ϱ is a linear representation of $G^r(n)$ with representation space $\mathscr{V}$; that is, ϱ is a continuous (or holomorphic) homomorphism from the Lie group $G^r(n)$ into the Lie group of linear automorphisms of the vector space $\mathscr{V}$. Suppose moreover that $\Theta : G^r(n) \to \mathscr{V}$ is a real (or complex) analytic mapping such that

$$\Theta(\xi \circ \eta) = \varrho(\eta)^{-1} \Theta(\xi) + \Theta(\eta) \tag{4}$$

for all elements ξ, η of $G^r(n)$; note that as a consequence of (4), $\Theta(\varepsilon) = 0$ for the identity element ε of $G^r(n)$ and $\Theta(\xi^{-1}) = -\varrho(\xi)\Theta(\xi)$. Thus the subset $H^r(\Theta) \subset G^r(n)$ defined by

$$H^r(\Theta) = \{\xi \in G^r(n); \Theta(\xi) = 0\} \tag{5}$$

is a Lie subgroup of $G^r(n)$. For subgroups of this form it follows readily from (4) that condition (2) can be rewritten:

$$\Theta(\xi^r_{\alpha\beta}) + \varrho(\xi^r_{\alpha\beta})^{-1} \Theta(\eta_\alpha) - \Theta(\eta_\beta) = 0 . \tag{6}$$

Considering the terms $\lambda_\alpha = \Theta(\eta_\alpha)$ as the primitive unknowns rather than the terms η_α, the two components of the reduction problem then take the simpler forms: (a) the problem of the existence of C^∞ (or holomorphic) mappings $\lambda_\alpha : U_\alpha \to \mathscr{V}$ such that in each nonempty intersection $U_\alpha \cap U_\beta$

$$\Theta(\xi^r_{\alpha\beta}) + \varrho(\xi^r_{\alpha\beta})^{-1} \lambda_\alpha - \lambda_\beta = 0 ; \tag{7}$$

and (b) the problem of the existence of C^∞ (or holomorphic) mappings $G_\alpha: V_\alpha \to \boldsymbol{R}^n$ (or $\boldsymbol{C}^n$) such that

$$\lambda_\alpha(p) = \Theta(D^r G_\alpha(F_\alpha p)) \quad \text{for all} \quad p \in U_\alpha . \tag{8}$$

A set of mappings $\{\lambda_\alpha\}$ satisfying (7) will be called a *connection* for the pseudogroup structure $H^r(\Theta)$ on the manifold M. The existence of mappings G_α satisfying (8) amounts to the integrability of the system of partial differential equations involved; the connection $\{\lambda_\alpha\}$ will be called *torsionless* if the system of equations (8) is integrable. It should be remarked that for real or complex analytic connections λ_α it is generally not difficult to derive explicit integrability conditions for the system (8), and to express these conditions in the form $D^* \lambda_\alpha = 0$ for a suitable differential operator D^*; the extension of such results to the C^∞ case may present considerable difficulties. In summary, the pseudogroup structure of M can be reduced to $H^r(\Theta)$ precisely when there is a torsionless connection for the pseudogroup structure $H^r(\Theta)$ on M.

The question of the existence of at least one connection for a pseudogroup structure $H^r(\Theta)$ is now quite easy to handle. Since the functions $\xi^r_{\alpha\beta}$ are the transition functions for the fibre bundle $\Xi^r(M)$, it follows immediately that the functions $\varrho(\xi^r_{\alpha\beta})$ are the transition functions for a vector bundle over M with fibre $\mathscr{V}$; this bundle will be denoted by $\varrho\Xi^r(M)$. Furthermore, it follows easily from (4) that in any non-empty triple intersection $U_\alpha \cap U_\beta \cap U_\gamma$ the following relation holds: $\Theta(\xi^r_{\alpha\gamma}) = \varrho(\xi^r_{\beta\gamma})^{-1} \Theta(\xi^r_{\alpha\beta}) + \Theta(\xi^r_{\beta\gamma})$; that is to say, the set of functions $\{\Theta(\xi^r_{\alpha\beta})\}$ defined in the intersections $U_\alpha \cap U_\beta$ form a one-cocycle on the nerve of the covering $\{U_\alpha\}$ with coefficients in the sheaf of germs of C^∞ (or holomorphic) cross-sections of the vector bundle $\varrho\Xi^r(M)$, [7]. Equation (7) is then just the condition that the mappings $\{\lambda_\alpha\}$ form a zero-cochain for the same covering and sheaf, and that the coboundary of this cochain is the one-cocycle $\{\Theta(\xi^r_{\alpha\beta})\}$. Since the choice of the covering $\{U_\alpha\}$ is quite free, a sufficient condition for the existence of a connection is the vanishing of the cohomology group $H^1(M, \varrho\Xi^r(M))$; in this context, $\varrho\Xi^r(M)$ denotes the sheaf of germs of C^∞ (or of holomorphic) cross-sections of the vector bundle. For sheaves of continuous or C^∞ cross-sections of vector bundles, all the psositive-dimensional cohomology groups vanish; and for sheaves of holomorphic cross-sections of complex vector bundles on a Stein manifold, all the positive-dimensional cohomology groups also vanish, [6]. Therefore *in the C^∞ case every manifold admits a connection for the pseudogroup structure $H^r(\Theta)$*; and *in the complex analytic case every Stein manifold admits a connection for the pseudogroup structure $H^r(\Theta)$*. The case of compact complex manifolds requires further study of the cohomology groups $H^1(M, \varrho\Xi^r(M))$.

3. Examples

For any germ $F(x) = (f_1(x), \ldots, f_n(x))$ of a C^∞ automorphism in $\boldsymbol{R}^n$ (or of a complex analytic automorphism in $\boldsymbol{C}^n$) write $\xi^i_j = \partial f_i / \partial x_j$, $\xi^i_{j_1 j_2} = \partial^2 f_i / \partial x_{j_1} \partial x_{j_2}$, and so on. Thus the elements of $G^r(n)$ are represented in the form $\xi = (\xi^i_j, \xi^i_{j_1 j_2}, \ldots, \xi^i_{j_1 \ldots j_r})$, where (ξ^i_j) is a non-singular matrix and the remaining entries are fully symmetric in their lower indices; and the group operation can be read off from the chain rule for differentiation.

(a) Let $\Theta: G^2(n) \to \mathscr{V}$ be the mapping defined by

$$\Theta(\xi) = (\Theta^i_{j_1 j_2}(\xi)) = (\Sigma_k \tilde{\xi}^i_k \xi^k_{j_1 j_2}), \tag{9}$$

where $(\tilde{\xi}^i_j) = (\xi^i_j)^{-1}$ denotes the inverse matrix to (ξ^i_j); here $\mathscr{V}$ is the vector space of dimension $n^2(n+1)/2$ consisting of all tensors $T^i_{j_1 j_2}$ fully symmetric in j_1 and j_2. A simple calculation shows that

$$\Theta^i_{j_1 j_2}(\xi \circ \eta) = \Sigma_{k k_1 k_2} \tilde{\eta}^i_k \Theta^k_{k_1 k_2}(\xi) \eta^{k_1}_{j_1} \eta^{k_2}_{j_2} + \Theta^i_{j_1 j_2}(\eta); \tag{10}$$

this equation is of the form (4), where ϱ is the tensor representation of the signature $(1) \otimes (-2)$ in the sense of [*10*]. Writing out equation (7) explicitly for this case, we see that a connection for the pseudogroup structure defined by (9) is given by mappings $\lambda_\alpha = (\lambda^i_{\alpha j_1 j_2})$ from the coordinate neighborhoods U_α into the vector space $\mathscr{V}$ such that in $U_\alpha \cap U_\beta$,

$$\frac{\partial^2 x^\alpha_i}{\partial x^\beta_{j_1} \partial x^\beta_{j_2}} + \Sigma_{k_1 k_2} \lambda^i_{\alpha k_1 k_2} \frac{\partial x^\alpha_{k_1}}{\partial x^\beta_{j_1}} \frac{\partial x^\alpha_{k_2}}{\partial x^\beta_{j_2}} - \Sigma_k \frac{\partial x^\alpha_i}{\partial x^\beta_k} \lambda^k_{\beta j_1 j_2} = 0. \tag{11}$$

However, this is just a symmetric affine connection on the manifold M in the usual sense, [*4*]; that is, *a connection for the pseudogroup structure* (9) *is a symmetric affine connection on the manifold*. The condition that the connection be torsionless is that the usual torsion tensor associated to an affine connection vanish. It is clear from the form of the defining equation (9) that the pseudogroup defined by $H^2(\Theta)$ is in this case the pseudogroup of affine transformations.

(b) The representation ϱ of example (a) just above is reducible to a direct sum $\varrho = \varrho' \oplus \varrho''$, where $\varrho' = (-1)$ and $\varrho'' = (1, 0, \ldots, 0, -2)$ in the notation of [*10*]. To exhibit this decomposition explicitly, introduce the linear mapping $P: \mathscr{V} \to \mathscr{V}$ defined by

$$(PT)^i_{j_1 j_2} = (n+1)^{-1} \Sigma_k (\delta^i_{j_1} T^k_{k j_2} + \delta^i_{j_2} T^k_{k j_1}), \tag{12}$$

where δ^i_j is the Kronecker symbol; a simple calculation shows that the mapping P commutes with the representation ϱ, and is a projection in the sense that $P^2 = P$. The linear subspaces $\mathscr{V}' = P\mathscr{V}$ and $\mathscr{V}'' = (I - P)\mathscr{V}$ of $\mathscr{V}$ are stable under the representation ϱ; the restriction of ϱ to $\mathscr{V}'$ is the representation ϱ', and the restriction of ϱ to $\mathscr{V}''$ is the representation ϱ''. The spaces $\mathscr{V}'$ and $\mathscr{V}''$ can also be characterized by

$$\mathscr{V}' = \{T \in \mathscr{V};\ PT = T\}; \mathscr{V}'' = \{T \in \mathscr{V}; PT = 0\}. \tag{13}$$

It follows immediately from (4) and the above observations that the mappings $\Theta' = P\Theta\colon G^2(n) \to \mathscr{V}'$ and $\Theta'' = (I - P)\Theta\colon G^2(n) \to \mathscr{V}''$ also satisfy an equation of the form (4), with the representations ϱ' and ϱ'' respectively; hence these mappings determine pseudogroups as well. Moreover if λ_α is a connection for the pseudogroup structure $H^2(\Theta)$ on a manifold, then $\lambda'_\alpha = P\lambda_\alpha$ and $\lambda''_\alpha = (I - P)\lambda_\alpha$ are connections for the pseudogroup structures $H^2(\Theta')$ and $H^2(\Theta'')$ respectively; and conversely the direct sum $\lambda'_\alpha + \lambda''_\alpha$ of connections for the structures $H^2(\Theta')$ and $H^2(\Theta'')$ is a connection for the structure $H^2(\Theta)$.

To consider these new structures more concretely, suppose firstly that λ'_α is a connection for the pseudogroup structure $H^2(\Theta')$. Since λ'_α takes values in $\mathscr{V}'$, it follows from (12) and (13) that this connection function $\{\lambda'^{i}_{\alpha j_1 j_2}\}$ is actually fully determined by the differential from

$$\lambda^*_\alpha = \Sigma_{ij}\, \lambda'^{i}_{\alpha ij}\, dx^\alpha_j\,.$$

Now in any intersection $U_\alpha \cap U_\beta$,

$$\lambda^*_\alpha(x) = \lambda^*_\beta(x) + \Sigma_{ijk}\, \frac{\partial^2 x^\alpha_i}{\partial x^\beta_j\, \partial x^\beta_k}\, \frac{\partial x^\beta_j}{\partial x^\alpha_i}\, dx^\beta_k = \lambda^*_\beta(x) + d(\log \det(\partial x^\alpha_i/\partial x^\beta_j)). \tag{14}$$

The transition functions $k_{\alpha\beta}(x) = \det(\partial x^\alpha_i/\partial x^\beta_j)$ define a line bundle on M called the *canonical bundle*; and equation (14) is just the condition that $\lambda^*_\alpha(x)$ is the connection form for that bundle, in the usual sense, [*2*]. Therefore *a connection for the pseudogroup structure* $H^2(\Theta')$ *is equivalent to a connection for the canonical bundle of the manifold.* The condition that this connection be torsionless is just that the differential form λ^*_α be closed. It is not difficult to verify that the pseudogroup defined by $H^2(\Theta')$ consists of all transformations with constant Jacobian determinant. Next suppose that λ''_α is a connection for the pseudogroup structure $H^2(\Theta'')$; the connection functions $\{\lambda''^{i}_{\alpha j_1 j_2}\}$ are thus subject to the condition that $P\lambda''_\alpha = 0$. These, however, are just the *projective connections* as discussed in [*4*] and [*9*]. Therefore *a connection for the pseudogroup structure* $H^2(\Theta'')$ *is a projective connection on the manifold.* The condition that this connection be torsionless is that the torsion tensor associated to the connection vanish, as in [*4*]. It is also demonstrated in [*4*] that the pseudogroup defined by $H^2(\Theta'')$ consists of all projective transformations in n dimensions (generalized linear fractional transformations). Finally, the decomposition of connections for the pseudogroup structure $H^2(\Theta)$ can be restated as the theorem that *any* *affine connection on a manifold can be decomposed uniquely as the sum of a projective connection and a connection for the canonical bundle of the manifold.*

(c) It is not difficult to see that the three preceding examples are essentially the only subgroups $H^r(\Theta) \subset G^r(n)$, for $n > 1$ and arbitrary r, such that there are no restrictions on the values of the Jacobian matrices $D^1 F$ at any fixed point for the pseudogroup of mappings F satisfying $\Theta(D^r F) = 0$; hence all other such pseudogroups must impose some restrictions on the first-order terms $D^1(F)$, [5]. Now any subgroup $H^1 \subset G^1(n)$ such that the fibration $G^1(n) \to G^1(n)/H^1$ is a vector bundle can be written in the form $H^1(\Theta)$ for a suitable mapping Θ, and so the associated pseudogroup structure admits connections. For example, the orthogonal subgroup of $G^1(n, \mathbf{R})$ is defined by the mapping $\Theta(\xi) = I - {}^t\xi \circ \xi$, if we consider the elements $\xi \in G^1(n)$ as matrices, where $\mathscr{V}$ is the space of symmetric $n \times n$ matrices under the representation $\varrho = (-2)$; that is, equation (4) is of the form $\Theta(\xi \circ \eta) = {}^t\eta \circ \Theta(\xi) \circ \eta + \Theta(\eta)$. Upon writing equation (7) out in full in this case, it follows that a connection is determined by symmetric matrix-valued functions $\{\lambda_\alpha\}$ such that $(I - \lambda_\beta) = {}^t\xi^1_{\alpha\beta} \circ (I - \lambda_\alpha) \circ \xi^1_{\alpha\beta}$ in the intersection $U_\alpha \cap U_\beta$ of two coordinate neighborhoods; but this is just the condition that the form $(I - \lambda_\alpha)$ define a Riemannian metric, although not necessarily a positive definite one, on the manifold.

(d) Finally, if the coordinate structure of a manifold has already been reduced so that the transition functions $\xi^r_{\alpha\beta}$ lie in a subgroup $H^r \subset G^r(n)$, there arises the question of what further reductions may be possible; connections can be introduced paralleling the previous construction but replacing $G^r(n)$ by H^r. For example, in the case $r = 1$, consider a subgroup $H^1 \subset G^1(n)$ defined as follows: decompose the matrices $\xi \in G^1(n)$ into four blocks by writing

$$\xi = \begin{pmatrix} \xi^{11} & \xi^{12} \\ \xi^{21} & \xi^{22} \end{pmatrix},$$

and let $H^1 = \{\xi \in G^1(n); \xi^{12} = 0\}$. Introduce a mapping Θ from H^1 into matrix space by defining $\Theta(\xi) = (\xi^{22})^{-1}\xi^{21}$ for any $\xi \in H^1$; note that $\Theta(\xi \circ \eta) = (\eta^{22})^{-1}\Theta(\xi)\eta^{11} + \Theta(\eta)$, which is the analogue of equation (4), and that $H^1(\Theta) \subset H^1$ consists of the matrices

$$\xi = \begin{pmatrix} \xi^{11} & 0 \\ 0 & \xi^{22} \end{pmatrix}.$$

If M is a manifold for which the coordinate transition functions have already been reduced to the pseudogroup associated to $H^1 \subset G^1(n)$, then there is a connection for the sub-pseudogroup structure associated to $H^1(\Theta) \subset H^1$; this connection is given by matrix-valued functions $\{\lambda_\alpha\}$

defined in the coordinate neighborhoods $\{U_\alpha\}$, such that in each intersection $U_\alpha \cap U_\beta$,

$$(\xi^{22}_{\alpha\beta})^{-1}\xi^{21}_{\alpha\beta} + (\xi^{22}_{\alpha\beta})^{-1}\lambda_\alpha \xi^{11}_{\alpha\beta} - \lambda_\beta = 0\,. \tag{15}$$

In particular, if M is a vector bundle over a manifold N, then taking local coordinates $x^\alpha = (x^\alpha_i)$ in N and $v^\alpha = \{v^\alpha_a\}$ in the fibre (which is a linear space), the coordinate transition functions have the form

$$x^\alpha = F_{\alpha\beta}(x^\beta) \quad \text{and} \quad v^\alpha_a = \Sigma\ g^{\alpha\beta}_{ab}(x^\beta)\, v^b_b$$

where $g^{\alpha\beta}(x^\beta) = (g^{\alpha\beta}_{ab}(x^\beta))$ is a non-singular linear transformation on the fibre; thus the matrices $\xi^1_{\alpha\beta}$ determining the tangent bundle $\Xi^1(M)$ decompose as follows:

$$\xi = \begin{pmatrix} \dfrac{\partial x^\alpha_i}{\partial x^\beta_j} & 0 \\ \dfrac{\partial}{\partial x^\beta_j}\Sigma_b g^{\alpha\beta}_{ab}(x^\beta)\, v^\beta_b & g^{\alpha\beta}_{ab}(x^\beta) \end{pmatrix}.$$

The connection as introduced above is then given by matrix-valued functions $\lambda^\alpha = (\lambda^\alpha_{ai}(x^\alpha, v^\alpha))$ in terms of local coordinates (x^α, v^α); and introducing the differential forms $\Lambda^\alpha_a(x^\alpha, v^\alpha) = \Sigma_j \lambda^\alpha_{aj}(x^\alpha, v^\alpha)\, dx^\alpha_j$ and $dg^{\alpha\beta}_{ab}(x^\beta) = \Sigma_j(\partial g^{\alpha\beta}_{ab}/\partial x^\beta_j)\, dx^\beta_j$, equation (15) takes the form

$$\Sigma_{b,c}\, \tilde g^{\alpha\beta}_{ab}\, dg^{\alpha\beta}_{bc}\, v^\beta_c + \Sigma_b\, \tilde g^{\alpha\beta}_{ab}\, \Lambda^\alpha_b(x^\alpha, v^\alpha) - \Lambda^\beta_a(x^\beta, v^\beta) = 0\,, \tag{16}$$

where $\tilde g^{\alpha\beta} = (g^{\alpha\beta})^{-1}$. When $\Lambda^\alpha_a(x^\alpha, v^\alpha)$ is actually a linear function of v^α, then (16) reduces to the familiar notion of a linear connection in a vector bundle over N, [1, pp. 47—49].

References

[1] Chern, S. S.: Topics in differential geometry. Mimeographed lecture notes, Institute for Advanced Study, Princeton, 1951.

[2] — Lectures on complex manifolds. Mimeographed lecture notes, University of Chicago, 1956.

[3] Ehresmann, C.: Introduction à la théorie des structures infinitésimales et des pseudogroupes de Lie. Colloque de géométrie differentielle de Strasbourg. C. N. R. S., Paris, 1953, 97—110.

[4] Eisenhart, L. P.: Non-Riemannian geometry. Amer. Math. Soc. (New York) 1927.

[5] Gunning, R. C.: The defining groups of Lie pseudogroups (to appear).

[6] —, and H. Rossi: Functions of several complex variables. Englewood Cliffs, N. J.: Prentice-Hall 1965.

[7] Hirzebruch, F.: Neue topologische Methoden in der algebraischen Geometrie. Berlin-Göttingen-Heidelberg: Springer 1956.

[8] KOBAYASHI, S., and K. NOMIZU: Foundations of differential geometry I. New York: Interscience 1963.
[9] —, and T. NAGANO: On projective connections. J. Math. and Mech. 13, 215—236 (1964).
[10] WEYL, H.: The classical groups. Princeton: Princeton University Press 1946.

Department of Mathematics
Princeton University
Princeton, N. J.

A Fundamental Lemma on Point Modifications*, **

By

H. HIRONAKA

Introduction. A general point modification (proper) can be dominated by another point modification which has better properties than the given one. Some theorems of this nature have been announced in my paper [1] and applied to some problems in a joint paper [2]. For instance, the following is one of the most important ones.

Theorem. *Let $f: X \to Y$ be a proper morphism of complex varieties (reduced irreducible complex spaces) which induces an isomorphism $X - f^{-1}(y) \to Y - y$ for a unique point y of Y. Suppose $Y - y$ is non-singular. Then there exists a monoidal transformation $g: \tilde{Y} \to Y$ of Y with center D such that:*

(1) *D is a complex subspace (not necessarily reduced) of Y such that y is the only point of D,*

(2) *there exists a morphism (unique) $h: \tilde{Y} \to X$ such that $g = f \circ h$,*

(3) *$\tilde{Y}$ is non-singular, and*

(4) *the analytic set $g^{-1}(y)$ is a union of non-singular subvarieties of $\tilde{Y}$ of codimension* 1, *and it has only normal crossings.* (cf. *Corollary* 1 of Main Theorem II′, §7, Chap. 0, [1].)

The property (1) is equivalent to saying that *$\tilde{Y}$ can be realized as a complex subvariety of $Y \times P^N$ with a certain complex projective space P^N so that:*

(a) *the point modification $g: \tilde{Y} \to Y$ is induced by the projection morphism, and*

(b) *if $\boldsymbol{O}(1)$ is the invertible sheaf on $\tilde{Y}$ induced by the sheaf of sections of the line bundle on $Y \times P^N$ which is induced by the line bundle of hyper-*

* This work was supported in part by the Sloan Foundation.
** Received June 2, 1964.

planes on P^N, then we have:

(i) $\boldsymbol{O}(1)$ *is ample with respect to the morphism g, and*

(ii) *the restriction of $\boldsymbol{O}(1)$ to $Y - g^{-1}(y)$ is isomorphic to the structural sheaf $\boldsymbol{O}_Y$ of Y (globally).*

Here the ampleness in (i) means the following: If $\boldsymbol{F}$ is any coherent analytic sheaf on Y, then there exists an integer $m = m(\boldsymbol{F}) > 0$ such that, if $\boldsymbol{O}(m)$ denotes the m-th tensor power of $\boldsymbol{O}(1)$, then $\boldsymbol{F} \otimes \boldsymbol{O}(m)$ is generated by its direct image $R^0 g(\boldsymbol{F} \otimes \boldsymbol{O}(m))$ on Y and $R^q g(\boldsymbol{F} \otimes \boldsymbol{O}(m))$ is the zero sheaf on Y for all $q > 0$.

In the application of the above result in [2], the properties (i) and (ii) played essential roles.

The proof of the theorem is based upon a lemma (in addition to the resolution theorems proven in [1]) which I stated in [1] without any proof. This lemma is to take care of the part (1)—(2) of the theorem. To be precise:

Fundamental Lemma. *Let $f: X \to Y$ be a proper morphism of complex varieties, which induces an isomorphism $X - f^{-1}(y) \to Y - y$ for a unique point y of Y. Then there exists a monoidal transformation $g: \tilde{Y} \to Y$ of Y with center D such that*

(1) *y is the only point of D, and*

(2) *there exists a morphism (unique) $h: \tilde{Y} \to X$ such that $g = f \circ h$.*

The main purpose of this paper is to present a proof of this lemma. In the proof, the theory of *complex analytic Zariski spaces* plays the principal role. This theory is developed in this paper slightly more than we need for the proof of the Fundamental Lemma.

§ 1. Zariski spaces

By a *projective system of complex varieties*, we shall mean a set of complex varieties, $\mathscr{S}$, together with a set of morphisms (holomorphic maps) among those complex varieties in the set, $\mathscr{M}$, which satisfy the following conditions: i) $\mathscr{S}$ is non-empty, ii) if $p_X^{X'}: X' \to X$ and $p_{X'}^{X''}: X'' \to X'$ are both in $\mathscr{M}$, then $p_X^{X'} \circ p_{X'}^{X''}$ is in $\mathscr{M}$, iii) for every pair (X', X'') with both X' and X'' in $\mathscr{S}$, there exists at most one $p_{X'}^{X''}: X'' \to X'$ which is in $\mathscr{M}$, and iv) given any pair (X, X') with both X and X' in $\mathscr{S}$, there exists $X'' \in \mathscr{S}$ with $p_X^{X''}; X'' \to X$ and $p_{X'}^{X''}: X'' \to X'$ both in $\mathscr{M}$. In the category of local-ringed spaces over the field of complex numbers $\boldsymbol{C}$, in which the category of complex varieties is imbedded, every projective system has its limit. Let $\mathscr{B} = (\mathscr{S}, \mathscr{M})$ be a projective system of complex varieties as above, and let $\mathscr{Z} = \mathit{lim}\, \mathscr{B}$, the limit of $\mathscr{B}$. Then we have a canonical morphism $p_X: \mathscr{Z} \to X$ for every $X \in \mathscr{S}$ and, for every $p_X^{X'}$:

$X' \to X$ in $\mathscr{M}$, we have $p_X = p_X^{X'} \circ p_{X'}$. Here are listed some of the basic properties of the limit $\mathscr{Z}$ with those p_X, which will be used later.

a) Let $\mathscr{S} = \{X_a\}_{a \in L}$. Then for every consistent $\{x_a\}_{a \in L}$ with $x_a \in X_a$, there exists one and only one $z \in Z$ such that $x_a = p_a(z)$ for all $a \in L$.

b) Let W be a subset of $\mathscr{Z}$ (i.e. a set of points of $\mathscr{Z}$). Then W is open if and only if for every $z \in W$ there exists $X \in \mathscr{S}$ with an open subset U of X such that $z \in p_X^{-1}(U) \subset W$.

c) Let $z \in \mathscr{Z}$ and $x' = p_{X'}(z)$ for $X' \in \mathscr{S}$.

Then we have a homomorphism (local) $p_{X'}^*: \boldsymbol{O}_{X',x'} \to \boldsymbol{O}_{\mathscr{Z},z}$ (local rings, or, stalks of the structural sheaves). Similarly, for $x = p_X^{X'}(x') \in X \in \mathscr{S}$, we have $p_X^{X'*}: \boldsymbol{O}_{X,x} \to \boldsymbol{O}_{X',x'}$, and $p_X^* = p_{X'}^* \circ p_X^{X'*}$. For every $\xi \in \boldsymbol{O}_{\mathscr{Z},z}$, there exists $X \in \mathscr{S}$ and $\eta \in \boldsymbol{O}_{X,x}$ with $x = p_X(z)$ such that $p_X^*(\eta) = \xi$. For every $X \in \mathscr{S}$ and $\eta \in \boldsymbol{O}_{X,x}$ with $x = p_X(z)$, $p_X^*(\eta) = 0$ if and only if there exists $X' \in \mathscr{S}$ with $p_X^{X'}: X' \to X$ such that

$$p_X^{X'*}(\eta) = 0 .$$

Remark. For the existence of the limit and those properties of the limit stated above (and some stated below), it is not necessary to restrict oneself to complex varieties; for instance, we may consider any complex spaces, reduced or not, and irreducible or not. However, in this paper, we are exclusively interested in the case of complex varieties.

Let X be a complex variety, and let Φ be a set of closed analytic subsets of X. This set will be assumed to have the property that *if* $D \in \Phi$ *and* E *is a closed analytic subset of* D, *then* $E \in \Phi$; this will be referred to as *the property* **C**. Let L_Φ be the set of all coherent sheaves of ideals on X, say $\boldsymbol{J}$, such that *Supp* $(\boldsymbol{O}_X/\boldsymbol{J}) \in \Phi$. Then for each $\boldsymbol{J} \in L_\Phi$, take the monoidal transformation $p_X^{X'}: X' \to X$, of X with center defined by $\boldsymbol{J}$. Let $\mathscr{S}_\Phi$ be the set of those complex varieties X' obtained in this manner, which are indexed by the elements of L_Φ. Let $p_X^{X'}: X' \to X$ (resp. $p_X^{X''}: X'' \to X$) be the monoidal transformations with center defined by $\boldsymbol{J}' \in L$ (resp. $\boldsymbol{J}'' \in L$). If there exists $\boldsymbol{J} \in L$ such that $\boldsymbol{J}'' = \boldsymbol{J}'\boldsymbol{J}$ in $\boldsymbol{O}_X$, then one knows that there exists a canonical morphism $p_{X'}^{X''}: X'' \to X'$ such that $p_X^{X''} = p_X^{X'} \circ p_{X'}^{X''}$. Let $\mathscr{M}_\Phi$ be the set of those morphisms $p_{X'}^{X''}$ which are obtained in this manner. It is easily seen that $\mathscr{B}_\Phi = (\mathscr{S}_\Phi, \mathscr{M}_\Phi)$ is a projective system of complex varieties.

Definition 1.1. *The limit of* $\mathscr{B}_\Phi$ *will be denoted by* $\mathscr{Z}_\Phi(X)$ *and called Zariski space of* X *with center* Φ. *In particular, if* Φ *contains all the analytic subsets of* X *other than the entire* X, $\mathscr{Z}_\Phi(X)$ *will be denoted by* $\mathscr{Z}(X)$ *and called Zariski space of* X.

Let $\mathscr{B}_1 = (\mathscr{S}_1, \mathscr{M}_1)$ and $\mathscr{B}_2 = (\mathscr{S}_2, \mathscr{M}_2)$ be projective systems of complex varieties. Then a *morphism* $F: \mathscr{B}_1 \to \mathscr{B}_2$ is a covariant functor from $\mathscr{B}_1$ into $\mathscr{B}_2$, i. e., a set of morphisms of complex varieties $f_Y^X: X \to Y$

with $X \in \mathscr{S}_1$ and $Y \in \mathscr{S}_2$, having the properties: i) for each $X \in \mathscr{S}_1$, there is a unique $Y \in \mathscr{S}_2$ and a unique $f_Y^X : X \to Y$ which is in F, ii) if $p_X^{X'} : X' \to X$ is in $\mathscr{M}_1$ and if $f_{Y'}^{X'} : X' \to Y'$ and $f_Y^X : X \to Y$ are both in F, then there is $p_Y^{Y'} : Y' \to Y$ in $\mathscr{M}_2$ such that $f_Y^X \circ p_X^{X'} = p_Y^{Y'} \circ f_{Y'}^{X'}$, and iii) for every $Y \in \mathscr{S}_2$, there exists $f_{Y'}^{X'} \in F$ with $p_Y^{Y'} \in \mathscr{M}_2$. If $F : \mathscr{B}_1 \to \mathscr{B}_2$ is a morphism as above, then there exists a canonical morphism $\hat{f} : \mathscr{Z}_1 \to \mathscr{Z}_2$, where $\mathscr{Z}_i = \lim \mathscr{B}_i (i = 1, 2)$, such that for every $f_Y^X : X \to Y$ in F, $p_Y \circ \hat{f} = f_Y^X \circ p_X$. For example, suppose there is given a proper morphism of complex varieties $f : X \to Y$. Let Ψ be a set of analytic subsets of Y with the property C. Let us denote by $f^{-1}\Psi$ the smallest set of closed analytic subsets of X which has the property C and which contains all $f^{-1}(D)$ with $D \in \Psi$. Let $\Phi = f^{-1}\Psi$. Then one gets a canonical morphism $F : \mathscr{B}_\Phi \to \mathscr{B}_\Psi$ and a canonical morphism $\hat{f} : \mathscr{Z}_\Phi(X) \to \mathscr{Z}_\Psi(Y)$. In particular, if f is bimeromorphic (hence, surjective) and if Ψ is the set of all closed analytic subsets of Y other than the entire Y, then Φ is the set of all closed analytic subsets of X other than the entire X, and one gets a canonical morphism $\hat{f} : \mathscr{Z}(X) \to \mathscr{Z}(Y)$.

Here I propose the following

Conjecture: Let $f : X \to Y$ be a proper bimeromorphic morphism of complex varieties. Let D be the set of those points of Y at which f is not isomorphic. Let Ψ be the set of all closed analytic subsets of D (including D) and let $\Phi = f^{-1}\Psi$. Then the canonical morphism $f : \mathscr{Z}_\Phi(X) \to \mathscr{Z}_\Psi(Y)$ is an isomorphism of local-ringed spaces. In this paper I give an affirmative answer to this conjecture in a certain special but interesting case, i. e., the case of point modification.

§ 2. Analytic sets in Zariski spaces

Let $\mathscr{B} = (\mathscr{S}, \mathscr{M})$ be a projective system of complex spaces. It will be said to be *proper* if all the morphisms in $\mathscr{M}$ are proper. Let $\mathscr{Z} = \lim \mathscr{B}$ with $p_X : \mathscr{Z} \to X$ for every $X \in \mathscr{S}$ as before. Note that if $\mathscr{B}$ is proper then p_X is proper for all $X \in \mathscr{S}$. By an analytic subset of a complex variety will be meant always a closed analytic subset.

Definition 2.1. *A set of points of the local-ringed space $\mathscr{Z} = \lim \mathscr{B}$, say $\hat{A}$, is said to be analytic (or called an analytic subset of $\mathscr{Z}$) if there can be found an analytic subset A_X of each $X \in \mathscr{S}$ such that $\hat{A} = \cap\, p_X^{-1}(A_X)$.*

Obviously, an analytic subset of $\mathscr{Z}$ is closed.

Lemma 2.2. *Assume that the projective system $\mathscr{B}$ is proper. Let $\mathscr{Z} = \lim \mathscr{B}$ as above. Then a closed subset $\hat{A}$ of $\mathscr{Z}$ is analytic if and only if $p_X(\hat{A})$ is an analytic subset of X for all $X \in \mathscr{S}$, where $\mathscr{B} = (\mathscr{S}, \mathscr{M})$.*

Proof. Suppose $\hat{A}$ is analytic. Then, by definition, one can choose an analytic subset A_X of each $X \in \mathscr{S}$ so that $A = \bigcap_{X \in \mathscr{S}} p_X^{-1}(A_X)$. Now, for each $X \in \mathscr{S}$, let $B_X = \bigcap p_X^{X'}(A_{X'})$ where X' runs through all those $X' \in \mathscr{S}$ with $p_X^{X'}: X' \to X$ in $\mathscr{M}$. Since all the $p_X^{X'}$ are proper, B_X is an analytic subset of X. One can check that $B_X = p_X(\hat{A})$ for all $X \in \mathscr{S}$. Thus the only-if part of the lemma is proven. The converse is immediate.

Q. E. D.

Lemma 2.3. *Let $\mathscr{B}$ be a projective system of complex varieties, not necessarily proper, and let $\mathscr{X} = \lim \mathscr{B}$. Then every finite union of analytic subsets of $\mathscr{X}$ is analytic, and an arbitrary intersection of analytic subsets of $\mathscr{X}$ is analytic.*

The proof is immediate from the definitions.

Definition 2.4. *An analytic subset of $\mathscr{X}$, say $\hat{A}$, is said to be reducible if $\hat{A} = \hat{A}_1 \bigcup \hat{A}_2$ where $\hat{A}_i$ is analytic and different from $\hat{A}$ for each $i = 1, 2$. Otherwise it will be said to be irreducible.*

Lemma 2.5. *Let $\mathscr{B}$ be proper, and let $\mathscr{X} = \lim \mathscr{B}$. Let $\hat{A}$ be an analytic subset of $\mathscr{X}$. Then $\hat{A}$ is irreducible if and only if $p_X(\hat{A})$ is an irreducible analytic subset of X for all $X \in \mathscr{S}$.*

Proof. Suppose $\hat{A}$ is reducible. Let $\hat{A} = \hat{A}_1 \bigcup \hat{A}_2$ with analytic $\hat{A}_2 (i = 1, 2)$ such that $\hat{A} \neq \hat{A}_1$ and $\hat{A} \neq \hat{A}_2$. Take a point z_i of $\hat{A} - \hat{A}_i$ for $i = 1, 2$. Since $\hat{A}_i$ is closed, there exists $X_i \in \mathscr{S}$ and an open subset U_i of X_i such that $\hat{A}_i \bigcap p_{X_i}^{-1}(U_i) = \Phi$ and $z_i \in p_{X_i}^{-1}(U_i)$. Take $X \in \mathscr{S}$ with $p_{X_i}^X : X \to X_i$ in $\mathscr{M}$ for both $i = 1, 2$. It is clear that $p_X(z_i) \in p_X(\hat{A}) - p_X(\hat{A}_i)$ for each $i = 1, 2$, and that $p_X(\hat{A}) = p_X(\hat{A}_1) \bigcup p_X(\hat{A}_2)$. Hence $p_X(\hat{A})$ is reducible. Conversely, suppose $p_X(\hat{A})$ is reducible for some $X \in \mathscr{S}$. So $p_X(\hat{A}) = A_1 \bigcup A_2$ with analytic $A_i \neq p_X(\hat{A})$ for $i = 1, 2$. Let $\hat{A}_i = \hat{A} \bigcap p_X^{-1}(A_i)$. Then $\hat{A} = \hat{A}_1 \bigcup \hat{A}_2$, $\hat{A}_i$ analytic and $\neq \hat{A}$ for $i = 1, 2$ (cf. Lemma 2.3.). Q. E. D.

Lemma 2.6. *Assume that $\mathscr{B}$ is proper. Let $\hat{A}$ be any analytic subset of $\mathscr{X}$. Let $X \in \mathscr{S}$ and let B be an irreducible analytic subset of X which is contained in $p_X(\hat{A})$. Then there exists an irreducible analytic subset $\hat{B}$ of $\mathscr{X}$ such that $p_X(\hat{B}) = B$ and $\hat{B} \subseteq \hat{A}$.*

Proof. Let $\mathscr{S}_0$ be the subset of $\mathscr{S}$ of those X' with $p_X^{X'} \in \mathscr{M}$. For each $X' \in \mathscr{S}_0$, let $T_{X'}$ denote the set of those irreducible analytic subsets of X' which are contained in $p_{X'}(\hat{A})$ and mapped onto B by $p_X^{X'}$. One can see that $T_{X'}$ is not empty, that if $p_{X'}^{X''} \in \mathscr{M}$ then $p_{X'}^{X''}$ induces a map $T_{X''} \to T_{X'}$ in the obvious manner, and that this map is surjective. It then follows that one can choose one $B_{X'}$ in each $T_{X'}$ in such a way that

$p_{X'}^{X''}(B_{X''}) = B_{X'}$ for all $p_{X'}^{X''} \in \mathscr{M}$ and $X' \in \mathscr{S}_0$. Let $\hat{B} = \bigcap_{X' \in \mathscr{S}_0} p_{X'}^{-1}(B_{X'})$. Then $\hat{B}$ is an analytic subset of $\mathscr{X}$ and $p_{X'}(\hat{B}) = B_{X'}$ for all $X' \in \mathscr{S}_0$. It follows that $\hat{B}$ is irreducible. Moreover, it is clear that $\hat{B} \subseteqq \hat{A}$ and that $p_X(\hat{B}) = B$. Q. E. D.

Let B be any subset of a complex variety X. Then there exists the smallest analytic subset A of X that contains B. This is due to the fact that an arbitrary intersection of analytic subsets is an analytic subset.

Definition 2.7. *Let $\hat{A}$ be an analytic subset of $\mathscr{X}$. Let A_X be the smallest analytic subset of X that contains $p_X(\hat{A})$, for each $X \in \mathscr{S}$. Then the dimension of $\hat{A}$, dim $\hat{A}$ in symbol, is*

$$\sup_{X \in \mathscr{S}} \left\{ \inf_{p_X^{X'} \in \mathscr{M}} (\dim A_{X'}) \right\}$$

Lemma 2.8. *Suppose $\mathscr{B}$ is proper. Then for an analytic subset $\hat{A}$ of $\mathscr{X}$, dim $\hat{A}$ is equal to sup $\{\dim p_X(\hat{A})\}$. Moreover, if dim $\hat{A}$ is finite, then there exists $X \in \mathscr{S}$ such that dim $\hat{A} = \dim p_{X'}(\hat{A})$ for all $X' \in \mathscr{S}$ with $p_X^{X'} \in \mathscr{M}$.*

The proof is immediate from the definition and from the fact that if $p : X' \to X$ is a proper morphism and if A' is an analytic subset of X', then $\dim A' \geqq \dim p(A')$.

Corollary. *Let A' be another analytic subset of $\mathscr{X}$. Suppose $\hat{A} \subseteqq \hat{A}'$ and that $\hat{A}'$ is irreducible. Then dim $\hat{A}$ = dim $\hat{A}'$ implies $\hat{A} = \hat{A}'$, provided dim $\hat{A}'$ is finite.*

Proof. By Lemma 2.5, $p_X(\hat{A}')$ is irreducible for all $X \in \mathscr{S}$. Take any $X \in \mathscr{S}$ such that $\dim \hat{A} = \dim p_X(\hat{A})$ and $\dim \hat{A}' = \dim p_X(\hat{A}')$. Then $\dim \hat{A} = \dim \hat{A}'$ (finite) implies $p_X(\hat{A}) = p_X(\hat{A}')$ because of the irreducibility of the second set. Now, $\hat{A} = \bigcap_{X \in \mathscr{S}} p_X^{-1}(p_X(\hat{A})) = \bigcap_{X \in \mathscr{S}} p_X^{-1}(p_X(\hat{A}')) = \hat{A}'$. Q. E. D.

Definition 2.9. *Let $\hat{A}$ be an analytic subset of $\mathscr{X} = \lim \mathscr{B}$. Then an irreducible component of $\hat{A}$ is an irreducible analytic subset $\hat{B}$ of $\mathscr{X}$ such that $\hat{B} \subseteqq \hat{A}$ and that there exists no irreducible analytic subset $\hat{B}'$ of $\mathscr{X}$ with $\hat{B} \subset \hat{B}' \subseteqq \hat{A}$ (other than $\hat{B}$ itself). An irreducible component $\hat{B}$ of $\hat{A}$ is said to be primary if there exists $X \in \mathscr{S}$ such that for every X' with $p_X^{X'} \in \mathscr{M}$, the smallest analytic subset of X' containing $p_{X'}(\hat{B})$ is an irreducible component of the smallest analytic subset of X' containing $p_{X'}(\hat{A})$. An irreducible component of $\hat{A}$ is said to be secondary if it is not primary.*

Recall that, if $\mathscr{B}$ is proper, $p_{X'}(\hat{B})$ and $p_{X'}(\hat{A})$ are both analytic and $p_{X'}(\hat{B})$ is irreducible for all $X' \in \mathscr{S}$.

Example. 2.10. Let X_0 be a complex manifold (i. e. a non-singular complex variety) of dimension $n \geqq 2$. Let us define a sequence of complex varieties X_i with morphisms $h_i: X_i \to X_{i-1} (i = 1, 2, \ldots)$ as follows: Let x_0 be a point of X_0 and let $h_1: X_1 \to X_0$ be the monoidal transformation of X_0 with center x_0. Let $E_1 = h_1^{-1}(x_0)$. Take a point x_1 of E_1 and let $h_2: X_2 \to X_1$ be the monoidal transformation of X_1 with center x_1. Let $E_2 = h_2^{-1}(x_1)$. Take a point x_2 of E_2 which is not on the strict transform of E_1 on X_2 (i. e., the closure of $h_2^{-1}(E_1 - x_1)$ in X_2). And repeat the process so that, for each i, $h_{i+1}: X_{i+1} \to X_i$ is the monoidal transformation of X_i with center x_i and x_{i+1} is a point of $E_{i+1} = h_{i+1}^{-1}(x_i)$ which is not on the strict transform of E_i on X_{i+1}. Let $\mathscr{B} = (\mathscr{S}, \mathscr{M})$ with $\mathscr{S}$ = the set of those X_i and $\mathscr{M}$ = the set of those h_i and their compositions. Let $\mathscr{X} = \lim \mathscr{B}$ with $p_i: \mathscr{X} \to X_i$ for all i. Then $\hat{A} = p_0^{-1}(x_0)$ is an analytic subset of $\mathscr{X}$. Let $\hat{B}$ be the point of $\mathscr{X}$ such that $p_i(\hat{B}) = x_i$ for all i. Then $\hat{B}$ is an irreducible component of $\hat{A}$, which is secondary. Note that all the other irreducible components of $\hat{A}$ are primary and have dimension one.

Lemma 2.11. *Suppose $\mathscr{B}$ is proper. Let $\hat{A}$ be any analytic subset of $\mathscr{X} = \lim \mathscr{B}$. Assume that $\dim \hat{A}$ is finite. Then, for every $X \in \mathscr{S}$, every irreducible component of $p_X(\hat{A})$ is the image of some primary irreducible component of $\hat{A}$. In particular, if A denotes the union of all primary irreducible components of $\hat{A}$, then $p_X(A) = p_X(\hat{A})$ for all $X \in \mathscr{S}$ and $\hat{A}$ is equal to the closure of A.*

Proof. I shall prove the first assertion. The rest follows immediately. So take an arbitrary $X \in \mathscr{S}$ and an arbitrary irreducible component B_X of $p_X(\hat{A})$. For every $X' \in \mathscr{S}$ with $p_X^{X'} \in \mathscr{M}$, there exists at least one irreducible component $B_{X'}$ of $p_{X'}(\hat{A})$ which is mapped onto B_X. Consider all such $B_{X'}$ and let r be the maximum of $\dim B_{X'}$, which is an integer because $\dim \hat{A}$ is finite. Choose an X' and a $B_{X'}$ such that $\dim B_{X'} = r$. Then, by Lemma 2.6., there exists an irreducible analytic subset $\hat{B}$ of $\mathscr{X}$ such that $\hat{B} \subsetneqq \hat{A}$ and $p_{X'}(\hat{B}) = B_{X'}$. Let $X'' \in \mathscr{S}$ with $p_{X'}^{X''} \in \mathscr{M}$. Then $p_{X''}(\hat{B})$ is irreducible and mapped onto $B_{X'}$. Hence it has dimension $\geqq r$. Therefore, by the above selection of r, $p_{X''}(\hat{B})$ must be an irreducible component of $p_{X''}(\hat{A})$. Hence, $\hat{B}$ is an irreducible component of $\hat{A}$, which is primary. Clearly $p_X(\hat{B}) = B_X$. It follows that $p_X(\hat{A}) = p_X(A)$.

Q. E. D.

Remark 2.12. One can show by examples that the finiteness of $\dim \hat{A}$ is essential. In fact, there is an example of an analytic subset $\hat{A}$ of

$\mathscr{Z} = lim\, \mathscr{B}$ with a proper $\mathscr{B}$, such that all the irreducible components of $\hat{A}$ are secondary (necessarily $dim\, \hat{A} = \infty$).

§ 3. Morphisms of Zariski spaces

Let $\mathscr{B}_i = (\mathscr{S}_i, \mathscr{M}_i)$ be a projective system of complex varieties for $i = 1, 2$. Let $\mathscr{Z}_i = lim\, \mathscr{B}_i$ for $i = 1, 2$. The canonical morphism $\mathscr{Z}_i \to \to S\,(S \in \mathscr{S}_i)$ will be denoted by p_S. A morphism $F: \mathscr{B}_1 \to \mathscr{B}_2$ is said to be *proper* if all the morphisms of complex varieties in F are proper. A morphism $\hat{f}: \mathscr{Z}_1 \to \mathscr{Z}_2$ (local-ringed spaces over the complex number field) is said to be *proper* if every compac tsubset K of $\mathscr{Z}_2$ has a compact inverse image $f^{-1}(K)$.

Lemma 3.1. *Let $F: \mathscr{B}_1 \to \mathscr{B}_2$ be a morphism and $\hat{f}: \mathscr{Z}_1 \to \mathscr{Z}_2$ the induced morphism. If $\mathscr{B}_1$ and F are both proper, then $\hat{f}$ is proper.*

Proof. For every compact subset K of $\mathscr{Z}_2$, the assumptions imply that $H = \bigcap_{X \in \mathscr{S}_1} p_X^{-1}\{(f_Y^X)^{-1}(p_Y(K))\}$ is compact, where $f_Y^X \in F$ for each $X \in \mathscr{S}_1$. Clearly, $H = \hat{f}^{-1}(K)$. Q. E. D.

A morphism $F: \mathscr{B}_1 \to \mathscr{B}_2$ is said to be *surjective* if all the morphisms in F are surjective.

Lemma 3.2. *If $F: \mathscr{B}_1 \to \mathscr{B}_2$ is surjective and if $\mathscr{B}_1$ and F are proper, then the induced morphism $\hat{f}: \mathscr{Z}_1 \to \mathscr{Z}_2$ is proper and surjective.*

Proof. Take any point z of $\mathscr{Z}_2$. Then the assumptions imply that $\bigcap_{X \in \mathscr{S}_1} p_X^{-1}\{(f_Y^X)^{-1}(p_Y(z))\}$ is not empty, because all the sets appearing in the intersection are compact and have the finite intersection property. This set is, on the other hand, equal to $\hat{f}^{-1}(z)$. Q. E. D.

Lemma 3.3. *Suppose both $\mathscr{B}_1$ and $F: \mathscr{B}_1 \to \mathscr{B}_2$ are proper. Then $\hat{f}: \mathscr{Z}_1 \to \mathscr{Z}_2$ has the property that the image of an analytic subset is analytic.*

Proof. Let $\hat{S}$ be an analytic subset of $\mathscr{Z}_1$. Then for each $X \in \mathscr{S}_1$, $S_X = p_X(\hat{S})$ is an analytic subset of X and hence $T_Y = f_Y^X(S_X)$ with $f_Y^X \in F$ is an analytic subset of Y. It is clear that $T_Y = p_Y(\hat{T})$ where $\hat{T} = \hat{f}(\hat{S})$. Thus the closure of $\hat{T}$ in $\mathscr{Z}_2$ is equal to the analytic subset $\bigcap p_Y^{-1}(T_Y)$ where Y runs through all those with $f_Y^X \in F$. But $\hat{T}$ is closed because $\hat{f}: \mathscr{Z}_1 \to \mathscr{Z}_2$ is proper by *Lemma* 3.1. Q. E. D.

Lemma 3.4. *Suppose both $\mathscr{B}_1$ and F are proper. Then for every analytic subset $\hat{A}$ of $\mathscr{Z}_1$, $dim\, \hat{A} \geq dim\, \hat{f}(\hat{A})$.*

Proof. Let $\hat{B} = \hat{f}(\hat{A})$. For every $Y \in \mathscr{S}_2$, there exists $f_{Y'}^{X'} \in F$ with $p_Y^{Y'} \in \mathscr{M}_2$. Then $dim\, \hat{A} \geq dim\, p_{X'}(\hat{A}) \geq p_{Y'}(\hat{B})$, where the second inequality is due to $p_{Y'} \circ \hat{f} = f_{Y'}^{X'} \circ p_{X'}$ and to the properness of these morphisms. It is then immediate from the definition of dimension that $dim\ \hat{A} \geqq sup(dim\ p_{Y'}(\hat{B})) \geqq dim\ \hat{B}$. Q. E. D.

Lemma 3.5. *Suppose* $\mathscr{B}_1$ *is proper. Let* $\hat{f}: \mathscr{X}_1 \to \mathscr{X}_2$ *be a morphism, where* $\mathscr{X}_i = lim\ \mathscr{B}_i$ *for* $i = 1, 2$. *Let* $\mathscr{B}_i = (\mathscr{S}_i, \mathscr{M}_i)$, $X \in \mathscr{S}_1$ *and* $Y \in \mathscr{S}_2$. *Then the image* $\tilde{X}$ *of* $p_X \times (p_Y \circ \hat{f}) : \mathscr{X}_1 \to X \times Y$ *is an irreducible analytic set and the projection of* $X \times Y$ *to* X *induces a proper morphism* $h : \tilde{X} \to X$. *Moreover, given* $Y \in \mathscr{S}_2$, *one can choose* $X \in \mathscr{S}_1$ *so that* h *is a bimeromorphic morphism.*

Proof. Since $\mathscr{B}_1$ is proper, $p_X : \mathscr{X}_1 \to X$ is proper for all $X \in \mathscr{S}_1$. Let $d = p_X \times (p_Y \circ \hat{f})$. Then $d : \mathscr{X}_1 \to X \times Y$ is proper. It follows that the image $\tilde{X}$ of d is a closed subset of $X \times Y$. To prove that $\tilde{X}$ is analytic, take any point $\tilde{x} = (x, y)$ of $\tilde{X}$. Let U be a relatively compact open neighborhood of x in X. Then $p_X^{-1}(\bar{U})$ is compact is $\mathscr{X}_1$, and therefore one can find $X' \in \mathscr{S}_1$ and an open neighborhood V' of $(p_X^{X'})^{-1}(\bar{U})$ in X' such that there exists a morphism $f : V' \to Y$ such that $f \circ p_{X'} = p_Y \circ \hat{f}$ within the open subset $p_{X'}^{-1}(V')$ of $\mathscr{X}_1$. (Note that the last condition determines f uniquely whenever it exists.) Now, let $\tilde{X}'$ be the image of $p_{X'} \times (p_Y \circ \hat{f})$. Then $\tilde{X}$ is the image of $\tilde{X}'$ by the morphism $p_X^{X'} \times id_Y$. The existence of f, however, shows that $\tilde{X}' \cap V' \times Y$ is the graph of the morphism f and hence an analytic subset of $V' \times Y$. Since $p_X^{X'} \times id_Y$ is proper and $V' \times Y$ contains the preimage of $U \times Y$, $\tilde{X} \cap U \times Y$ is an analytic subset of $U \times Y$. In particular, $\tilde{X}$ is analytic in a neighborhood of $\tilde{x}$. One concludes that $\tilde{X}$ is analytic. $\tilde{X}$ is irreducible because it is the image of $\mathscr{X}_1$ by a morphism d. The morphism $h : \tilde{X} \to X$ is proper because p_X is proper. Finally, if X is replaced by X' as above, then the morphism $h' : \tilde{X}' \to X'$ is clearly bimeromorphic. Q. E. D.

Lemma 3.6. *Suppose that* $\mathscr{B}_1$, $\mathscr{B}_2$ *and* F *are all proper. Let* $\hat{B}$ *be an irreducible analytic subset of* $\mathscr{X}_2$, *and let* $\hat{D} = \hat{f}^{-1}(\hat{B})$. *Suppose* $dim\ \hat{D}$ *is finite. If* $\hat{f}(\hat{D}) = \hat{B}$, *then there exists at least one primary irreducible component* $\hat{A}$ *of* $\hat{D}$ *such that* $\hat{f}(\hat{A}) = \hat{B}$.

Proof. By *Lemma* 3.4, $dim\ \hat{D} \geqq dim\ \hat{B}$. In particular, $dim\ \hat{B}$ is finite. By Lemmas 2.2 and 2.8, there exists $Y \in \mathscr{S}_2$ such that if $B_Y = p_Y(\hat{B})$, $dim\ \hat{B} = dim\ B_Y$. Here Y may be replaced by any $Y' \in \mathscr{S}_2$ with $p_Y^{Y'} \in \mathscr{M}_2$. Hence one can assume that there exists $X \in \mathscr{S}_1$ and $f_Y^X \in F$. Let $D_X = p_X(\hat{D})$. Since $\hat{f}(\hat{D}) = \hat{B}$, we have $f_Y^X(D_X) = B_Y$. It follows that there exists an irreducible component A_X of D_X such that $f_Y^X(A_X) = B_Y$. By *Lemma* 2.11, there exists a primary irreducible component $\hat{A}$ of $\hat{D}$ such that $p_X(\hat{A}) = A_X$. Now I claim that $\hat{A}$ has the required property: $\hat{f}(\hat{A}) = \hat{B}$. In fact, $\hat{f}(\hat{A}) \subseteq \hat{B}$ and $p_Y(\hat{f}(\hat{A})) = f_Y^X(p_X(\hat{A})) = f_Y^X(A_X) = B_Y$. Hence $dim\, \hat{f}(\hat{A}) \geqq dim\, B_Y = dim\, \hat{B}$. By the corollary of *Lemma* 2.8, $\hat{f}(\hat{A}) = \hat{B}$. Q. E. D.

§ 4. The main theorem on irregular correspondences

A complex space X is called a complex S-space, meaning that a morphism of complex spaces $h: X \to S$ is specified. (Or, more logically, a complex S-space is a pair (X, h).) Let Y be another complex S-space with $g: Y \to S$. Then a morphism of complex S-spaces $f: X \to Y$, or simply an S-morphism, is a morphism of complex spaces such that $h = g \circ f$. In the category of complex S-spaces for a fixed complex space S, the product exists. This means that if X and Y are complex S-spaces as above, then there exists a complex S-space P with S-morphisms $p: P \to X$ and $q: P \to Y$ having the following universal mapping property: For any pair of S-morphisms $f': T \to X$ and $g': T \to Y$, there exists a unique S-morphism $h': T \to P$ such that $f' = p \circ h'$ and $g' = q \circ h'$. The above product P is also called *the fibre product of X and Y over S and denoted by $X \times_S Y$*. The morphisms p and q are called *projection morphisms*, or just *projections*. The fibre product $X \times_S Y$ is obtained as a complex subspace of the product of complex spaces, $X \times Y$ (the product in the category of complex spaces). Let $x \in X$ and $y \in Y$ such that $f(x) = g(y) = s \in S$. Choose a coordinate system $(x_1, \ldots, x_n)$ of X at x and $(y_1, \ldots, y_m)$ of Y at y. Suppose X is locally defined by the equations $a_1(x) = a_2(x) = \cdots = a_i(x) = 0$ in $\boldsymbol{C}^n$ and Y by $b_1(y) = \cdots = b_j(y) = 0$ in $\boldsymbol{C}^m$. Then $X \times Y$ is locally defined by the equations $a_1(x) = \cdots = a_i(x) = b_1(y) = \cdots = b_j(y) = 0$ in $\boldsymbol{C}^{n+m} = \boldsymbol{C}^n \times \boldsymbol{C}^m$. The fibre product $X \times_S Y$ is locally defined by these equations plus the equations: $f_\alpha(x) = g_\alpha(y)$ for $\alpha = 1, 2, \ldots, t$, where the morphisms f and g are written, in terms of a local coordinate system $(s_1, \ldots, s_t)$ of S at s, as $s_\alpha = f_\alpha(x)$ and $s_\alpha = g_\alpha(y)$ for $\alpha = 1, 2, \ldots, t$ respectively.

Let $\mathscr{B}_1 = (\mathscr{S}_1, \mathscr{M}_1)$ and $\mathscr{B}_2 = (\mathscr{S}_2, \mathscr{M}_2)$ be projective systems of complex varieties. A morphism $F: \mathscr{B}_1 \to \mathscr{B}_2$ being given, I shall say that *$\mathscr{B}_1$ is induced from $\mathscr{B}_2$ (by the morphism F)* if for every $p_X^{X'} \in \mathscr{M}_1$ and $p_Y^{Y'} \in \mathscr{M}_2$ such that $f_{Y'}^{X'} \in F$ and $f_Y^X \in F$, there exists an imbedding of X' into $X \times_Y Y'$ which gives $f_{Y'}^{X'}$ and $p_X^{X'}$ by means of projection morphisms. In this paper, I am primarily interested in the case where all the morphisms in $\mathscr{M}_1$ and $\mathscr{M}_2$ are proper bimeromorphic and all the morphisms in F are proper surjective. In this case, if $\mathscr{B}_1$ is induced from $\mathscr{B}_2$, then any one of the morphisms in F "essentially" determines B_1 and F. Namely, $p_X^{X'}$, $p_Y^{Y'}$, $f_{Y'}^{X'}$ and f_Y^X being as above, $p_X^{X'}$ and $f_{Y'}^{X'}$ are uniquely determined by $p_Y^{Y'}$ and f_Y^X through an isomorphism of X' to the unique irreducible component of $X \times_Y Y'$ which projects surjectively to X and Y'.

Let $f: X \to Y$ be a morphism of complex varieties. Let x be a point of X and $y = f(x)$. Then one can prove that the following conditions are equivalent to one another:

(i) *The local ring* $\boldsymbol{O}_{X,x}$ *is an integral extension of* $\boldsymbol{O}_{Y,y}$ *by means of* f, i. e., *every element of* $\boldsymbol{O}_{X,x}$ *safisfies a monic polynomial equation with coefficients in* $\boldsymbol{O}_{Y,y}$. (It is not required that $O_{Y,y} \to O_{X,x}$ is injective.)

(ii) *The maximal ideal of* $\boldsymbol{O}_{Y,y}$ *generates a primary ideal in* $\boldsymbol{O}_{X,x}$ *belonging to the maximal ideal.*

(iii) *The analytic subset* $f^{-1}(y)$ *of* X *has* x *as an isolated point.*

(The implication (ii) $\Rightarrow$ (i) is proved by the Weierstrass preparation theorem.) If one (hence all) of these conditions is satisfied, then f is said to be *integral at* x. It is easy to see that f is integral at x if and only if:

(iv) f *is integral at every point in a neighborhood of* x *in* X.

Given a point y of Y, I say that f is *integral at* y if it is so at every point of $f^{-1}(y)$; for instance, $f^{-1}(y)$ is empty. The morphism f is said to be *integral* if it is so at every point of X. It is known that the set of points $x \in X$ at which f is integral is the complement of an analytic subset of X.

Let $\mathscr{B}_1$ and $\mathscr{B}_2$ be projective systems of complex varieties, and let $\mathscr{Z}_i = \lim \mathscr{B}_i$ for $i = 1, 2$. A morphism $\hat{f}: \mathscr{Z}_1 \to \mathscr{Z}_2$ will be said to be *integral at* $z_1 \in \mathscr{Z}_1$ if $\boldsymbol{O}_{\mathscr{Z}_1, z_1}$ is an integral extension of $\boldsymbol{O}_{\mathscr{Z}_2, z_2}$ (by means of $\hat{f}$), where $z_2 = \hat{f}(z_1)$. I shall say that $\hat{f}$ is integral at $z_2 \in \mathscr{Z}_2$, if it is so at every point of $\hat{f}^{-1}(z_2)$; for instance, $\hat{f}^{-1}(z_2)$ is empty.

Lemma 4.1. *Suppose* $\mathscr{B}_1$ *is induced from* $\mathscr{B}_1$ *by a morphism* $F: \mathscr{B}_1 \to$ $\to \mathscr{B}_2$. *Let* $\hat{f}: \mathscr{Z}_1 \to \mathscr{Z}_2$, *with* $\mathscr{Z}_i = \lim \mathscr{B}_i$, *be the morphism induced by* F. *Let* $z_1 \in \mathscr{Z}_1$. *Then* $\hat{f}$ *is integral at* z_1 *if and only if there exists* $X \in \mathscr{S}_1$ *such that* $f_Y^X \in F$ *is integral at* $p_X(z) \in X$.

Proof. Suppose $\hat{f}$ is integral at $z_1 \in \mathscr{Z}_1$. Let $z_2 = \hat{f}(z_1)$. Let $\hat{\boldsymbol{O}}_i = \boldsymbol{O}_{\mathscr{Z}_i, z_i}$ for $i = 1, 2$. Take any $X \in \mathscr{S}_1$. Let $x = p_X(z_1)$. Let $(x_1, \ldots, x_n)$ be the coordinate functions of X at x. Let ξ_i be the image of x_i in $\hat{\boldsymbol{O}}_1$. Then each ξ_i satisfies a monic polynomial equation with coefficients in $\hat{\boldsymbol{O}}_2$, say $H_i(\xi_i) = 0$. There exists $Y \in \mathscr{S}_2$ such that if $y = p_Y(z_2)$ then all the coefficients of the polynomials $H_i (1 \leqq i \leqq n)$ belong to the image of $\boldsymbol{O}_{Y,y}$ in $\hat{\boldsymbol{O}}_2$. Now, take $X' \in \mathscr{S}_1$ with $p_X^{X'} \in \mathscr{M}_1$, $f_{Y'}^{X'} \in F$ and $p_Y^{Y'} \in \mathscr{M}_2$. Let $x' = p_{X'}(z_1)$, and let $(x_1', \ldots, x_n')$ be the image of $(x_1, \ldots, x_n)$ in the local ring $\boldsymbol{O}_{X',x'}$. For each coefficient of the polynomial H_i, firstly take one of its representatives in $\boldsymbol{O}_{Y,y}$ and then take its image in $\boldsymbol{O}_{X',x'}$ by means of $p_Y^{Y'} \circ f_{Y'}^{X'}$. Let H_i' be the polynomial obtained by replacing the coefficients of H_i by the so obtained elements of $\boldsymbol{O}_{X',x'}$. Consider the elements $H_i'(x_i')$ of $O_{X',x'}$ for $1 \leqq i \leqq n$. Clearly these elements are mapped to zero by the homomorphism $\hat{O}_{X',x'} \to O_1$. Therefore, by replacing X' if necessary, one can assume that $H_i'(x_i') = 0$ for all i. Since $f_{Y'}^{X'}$ is induced by an embedding of X' into the fibre product $X \times_Y Y'$ by assumption, it follows that $f_{Y'}^{X'}$ is integral at the point x'. This shows the only-if part of the assertion. The if-part is immediate in view of the assumption that $\mathscr{B}_1$ is induced from $\mathscr{B}_2$. Q. E. D.

Corollary 1. *The assumptions being the same as in the lemma, there exists an analytic subset $\hat{A}$ of $\mathscr{Z}_1$ such that $\hat{f}$ is integral at a point $z_1 \in \mathscr{Z}_1$ if and only if $z_1 \in \mathscr{Z}_1 - \hat{A}$.*

Proof. For each $X \in \mathscr{S}_1$, let A_X be the analytic subset of X such that $f_Y^X \in F$ is integral at $x \in X$ if and only if $x \in X - A_X$. Then let $\hat{A} = \bigcap_{X \in \mathscr{S}_1} p_X^{-1}(A_X)$. By *Lemma* 4.1, it is enough to show that this $\hat{A}$ has the required property. Q. E. D.

Corollary 2. *Suppose, in addition, that $\mathscr{B}_1$ and F are proper. Then $\hat{f}$ is integral at $z_2 \in \mathscr{Z}_2$ if and only if there exists $X \in \mathscr{S}_1$ such that $f_Y^X \in F$ is integral at $p_Y(z_2) \in Y$. Moreover, there exists an analytic subset $\hat{B}$ of $\mathscr{Z}_2$ such that $\hat{f}$ is integral at $z_2 \in \mathscr{Z}_2$ if and only if $z_2 \in \mathscr{Z}_2 - \hat{B}$.*

Proof. Let $\hat{A}$ be the analytic subset of $\mathscr{Z}_1$ such that $\hat{f}$ is integral at $z_1 \in \mathscr{Z}_1$ if and only if $z_1 \in \mathscr{Z}_1 - \hat{A}$. By *Lemma* 3.1, $\hat{f}$ is proper. Therefore $\hat{f}^{-1}(z_2)$ is compact. Let A_X be the analytic subset of $X \in \mathscr{S}_1$ of those points at which $f_Y^X \in F$ is integral. Let $y = p_Y(z_2) \in Y$. Then we have $\hat{A} = \bigcap p_X^{-1}(A_X)$ and $\hat{f}^{-1}(z_2) = \bigcap p_X^{-1}(f_Y^X)^{-1}(y)$, where the first (resp. the second) intersection is for all $X \in \mathscr{S}_1$ (resp. all $f_Y^X \in F$). Here note that all the $(f_Y^X)^{-1}(y)$ are compact. Now, we can see that $\hat{f}^{-1}(z_2)$ does not meet $\hat{A}$ (or, equivalently, $\hat{f}$ is integral at z_2) if and only if there exists $f_Y^X \in F$ such that $(f_Y^X)^{-1}(y)$ does not meet A_X (or, equivalently, f_Y^X is integral at y). The second assertion is clear from *Lemma* 3.3, if we take $\hat{B} = \hat{f}(\hat{A})$. Q. E. D.

Lemma 4.2. *Let $f : X \to Y$ be a proper morphism of complex varieties. Let A be an irreducible analytic subset of X and let $B = f(A)$. If f is not integral at any point of A, then every irreducible component A' of $f^{-1}(B)$ with $A' \supseteqq A$ has dim $A' >$ dim B.*

Proof. Take any irreducible component A' of $f^{-1}(B)$ with $A' \supseteqq A$. We have then $f(A') = B$. Suppose *dim* A' = *dim* B. Then there exists a point $x \in A'$ such that $f^{-1}(f(x)) \cap A'$ contains x as an isolated point. We can choose such a point x so that it has the further property that A' is the only irreducible component of $f^{-1}(B)$ containing x. It then follows that $f^{-1}(f(x)) \cap f^{-1}(B)$, which is obviously $f^{-1}(f(x))$, contains x as an isolated point. This implies that f is integral at x, which contradicts the assumption. Hence we must have $dim\ A' > dim\ B$. Q. E. D.

Let A be an analytic subset of a complex variety and x a point of A. Then A determines a germ of analytic set at x. The dimension of this germ will be denoted by $dim_x A$.

Lemma 4.3. *Let X be a complex variety of dimension n. Let A be an analytic subset of X and x a point of A. Let $r = dim_x\ A$. Let s be an integer such that $n \geqq s \geqq n - r$. Then there exists an analytic subset T_s of X in a neighborhood of x which has the following properties:* (i) *$dim_x\ T_s = s$,* (ii) *$dim_x(A \cap T_s) = r + s - n$, and* (iii) *for every analytic subset D of X in a neighborhood of x, $dim_x(D \cap T_s) \geqq dim_x\ D + s - n$.*

Proof. If $n = s$, then take $T_s = X$. The proof in the general case is done by induction. Suppose we have obtained T_{s+1} for some $s \geqq n - r$. Let $\boldsymbol{O}_x$ be the local ring of X at x, and J the ideal of $A \cap T_{s+1}$ in $\boldsymbol{O}_x$. Then $r + s - n + 1$ is the Krull dimension of $\boldsymbol{O}_x/J$. Let I be the ideal of T_{s+1} in $\boldsymbol{O}_x$. Then $s + 1$ is the Krull dimension of $\boldsymbol{O}_x/I$. Choose an element h of the maximal ideal of $\boldsymbol{O}_x$ which is not contained in any prime ideal of either I or J. (Note that I and J are intersections of prime ideals.) Then, by a theorem of Krull, $\boldsymbol{O}_x/(J, h)$ has dimension $r + s - n$ and $\boldsymbol{O}_x/(I, h)$ has dimension s. By the same theorem, if H is any ideal in $\boldsymbol{O}_x$ then the dimension of $\boldsymbol{O}_x/(H, h)$ is at least that of $\boldsymbol{O}_x/H$. The three properties of T_s are then verified if we define it as the subset of T_{s+1} of those points satisfying the equation $h = 0$ in a suitable neighborhood of x. Q.E.D.

Lemma 4.4. *Let X be a complex variety, A an analytic subset of X, and x a point of A. Then there exists a compact subset E of X such that $A \cap E$ is empty and that $A^* \cap E$ is not empty for any analytic subset A^* of X with $dim_x\ A^* > dim_x\ A$.*

Proof. Let $dim_x\ A = r$. Take T_{n-r} which has the properties in *Lemma* 4.3. With reference to a local embedding of X into a complex number space, take the intersections of T_{n-r} with balls of radii r_1 and r_2 with $0 < r_1 < r_2$. Call these D_1 and D_2. We choose r_2 so small that $\bar{D}_2$ is compact and that within a neighborhood of $\bar{D}_2$, x is the only common point of A and T_{n-r}. Now, let $E = \bar{D}_2 - D_1$. This set has the required property, in virtue of the property (iii) of T_{n-r} and by the fact that D_2 is a Stein space. Q.E.D.

I say that a projective system of complex varieties $\mathscr{B} = (\mathscr{S}, \mathscr{M})$ is *surjective*, if all the morphisms in $\mathscr{M}$ are surjective. If $\mathscr{B}$ is surjective, then $p_X : \mathscr{X} \to X$ is surjective for all $X \in \mathscr{S}$, where $\mathscr{X} = lim\ \mathscr{B}$.

Theorem 4.5. *Let $F : \mathscr{B}_1 \to \mathscr{B}_2$ be a morphism of projective systems of complex varieties, and let $\hat{f} : \mathscr{X}_1 \to \mathscr{X}_2$ be the induced morphism, where $\mathscr{X}_i = lim\ \mathscr{B}_i$ for $i = 1, 2$. Suppose that $\mathscr{B}_1$, $\mathscr{B}_2$ and F are all proper, that*

$\mathscr{B}_1$ is induced from $\mathscr{B}_2$ by F, and that $\mathscr{B}_1$ is surjective. Let $\hat{T}$ be the set of those points of $\mathscr{X}_1$ at which $\hat{f}$ is not integral, and let $\hat{S} = \hat{f}(\hat{T})$. Then we have:

I) *$\hat{T}$ (resp. $\hat{S}$) is an analytic subset of $\mathscr{X}_1$ (resp. $\mathscr{X}_2$).*

II) *Let $\hat{B}$ be an irreducible analytic subset of $\mathscr{X}_2$ which is non-empty and finite-demensional. Then either one of the following is true:*

(a) *$\hat{B}$ is not contained in $\hat{S}$.*

(b) *There exists a primary irreducible component $\hat{C}$ of $\hat{f}^{-1}(\hat{B})$ such that $\hat{C} \subseteqq \hat{T}$ and $\hat{f}(\hat{C}) = \hat{B}$.*

III) *$\hat{B}$ being as above, if $\hat{A}$ is any primary irreducible component of $\hat{f}^{-1}(\hat{B})$ such that $\hat{f}(\hat{A}) = \hat{B}$, then either one of the following is true:*

(c) *$\hat{A}$ is not contained in $\hat{T}$ and $\dim \hat{A} = \dim \hat{B}$.*

(d) *$\hat{A}$ is contained in $\hat{T}$ and $\dim \hat{A} > \dim \hat{B}$.*

Proof. I) had been proved in the Corollaries of *Lemma* 4.1. Let $\hat{B}$ be the same as in II). Since $\mathscr{B}_1$ is induced from $\mathscr{B}_2$ by F, $\dim \hat{B} < \infty$ implies $\dim \hat{f}^{-1}(\hat{B}) < \infty$. In fact, as is easily seen, if $X \in \mathscr{S}_1$ and $f_Y^X \in F$, then $\dim \hat{f}^{-1}(\hat{B}) - \dim \hat{B} \leqq \dim (f_Y^X)^{-1}(B_Y) - \dim B_Y$, where $B_Y = p_Y(\hat{B})$. Now, suppose $\hat{B}$ is contained in $\hat{S}$. Let r be the maximum of the dimensions of those irreducible components $\hat{C}$ of $\hat{f}^{-1}(\hat{B})$ such that $\hat{f}(\hat{C}) = \hat{B}$. Then obviously $\dim \hat{B} \leqq r \leqq \dim \hat{f}^{-1}(\hat{B})$. Let us first prove that every irreducible component $\hat{C}$ of $\hat{f}^{-1}(\hat{B})$ with $\hat{f}(\hat{C}) = \hat{B}$ and $\dim \hat{C} = r$ is primary. In fact, choose $X \in \mathscr{S}_1$ and $f_Y^X \in F$ so that $\dim p_X(\hat{C}) = r$ and $\dim p_Y(\hat{B}) = \dim \hat{B}$. Take any $X' \in \mathscr{S}_1$ with $p_X^{X'} \in \mathscr{M}_1$. I claim that $p_{X'}(\hat{C})$ is an irreducible component of $p_X'(\hat{f}^{-1}(\hat{B}))$. Suppose it is not. Then there exists an irreducible component $C_{X'}$ of $p_{X'}(\hat{f}^{-1}(\hat{B}))$ such that $C_{X'} \supseteqq p_X'(\hat{C})$. By *Lemma* 2.11, there exists a primary irreducible component $\hat{C}'$ of $\hat{f}^{-1}(\hat{B})$ such that $p_{X'}(\hat{C}') = C_{X'}$. Then $\hat{f}(\hat{C}') \subseteqq \hat{B}$ and, if $f_{Y'}^{X'} \in F$, $p_{Y'}(\hat{f}(\hat{C}')) = f_{Y'}^{X'}(p_{X'}(\hat{C}')) \supseteqq f_{Y'}^{X'}(p_{X'}(\hat{C})) = p_{Y'}(\hat{B})$, so that $\dim \hat{f}(\hat{C}') \geqq \dim p_{Y'}(\hat{f}(\hat{C}')) \geqq \dim p_{Y'}(\hat{B}) \geqq \dim p_Y(\hat{B}) = \dim \hat{B}$. It follows, by the Corollary of *Lemma* 2.8, that $\hat{f}(\hat{C}') = \hat{B}$. However, $\dim \hat{C}' \geqq \dim p_{X'}(\hat{C}') > \dim p_{X'}(\hat{C}) = r$. This contradicts the above selection of r. Therefore, $p_{X'}(\hat{C})$ is an irreducible component of $p_{X'}(\hat{f}^{-1}(\hat{B}))$. Thus $\hat{C}$ is primary. We are now ready to prove (b). If $r > \dim B$, then every irreducible component $\hat{C}$ of $\hat{f}^{-1}(\hat{B})$ with $\hat{f}(\hat{C}) = \hat{B}$ and $\dim \hat{C} = r$ (which was proved to be primary) must be contained in $\hat{T}$. In fact, otherwise $\hat{f}$ is integral at some point of $\hat{C}$ and one gets $\dim \hat{C} = \dim \hat{f}(\hat{C})$, which is not the case. (To see this, Lemma 4.1. may be referred to.) So $\hat{C}$ has the properties required in (b). Now, consider the case of $r = dim\ \hat{B}$. In this case, take an irreducible component $\hat{C}$ of $\hat{f}^{-1}(\hat{B}) \cap \hat{T}$ such that $\hat{f}(\hat{C}) = \hat{B}$. Such a $\hat{C}$ exists because $\hat{f}(\hat{T}) = \hat{S} \supseteqq \hat{B}$. We have $dim\ \hat{C} \geqq dim\ \hat{B} = r$ and hence $dim\ \hat{C} = r$, so that $\hat{C}$ is a primary irreducible component of $\hat{f}^{-1}(\hat{B})$. Again this $\hat{C}$ has the required properties in (b). We shall next prove III). It suffices to prove that if $\hat{A}$ is contained in $\hat{T}$, then $dim\ \hat{A} > dim\ \hat{B}$. Choose $X \in \mathscr{S}_1$ such that $dim\ p_X(\hat{A}) = dim\ \hat{A}$ and, if $f_Y^X \in F$, $dim\ p_Y(\hat{B}) = dim\ \hat{B}$, and such that $p_X(\hat{A})$ is an irreducible component of $p_X(\hat{f}^{-1}(\hat{B}))$. Take a point $z_1 \in \hat{A}$ such that $p_X(z_1)$ is not contained in any irreducible component of $p_X(\hat{f}^{-1}(\hat{B}))$ other than $p_X(\hat{A})$. Let us write A for $p_X(\hat{A})$, B for $p_Y(\hat{B})$ and x for $p_X(z_1)$. By *Lemma* 4.4, choose a compact subset E of X such that $E \cap p_X(\hat{f}^{-1}(\hat{B}))$ is empty, but for any analytic subset A^* of X with $dim_x A^* > dim_x(p_X(\hat{f}^{-1}(\hat{B}))) = dim_x A$, $E \cap A^*$ is not empty. Let $\hat{E} = p_X^{-1}(E)$. Then $\hat{E}$ is a compact subset of $\mathscr{X}_1$ (by the properness of $\mathscr{B}_1$) and $E \cap \hat{f}^{-1}(\hat{B})$ is empty. Let $\hat{D} = \hat{f}(\hat{E})$. Then $\hat{D}$ is a compact subset of $\mathscr{X}_2$ and $\hat{D} \cap \hat{B}$ is empty. Since $\hat{B}$ is closed, it follows that there exists $Y' \in \mathscr{S}_2$ and an open subset U' of Y' such that $p_{Y'}^{-1}(U') \supset \hat{D}$ and $p_{Y'}^{-1}(U') \cap \hat{B}$ is empty. It is clear that Y' may be replaced by any $Y'' \in \mathscr{S}_2$ with $p_{Y'}^{Y''} \in \mathscr{M}_2$ and accordingly U' by $(p_{Y'}^{Y''})^{-1}(U')$. Therefore, we may assume that there exists $X' \in \mathscr{S}_1$ with $p_X^{X'} \in \mathscr{M}_1$ and with $f_{Y'}^{X'} \in F$. Now, let us assume that $\hat{A}$ is contained in $\hat{T}$. We want to prove: $dim\ \hat{A} > dim\ \hat{B}$. Let $A' = p_{X'}(\hat{A})$, $B' = p_{Y'}(\hat{B})$ and $x' = p_{X'}(z_2)$. The assumption implies that $f_{Y'}^{X'}$ is not integral at any point of A', by *Lemma* 4.1. Therefore, by *Lemma* 4.2, there exists an irreducible component C' of $(f_{Y'}^{X'})^{-1}(B')$ which contains A' (hence $f_{Y'}^{X'}(C') = B'$) and such that $dim\ C' > dim\ B'$. Let $C = p_X^{X'}(C')$. Then C contains A. Now, by the assumption that $\mathscr{B}_1$ is induced from $\mathscr{B}_2$ by F, we must have $dim\ C' - dim\ B' \leqq dim\ C - dim\ B$: Therefore $dim\ C > dim\ B$. I now claim that $dim\ C = dim\ A$. This will imply that $dim\ A > dim\ B$, hence, $dim\ \hat{A} > dim\ \hat{B}$, and the proof of he theorem will be completed. Suppose $dim\ A \neq dim\ C$. This implies $dim_x A < dim_x C$ because both A and C are irreducible and $C \supset A$. Then $C \cap E$ is not empty. Hence $C' \cap (p_X^{X'})^{-1}(E)$ is not empty. Since $\mathscr{B}_1$ is assumed to be surjective, $p_{X'}^{-1}(C') \cap \hat{E}$ is not empty. Take a point $\hat{w}_1$ in this set. Let $\hat{w}_2 = \hat{f}(\hat{w}_1)$. Then $p_{Y'}(\hat{w}_2) = f_{Y'}^{X'}(p_{X'}(\hat{w}_1)) \in f_{Y'}^{X'}(C') = B'$. On the other hand, $\hat{w}_1 \in \hat{E}$ and hence $\hat{w}_2 \in \hat{D} \subset p_{Y'}^{-1}(U')$. But $\hat{B} \cap p_{Y'}^{-1}(U')$ is empty and hence $B' \cap U'$ is empty. $p_{Y'}(w_1) \in B' \cap U'$ contradicts this statement. Thus we conclude that $dim\ C = dim\ A$, and complete the proof. Q.E.D.

§ 5. Point modifications

In this section, I shall prove the fundamental lemma which was announced in the introduction. We are given a proper bimeromorphic morphism $f: X \to Y$ of complex varieties, such that there is a point $y \in Y$ and f induces an isomorphism of $X - f^{-1}(y)$ to $Y - y$. The fundamental lemma then asserts that there exists a complex subspace D (not necessarily reduced) such that y is the only point of D and that if $\tilde{f}: \tilde{X} \to Y$ is the monoidal transformation of Y with center D, then there exists a morphism $g: \tilde{X} \to X$ (necessarily unique) with $\tilde{f} \circ g = f$. The ideal sheaf $\boldsymbol{J}$ of D in $\boldsymbol{O}_Y$ is then such that the stalk $\boldsymbol{J}_y$ is a primary ideal belonging to the maximal ideal of $\boldsymbol{O}_{Y,y}$ and $\boldsymbol{J}_{y'} = O_{Y,y'}$ for all points $y' \in Y - y$. Thus what we seek for is a primary ideal $J = \boldsymbol{J}_y$ belonging to the maximal ideal of $\boldsymbol{O}_{Y,y}$ which has the above property with reference to the given point modification $f: X \to Y$.

Let Φ be the set consisting of a single element which is the one-point analytic subset y of Y. In the sense of section 1, we obtain the Zariski space $\mathscr{Z}_\Phi(Y)$ of Y with center Φ. (cf. *Definition* 1.1.)

Let us write $\mathscr{S}_2$ for $\mathscr{S}_\Phi$, $\mathscr{M}_2$ for $\mathscr{M}_\Phi$, $\mathscr{B}_2$ for $\mathscr{B}_\Phi = (\mathscr{S}_\Phi, \mathscr{M}_\Phi)$ and $\mathscr{Z}_2$ for $\mathscr{Z}_\Phi = \lim \mathscr{B}_\Phi$. These are obtained as follows. Let L be the set of all primary ideals belonging to the maximal ideal of the local ring $\boldsymbol{O}_{Y,y}$. Each $J \in L$ defines a complex subspace D_J of Y which has one and only one point y. Let $p_J: Y_J \to Y$ be the monoidal transformation of Y with center D_J. Then $\mathscr{S}_2$ is the set of complex varieties Y_J with $J \in L$, and $\mathscr{M}_2$ is the set of all Y-morphisms among those Y_J, i.e., morphisms $p_J^{J'}: Y_{J'} \to Y_J$ such that $p^{J'} = p^J \circ p_J^{J'}$. Note that $p_J^{J'}$ exists if and only if the ideal J generates a locally principal ideal sheaf on $Y_{J'}$, and that it is unique if it exists at all. Also note that $\mathscr{B}_2 = (\mathscr{S}_1, \mathscr{M}_2)$ has the property to be a projective system of complex varieties, and that $\mathscr{B}_2$ is proper and surjective.

Given a point modification $f: X \to Y$ as above, one can construct another projective system of complex varieties, $\mathscr{B}_1 = (\mathscr{S}_1, \mathscr{M}_1)$, and a morphism $F: \mathscr{B}_1 \to \mathscr{B}_2$ as follows. For each $J \in L$, take the ideal sheaf $\boldsymbol{I}$ in $\boldsymbol{O}_X$ generated by J, which is coherent and has $supp\ (\boldsymbol{O}_X/\boldsymbol{I}) = f^{-1}(y)$. Let $q^J: X_J \to X$ be the monoidal transformation of X with the center defined by the ideal sheaf I. Then $\mathscr{S}_1$ is the set of complex varieties X_J with $J \in L$. The ideal sheaf $\boldsymbol{I}$ on X, and hence the ideal J, generates a locally principal ideal sheaf on X_J. Therefore there exists a unique morphism of complex varieties $f_J: X_J \to Y_J$ such that $p^J \circ f_J = f \circ q^J$. Let (J', J) be a pair of ideals in L, such that there exists a Y-morphism $p_J^{J'}: Y_{J'} \to Y_J$ as above. This implies that J generates a locally principal ideal sheaf on $Y_{J'}$. Hence, by $p^{J'} \circ f_{J'} = f \circ q^{J'}$, the ideal sheaf $\boldsymbol{I}$ on X generated by J should generate a locally principal ideal sheaf on $X_{J'}$.

Hence there exists a morphism $q_J^{J'}: X_{J'} \to X_J$ such that $q^J \circ q_J^{J'} = q^{J'}$. Now, $\mathcal{M}_1$ is the set of those morphisms $q_J^{J'}$ for the pairs (J', J) with $p_J^{J'}$. It is seen that $\mathcal{B}_1 = (\mathcal{S}_1, \mathcal{M}_1)$ is a projective system of complex varieties. Moreover, if F is the set of those morphisms $f_J: X_J \to Y_J$ for $J \in L$, then F can be viewed as a morphism of projective systems $\mathcal{B}_1 \to \mathcal{B}_2$. It is easy to see that the morphisms f_J and q^J are obtained from an imbedding of X_J in the fibre product $X \times_Y Y_J$ (where the image is the unique irreducible component (reduced) of $X \times_Y Y_J$ which is projected onto X and onto Y_J).

We have thus constructed projective systems $\mathcal{B}_1$ and $\mathcal{B}_2$, and a morphism $F: \mathcal{B}_1 \to \mathcal{B}_2$, starting from the given point modification $f: X \to Y$. It should be noted that *$\mathcal{B}_1$, $\mathcal{B}_2$ and F are all proper and surjective* and that *$\mathcal{B}_1$ is induced from $\mathcal{B}_2$ by F.*

Theorem 5.1. *The induced morphism $\hat{f}: \mathcal{X}_1 \to \mathcal{X}_2$ by $F: \mathcal{B}_1 \to \mathcal{B}_2$ is an isomorphism of local-ringed spaces, where $\mathcal{X}_i = \lim \mathcal{B}_i$ for $i = 1$ and 2.*

The proof of this theorem is done by induction on the integer $n = \dim X = \dim Y$. The assertion is trivially true for $n = 0$. Let us assume that $n > 0$. I shall first prove that $\hat{f}$ is integral (at every point of $\mathcal{X}_1$). Suppose $\hat{f}$ is not integral at some point. Referring to *Theorem* 4.5, let $\hat{T}$ be the analytic subset of $\mathcal{X}_1$ of those points at which $\hat{f}$ is not integral, let $\hat{S} = \hat{f}(\hat{T})$ which is an analytic subset of $\mathcal{X}_2$, let $\hat{B}$ be an irreducible component of $\hat{S}$, and let $\hat{A}$ be a primary irreducible component of $\hat{f}^{-1}(\hat{B})$ such that $\hat{f}(\hat{A}) = \hat{B}$ and $\hat{A} \subseteqq \hat{T}$. Then by the same theorem, $\dim \hat{A} > \dim \hat{B}$. From now on, all we need is the existence of an irreducible anlytic subset $\hat{A}$ of $\mathcal{X}_1$, and $\hat{B}$ of $\mathcal{X}_2$, such that $\hat{f}(\hat{A}) = \hat{B}$ and $\dim \hat{A} > \dim \hat{B}$. This will lead to a contradiction.

We write q_J (resp. q) for the canonical morphism $\mathcal{X}_1 \to X_J$ (resp. $\mathcal{X}_1 \to X$), and p_J (resp. p) for the canonical morphism $\mathcal{X}_2 \to Y_J$ (resp. $\mathcal{X}_2 \to Y$). We shall speak of local rings of these local-ringed spaces and ideals in them, and here we make once for all the following general agreement: Suppose $h: V \to W$ is a "canonical" morphism, i.e., a morphism obtained by composing some of the morphisms $\hat{f}$, p_J, $p_J^{J'}$, p^J, q_J, $q_J^{J'}$, q^J, f and f_J. Let $t \in V$ and $u = h(t)$. Then I say that an ideal in $\boldsymbol{O}_{V,t}$ contains an ideal (or an element) of $\boldsymbol{O}_{W,u}$, meaning that the image of the latter in $\boldsymbol{O}_{V,t}$ is contained in the former, with reference to the homomorphism $\boldsymbol{O}_{W,u} \to \boldsymbol{O}_{V,t}$ associated with h. Moreover, if $\boldsymbol{a}$ is an ideal in $\boldsymbol{O}_{V,t}$, I write $\boldsymbol{a} \cap \boldsymbol{O}_{W,n}$, meaning the preimage of $\boldsymbol{a}$ in $\boldsymbol{O}_{W,u}$, which is an ideal.

Let us pick a point z_1 of $\hat{A}$, and let $\hat{\boldsymbol{O}}_1$ be the local ring of $\mathcal{X}_1$ at z_1. Let $\hat{\boldsymbol{a}}$ be the ideal of $\hat{A}$ in $\hat{\boldsymbol{O}}_1$, which may be defined as follows: For each

$J \in L$, take the ideal $\boldsymbol{a}_J$ of $q_J(\hat{A})$ in the local ring of X_J at the point $q_J(z_1)$. Then the union of the images of those $\boldsymbol{a}_J$ in $\hat{\boldsymbol{O}}_1$ is an ideal, and this ideal is the ideal $\hat{\boldsymbol{a}}$ of $\hat{A}$ in $\hat{\boldsymbol{O}}_1$. It is easy to see that $\hat{\boldsymbol{a}}$ is the unique ideal in $\hat{\boldsymbol{O}}_1$ such that for every $J \in L$, $\hat{\boldsymbol{a}} \cap \boldsymbol{O}_{1J}$ is the ideal of $q_J(\hat{A})$ in the local ring $\boldsymbol{O}_{1J}$ of X_J at $q_J(z_1)$. Now, take a minimal prime ideal $\hat{\boldsymbol{d}}$ of $\hat{\boldsymbol{a}}$, i.e. a prime ideal in $\hat{\boldsymbol{O}}_1$ containing $\hat{\boldsymbol{a}}$ such that if $\hat{\boldsymbol{d}}^*$ is any prime ideal with $\hat{\boldsymbol{d}} \supseteqq \hat{\boldsymbol{d}}^* \supseteqq \hat{\boldsymbol{a}}$, then $\hat{\boldsymbol{d}} = \hat{\boldsymbol{d}}^*$. We need the following fact about such an ideal $\hat{\boldsymbol{d}}$:

Lemma (a). *For each $J \in L$, $\boldsymbol{O}_{1J}$ denotes the local ring of X_J at the point $x_J = q_J(z_1)$, and $\boldsymbol{a}_J$ denotes the ideal of $A_J = q_J(\hat{A})$ in $\boldsymbol{O}_{1J}$, so that $\hat{\boldsymbol{a}}_J = \boldsymbol{a} \cap \boldsymbol{O}_{1J}$. Then $\boldsymbol{d}_J = \hat{\boldsymbol{d}} \cap \boldsymbol{O}_{1J}$ is a minimal prime ideal of $\boldsymbol{a}_J$.*

Proof. It is essential that $\hat{A}$ is irreducible, so that A_J is irreducible for all $J \in L$. Suppose $\boldsymbol{d}_J$ is not a minimal prime ideal of $\boldsymbol{a}_J$ for some $J \in L$. Then take a minimal prime ideal $\boldsymbol{d}_J^*$ of $\boldsymbol{a}_J$ which is contained in $\boldsymbol{d}_J$. For each $q_{J'}^{J''} \in \mathscr{M}_1$, if $\boldsymbol{d}''$ is a minimal prime ideal of $\boldsymbol{a}_{J''}$ then $\boldsymbol{d}'' \cap \boldsymbol{O}_{1J'}$ is a minimal prime ideal of $\boldsymbol{a}_{J'}$, and every minimal prime ideal of $\boldsymbol{a}_{J'}$ is obtained in this manner. Here the fact is needed that $A_{J''}$ and $A_{J'} = q_{J'}^{J''}(A_{J''})$ are both irreducible. It follows, by the standard argument, that one can choose a minimal prime ideal $\boldsymbol{d}_{J'}^*$ of $\boldsymbol{a}_{J'}$ for each J' with $q_J^{J'} \in \mathscr{M}_1$, such that $\boldsymbol{d}_{J'} \supset \boldsymbol{d}_{J'}^*$ for every J' and that $\boldsymbol{d}_{J''}^* \cap \boldsymbol{O}_{1J'} = \boldsymbol{d}_{J'}^*$ for every $q_{J'}^{J''} \in \mathscr{M}_1$. Then, as is easily seen, the union $\hat{\boldsymbol{d}}^*$ of the images in $\hat{\boldsymbol{O}}_1$ of those $\boldsymbol{d}_{J'}^*$ is a prime ideal which contains $\hat{\boldsymbol{a}}$ and which is strictly contained in $\hat{\boldsymbol{d}}$. This contradicts the assumption that $\hat{\boldsymbol{d}}$ is a minimal prime ideal of $\hat{\boldsymbol{a}}$. Q.E.D.

It is immediate from Lemma (a), that the Krull dimension of the localization $(\hat{\boldsymbol{O}}_1)_{\hat{\mathfrak{d}}}$ is at least one. Let $z_2 = \hat{f}(z_1)$, and $\hat{\boldsymbol{O}}_2$ the local ring of $\mathscr{X}_2$ at the point z_2. We shall use the notions: $x_J = q_J(z_1)$, $y_J = p_J(z_2)$, $\boldsymbol{O}_{1J}$ = the local ring of X_J at x_J, $\boldsymbol{O}_{2J}$ = the local ring of Y_J at y_J, etc. An element η of $\hat{\boldsymbol{O}}_1/\hat{\boldsymbol{e}}$ is said to *satisfy an analytic equation with coefficients in $\hat{\boldsymbol{O}}_2/\hat{\boldsymbol{e}} \cap \hat{\boldsymbol{O}}_2$*, where $\hat{\boldsymbol{e}}$ is a prime ideal in $\hat{\boldsymbol{O}}_1$, if there exists $J \in L$ such that η is an element of $\boldsymbol{O}_{1J}/\hat{\boldsymbol{e}} \cap \boldsymbol{O}_{1J}$ and in this ring the following equality holds:

$$\sum_{m=0}^{\infty} k_m \eta^m = 0 \quad (k_m \neq 0 \text{ for some } m)$$

where $k_m \in \boldsymbol{O}_{2J}/\hat{\boldsymbol{e}} \cap \boldsymbol{O}_{2J}$ for all m. (It is here important that there is one J such that $\boldsymbol{O}_{2J}/\hat{\boldsymbol{e}} \cap \boldsymbol{O}_{2J}$ contains all the coefficients of the convergent power series in η. Moreover, the convergence makes sense because $\boldsymbol{O}_{1J}/\hat{\boldsymbol{e}} \cap \boldsymbol{O}_{1J}$ can be canonically realized as a local ring of a complex variety.)

Lemma (b). *The ideal* $\hat{d}$ *has the property that there exists at least one (in fact, infinitely many) element of* $\hat{O}_1/\hat{d}$ *which satisfies no analytic equation with coefficients in* $\hat{O}_2/\hat{d} \cap \hat{O}_2$.

Proof. First note that $\hat{d} \cap \hat{O}_2$ is a minimal prime ideal of the ideal $\hat{b}$ of $\hat{B}$ in $\hat{O}_2$. In fact, by *Lemma* (a), for every $J \in L$, $d_J = \hat{d} \cap O_{1J}$ is a minimal prime ideal of the ideal a_J of $A_J = q_J(\hat{A})$ in O_{1J}. If $B_J = p_J(\hat{B})$, then $B_J = f_J(A_J)$. Since A_J is irreducible, $d_J \cap O_{2J}$ is a minimal prime ideal of the ideal b_J of B_J in O_{2J}. We have $b_J = \hat{b} \cap O_{2J}$. Thus, for every $J \in L$, $\hat{d} \cap O_{2J}$ is a minimal prime ideal of $\hat{b} \cap O_{2J}$. It follows that $\hat{d} \cap \hat{O}_2$ is a minimal prime ideal of $\hat{b}$. Now, choose $J \in L$ such that $\dim \hat{A} = \dim A_J$ and $\dim \hat{B} = \dim B_J$, so that $\dim(O_{1J}/\hat{d} \cap O_{1J}) = \dim \hat{A}$ and $\dim(O_{2J}/\hat{d} \cap O_{2J}) = \dim \hat{B}$. These equalities remain true if we replace J by any J' such that $q_J^{J'}$ (and $p_J^{J'}$) exists. Then, for every J' with $q_J^{J'}$, every element of $O_{2J'}/\hat{d} \cap O_{2J'}$ satisfies an analytic equation with coefficients in $O_{2J}/d \cap O_{2J}$. However, in view of the difference in Krull dimensions, there exists an element η in $O_{1J}/\hat{d} \cap O_{1J}$ which satisfies no analytic equation with coefficients in $O_{2J}/\hat{d} \cap O_{2J}$. Therefore, for any J' with $q_J^{J'}$, the image of η in $O_{1J'}/\hat{d} \cap O_{1J'}$ satisfies no analytic equation with coefficients in $O_{2J'}/\hat{d} \cap O_{2J'}$. This η, viewed as an element of $\hat{O}_1/\hat{d}$, satisfies no analytic equation with coefficients in $\hat{O}_2/\hat{d} \cap \hat{O}_2$ in the sense defined above. Q.E.D.

Now, the proof of the *Theorem* 5.1. proceeds with a minimal prime ideal $\hat{d}$ of $\hat{a}$, fixed once for all. We shall consider the following two cases separately.

Case I. *Every prime ideal* $\hat{e}$ *in* $\hat{O}_1$ *such that*

(i) $\hat{e} \subseteqq \hat{d}$, *and*

(ii) $\dim(\hat{O}_1)_{\hat{e}} \geqq 1$,

has the property:

(iii) $\hat{e} \supset m$, *where* m *denotes the maximal ideal of the local ring* O_2 *of* Y *at the point* y.

Case II. *There exists at least one prime ideal* $\hat{e}$ *in* $\hat{O}_1$ *which has the properties* (i) *and* (ii) *but not* (iii).

Let us first consider Case I. By Lemma (b), choose an element η of $\hat{O}_1/\hat{d}$ which satisfies no analytic equation with coefficients in $\hat{O}_2/\hat{d} \cap \hat{O}_2$. Pick an element t of $\hat{O}_1$ which represents the residue class η. Then choose $J \in L$ such that this element t is the image of an element of the local ring O_{1J} of X_J at the point $x_J = q_J(z_1)$. Let us pick such an element of O_{1J}, and denote it by t again. From now on, we shall use the symbol t to denote the image of this t in any one of the local rings $O_{1J'}$ for $J' \in L$ with $q_J^{J'} \in \mathscr{M}_1$. Since the canonical morphism $f \circ q^J : X_J \to Y$ is bimeromorphic, the element t satisfies an analytic equation with coefficients in the local ring O_2 of Y at y. Let us take such an equation and write it as

$$\sum_{m=0}^{\infty} k_m t^m = 0 \quad \text{with} \quad k_m \in O_2 .$$

Here, as it is easily seen, we can choose this equation so that at least one of the k_m has a non-zero image in the ring $\hat{O}_1$. Let J_0 be the ideal in O_2 generated by the k_m for all $m \geqq 0$, which may not be in L. Let $J_1 = J_0(\hat{O}_1)_{\hat{d}} \cap O_2$. Then the assumption in *Case* I assures me that J_1 is a primary ideal belonging to the maximal ideal m of O_2. (Note that J_0, hence J_1, is not a unit ideal, because otherwise the Weierstrass preparation theorem tells us that t is integral over O_2, which is obviously not the case.) Now, let $J' = JJ_1$, which is in L. Then we have $q_J^{J'}$ and hence t is an element of $O_{1J'}$. Moreover, by the property of monoidal transformation, $J'O_{2J'}$ is principal. Hence $J_1 O_{2J'}$ is principal. Let h be an element of J_1 which generates this principal ideal. Then, since $J_0 \subseteqq J_1$, we can write $k_m = k_m^* h$ with $k_m^* \in O_{2J'}$. (The equality makes sense in $O_{2J'}$.) We thus obtain an analytic equation of the form $\sum_{m=0}^{\infty} k_m^* t^m = 0$. Let J^* be the ideal in $O_{2J'}$ generated by those k_m^* for $m \geqq 0$. Then $hJ^*(\hat{O}_1)_{\hat{d}} = J_0(\hat{O}_1)_{\hat{d}} = J_1(\hat{O}_1)_{\hat{d}} = h(\hat{O}_1)_{\hat{d}}$. Therefore $J^*(\hat{O}_1)_{\hat{d}}$ is the unit ideal. Namely, at least one of the k_m^* with $m \geqq 0$ is a unit in the local ring $(\hat{O}_1)_{\hat{d}}$. Equivalently, if u_m denotes the residue class of k_m^* in $\hat{O}_1/\hat{d}$, then at least one of the u_m is not zero. However, we arrive at the equation $\sum_{m=0}^{\infty} u_m \eta^m = 0$. This contradicts the selection of η.

Next, let us consider *Case* II. Let us pick once for all a prime ideal $\hat{e}$ in $\hat{O}_1$ such that $\hat{e} \subseteqq \hat{d}$, $\dim (\hat{O}_1)_{\hat{e}} \geqq 1$ and $\hat{e} \not\supset m$. Since $\hat{d} \supset m$, $\hat{d} \neq \hat{e}$. Let $e_2 = \hat{e} \cap O_2$. Then e_2 is a prime ideal in O_2 and defines an irreducible subvariety $\bar{Y}$ of Y in a certain open neighborhood U of the point y. Since the given morphism $f: X \to Y$ is isomorphic outside the point y, there exists a unique irreducible subvariety $\bar{X}$ of X in the open subset $f^{-1}(U)$. f induces a proper morphism $\bar{f}: \bar{X} \to \bar{Y}$, which is again a point modification with the same center y as before. Now, $\dim \bar{X} = \dim \bar{Y}$ is strictly smaller than $n = \dim X = \dim Y$. Hence, by induction assumption, there exists a primary ideal $\bar{J}$ in the local ring $\bar{O}_2$ of $\bar{Y}$ at y such that, if D is the complex subspace (not necessarily reduced) of $\bar{Y}$ defined by $\bar{J}$, then the monoidal transformation $p_{\bar{J}}: \bar{Y}_{\bar{J}} \to \bar{Y}$ of $\bar{Y}$ with center D *dominates* $\bar{f}$, i. e., there exists a morphism $\bar{h}: \bar{Y}_{\bar{J}} \to \bar{X}$ such that $p_{\bar{J}}$

$=\bar{f}\circ\bar{h}$. Let J be the ideal in $\boldsymbol{O}_2$ obtained as the preimage of $\bar{J}$ by the natural homomorphism $\boldsymbol{O}_2\to\bar{\boldsymbol{O}}_2=\boldsymbol{O}_2/\boldsymbol{e}_2$. Then this J belongs to L. There is then a unique imbedding (closed in $p_{\bar{J}}^{-1}(U)$) $\bar{Y}_{\bar{J}}\to Y_J$ which is compatible with $p_{\bar{J}}$, p_J and the imbedding $\bar{Y}\to Y$ (closed in U). In the local ring $\boldsymbol{O}_{2J}$ of Y_J at $y_J=p_J(z_2)$, the prime ideal $\hat{\boldsymbol{e}}\cap\boldsymbol{O}_{2J}$ is a minimal prime ideal of the ideal of $\bar{Y}_{\bar{J}}$. Let $\bar{X}_{\bar{J}}$ be the unique irreducible subvariety of X_J in an open subset coming from U, which is mapped onto $\bar{X}$. Then in the local ring $\boldsymbol{O}_{1J}$ of X_J at $x_J=q_J(z_1)$, the prime ideal $\hat{\boldsymbol{e}}\cap\boldsymbol{O}_{1J}$ is a minimal prime ideal of the ideal of X_J. Since X_J is canonically imbedded in the fibre product of X and Y_J over Y, $\bar{X}_{\bar{J}}$ is canonically imbedded in the fibre product of $\bar{X}$ and $\bar{Y}_{\bar{J}}$ over $\bar{Y}$. Then the existence of the morphism $\bar{h}:\bar{Y}_{\bar{J}}\to\bar{X}$ implies that the morphism $f_{\bar{J}}:\bar{X}_{\bar{J}}\to\bar{Y}_{\bar{J}}$, induced by $f_J:X_J\to Y_J$, is an isomorphism. The above arguments remain valid without any change, if we replace J by a multiple by any $J'\in L$. Therefore we conclude that $\hat{\boldsymbol{O}}_1/\hat{\boldsymbol{e}}$ is a trivial extension of $\hat{\boldsymbol{O}}_2/\hat{\boldsymbol{e}}\cap\hat{\boldsymbol{O}}_2$. This is a contradiction, because $\hat{\boldsymbol{d}}\supset\hat{\boldsymbol{e}}$ and $\hat{\boldsymbol{O}}_1/\hat{\boldsymbol{d}}$ is not a trivial extension of $\hat{\boldsymbol{O}}_2/\hat{\boldsymbol{d}}\cap\boldsymbol{O}_2$.

We can thus conclude that the existence of $\hat{A}$ and $\hat{B}$ was false. Namely, the morphism $\hat{f}:\mathscr{X}_1\to\mathscr{X}_2$ is integral at every point.

Now, by *Lemma* 4.1. and by the compactness of the preimage in $\mathscr{X}_1$ of the point $y\in Y$, there can be found $J\in L$ such that $f_J:X_J\to Y_J$ is integral everywhere. Since f_J is bimeromorphic, if $\widetilde{Y}_J$ is the normalization of Y_J, then the canonical morphism $\widetilde{Y}_J\to Y_J$ is factored by f_J. Moreover, since f_J is isomorphic outside $(p^J)^{-1}(y)$, the direct image $\widetilde{\boldsymbol{O}}$ of $\boldsymbol{O}_{X_J}$ by f_J is a coherent sheaf of fractional ideals on Y_J which coincides with $\boldsymbol{O}_{Y_J}$ outside $(p^J)^{-1}(y)$. Now, let $\boldsymbol{O}(1)$ be the invertible ideal sheaf generated by J on X_J, and $\boldsymbol{O}(m)$ its m-th power. Then, for all sufficiently large integers m, there exists an ideal sheaf $\boldsymbol{J}(m)$ in $\boldsymbol{O}_Y$ which generates the sheaf of fractional ideals $\boldsymbol{O}(m)\widetilde{\boldsymbol{O}}$. Obviously $J(m)=\boldsymbol{J}(m)_y$ belongs to L, and the morphism $p^{J(m)}:Y_{J(m)}\to Y$ dominates $X_J\to Y$ (in fact, they are the same). It follows that $f_{J'}:X_{J'}\to Y_{J'}$ is an isomorphism for all J' such that $p^{J'}_{J(m)}$ exists. It implies that $\hat{f}:\mathscr{X}_1\to\mathscr{X}_2$ is an isomorphism. We have completed the proof of *Theorem* 5.1.

Theorem 5.1. has an immediate consequence as follows.

Corollary 5.2. *Given a point modification $f:X\to Y$ which induces an isomorphism of $X-f^{-1}(y)$ to $Y-y$, there exists a complex subspace (not reduced, in general) D of Y such that y is the only point of D and such that if $g:\widetilde{Y}\to Y$ is the monoidal transformation of Y with center D, then there exists a unique morphism $h:\widetilde{Y}\to X$* with $g=f\circ h$.

Throughout this section, I kept the assumption that $f: X \to Y$ induces an isomorphism outside the unique point y. I can drop this assumption and prove, by the same argument as given above, the following generalization.

Theorem 5.2. *Let $f: X \to Y$ be any proper bimeromorphic morphism of complex varieties. Let S be the analytic subset of Y of those points at which f is not isomorphic. Then there exists an open neighborhood U of S and a complex subspace D of U (not necessarily reduced) such that $U \cap S$ is exactly the set of points of D and such that if $g: \widetilde{U} \to U$ is the monoidal transformation of U with center D, then there exists a unique morphism $h: \widetilde{U} \to X$ with $g = f \circ h$.*

It is not true that any D as in this theorem can be extended to a complex subspace of Y throughout. Here is an open question whether we can choose D of *Theorem* 5.2 so that it is a restriction of a complex subspace of Y and that the monoidal transformation g can be extended through the entire variety Y.

References

[1] Hironaka, H.: Resolution of singularities of an algebraic variety over a field of characteristic zero. Ann. Math. **79**, 109—326 (1964).

[2] —, and H. Rossi: On the equivalence of imbeddings of exceptional complex spaces. (Forthcoming in Math. Ann.)

Department of Mathematics
Brandeis University
Waltham, Mass.

Transmission Problems for Holomorphic Fiber Bundles*,**

By

H. Röhrl

With 2 Figures

In a previous paper ([9]) the author considered transmission problems for holomorphic fiber bundles over Riemann surfaces and families of Riemann surfaces. In this context the question arose as to how to treat analogous problems for fiber bundles over complex spaces X. The situation in higher dimensional complex spaces is by necessity different

* This research was supported by the Air Force Office of Scientific Research.
** Received June 15, 1964.

from the one dimensional case since only in more than one complex dimension the phenomenon of pseudoconvexity appears. At the same time it turned out to be desirable to deal with a more general geometric and analytic situation as in [9], thus making it possible to interpret "topologically correct" transmission problems as cycles in a certain homology theory, so that two such topological transmission problems are isomorphic if the corresponding cycles are homologous.

In section 1 topological preparations pertaining to the notion of "boundary values on a hypersurface by approach from various sides" are carried out. For this purpose the subset P of X along which the transmission occurs is assumed to be an oriented polyhedron imbedded in X. To the given oriented polyhedron P two maps are constructed, one of which relates the various wedges determined by P in a given point of P, while the other one turns up in the exponent of the transmission problem (see formula (7)). Section 2 deals with certain analytical preparations, mainly with holomorphic bundles $\mathfrak{L}$ of complex Lie groups acting on holomorphic fiber bundles (cf. [2]). In section 3 the transmission problems we are dealing with are stated. To each holomorphically correct transmission function a holomorphic fiber bundle $\mathfrak{B}_\tau$ over X is constructed such that the set of solutions of the transmission problem τ (with the given transmission function) corresponds bijectively to the set of holomorphic sections in $\mathfrak{B}_\tau$. Analogous theorems concerning topologically correct transmission functions are stated. In section 4 topologically correct transmission functions are discussed and related to a certain homology theory, assuming that P possesses a sufficiently well behaved neighborhood. For this homology theory, a functorial map $H_{n-1}(P; \mathfrak{L}) \to$ $\to H^1(X, \mathfrak{L}_c)$ is constructed which under suitable hypotheses turns out to be bijective. Section 5 deals with holomorphically correct transmission functions. The main theorem of this section states that a germ of a holomorphic section in $\mathfrak{L}$ at $x \in P$ is holomorphically correct if (and only if) it is topologically correct. The one dimensional case is discussed briefly only, since the essential part of this case has been treated in [9] already. For complex spaces X of dimension bigger than one, holomorphic correctness is investigated for such polyhedra whose simplices of top dimension are holomorphic families of strongly pseudoconvex surfaces.

1. Topological preparations

I. Let X be a topological space and S a closed subset of X. Denoting the neighborhood filter of $x \in X$ by $\mathfrak{U}_x$, we consider the set X_S of all pairs $(x, \mathfrak{F})$ such that

(i) $x \in X$ and $\mathfrak{F}$ is a filter on $X - S$ that is finer than $(\mathfrak{U}_x)_{X-S}$;[1]

(ii) $\mathfrak{F}$ is generated by connected components of elements contained in $(\mathfrak{U}_x)_{X-S}$.

In the set X_S we introduce a topology as follows. Suppose that $(x_0, \mathfrak{F}_0)$ is in X_S. Then for every $F_0 \in \mathfrak{F}_0$ we can form the set $\hat{F}_0 = \{(x, \mathfrak{F}) : F_0 \in \mathfrak{F}\}$. With the elements $(x_0, \mathfrak{F}_0)$ we can associate the filter $\mathfrak{U}_{(x_0, \mathfrak{F}_0)}$ on X_S that is generated by the set $\{\hat{F}_0 : F_0 \in \mathfrak{F}_0\}$. It is easy to check ([*1*], Chap. I, § 1) that there is a topology on X_S such that for every point $(x_0, \mathfrak{F}_0)$ the neighborhood filter coincides with $\mathfrak{U}_{(x_0, \mathfrak{F}_0)}$.

Let $\pi : X_S \to X$ be defined by $\pi(x, \mathfrak{F}) = x$ for all $(x, \mathfrak{F})$ in X_S. Then the following statements can be shown easily:

1) $\pi : X_S \to X$ is a continuous map;

2) for every open subset U of X there is a canonical bijective map j: $U_{U \cap S} \to \pi^{-1}(U)$ such that $\pi \circ j = \pi'$ holds where $\pi' : U_{U \cap S} \to U$ is the previously defined projection; by means of j, $U_{U \cap S}$ shall be regarded as a subset of X_S;

3) $\pi(X_S) \subset \overline{X - S}$;

4) if every point x of S has a basis of neighborhoods U_i, $i \in I$, such that for every i there is a connected component, V, of $U_i - U_i \cap S$ satisfying $x \in \bar{V}$, then $S \subset \pi(X_S)$;

5) if X is locally connected, then $X - S \subset \pi(X_S)$;

6) if X is locally connected, then for every connected component X_λ of $X - S$, $\pi^{-1}(X_\lambda)$ is a connected component of X_S;

7) if X is hausdorff, then X_S is hausdorff;

8) if X is hausdorff and locally connected, then π maps $(X - S)_\emptyset = \pi^{-1}(X - S)$ homeomorphically onto $X - S$.

Let S be a closed subset of X and $p : T \to X$ be a continuous map. Then we can form the pull-back[2] $\pi^*(p) : \pi^*(T) \to X_S$. Suppose that s

---

[1] For a filter $\mathfrak{G}$ on X and the subset A of X, $\mathfrak{G}_A$ denotes the trace of $\mathfrak{G}$ on A, i. e. the filter on A generated by $\{G \cap A : G \in \mathfrak{G}\}$.

[2] For every pair of continuous maps $p_i : T_i \to X$, $i = 1, 2$, we denote by $T_1 \times_X T_2$ the subspace $\{(t_1, t_2) : p_1(t_1) = p_2(t_2)\}$ of $T_1 \times T_2$. Given another such pair $p_i' : T_i' \to X$ together with maps $q_i : T_i' \to T_i$ satisfying $p_i \circ q_i = p_i'$, there is a map $q_1 \times_X q_2 : T_1' \times_X T_2' \to T_1 \times_X T_2$ that is defined by $q_1 \times_X q_2(t_1', t_2') = (q_1(t_1'), q_2(t_2'))$.

Sometimes it is advisable to relinquish the symmetry of this notation. Then, given the maps $p : T \to X$ and $\pi : X' \to X$, we denote by $\pi^*(T)$ the space $X' \times_X T$ and by $\pi^*(p)$ the restriction $pr_1 | X' \times_X T$, pr_1 being the first projection of $X' \times T$ onto X'. The second projection restricted to $X' \times_X T$ defines a map $\pi_* : \pi^*(T) \to T$ which satisfies the obvious commutativity property. Given a continuous section s in $p : T \to X$ over the open subset U of X, the map $\pi^*(s)$ of $\pi^{-1}(U)$ into $\pi^*(T)$ that sends $x' \in \pi^{-1}(U)$ into $(x', s(\pi(x')))$ is a continuous section in $\pi^*(p) : \pi^*(T) \to X'$.

In the complex analytic case, the same considerations, starting out from complex spaces and holomorphic maps, lead again to complex spaces and holomorphic maps, provided $T_1 \times_X T_2$ is equipped with the induced structure.

is a continuous section in $p : T \to X$ over $X - S$. Then s is said to have a *continuous extension to* S if there is a global continuous section $\hat{s}$ in $\pi^*(p) : \pi^*(T) \to X_S$ that coincides with $\pi^*(s)$ on $\pi^{-1}(X - S)$. Intuitively speaking, this definition amounts to the following: if S does not separate X in the neighborhood of $x \in S$, then continuous extendability in our sense is just continuous extendability into x in the usual sense; however, if S separates X in the neighborhood of $x \in S$, then continuous extendability in our sense means the existence of "boundary values" from different "sides" of S at x, the boundary value at the side specified by $\hat{x} \in \pi^{-1}(S)$ being $\pi_*(\hat{s}(\hat{x}))$. It is convenient to denote this boundary value by $s(\hat{x})$.

II. From now on X is assumed to be a (not necessarily reduced) complex space. We impose the following requirements, which shall be referred to as the *geometric hypothesis*:

(i) X is reduced and normal, and has complex dimension n in all points of the closed subset S of X;

(ii) the subset S is equipped with the structure of a locally finite, oriented polyhedron P of dimension $2n - 1$ such that:

> (ii′) every simplex of P is contained in some $(2n - 1)$-dimensional simplex of P;
>
> (ii″) the intersection of S with the singular locus of X is a sub-polyhedron of P.

Here, the terminology is the one used in [*5*].

It should be remarked that the condition (ii″) is no essential restriction, provided the subspace S of X has countable topology: in this case there is an open neighborhood V of S that has also countable topology, whence the results of [*3*] lead to a subdivision satisfying the desired requirement (ii″) (see also section 3, Remark 1).

Under the geometric hypothesis it follows immediately from the statements in I. that $\pi : X_{|P|} \to X$ is surjective, $|P|$ being the carrier of P.

Under the geometric hypothesis, the *polyhedron* P is said to be *of class* C^k, if for every simplex σ^q of P and the map $\varphi_\sigma : \Delta^q \to |\sigma^q|$ of the standard q simplex $\Delta^q \subset \boldsymbol{R}^q$ onto $|\sigma^q|$ that defines σ^q, the following statement is true:

every point $x \in |\sigma^q|$ possesses an open neighborhood U of $\varphi_\sigma^{-1}(x)$ in $\boldsymbol{R}^q$ and an open neighborhood V of x in X together with a biholomorphic map ψ of V onto some analytic subset of (a domain in) $\boldsymbol{C}^N$ such that the map $\psi \circ (\varphi_\sigma | \Delta^q \cap U)$ can be extended to a map of class C^k (in the usual sense) from U to $\boldsymbol{C}^N$.

The orientation on P gives rise in the well-known way (cf. [*5*]) to the incidence number $[\sigma, \sigma']$, for any two simplices σ and σ' of P. Let us consider the (possibly infinite) integral $(2n-1)$-chain

$$\gamma = \sum \{\sigma^{2n-1} : \sigma^{2n-1} \in P\}$$

on P. Since, under the geometric hypothesis, P is locally finite, we can form the boundary of γ:

$$\begin{aligned}\partial\gamma &= \sum \{\partial\gamma(\sigma^{2n-2}) \cdot \sigma^{2n-2} : \sigma^{2n-2} \in P\} \\ &= \sum \{\sum [\sigma^{2n-1}, \sigma^{2n-2}] : \sigma^{2n-1} \in P\} \cdot \sigma^{2n-2} : \sigma^{2n-2} \in P\}.\end{aligned}$$

$\partial\gamma$ will play an important role later on. Right now, however, we shall assign to P two maps with domain

$$\hat{S} = \pi^{-1}(\bigcup\{\mathring{\sigma}^{2n-1} : \sigma^{2n-1} \in P\} \cup \bigcup\{\mathring{\sigma}^{2n-2} : \sigma^{2n-2} \in P\}),$$

$\mathring{\sigma}$ denoting the interior of the simplex σ.

It is easy to see that for every $(2n-1)$-simplex σ^{2n-1} in P there exists an open subset $U = U(\sigma^{2n-1})$ of X of the topological type of the $2n$-cell such that

(i) $U \cap S = \mathring{\sigma}^{2n-1}$;

(ii) $U - \mathring{\sigma}^{n-1}$ consists of two connected components, each of which is of the topological type of the $2n$-cell.

Of these two connected components of $U - \mathring{\sigma}^{2n-1}$ there is precisely one, W_1, with the following property:

to any point x_0 in W_1 there exists a $2n$-simplex contained in W_1 that has vertex x_0 and $(2n-1)$-face σ^{2n-1} such that the orientation given by $x_0\,\sigma^{2n-1}$ coincides with the orientation induced by the ambient space.

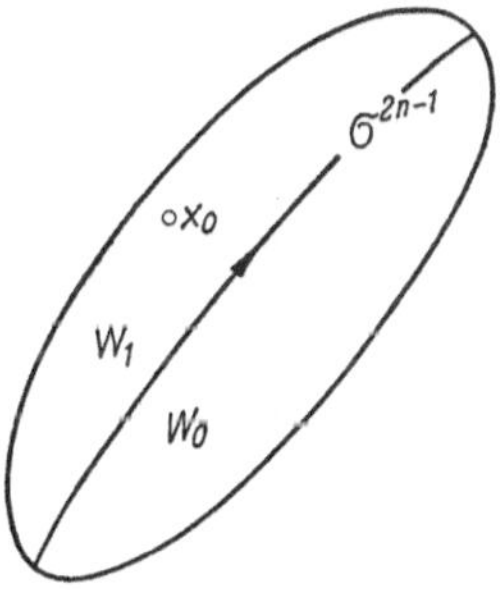

Fig. 1

The other component of $U - \mathring{\sigma}^{2n-1}$ shall be denoted by W_0. For every point x in $\mathring{\sigma}^{2n-1}$, the set $\pi^{-1}(x)$ consists of exactly two points, $x_\varkappa = (x, \mathfrak{F}_\varkappa)$, $\varkappa = 0, 1$. It is no restriction to assume that $W_\varkappa \in \mathfrak{F}_\varkappa$ holds. Now we define, counting indices mod 2,

$$\beta(x_\varkappa) = x_{\varkappa+1} \quad \text{and} \quad \gamma(x_0) = -\gamma(x_1) = 1. \tag{1}$$

Obviously, the definitions (1) do not depend on the choice of U in the above construction. Since the interiors of different $(2n-1)$-simplices are mutually disjoint, β and γ have now been defined on

$$\pi^{-1}(\bigcup\{\mathring{\sigma}^{2n-1} : \sigma^{2n-1} \in P\}).$$

Due to the geometric hypothesis, for every simplex σ^{2n-2} in P the intersection of $\mathring{\sigma}^{2n-2}$ with the singular locus of X will be empty. Hence it is clear that for every point x in $\mathring{\sigma}^{2n-2}$ there is a neighborhood U of x

in X and a homeomorphism χ of U onto $\boldsymbol{R}^{2n}$ (with coordinates $\xi_1, \ldots, \xi_{2n}$) such that χ maps $U \cap S$ onto the union of the half planes $H_\varkappa$, $\varkappa = 0, \ldots, k-1$, that are given by the relations

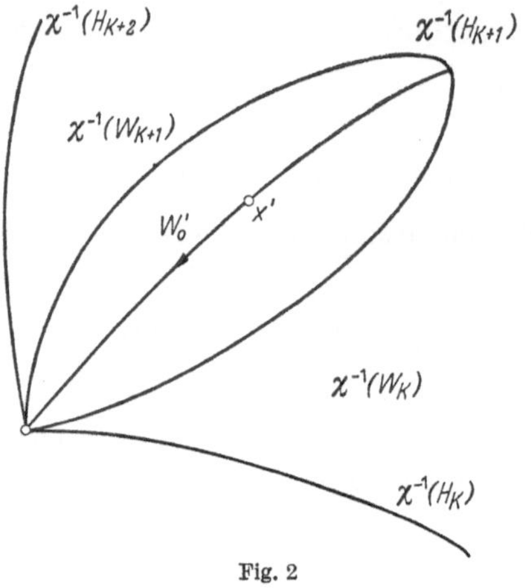

Fig. 2

$$\xi_1 = t \cos \frac{2\pi\varkappa}{k},$$

$$\xi_2 = t \sin \frac{2\pi\varkappa}{k}, \quad t \geqq 0.$$

k, of course, denotes the number of $(2n-1)$-simplices that contain σ^{2n-2} as a face. In particular, $U \cap \mathring{\sigma}^{2n-2}$ is mapped by χ onto $\xi_1 = \xi_2 = 0$. The "wedge" $W_\varkappa$, given by

$$\xi_1 = t \cos \varphi, \quad \xi_2 = t \sin \varphi,$$

$$t > 0, \quad \frac{2\pi\varkappa}{k} < \varphi < \frac{2\pi(\varkappa+1)}{k}$$

has boundary $H_\varkappa \cup H_{\varkappa+1}$ (indices being taken mod k). Clearly, for every point x of $\mathring{\sigma}^{2n-2}$, $\pi^{-1}(x)$ consists of k points $x_\varkappa = (x, \mathfrak{F}_\varkappa)$, $\varkappa = 0, \ldots, k-1$. The indexing can be chosen such that $\chi^{-1}(W_\varkappa) \in \mathfrak{F}_\varkappa$, $\varkappa = 0, \ldots, k-1$, is satisfied. With these notations we define (again indices being taken mod k)

$$\beta(x_\varkappa) = x_{\varkappa+1} \quad \text{and} \quad \gamma(x_\varkappa) = \delta_\varkappa \tag{2}$$

where $\delta_\varkappa = +1$ resp. -1 depending on whether for $x' \in \chi^{-1}(H_{\varkappa+1})$ and $\pi^{-1}(x') = \{(x', \mathfrak{F}_0'), (x', \mathfrak{F}_1')\}$ the relation $\chi^{-1}(W_\varkappa) \in \mathfrak{F}_0'$ resp. $\chi^{-1}(W_\varkappa) \in \mathfrak{F}_1'$ holds. (In Fig. 2 the case $\delta_\varkappa = -1$ is sketched.) Again it is easy to see that the definitions (2) do not depend on the choice of U, χ, and x'.

2. Analytic preparations

Let X be a complex space and $\mathfrak{L} = (T_\mathfrak{L}, p_\mathfrak{L}, X)$ be a *holomorphic fiber bundle of complex Lie groups* over X. The latter shall mean (cf. [2]) that there is a complex Lie group L, an open covering U_i, $i \in I$, of X, and biholomorphic maps $\psi_i: U_i \times L \to p_\mathfrak{L}^{-1}(U_i)$ satisfying

(i) $p_\mathfrak{L} \circ \psi_i = pr_1$, pr_1 denoting the projection of $U_i \times L$ onto U_i

(ii) $\psi_i^{-1} \circ \psi_j(x, l) = (x, l_{ij}(x, l))$ in $U_{ij} = U_i \cap U_j$,

where for every x in U_{ij} the map $l \to l_{ij}(x, l)$ is an automorphism of L. The last condition makes it possible to define a canonical holomorphic map $\mu: T_\mathfrak{L} \times_X T_\mathfrak{L} \to T_\mathfrak{L}$ as follows: if $m: L \times L \to L$ is the multiplication in L, i.e. $m(l', l'') = l' l''$, then for every U_i, $\mu \mid p_\mathfrak{L}^{-1}(U_i)$ is defined as the composition of maps[3]

$$p_{\mathfrak{L}}^{-1}(U_i)\times_{U_i} p_{\mathfrak{L}}^{-1}(U_i) \xrightarrow{\varphi_i^{-1}\times_{U_i}\varphi_i^{-1}} (U_i\times L)\times_{U_i}(U_i\times L) \xrightarrow{\simeq} U_i\times L\times \times L \xrightarrow{U_i\times m} U_i\times L \xrightarrow{\varphi_i} p_{\mathfrak{L}}^{-1}(U_i)$$

where $\xrightarrow{\simeq}$ stands for the canonical homeomorphism between $(U_i\times L)\times \times_{U_i}(U_i\times L)$ and $U_i\times L\times L$. Compatibility of this definition is an immediate consequence of (ii). The restriction μ_x of μ to $L_x\times L_x$, $L_x = p_{\mathfrak{L}}^{-1}(x)$ being the fiber in $\mathfrak{L}$ over x, defines a group structure on L_x that is isomorphic to the group structure on L. The holomorphic sections in $\mathfrak{L}$ over a fixed open subset U of X form a group, the product of two sections being defined point wise. The neutral element e_U of this group is the section which sends every $x \in U$ into the neutral element e_x of L_x.

Given two such fiber bundles, $\mathfrak{L}$ and $\mathfrak{L}'$, with associated maps μ and μ', a holomorphic map $\lambda: T_{\mathfrak{L}} \to T_{\mathfrak{L}}'$ is called a *homomorphism* from $\mathfrak{L}$ to $\mathfrak{L}'$ if both conditions,

$$p_{\mathfrak{L}} = p_{\mathfrak{L}}' \circ \lambda \quad \text{and} \quad \mu = \mu' \circ (\lambda \times_X \lambda)$$

are fulfilled. Obviously, these conditions state that λ restricted to L_x is a homomorphism of groups into L_x' for every $x \in X$.

In case $\mathfrak{L}$ is globally trivial, i.e. isomorphic to $(X\times L, pr_1, X)$, we shall denote $\mathfrak{L}$ also by L.

Next, let us assume that we are given a holomorphic fiber bundle $\mathfrak{B} = (T_{\mathfrak{B}}, p_{\mathfrak{B}}, X)$ over X with structural group the complex Lie group K and fiber the complex space F, K left-acting on F. It is no restriction to assume that $\mathfrak{B}$ be trivial over the previously given open covering U_i, $i \in I$, of X. Then we get biholomorphic maps $\varphi_i: U_i\times F \to p_{\mathfrak{B}}^{-1}(U_i)$ such that

(i') $p_{\mathfrak{B}} \circ \varphi_i = pr_1$

(ii') $\varphi_i^{-1} \circ \varphi_j(x, f) = (x, k_{ij}(x)\cdot f)$ in U_{ij},

where $k_{ij}: U_{ij} \to K$ is a holomorphic map and $(k, f) \to k\cdot f$ denotes the action of K on F.

As usual, the restriction of φ_i to $\{x\}\times F$ shall be denoted by $\varphi_{i,x}$, or shorter, by φ_x.

The holomorphic map $\alpha: T_{\mathfrak{B}}\times_X T_{\mathfrak{L}} \to T_{\mathfrak{L}}$ is said to be an *action of* $\mathfrak{L}$ *on* $\mathfrak{B}$ if the following two diagrams

$$\begin{array}{ccc} T_{\mathfrak{B}}\times_X T_{\mathfrak{L}} & \xrightarrow{\alpha} & T_{\mathfrak{B}} \\ {\scriptstyle p_{\mathfrak{B}}\times_X p_{\mathfrak{L}}}\downarrow & & \downarrow{\scriptstyle p_{\mathfrak{B}}} \\ X\times_X X & \xleftrightarrow{\Delta} & X \end{array} \qquad \begin{array}{ccc} T_{\mathfrak{B}}\times_X T_{\mathfrak{L}}\times_X T_{\mathfrak{L}} & \xrightarrow{\alpha\times_X T_{\mathfrak{L}}} & T_{\mathfrak{B}}\times_X T_{\mathfrak{L}} \\ {\scriptstyle T_{\mathfrak{B}}\times_X \mu}\downarrow & & \downarrow{\scriptstyle \alpha} \\ T_{\mathfrak{B}}\times_X T_{\mathfrak{L}} & \xrightarrow{\alpha} & T_{\mathfrak{B}} \end{array} \tag{3}$$

[3] Conforming to a usage in categorical algebra, the identity map of the set X onto itself shall be denoted by X.

commute, Δ being the canonical map between X and the diagonal $X \times_X X$ in $X \times X$, and if, in addition,

$$\alpha(T_{\mathfrak{B}} \times_X \{e_x : x \in X\}) = T_{\mathfrak{B}} \tag{3'}$$

holds. In the previous fiber coordinates, these conditions can be expressed as follows. Let α_i' denote the composition of maps

$$U_i \times F \times L \xrightarrow{\cong} (U_i \times F) \times_{U_i} (U_i \times L) \xrightarrow{\varphi_i \times_{U_i} \psi_i} p_{\mathfrak{B}}^{-1}(U_i) \times_{U_i} p_{\mathfrak{B}}^{-1}(U_i)$$
$$\xrightarrow{\alpha} p_{\mathfrak{L}}^{-1}(U_i) \xrightarrow{\psi_i^{-1}} U_i \times F.$$

Then commutativity of the first diagram (3) means that $pr_1 \circ \alpha_i' = pr_1$. Hence there is a holomorphic map $\alpha_i'' : U_i \times F \times L \to F$ such that $\alpha_i'(x, f, l) = (x, \alpha_i''(x, f, l))$. It is convenient to write $f \cdot \alpha_i(x, l)$ instead of $\alpha_i''(x, f, l)$. With this notation, the second condition (3) and (3′) can be rewritten as

$$f \cdot \alpha_i(x, l' l'') = (f \cdot \alpha_i(x, l')) \cdot \alpha_i(x, l''), \qquad f \cdot \alpha_i(x, e_x) = f. \tag{4}$$

for all $x \in U_i$, $f \in F$, $l' \in L$, and $l'' \in L$. This means that for every point x of U_i the action α gives rise (via the fiber coordinates φ_i and ψ_i) to an action, α_i, of L on F.

An easy computation shows that the local actions α_i of L on F are interconnected by the relations

$$(k_{ij}(x) \cdot f) \cdot \alpha_i(x, l_{ij}(x, l)) = k_{ij}(x) \cdot (f \cdot \alpha_j(x, l)) \quad \text{in} \quad U_{ij} \times F \times L. \tag{5}$$

Conversely, whenever local actions α_i are given such that (4) and (5) are satisfied, then they match up to a global action.

The previously constructed holomorphic map $\mu: T_{\mathfrak{L}} \times_X T_{\mathfrak{L}} \to T_{\mathfrak{L}}$ shall be called the *canonical action of* $\mathfrak{L}$ *on itself*. It is not an action in the sense of our definition since the holomorphic fiber bundle $\mathfrak{L}$ lacks a structural group. Yet the obvious analogues to (4) and (5) are still valid. Therefore, from now on we mean by an action either an action in the previous sense or the canonical action of $\mathfrak{L}$ on itself.

Given a map $\pi: X' \to X$ and an action α of $\mathfrak{L}$ on the fiber bundle $\mathfrak{B}$ over X, there is in an obvious way an action of the pull-back $\pi^*(\mathfrak{L})$ on the pull-back $\pi^*(\mathfrak{B})$ which is called the pull-back of α and denoted by $\pi^*(\alpha)$. In case all maps involved are holomorphic, $\pi^*(\alpha)$ is also holomorphic.

Let us assume that S is a closed subset of the complex space X and that $\mathfrak{B}$ is a holomorphic fiber bundle over X. Then we say that the holomorphic fiber bundle $\mathfrak{L}$ of complex Lie groups *acts on* $\mathfrak{B}$ *over* S, if there is given an open neighborhood V of S, a bundle $\mathfrak{L}$ over V, and an action of $\mathfrak{L}$ on the restriction $\mathfrak{B} \mid V$.

The following assumption shall be referred to as *analytic hypothesis*: There is a complex Lie transformation group G left-acting on F and an antihomomorphism $\lambda: \mathfrak{L} \to G$ (i.e. a homomorphism $\mathfrak{L} \to G^{op}$) such that

(i) K is a Lie transformation subgroup of G

(ii) for every x in X, the diagram

$$\begin{array}{ccc} F_x \times L_x & \xrightarrow{\alpha_x} & F_x \\ {\scriptstyle \varphi_x^{-1} \times \lambda_x}\downarrow & & \downarrow{\scriptstyle \varphi_x^{-1}} \\ F \times G & \longrightarrow & F \end{array}$$

commutes, $F \times G \to F$ being the map that sends (f, g) into $g \cdot f$. In terms of the previous fiber coordinates, this means that

$$f \cdot \alpha_i(x, l) = \lambda_i(x, l) \cdot f \qquad \text{in} \quad U_i \times F \times L, \tag{6}$$

$\lambda_i: U_i \times L \to G$ being the composition of maps

$$U_i \times L \xrightarrow{\varphi_i} p_{\mathfrak{B}}^{-1}(U_i) \xrightarrow{\lambda} U_i \times G.$$

Obviously, we get $\alpha_i'(x, f, l) = (x, \lambda_i(x, l) \cdot f)$ in $U_i \times F \times L$.

Intuitively speaking, the analytic hypothesis means that both, the action of the structural group K on F and the local actions α_i of L on F, can simultaneously be described by means of a complex Lie transformation group G. The analytic hypothesis is certainly satisfied, provided the fiber F is a compact complex space, since in this case the group of all automorphisms is a complex Lie group (cf. [*6*]). In general, however, one has only a necessary condition for the analytic hypothesis to be fulfilled: the Lie algebra of tangent fields on F generated by K and by all local actions α_i of L must be finite dimensional. The latter condition is also sufficient in case both, K and L, are simply connected.

The analytic hypothesis is also satisfied in the "classical case" where both, $\mathfrak{B}$ and $\mathfrak{L}$, are trivial bundles and the action α is the "product action". In this case, G may be chosen to be L.

3. The main theorem

Let X be a complex space and P a polyhedron imbedded in X, both subject to the geometric hypothesis (cf. section 1). Let furthermore $\mathfrak{B}$ be a holomorphic fiber bundle over X and $\mathfrak{L}$ a holomorphic fiber bundle of complex Lie groups acting on $\mathfrak{B}$ over $|P|$. Moreover, denoting by $\mathfrak{L}_c$ the sheaf of germs of continuous sections in $\mathfrak{L}$, let t be a map that assigns to each $(2n-1)$-simplex σ^{2n-1} of P an element, $t(\sigma^{2n-1})$, in $\Gamma(|\sigma^{2n-1}|, \mathfrak{L}_c)$ such that[4] for every $(2n-2)$-simplex σ^{2n-2} and all σ^{2n-1} satisfying $[\sigma^{2n-2}, \sigma^{2n-1}] \neq 0$ the sections $t(\sigma^{2n-1})|\sigma^{2n-2}|$ mutually commute. With these data we can associate the following *transmission problem* $\tau = \tau(X, P; \mathfrak{B}, \mathfrak{L}; t)$:

[4] The geometric meaning of the following condition shall be given in section 4.

to find holomorphic sections s in $\mathfrak{B}$ over $X - |P|$ that have a continuous extension to $|P|$ such that for every $\hat{x} \in \pi^{-1}(\bigcup \{\mathring{\sigma}^{2n-1}: \sigma^{2n-1} \in P\})$

$$s(\beta(\hat{x})) = \alpha(s(\hat{x}), t(\pi(\hat{x}))^{\nu(\hat{x})}) \tag{7}$$

holds, where $t(x)$, x being in $\mathring{\sigma}^{2n-1}$, abbreviates $t(\sigma^{2n-1})(x)$.

A holomorphic section s with these properties is called a *solution* of the transmission problem τ. The set of all solutions of τ shall be denoted by Σ_τ. The aim of this paper is to get informations concerning Σ_τ. For this purpose we need several preliminary remarks.

Remark 1. Suppose that the closed subset S is given in two ways the structure of an oriented polyhedron such that in either case the geometric hypothesis is satisfied. Denoting these two polyhedra by P_1 and P_2, we assume that P_2 is a subdivision of P_1. Then we get a canonical chain map $\varrho: C^{2n-1}(P_1) \to C^{2n-1}(P_2)$ for the integral $(2n-1)$-chains. It is easy to verify that under this assumption there is a natural bijective map $\Sigma_\tau \to \Sigma_{\varrho(\tau)}$ for every $\tau = \tau(X, P_1; \mathfrak{B}, \mathfrak{L}; t)$ and $\varrho(\tau) = \tau(X, P_2; \mathfrak{B}, \mathfrak{L}; t_\varrho)$ where t_ϱ is given by $t_\varrho(\sigma_2^{2n-1}) = t(\sigma_1^{2n-1})^{(\varrho\gamma)(\sigma_2^{2n-1})}$, σ_1^{2n-1} being the unique simplex of P_1 containing σ_2^{2n-1}.

Remark 2. Let $h: X' \to X$ be a holomorphic map. Then the action α of $\mathfrak{L}$ on $\mathfrak{B}$ over S gives rise (cf. section 2) to an action $h^*(\alpha)$ of $h^*(\mathfrak{L})$ on $h^*(\mathfrak{B})$ over $h^{-1}(S)$. We assume that the geometric hypothesis is satisfied for X and S as well as X' and $S' = h^{-1}(S)$, denoting the polyhedron on S' by P'. Furthermore we assume that the restriction of h to P' is a simplicial map and that h preserves orientations. Then the pull-back on the level of sections gives rise to a canonical map, h^*, from Σ_τ to $\Sigma_{h^*(\tau)}$ for every $\tau = \tau(X, P; \mathfrak{B}, \mathfrak{L}; t)$ and $h^*(\tau) = \tau(X', P'; h^*(\mathfrak{B}), h^*(\mathfrak{L}); t')$ where t' is given by $t'(\sigma'^{2n'-1}) = h^*(t(h\sigma'^{2n'-1}))$, n' being the complex dimension of X'. The assignment of $h^*: \Sigma_\tau \to \Sigma_{h^*(\tau)}$ to h, subject to the above conditions, constitutes a contravariant functor.

Remark 3. Let $b: \mathfrak{B} \to \mathfrak{B}'$ be a holomorphic fiber map and $\lambda: \mathfrak{L} \to \mathfrak{L}'$ be a homomorphism such that for the actions, α, of $\mathfrak{L}$ on $\mathfrak{B}$ and α', of $\mathfrak{L}'$ on $\mathfrak{B}'$, the following diagram is commutative

$$\begin{array}{ccc} T_{\mathfrak{B}} \times_X T_{\mathfrak{L}} & \xrightarrow{\alpha} & T_{\mathfrak{B}} \\ b \times_X \lambda \downarrow & & \downarrow b \\ T_{\mathfrak{B}'} \times_X T_{\mathfrak{L}'} & \xrightarrow{\alpha'} & T_{\mathfrak{B}'} \end{array}$$

Then the map on the level of sections, that is induced by b, gives rise to a canonical map, b_*, from Σ_τ to $\Sigma_{b_*(\tau)}$ for every $\tau = (X, P; \mathfrak{B}, \mathfrak{L}; t)$ and $b_*(\tau) = \tau(X, P; \mathfrak{B}', \mathfrak{L}'; \lambda \circ t)$. Again it is clear that the assignment of $b_*: \Sigma_\tau \to \Sigma_{b_*(\tau)}$ to b constitutes a covariant functor in both, b and λ.

Remark 4. Given an open subset U of X and the closed subset $|P|$ of X where X and P satisfy the geometric hypothesis. Then $|P| \cap U$ is an open subset of $|P|$ and can therefore be equipped with the structure of an oriented polyhedron P' such that

(i) for every $q \geqq 0$ the carrier of the q-skeleton P'^q of P' contains $|P^q| \cap U$;

(ii) if $\sigma'^q \in P'$ is contained in $\sigma^q \in P$, then the orientation of σ'^q coincides with the one induced from σ^q.

The condition (ii) makes sense because (i) implies that for every q-dimensional simplex σ'^q in P' there is precisely on q-dimensional simplex σ^q containing σ'^q. Denoting this simplex by $\sigma(\sigma'^q)$, we put $t'(\sigma'^{2n-1}) = t(\sigma(\sigma'^{2n-1}))| |\sigma'^{2n-1}|$ and associate with $\tau = \tau(X, P; \mathfrak{B}, \mathfrak{L}; t)$ the restriction $\tau | U = \tau(U, P'; \mathfrak{B}| U, \mathfrak{L} | U; t')$. Due to Remark 1, $\Sigma_{\tau|U}$ depends only on U and τ, and not on the choice of P'. Therefore, though $\tau | U$ does depend on the choice of P', we may speak of *the restriction* of τ to U. In particular, we can speak of the germ of a transmission problem τ at the point x.

Remark 5. A last preliminary construction is to associate with every transmission problem $\tau(X, P; \mathfrak{B}, \mathfrak{L}; t)$ a *principal transmission problem.* By that we mean the transmission problem $\tau(X, P; \mathfrak{L}, \mathfrak{L}; t)$ where $\mathfrak{L}$ acts on $\mathfrak{L}$ canonically (cf. section 2). Locally, a principal transmission problem is of the form $\tau(U, P; L, L; t)$.

With these notations we have

Theorem 3.1. *Let $\tau = \tau(X, P; \mathfrak{B}, \mathfrak{L}; t)$ be a transmission problem such that for every point x in $|P|$ the germ at x of the associated principal transmission problem has a solution. Assume that both, the analytic and the geometric hypothesis, are fulfilled. Then there is a holomorphic fiber bundle $\mathfrak{B}_\tau$ over X with fiber F and structural group G such that the set Σ_τ of solutions of τ corresponds bijectively to the set of global holomorphic sections in $\mathfrak{B}_\tau$.*

Proof. We choose an open covering U_i, $i \in I$, of X in the following way: first we cover $X - |P|$ by open sets U_{i_1}, $i_1 \in I_1$, such that $\mathfrak{B}$ is trivial over each U_{i_1}. Then we select open sets U_{i_2}, $i_2 \in I_2$, such that this family of open subsets of X covers $|P|$ and has the following properties:

(0) U_{i_2} is contained in the base space V of $\mathfrak{L}$;

(i) both, $\mathfrak{B}$ and $\mathfrak{L}$, are trivial over each U_{i_2};

(ii) the restriction to each U_{i_2} of the principal transmission problem associated with τ admits a solution;

(iii) every U_{i_2} is a normal subspace of X.

The fact that condition (iii) can be satisfied, follows immediately from the geometric hypothesis. Having found two such families, we take as family U_i, $i \in I$, the disjoint union of them.

Next we assign to each $i \in I$ a holomorphic map $t_i: U_i - U_i \cap |P| \to L$. In case $U_i \cap |P| = \emptyset$, we choose as t_i the constant map that sends U_i into the neutral element e of L; it is convenient, in this case, to put $t_i'(x) = (x, t_i(x))$. In case $U_i \cap |P| \neq \emptyset$, let the section t_i' be a solution of the restriction to U_i of the associated principal transmission problem $\tau_i = \tau(U_i, P_i; L, L; \psi_i^{-1} \circ t | U_i)$ where P_i is constructed according to Remark 4, and define t_i by the relation $t_i'(x) = (x, t_i(x))$.

Suppose that the transition functions for the bundle $\mathfrak{B}$ are k_{ij}. Then we define $\hat{k}_{ij}: U_{ij} - U_{ij} \cap |P| \to G$ by

$$(\lambda_i(x, t_i(x)))^{-1} \cdot k_{ij}(x) \cdot \lambda_j(x, t_j(x)) . \tag{8}$$

Obviously, $\hat{k}_{ij}$ constitutes a holomorphic 1-cocycle on $X - |P|$ with values in G. If we can show that each $\hat{k}_{ij}$ can be extended holomorphically to all of U_{ij}, then we get in fact a holomorphic 1-cocycle on all of X. In order to show that $\hat{k}_{ij}$ can be extended holomorphically we recall that G is a Lie transformation group on F. Then a straightforward computation, using formulas (4), (6), and (8), leads to

$$\hat{k}_{ij}(x) = \lambda_i(x, l_{ij}(x, t_j(x) \cdot t_i^{-1}(x))) \cdot k_{ij}(x). \tag{8'}$$

Now, let $\hat{x} \in \pi^{-1}(x)$ where x is a point of $U_{ij} \cap \bigcup \{\mathring{\sigma}^{2n-1} : \sigma^{2n-1} \in P\}$. Then another straightforward computation, using the fact that the map $l \to l_{ij}(x, l)$ is an automorphism of L and that, in any case, t_i' is a solution of the associated local principal transmission problem, results in the relation

$$l_{ij}(x, t_j(\beta(\hat{x}))) \cdot t_i(\beta(\hat{x}))^{-1} = l_{ij}(\hat{x}, t_j(\hat{x})) \cdot t_i(\hat{x})^{-1} .$$

That is, the boundary values from various sides at x coincide for the function $l_{ij}(x, t_j(x)) \cdot t_i(x)^{-1}$ which therefore extends to a continuous function in all of U_{ij}. Outside of the singular locus of X, $l_{ij}(x, t_j(x)) \cdot t_i(x)^{-1}$ is thus holomorphic in U_{ij}, as can be taken immediately from Hartog's theorem. Since, according to hypothesis, the structural sheaf of X is normal on U_{ij}, $l_{ij}(x, t_j(x)) \cdot t_i(x)^{-1}$ is also holomorphic in the intersection of U_{ij} and the singular locus of X. Consequently, by (8'), $\hat{k}_{ij}(x)$ has been extended to a holomorphic map from U_{ij} to G.

The holomorphic fiber bundle over X with fiber F that is defined by the 1-cocycle $\hat{k}_{ij}$ is denoted by $\mathfrak{B}_\tau$. We want to show that to each global holomorphic section in $\mathfrak{B}_\tau$ there corresponds an element in Σ_τ, and vice versa. Let s be such a section. It can be described in the previous fiber coordinates by holomorphic maps $\hat{s}_i: U_i \to F$ satisfying

$$\hat{s}_i(x) = \hat{k}_{ij}(x) \cdot \hat{s}_j(x) \quad \text{in} \quad U_{ij} .$$

In order to construct an element in Σ_τ, take the collection of holomorphic maps $s_{i_1}(x) = \hat{s}_{i_1}(x)$, $i_1 \in I_1$. Due to (8) and the definition of t_{i_1}, we have

$$s_{i_1}(x) = k_{i_1 j_1}(x) \cdot s_{j_1}(x), \qquad i_1, j_1 \in I_1.$$

Therefore, the s_{i_1}, $i_1 \in I_1$, define a holomorphic section, s, in $\mathfrak{B}$ over $X - |P|$. In case $i_1 \in I_1$ and $i_2 \in I_2$, we get

$$s_{i_1}(x) = \widehat{k}_{i_1 i_2}(x) \cdot \hat{s}_{i_2}(x) = k_{i_1 i_2}(x) \cdot (\lambda_{i_2}(x, t_{i_2}(x)) \cdot \hat{s}_{i_2}(x)). \tag{9}$$

Since, according to construction, $t_{i_2}(x)$ can be extended continuously onto $U_{i_2} \cap |P|$, the same is true for $\lambda_{i_2}(x, t_{i_2}(x)) \cdot s_{i_2}(x)$. Hence (9) shows that the section s can be extended continuously onto $|P|$. For $x \in \pi^{-1}(U_{i_2}) \cap |P|$ we get then, denoting $\pi(\hat{x})$ by x,

$$\begin{aligned}\lambda_{i_2}(x, t_{i_2}(\beta(\hat{x}))) \cdot \hat{s}_{i_2}(x) &= \lambda_{i_2}(x, t_{i_2}(\hat{x}) \cdot pr_2(\psi_{i_2}^{-1} \circ t)^{\gamma(\hat{x})}) \cdot \hat{s}_{i_2}(x) \\ &= \lambda_{i_2}(x, pr_2(\psi_{i_2}^{-1} \circ t))^{\gamma(\hat{x})} \cdot (\lambda_{i_2}(x, t_{i_2}(\hat{x})) \cdot \hat{s}_{i_2}(x)) \\ &= (\lambda_{i_2}(x, t_{i_2}(\hat{x})) \cdot \hat{s}_{i_2}(x)) \cdot \alpha_{i_2}(x, pr_2(\psi_{i_2}^{-1} \circ t))^{\gamma(\hat{x})}\end{aligned}$$

and thus

$$s(\beta(\hat{x})) = \alpha(s(\hat{x}), t(\pi(\hat{x}))^{\gamma(\hat{x})})$$

as required. Tracing this argument in the opposite direction, one finds to each element s in Σ_τ a global holomorphic section s in $\mathfrak{B}_\tau \to X$. Since the thus described map is evidently bijective, our theorem is proved.

Corollary 3.2. *The assignment of $\mathfrak{B}_\tau \to X$ to τ has the same functorial properties as described in Remarks 2 and 3. With the notation of Remark 1, $\mathfrak{B}_\tau \to X$ and $\mathfrak{B}_{\varrho(\tau)} \to X$ are naturally isomorphic. Moreover, $\mathfrak{B}_{\tau|U} \to U$ is naturally isomorphic to the restriction of $\mathfrak{B}_\tau \to X$ to U.*

Proof: Given the transmission problem τ, we can form the sheaf S_τ of germs of solutions of τ; here $S_\tau(U) = \Sigma_{\tau|U}$, while for $V \subset U$ the map $S_\tau(U) \to S_\tau(V)$ is the one described in Remark 4. According to the construction of $\mathfrak{B}_\tau \to X$, $S_\tau(U)$ is naturally isomorphic to the set $\Gamma(U, \mathfrak{B}_\tau)$ of holomorphic sections in $\mathfrak{B}_\tau \to X$ over U. Hence the corollary follows from the Remarks 1—4.

In case the fiber bundle $\mathfrak{B}$ is a vector bundle of rank q, the fiber bundle $\mathfrak{L}$ has group L a closed subgroup of $GL(q, \mathbf{C})$, and the group G equals $GL(q, \mathbf{C})$, we call a transmission problem to these data a *linear transmission problem.* For linear transmission problems $\tau_i = \tau(X, P; \mathfrak{B}_i, \mathfrak{L}_i; t_i)$, $i = 1, 2$, one can form

$$\tau_1 \oplus \tau_2 = \tau(X, P; \mathfrak{B}_1 \oplus \mathfrak{B}_2, \mathfrak{L}_1 \oplus \mathfrak{L}_2; t_1 \oplus t_2)$$

and

$$\tau_1 \otimes \tau_2 = \tau(X, P; \mathfrak{B}_1 \otimes \mathfrak{B}_2, \mathfrak{L}_1 \otimes \mathfrak{L}_2; t_1 \otimes t_2)$$

where the group of $\mathfrak{L}_1 \oplus \mathfrak{L}_2$ is $L_1 \oplus L_2$ [5], the transition functions for

[5] For the subgroups L_i of $GL(q_i, \mathbf{C})$, $i = 1, 2$, $L_1 \oplus L_2$ the subgroup of $GL(q_1 + q_2, \mathbf{C})$ consisting of all matrices $\begin{pmatrix} l_1 & 0 \\ 0 & l_2 \end{pmatrix}$, $l_1 \in L_1$ and $l_2 \in L_2$. $L_1 \otimes L_2$ is defined analogously.

$\mathfrak{L}_1 \oplus \mathfrak{L}_2$ are given by $(l_1, l_2) \to l_{ij}^{(1)}(x, l_1) \oplus l_{ij}^{(2)}(x, l_2)$, $x \in U_{ij}$, and the action, $\alpha_1 \oplus \alpha_2$, of $\mathfrak{L}_1 \oplus \mathfrak{L}_2$ on $\mathfrak{B}_1 \oplus \mathfrak{B}_2$ is described by

$$(\alpha_1 \oplus \alpha_2)_i'(x, (f_1, f_2), (l_1, l_2)) = \alpha_i^{(1)}(x, f_1, l_1) \oplus \alpha_i^{(2)}(x, f_2, l_2) ;$$

correspondingly in the case of $\otimes$ instead of $\oplus$.

With these notations one derives readily from the definition of $\mathfrak{B}_\tau$ the

Corollary 3.3. $\mathfrak{B}_{\tau_1 \oplus \tau_2}$ *is canonically isomorphic to* $\mathfrak{B}_{\tau_1} \oplus \mathfrak{B}_{\tau_2}$, $\mathfrak{B}_{\tau_1 \otimes \tau_2}$ *is canonically isomorphic to* $\mathfrak{B}_{\tau_1} \otimes \mathfrak{B}_{\tau_2}$.

It is of interest to know the "range" of the map $\tau \to \mathfrak{B}_\tau$. Here we shall give a result that leads to an interesting application in the theory of singular integral operators ([*10*]). For this purpose we need a preliminary definition.

Suppose that the geometric hypothesis is satisfied. Then for a given complex Lie group G we can form the sheaf $G_{\omega, P}$ of germs of those continuous maps of X_P into G that are holomorphic on $\pi^{-1}(X - |P|)$ and take values in the center of G on $\pi^{-1}(\bigcup\{|\sigma^{2n-2}| : \sigma^{2n-2} \in P\})$. The holomorphic fiber bundles $\mathfrak{B}$ and $\mathfrak{B}'$ over X (with structural group G) are said to be *holomorphically isomorphic over* $X_{|P|}$ if the pull-back $\pi^*(\mathfrak{B})$ is isomorphic to $\pi^*(\mathfrak{B}')$ via coordinate changes that are sections in $G_{\omega, P}$. With this notation we get

Proposition 3.4. *Suppose that for X and P the geometric hypothesis is satisfied. Let furthermore $\mathfrak{B}$ and $\mathfrak{B}'$ be holomorphic fiber bundles over X with structural group G, such that $\mathfrak{B}$ and $\mathfrak{B}'$ are holomorphically isomorphic over $X_{|P|}$. Assume in addition that $\mathfrak{B}$ is holomorphically trivial over some neighborhood of $|P|$. Then there is a transmission problem $\tau = \tau(X, P; \mathfrak{B}, G^{op}; t)$ such that $\mathfrak{B}'$ is holomorphically isomorphic to $\mathfrak{B}_\tau$.*

Proof. Using the notation adopted in the proof of Theorem 3.1, we may assume that both, $\mathfrak{B}$ and $\mathfrak{B}'$, are trivial over each U_i, $i \in I_1 \cup I_2$. Denoting the transition functions by k_{ij} resp. k_{ij}', we can find sections t_i' in $G_{\omega, P}$ over $\pi^{-1}(U_i)$ such that

$$k_{ij}'(\pi(\hat{x})) = t_i'^{-1}(\hat{x}) \circ k_{ij}(\pi(\hat{x})) \cdot t_j'(\hat{x}) \quad \text{in} \quad U_{ij}$$

holds. It is no restriction to assume that $t_i'(x) = e (=$ neutral element in $G)$ for all $x \in \pi^{-1}(U_i)$ and all $i \in I_1$. Moreover we may assume that the fiber bundle $\mathfrak{B}$ is trivial over $\bigcup\{U_i : i \in I_2\}$. Therefore, for a suitable collection of holomorphic maps h_i, $i \in I_2$, of U_i into G we get for all i, j in I_2

$$k_{ij}'(\pi(\hat{x})) = t_i'^{-1}(\hat{x}) \cdot h_i^{-1}(\pi(\hat{x})) \cdot h_j(\pi(\hat{x})) \cdot t_j'(\hat{x}) = t_i^{-1}(\hat{x}) \cdot t_j(\hat{x}) \text{ in } U_{ij} \quad (10)$$

where $t_i(\hat{x}) = h_i(\pi(\hat{x})) \cdot t_i'(\hat{x})$. Given the $(2n-1)$-simplex σ^{2n-1} of P, we define for $x \in \mathring{\sigma}^{2n-1} \cap U_i$, using the notation of section 1, (1),

$$t_i(\sigma^{2n-1})(x) = t_i(x_0) \cdot t_i^{-1}(\beta(x_0)) = t_i(x_0) \cdot t_i^{-1}(x_1) .$$

Due to (10) we have in $\mathring{\sigma}^{2n-1} \cap U_{ij}$

$$t_j(\sigma^{2n-1})(x) = t_j(x_0) \cdot t_j^{-1}(x_1) = t_i(x_0) \cdot k'_{ij}(x) \cdot k'^{-1}_{ii}(x) \cdot t_i^{-1}(x_1)$$
$$= t_i(\sigma^{2n-1})(x)$$

whence the family of functions $t_i(\sigma^{2n-1})$ defines a continuous map of $\mathring{\sigma}^{2n-1}$ into G^{op}. Since the t'_i are continuous in $X_{|P|}$, this map can be extended to a continuous map, $t(\sigma^{2n-1})$, of $|\sigma^{2n-1}|$ into G^{op}. The collection of maps $t(\sigma^{2n-1})$ constitutes a map t from the set of $(2n-1)$-simplices of P into the set of sections in G_c^{op}. Since the t'_i take values in the center of G over $\pi^{-1}(\bigcup\{|\sigma^{2n-2}| : \sigma^{2n-2} \in P\})$, we conclude that t satisfies the requirements imposed on such maps in the beginning of this section. Finally, we let G^{op} operate canonically on $\mathfrak{B}$ over $|P|$, i. e. after the coordinate changes by the maps h_i the action of G^{op} on $U_i \times F$ is given by $\alpha'_i(x, f, g) = (x, g \cdot f)$ where $g \cdot f$ is the action of $g \in G$ on $f \in F$. One checks now easily that these data together with the canonical map $G^{op} \to G$ (replacing $\lambda : \mathfrak{L} \to G$) give rise to a transmission problem $\tau = \tau(X, P; \mathfrak{B}, G^{op}; t)$ that satisfies the hypotheses of Theorem 3.1. From the proof of Theorem 3.1 and the construction carried out here, it is evident that $\mathfrak{B}'$ is holomorphically isomorphic to $\mathfrak{B}_\tau$.

Remark 6. The proof of Theorem 4.1 does not cover the case of principal transmission problems $\tau = \tau(X, P; \mathfrak{L}, \mathfrak{L}; t)$ since neither $\mu : \mathfrak{L} \times_X \mathfrak{L} \to \mathfrak{L}$ is an action in the sense used in the proof, nor a Lie transformation group G gets involved. However, the idea of the proof carries over to principal transmission problems whence the statement of Theorem 3.1 is also valid for this case. Yet, the fiber bundle $\mathfrak{L}_\tau$ has an additional structural property: it is a $\mathfrak{L}$-principal holomorphic fiber bundle in the sense of [2].

Remark 7. In case we are dealing with a principal transmission problem τ of the form $\tau(X, P; L, L; t)$, that is if $\mathfrak{L}$ is globally trivial, then the holomorphic fiber bundle L_τ is a principal L-bundle in the usual sense.

The transmission problems we dealt with so far were stated for holomorphic fiber bundles, holomorphic sections etc., i. e. they were stated for the category of complex spaces. Obviously, one can make analogous definitions involving topological fiber bundles, continuous sections, continuous actions etc., i. e. involving the category of topological spaces. Whenever it is necessary from now on to distinguish between these two cases we speak of *topological transmission problems* as compared to *holomorphic transmission problems*. Clearly, every holomorphic transmission problem can be viewed as a topological transmission problem. Moreover, the statements proved so far for holomorphic transmission problems remain valid for topological transmission problems under the corresponding hypotheses[6], as can be seen easily from the proofs.

4. Topologically correct transmission functions

Let $\tau = \tau(X, P; \mathfrak{B}, \mathfrak{L}; t)$ be a topological (holomorphic) transmission problem. Then t is called the *transmission function* of τ. The transmission function t is said to be *topologically (holomorphically) correct at* $x \in |P|$ if the germ of the associated principal transmission problem $\tau(X, P; \mathfrak{L}, \mathfrak{L}; t)$ at x admits a solution. t is said to be *topologically (holomorphically) correct* if it is topologically (holomorphically) correct at every point of $|P|$. The importance of the notion of correctness of a transmission function is obvious from Theorem 3.1.

Since correctness is a local notion, we can speak of the sheaf $\mathfrak{T}^t(\mathfrak{T}^h)$ of germs of topologically (holomorphically) correct transmission functions on $|P|$ with values in $\mathfrak{L}$. The correct transmission functions are then precisely the global sections in $\mathfrak{T}^t(\mathfrak{T}^h)$. Therefore it is desirable to arrive at necessary and sufficient conditions for a germ of a (not necessarily continuous) section in $\mathfrak{L}$ to belong to $\mathfrak{T}^t(\mathfrak{T}^h)$.

In this section we shall deal with topologically correct transmission functions. We will get necessary and sufficient conditions for a transmission function to be topologically correct. These conditions can be best expressed using a certain homology theory which we shall define only in the dimension pertinent to our problem. By localizing these conditions we get then a complete description of $\mathfrak{T}^t$.

We assume that the geometric hypothesis is satisfied in the modified form given in footnote 6. Moreover, let $\mathfrak{L}$ be a fiber bundle of topological groups over some neighborhood V of $|P|$. Then a $(n-1)$-*chain on P with values in* $\mathfrak{L}$ is defined to be a map, t, which assigns to each $(n-1)$-simplex σ^{n-1} of P an element $t(\sigma^{n-1}) \in \Gamma(|\sigma^{n-1}|, \mathfrak{L}_c)$ such that for every $(n-2)$-simplex σ^{n-2} in P all sections

$$t(\sigma^{n-1}) \,|\, |\sigma^{n-2}| \tag{11}$$

mutually commute. The set of all $(n-1)$-chains on P with values in $\mathfrak{L}$ is denoted by $C_{n-1}(P; \mathfrak{L})$. It is a based set, the base point being the chain t_0 that assigns to each σ^{n-1} the unit section in $\mathfrak{L}_c$, i. e. the section defined by $t_0(\sigma^{n-1})(x) = e_x$ for all x in σ^{n-1}.

Analogously, a $(n-2)$-*chain on P with values in* $\mathfrak{L}$ is a map, t, which assigns to each $(n-2)$-simplex σ^{n-2} of P an element $t(\sigma^{n-2}) \in \Gamma(|\sigma^{n-2}|, \mathfrak{L}_c)$. The set of all $(n-2)$-chains on P is denoted by $C_{n-2}(P; \mathfrak{L})$; it is also a based set. Note that on the $(n-2)$-level there is no condition of the type

[6] For topological transmission problems, in the geometric hypothesis the condition (i) as well as (ii′′) is meaningless and has to be dropped, while in (ii′) the dimension $2n$ has to be replaced by the (topological) dimension n.

(11) required; this is due to the fact that we are only interested in dimension $n-1$.

Next we define a map $\partial : C_{n-1}(P; \mathfrak{L}) \to C_{n-2}(P; \mathfrak{L})$ as follows: for $t \in C_{n-1}(P; \gamma, \mathfrak{L})$ we set

$$(\partial t)(\sigma^{n-2}) = \prod \{(t(\sigma^{n-1}) \| \sigma^{n-2}|)^{(\partial\gamma)(\sigma^{n-2})} : \sigma^{n-1} \in P\}. \tag{12}$$

Due to (11), this product is independent of the order of the factors. The kernel of ∂ is denoted by $Z_{n-1}(P; \mathfrak{L})$ and called the set of $(n-1)$-*cycles on* P *with values in* $\mathfrak{L}$.

In order to get a homology set in dimension $n-1$, we have to define an equivalence relation on $C_{n-1}(P; \mathfrak{L})$. For that purpose, let t and t' be two $(n-1)$-chains on P with values in $\mathfrak{L}$. Then t and t' are called *homologous* if and only if there exists a neighborhood $V' \subset V$ of P and a continuous section s in $\mathfrak{L}$ over $V' - |P|$ that has a continuous extension to $|P|$ such that for all σ^{n-1} in P and $x \in \pi^{-1}(\mathring{\sigma}^{n-1})$

$$t'(\sigma^{n-1})(\pi(\hat{x}))^{\gamma(\hat{x})} = s(\hat{x}) \cdot t(\sigma^{n-1})(\pi(\hat{x}))^{\gamma(\hat{x})} \cdot s(\beta(\hat{x}))^{-1} \tag{13}$$

holds. This constitutes obviously an equivalence relation in $C_{n-1}(P; \mathfrak{L})$. Under suitable additional hypotheses (cf. Theorem 4.1), a chain that is homologous to a cycle is itself a cycle. The based set of equivalence classes of $(n-1)$-cycles on P with values in $\mathfrak{L}$ is denoted by $H_{n-1}(P; \mathfrak{L})$ and called the set of $(n-1)$-*homology classes on* P *with values in* $\mathfrak{L}$.

Theorem 4.1. *Suppose that* X *and* $|P|$ *satisfy the geometric hypothesis in the modified form given in footnote* 6 *and that* X *is a n-dimensional topological manifold in some neighborhood of* $|P|$. *Suppose furthermore that a fiber bundle* $\mathfrak{L}$ *of topological groups over some neighborhood of* $|P|$ *is given. Then a* $(n-1)$-*chain on* P *with values in* $\mathfrak{L}$ *is a topologically correct transmission function if and only if it belongs to* $Z_{n-1}(P; \mathfrak{L})$.

Proof. Suppose that t is topologically correct. Then for every $(n-2)$-simplex σ^{n-2} in P and every point $x \in \mathring{\sigma}^{n-2}$ there is a neighborhood U and a continuous section s in $\mathfrak{L}$ over $U - U \cap |P|$ that has a continuous extension to $U \cap |P|$ which satisfies (7) as formulated for principal transmission problems. Choosing the point $x_0 \in \pi^{-1}(x)$ in accordance with section 1, (2), and assuming that there are precisely k $(n-1)$-simplices in P that contain x, we get therefore

$$s(x_k) = t(\sigma_k)(x)^{\gamma(x_{k-1})} \cdot s(x_{k-1}) = \prod \{t(\sigma_\varkappa)(x)^{\gamma(x_{k-1})} : \varkappa = 1, \ldots, k\} \cdot s(x_0),$$

the product being taken in the canonical order. One checks easily that either $\gamma(x_\varkappa) = [\sigma_{\varkappa+1}^{n-1}, \sigma^{n-2}]$ for all $\varkappa$ or else $\gamma(x_\varkappa) = -[\sigma_{\varkappa+1}^{n-1}, \sigma^{n-2}]$ for all $\varkappa$ holds. Consequently, since $x_0 = x_k$ and since s takes values in a fiber bundle of groups, we get for all x in σ^{n-2} the relation

$$\prod \{t(\sigma_\varkappa)(x)^{(\partial\gamma)(\sigma^{n-2})} : \varkappa = 1, \ldots, k\} = e_x.$$

Due to continuity this relation holds on all of $|\sigma^{n-2}|$ which, according to definition, means that t belongs to $Z_{n-1}(P; \mathfrak{L})$.

Conversely, let t be a $(n-1)$-cycle on P with values in $\mathfrak{L}$. Since correctness is a local notion we may assume that $\mathfrak{L} = L$ holds. Then, however, a trivial extendability argument shows that t is topologically correct on $\bigcup\{\mathring{\sigma}^{n-1} : \sigma^{n-1} \in P\}$.

Now we have to discuss points x in $\mathring{\sigma}^q$, $q \leqq n-2$, for each σ^q in P. As in section 1 we can find to such a point a neighborhood U of the topological type of the cell and a homeomorphism, χ, of U onto $\boldsymbol{R}^n$ (with coordinates $\xi_1, \ldots, \xi_n$) such that

(i) $\chi(\mathring{\sigma}^q \cap U)$ equals the subspace $\xi_1 = \cdots = \xi_{n-q} = 0$;

(ii) whenever $\mathring{\sigma}^p \cap U$ is non empty and $p \geqq q$, then $\mathring{\sigma}^q \cap U$ is contained in $\mathring{\sigma}^p \cap U$ and $\chi(\mathring{\sigma}^p \cap U)$ can be described by

$$L_1(\xi) = 0, \ldots, L_{n-p}(\xi) = 0, \quad L_{n-p+1}(\xi) \geqq 0, \ldots, L_{n-q+1}(\xi) \geqq 0,$$

the L_ϱ being linear forms in $\xi_1, \ldots, \xi_{n-q}$.

Let $\sigma_1^n, \ldots, \sigma_l^n$ be the closures of the connected components of $\boldsymbol{R}^n - \chi(|P| \cap U)$ and denote $\chi(\mathring{\sigma}^p \cap U)$ again by σ^p for every $\sigma^p \in P$ such that $\mathring{\sigma}^p \cap U$ is non empty. *A chain of length r from $\sigma_{\lambda_1}^n$ to $\sigma_{\lambda_2}^n$* is a sequence $C = (\sigma_{\lambda_1}^n, \sigma_1^{n-1}, \ldots, \sigma_r^{n-1}, \sigma_{\lambda_2}^n)$ of wedges σ^n alternating with $(n-1)$-simplices of $\chi(P \cap U)$ such that any two consecutive elements of this sequence are incident; C is said to be incident with a given $\sigma^p \in \chi(P \cap U)$ if every element of C is incident with σ^p. A special chain is one that contains exactly one copy of each σ^n and each σ^{n-1} incident with a given σ^{n-2} (see section 1, (2)); a special chain that is incident with a given σ^{n-2} is unique up to cyclic permutation and reversal of the order. Given a chain C, the ϱth $(n-1)$-simplex σ_ϱ^{n-1} is preceded in C by some σ_μ^n and succeeded by some σ_ν^n. If $\mu = \nu$, then we put $\sigma(C, \varrho) = 0$; otherwise we denote by $\gamma(C, \varrho)$ the integer $\gamma(\hat{x})$ where $\pi(\hat{x})$ is contained in $\mathring{\sigma}^{n-1}$ and for $\hat{x} = (x, \mathfrak{F})$ the relation $\mathring{\sigma}_\mu^n \in \mathfrak{F}$ is satisfied (note that $\gamma(\hat{x})$ does not depend on the choice of $\hat{x} \in \pi^{-1}(\mathring{\sigma}_\varrho^{n-1})$).

We want to show by induction on l_0 that for every $l_0 \leqq l$ there are continuous maps $t_\lambda : |\sigma_\lambda^n| \to L$, $\lambda = 1, \ldots, l_0$, such that

(i) whenever $C = (\sigma_{\lambda_1}^n, \sigma_1^{n-1}, \sigma_{\lambda_2}^n)$, $\lambda_1, \lambda_2 \leqq l_0$, is a chain, then for all $\xi \in |\sigma_1^{n-1}|$

$$t_{\lambda_2}(\xi) = t_{\lambda_1}(\xi)\, t(\sigma_1^{n-1})\, (\chi^{-1}(\xi))^{\gamma(C,1)}.$$

(ii) whenever $\sigma^p, p < n-1$, is incident with the chain $C = (\sigma_{\lambda_1}^n, \sigma_1^{n-1}, \ldots, \sigma_r^{n-1}, \sigma_{\lambda_2}^n)$, $\lambda_1, \lambda_2 \leqq l_0$, then for all $\xi \in |\sigma^p|$

$$t_{\lambda_2}(\xi) = t_{\lambda_1}(\xi) \prod \{t(\sigma_\varrho^{n-1})\, (\chi^{-1}(\xi))^{\gamma(C,\varrho)} : \varrho = 1, \ldots, r\}.$$

It should be noted that, due to (11), the product $\prod\{\ldots\}$ is independent of the order of its factors.

Assuming that these two conditions are satisfied for $l_0 \geqq 1$, we construct $t_{l_0+1}: |\sigma_{l_0+1}^n| \to L$ as follows. If $C = (\sigma_\lambda^n, \sigma_1^{n-1}, \sigma_{l_0+1}^n)$, $\lambda \leqq l_0$, is a chain, then we define on $|\sigma_1^{n-1}|$ the value of t_{l_0+1} by

$$t_{l_0+1}(\xi) = t_\lambda(\xi)\, t(\sigma_1^{n-1})\, (\chi^{-1}(\xi))^{\gamma(C,1)}\,.$$

If $C' = (\sigma_{\lambda'}^n, \sigma_1'^{n-1}, \sigma_{l_0+1}^n)$, $\lambda' \leqq l_0$, is another such chain, then $|\sigma_1^{n-1}| \cap \cap |\sigma_1'^{n-1}|$ is a σ^{n-2} and we have, applying (ii) to the chain $(\sigma_\lambda^n, \sigma_1^{n-1}, \sigma_{l_0+1}^n, \sigma_1'^{n-1}, \sigma_{\lambda'}^n)$

$$t_{\lambda'}(\xi) = t_\lambda(\xi) \cdot t(\sigma_1^{n-1})\, (\chi^{-1}(\xi))^{\gamma(C,1)} \cdot t(\sigma_1'^{n-1})\, (\chi^{-1}(\xi))^{-\gamma(C',1)}$$

whence the function t_{l_0+1} is well defined on $|\sigma_1^{n-1}| \cup |\sigma_1'^{n-1}|$ and thus on the union of all $(n-1)$-faces of $\sigma_{l_0+1}^n$ that are also faces of some σ_λ^n, $\lambda \leqq l_0$. If, moreover, σ^p is a p-simplex incident with $\sigma_{l_0+1}^n$ and with some σ_λ^n, $\lambda \leqq l_0$, then we choose a chain $C = (\sigma_\lambda^n, \ldots, \sigma_{l_0+1}^n)$ of length r that is incident with σ^p and define t_{l_0+1} on $|\sigma^p|$ by

$$t_{l_0+1}(\xi) = t_\lambda(\xi) \prod \{t(\sigma_\varrho^{n-1})\, (\chi^{-1}(\xi))^{\gamma(C,\varrho)} : \varrho = 1, \ldots, r\}\,.$$

As before, it follows from (ii) that the value of $t_{l_0+1}(\xi)$ for $\xi \in |\sigma^p|$ does not depend on the choice of the chain C and that for $\xi \in |\sigma^p| \cap |\sigma'^{p'}|$ the value of t_{l_0+1} does not depend on whether σ^p or $\sigma^{p'}$ was used for its definition. Since every σ_λ^n, $\lambda = 1, \ldots, l$, as well as every $\sigma^p \in \chi(P \cap U)$ is a generalized cone with vertex σ^q, the (partially defined) map t_{l_0+1} can be extended to a continuous map of all of $|\sigma_{l_0+1}^n|$ to L. According to construction, t_{l_0+1} together with the previous t_λ, $\lambda \leqq l_0$, fulfills (i) and (ii).

In order to see that the induction can be started at all, we choose for t_1 the constant map given by $t_1(\xi) = e$. Since the validity of (ii) only has to be established, we need to verify that for every chain

$$C' = (\sigma_1^n, \sigma_1^{n-1}, \ldots, \sigma_r^{n-1}, \sigma_1^n)$$

that is incident with some σ^p, $p < n-1$, and for every point $\xi \in |\sigma^p|$ the relation

$$\prod \{t(\sigma_\varrho^{n-1})\, (\chi^{-1}(\xi))^{\gamma(C',\varrho)} : \varrho = 1, \ldots, r\} = e \tag{14}$$

is satisfied. This relation is certainly satisfied for every special chain since t is a cycle. It is also fulfilled for every chain C' that is incident with some σ^{n-2} but is not a special chain, as can be seen in Fig. 2. Since all factors in the last equation, whose validity has to be proved, commute with each other as a concequense of (11), our claim will be established if we can show that every chain $C' = (\sigma_1^n, \ldots, \sigma_1^n)$ that is incident with some σ^p, $p < n-2$, is a combination of chains that are incident with some σ^{n-2}.

The latter statement, however, is a purely combinatorial one and can thus be checked for $L = \boldsymbol{Z}$. It is possible to give a direct, elementary proof of this combinatorial statement. Yet, for reasons that become apparent in Proposition 4.2, we prefer another proof. In case $L = \boldsymbol{Z}$ our whole construction amounts to exhibiting explicitly, to a given integral $(n-1)$-cycle on $\chi(P \cap U)$, an integral n chain which bounds the given cycle. Necessary and sufficient for the existence of such an n-chain is that the induction which we described previously works. This induction, in turn, works if and only if the induction can be started, i. e., if and only if all the relations (14) are satisfied for the given cycle. Since $\chi(U) = \boldsymbol{R}^n$ is acyclic, all the relations (14) have to be satisfied for every integral $(n-1)$-cycle. Because the only relations which are simultaneously satisfied by all integral $(n-1)$-cycles are the relations (14) for chains that are incident with some σ^{n-2}, the required combinatorial statement is true. This finishes the proof of Theorem 4.1.

The crucial hypothesis in Theorem 4.1 is that some neighborhood of P be a n-dimensional topological manifold. However, an analysis of the proof shows that in fact only part of this hypothesis has been used, namely

(i) there exists a neighborhood of

$$\bigcup \{\mathring{\sigma}^{n-1} : \sigma^{n-1} \in P\} \cup \bigcup \{|\sigma^{n-2}| : \sigma^{n-2} \in P\}$$

that is a n-dimensional topological manifold (necessity part of Theorem 4.1).

(ii) every point in $\bigcup \{|\sigma^{n-2}| : \sigma^{n-2} \in P\}$ possesses a neighborhood U such that $P \cap U$ is acyclic (sufficiency part of Theorem 4.1).

In case X is a complex space and P is a polyhedron imbedded in X such that the geometric hypothesis is satisfied, (i) is fulfilled since in a normal complex space the singular locus has topological codimension $\geqq 4$. The fact that in this case (ii) is satisfied too, is an immediate consequence of [*3*]. Therefore we get.

Proposition 4.2. *Suppose that X is a complex space and P is a polyhedron imbedded in X satisfying the geometric hypothesis. Suppose furthermore that a topological fiber bundle $\mathfrak{L}$ of topological groups over some neighborhood of $|P|$ is given. Then a $(2n-1)$-chain on P with values in $\mathfrak{L}$ is a topologically correct transmission function if and only if it belongs to* $Z_{2n-1}(P;\mathfrak{L})$.

In the two geometric situations covered by Theorem 4.1 and Proposition 4.2 we get now a complete description of the sheaf $\mathfrak{T}^t$: a germ of a transmission function with values in $\mathfrak{L}$ belongs to $\mathfrak{T}^t$ if and only if it is locally a cycle.

After having discussed cycles we shall now turn to homology classes. Again we assume that the assumptions of either Theorem 4.1 or of Proposition 4.2 are satisfied. According to Remark 6, the topological principal transmission problem $\tau = \tau(X, P; \mathfrak{L}, \mathfrak{L}; t)$ gives rise to a topological $\mathfrak{L}$-principal fiber bundle $\mathfrak{L}_\tau$ for every topologically correct transmission function t. Hence, for fixed X, P, $\mathfrak{L}$, we get a canonical map $Z_{n-1}(P;\mathfrak{L}) \to H^1(X, \mathfrak{L}_c)$ resp. $Z_{2n-1}(P;\mathfrak{L}) \to H^1(X, \mathfrak{L}_c), H^1(X, \mathfrak{L}_c)$ being the set of (isomorphy classes of) topological $\mathfrak{L}$-principal bundles in the sense of [*2*]. In case $\mathfrak{L}$ is a principal L-bundle, it follows from Remark 7 that $H^1(X, \mathfrak{L}_c) = H^1(X, L_c)$, the latter being the cohomology set in dimension one for the topological group L. For the sake of convenience we phrase the following statements only for the case covered by Theorem 4.1; analogous results are valid for the case covered by Proposition 4.2.

Proposition 4.3. *Suppose that the hypotheses of Theorem* 4.1 *are satisfied. Then the canonical map* $Z_{n-1}(P;\mathfrak{L}) \to H^1(X, \mathfrak{L}_c)$ *factors through* $H_{n-1}(P;\mathfrak{L})$. *The induced map* $H_{n-1}(P;\mathfrak{L}) \to H^1(X, \mathfrak{L}_c)$ *is functorial.*

Proof: Suppose that t and t' are homologous cycles. Then, given a continuous solution t_i of the restriction to U_i of the principal transmission problem $\tau(X, P; \mathfrak{L}, \mathfrak{L}; t)$, the section $s \cdot s_i$ defined by $(s \cdot s_i)(\hat{x}) = s(\hat{x}) \cdot s_i(\hat{x})$ is a continuous solution t_i of the restriction to U_i of $\tau(X, P; \mathfrak{L}, \mathfrak{L}; t')$. Therefore we conclude from (8′) by an easy computation that the transition functions $\hat{k}_{ij}$ formed by means of the t_i coincide with the ones formed by means of the t_i'. Hence both, t and t', give rise to isomorphic $\mathfrak{L}$-principal bundles. The functorial properties of the induced map follow immediately from Corollary 3.2.

Corollary 4.4. *Suppose that the hypotheses of Theorem* 4.1 *are satisfied for* $\mathfrak{L} = L$. *If in addition* $H^1(X_{|P|}, L_c)$ *is trivial, then* $H_{n-1}(P;\mathfrak{L}) \to H^1(X, L_c)$ *is bijective.*

Proof. Surjectivity of this map is simply a restatement of Proposition 3.4 for topological transmission problems. Injectivity of this map is an obvious consequence of $H^1(X_{|P|}, L_c)$ being trivial.

5. Holomorphically correct transmission functions

It can be shown quickly that a transmission function with values in a holomorphic fiber bundle $\mathfrak{L}$ of complex Lie groups may be topologically correct without being holomorphically correct. We shall therefore give sufficient conditions for such a transmission function to be holomorphically correct at a given point.

Proposition 5.1. ([*9*]). *Suppose that the 1-dimensional complex space* X *and the polyhedron* P *of class* C^1 *imbedded in* X *satisfy the geometric hypothesis. Then the transmission function* $t \in Z_1(P;\mathfrak{L})$ *is holomorphically correct at* x *if there is a neighborhood* U *of* x *such that for each* $\sigma^1 \in P$ *containing* x, $t(\sigma^1) \,|\, U \cap |\sigma^1|$ *is Hölder continuous.*

Proof. This statement has been proved in [*9*] under the additional hypothesis that for all $\sigma^1 \in P$ containing x, $t(\sigma^1)(x)$ has the same value. However, our present, more general case can be reduced to the one treated in [*9*] by the following method. Let $\tilde{t}$ be the transmission function that is constant on each $|\sigma^1|$ and has values $t(\sigma^1)(x)$ for all σ^1 containing x. $\tilde{t}$ is a 1-cycle in the usual sense, with values in the abelian group generated by the element $t(\sigma^1)(x)$. Hence $\tilde{t}$ is locally a boundary. Therefore there exists a locally constant function s on $U - U \cap |P|$ with values in L, U being a suitable neighborhood of x, which constitutes a holomorphic solution of the holomorphic principal transmission problem $\tau(U, U \cap P; L, L; t)$. The map t' from the set of 1-simplices that contain x, defined by

$$t'(\sigma^1)(\pi(\hat{\xi}))^{\gamma(\hat{\xi})} = s(\beta(\hat{\xi}))^{-1} \cdot t(\sigma^1)(\pi(\hat{\xi}))^{\gamma(\hat{\xi})} \cdot s(\hat{\xi}), \quad \xi \in \pi^{-1}(|\sigma^1|),$$

is then a cycle for which $t'(\sigma^1)(x) = e$ for all σ^1 containing x. Since $t'(\sigma^1)$ is Hölder continuous in U whenever $t(\sigma^1)$ is, t' and thus t are holomorphically correct at x.

For additional details about the 1-dimensional case see [*9*] and the literature quoted there. It should also be noted that in this case the analogue of Corollary 4.4 is true, provided the section s appearing in (13) is assumed to be holomorphic.

The next question we want to answer is a question concerning the relation between germs of holomorphic sections in $\mathfrak{L}$ and germs of holomorphically correct transmission functions with values in $\mathfrak{L}$. For that purpose, let $\mathfrak{L}_\omega$ be the sheaf of germs of holomorphic sections in $\mathfrak{L}$ and $\mathfrak{T}^t$ resp. $\mathfrak{T}^h$ the sheaf of germs of topologically resp. holomorphically correct transmission functions with values in $\mathfrak{L}$. Then we have

Theorem 5.2. $(\mathfrak{L}_\omega \,|\, |P|) \cap \mathfrak{T}^t \subset \mathfrak{T}^h$.

Proof. Again this relation is obvious on $\bigcup \{\mathring{\sigma}^{2n-1} : \sigma^{2n-1} \in P\}$, since one has only to take the unit section on the one side of $\mathring{\sigma}^{2n-1}$ and the given local section in $\mathfrak{L}_\omega$ to the power ± 1 on the other side of $\mathring{\sigma}^{2n-1}$. In order to see that for every point $x \in |\sigma^{2n-2}|$, where σ^{2n-2} is in P, a holomorphic germ at x that at the same time is a cycle is holomorphically correct, we go back to the proof of Theorem 4.1 We claim that the induction proof given there works also in the present situation. As t_1 we choose the restriction of the given local section t in $\mathfrak{L}_\omega$ to σ_1^{2n-1}. If the t_λ, $\lambda \leq l_0$, are already constructed, then the formulas in the proof of Theorem 4.1 that were used in order to define t_{l_0+1} by induction, show that t_{l_0+1} can be chosen as the restriction to $\sigma_{l_0+1}^{2n-1}$ of a suitable power of t. Hence t is holomorphically correct.

From Theorem 4.1, Theorem 5.2, and the meaning of $\partial\gamma$ we conclude easily

Corollary 5.3. If γ is a cylce, then $\mathfrak{L}_\omega \,|\, |P| \subset \mathfrak{T}^h$.

In general, $(\mathfrak{L}_\omega| \,|P|) \cap \mathfrak{T}^t$ is a proper, and for that matter rather small, subsheaf of $\mathfrak{T}^h$. However, in order to determine $\mathfrak{T}^h$, we have to impose on the polyhedron P geometric restrictions which reflect the fact that we are dealing with functions of several complex variables. The hypotheses we impose are

(i) each point of P is a regular point of X;

(ii) the polyhedron P is of class C^2 and is strongly pseudoconvex.

Here, a polyhedron P of class C^2 in a complex manifold X' is called *strongly pseudoconvex* if for every $(2n-1)$-simplex σ^{2n-1} in P, $\mathring{\sigma}^{2n-1}$ is a strongly pseudoconvex hypersurface. This means that to each $\sigma = \mathring{\sigma}^{2n-1}$ there exists an open neighborhood V of $\mathring{\sigma}^{2n-1}$, containing $\mathring{\sigma}^{2n-1}$ as a closed subset, and a real valued function φ_σ of class C^2 in V such that $\mathring{\sigma}^{2n-1} = \{\xi : \varphi(\xi) = 0\}$, $d\varphi_\sigma$ vanishes nowhere in σ^{2n-1}, and the Levi-form $L(\varphi_\sigma)$ is positive definite on $\mathring{\sigma}^{2n-1}$. Given such a function φ_σ, σ^{2n-1} is strongly pseudoconvex from the side $\{\xi : \varphi(\xi) > 0\}$. This side shall be called the *positive side of* σ^{2n-1} as opposed to the *negative side of* σ^{2n-1} given by $\{\xi : \varphi(\xi) < 0\}$.

Suppose now that t is a holomorphically correct transmission function at x with values in $\mathfrak{L}$. Then we choose an open neighborhood U of x such that there is a holomorphic solution s of the restriction to U of the principal transmission problem with transmission function t. Take a $(2n-1)$-simplex σ^{2n-1} in P such that $\mathring{\sigma}^{2n-1}$ contains x. Then, since s can be extended continuously onto $|P|$, s can be extended holomorphically from the positive side of σ^{2n-1} across σ^{2n-1}: this is clear because, locally, $\mathfrak{L}$ equals L and because the exponential map is biholomorphic in sufficiently small neighborhoods of the neutral element of L. Consequently, $t(\sigma^{2n-1})$ has to be the boundary value of some holomorphic section in $\mathfrak{L}$ over the negative side of σ^{2n-1}. The condition of being the boundary value of some holomorphic section in $\mathfrak{L}$ over the negative side of σ^{2n-1} is trivially also sufficient for t to be holomorphically correct at every point of $\bigcup \{\mathring{\sigma}^{2n-1} : \sigma^{2n-1} \in P\}$[7]. As can be seen from examples, this

[7] Necessary and sufficient conditions for a continuous complex valued function on a strongly pseudoconvex hypersurface to be a boundary value from the negative side can be found, for $n = 2$, in [*7*] (see also [*8*]). This condition generalizes to arbitrary complex dimension and so do the corresponding proofs and statements of [*7*], as can be checked easily. The necessary and sufficient conditions in [*7*] can locally be carried over (via the exponential map) to functions with values in a given complex Lie group L. A modern way of phrasing these necessary and sufficient conditions is the following: Suppose that the strongly pseudoconvex hypersurface S is contained in a domain in $\boldsymbol{C}^n$ (with complex-euclidean coordinates $z_1, \ldots, z_n$), and denote the canonical injection $S \to \boldsymbol{C}^n$ by j; then the necessary and sufficient conditions, stated in [*7*], for the continuous function t on S to be a boundary value from the negative side are

(i) $t \cdot j^*(dz_1 \wedge \ldots \wedge dz_n)$ is a regular differential form (see [*12*]) on S

(ii) $d(t \cdot j^*(dz_1 \wedge \ldots \wedge dz_n) = 0$.

condition fails to be sufficient for points in $\bigcup\{\sigma^{2n-2}:\sigma^{2n-2}\in P\}$. Summing up we get

Proposition 5.4. *Suppose that X and P satisfy the geometric hypothesis and that P is a strongly pseudoconvex polyhedron of class* C^2. *Then for every* σ^{2n-1} *in P, the restriction* $\mathfrak{T}^h\,|\,\mathring{\sigma}^{2n-1}$ *consists of all germs of boundary values of holomorphic sections in* $\mathfrak{L}$ *over the negative side of* σ^{2n-1}.

It should be remarked that, without assuming strong pseudoconvexity of the polyhedron P, germs of one sided boundary values of holomorphic sections in $\mathfrak{L}$ over $\bigcup\{\mathring{\sigma}^{2n-1}:\sigma^{2n-1}\in P\}$ belong to $\mathfrak{T}^h$.

By an obvious argument one gets from Proposition 5.4 the

Corollary 5.5. *Let* $|P|$ *be the union of mutually disjoint, closed, strongly pseudoconvex hypersurfaces in X and assume that all simplices of P are coherently oriented. Then* $\mathfrak{T}^h$ *consists of all germs of boundary values of holomorphic sections in* $\mathfrak{L}$ *over the negative side of the appropriate connected component of* $|P|$.

It seems to be quite difficult to describe $\mathfrak{T}^h\,|\bigcup\{|\sigma^{2n-2}|:\sigma^{2n-2}\in P\}$ in general. A special case that can be handled and that is closely related to the one described in Corollary 5.5 is the following. Suppose that all simplices of P are coherently oriented. Assume furthermore that a given point x in $\bigcup\{|\sigma^{2n-2}|:\sigma^{2n-2}\in P\}$ possesses a neighborhood U such that $U\cap|P|$ is the union of finitely many strongly pseudoconvex hypersurfaces of class C^2 in U such that for each of these hypersurfaces h_μ, $\mu=1,\ldots,m$, there exist precisely two $(2n-1)$-simplices, ${}^+\sigma^{2n-1}$ and ${}^-\sigma^{2n-1}$, such that both relations,

$$h_\mu\cap|{}^+\sigma^{2n-1}|\neq\emptyset \quad\text{and}\quad h_\mu\cap|{}^-\sigma^{2n-1}|\neq\emptyset,$$

are satisfied. Then the germ at x of a transmission function t with values in the fiber bundle $\mathfrak{A}$ of complex abelian groups is holomorphically correct, provided

(i) for every μ, $t({}^+\sigma_\mu^{2n-1})$ and $t({}^-\sigma_\mu^{2n-1})$ are equal on $|{}^+\sigma_\mu^{2n-1}|\cap|{}^-\sigma_\mu^{2n-1}|$;

(ii) for every μ, the continuous section in $\mathfrak{A}$ over a suitable neighborhood of x in h_μ that is given by $t({}^+\sigma_\mu^{2n-1})$ and $t({}^-\sigma_\mu^{2n-1})$ is the boundary value of some holomorphic section in $\mathfrak{A}$ over the negative side of h_μ.

In order to see that this is true, we choose for every μ the section s_μ in $\mathfrak{A}$ over $V-V\cap h_\mu$, V a suitable neighborhood of x, that is the unit section over the positive side of h_μ and $\pm$ the holomorphic section over the negative side of h_μ whose boundary values are given by $t({}^+\sigma_\mu^{2n-1})$ and $t({}^-\sigma_\mu^{2n-1})$, $\pm$ being the appropriate sign. Then a straightforward computation shows that the section $\Sigma\{s_\mu:\mu=1,\ldots,m\}$ in $\mathfrak{A}$ over $V-V\cap|P|$ satisfies the conditions that are required for t to be holomorphically correct at x.

With respect to analytic properties, strongly pseudoconvex hypersurfaces in a complex manifold are one extreme. The other extreme has been treated in [9]: *local holomorphic families of 1-dimensional polyhedra of class* C^1. By this we mean (see [9]) a closed subset $|P|$ of a complex space X such that every point $x \in |P|$ possesses a neighborhood U and a biholomorphic map χ of U onto the n-dimensional polydisk

$$D^n = \{(z_1, \ldots, z_n) : |z_\nu| < 1, \quad \nu = 1, \ldots, n\}$$

satisfying

(i) $\chi(U \cap |P|)$ can be written as

$$\bigcup \{(z_1, \ldots, z_{n-1}, \varphi_{\varkappa, n}(z_1, \overline{z_1}, \ldots, z_{n-1}, \overline{z_{n-1}}, t)) : (z_1, \ldots, z_{n-1}) \in D^{n-1}, \quad t \in [0,1], \varkappa = 1, \ldots, k\};$$

(ii) $\varphi_{\varkappa, n}(z_1, \overline{z_1}, \ldots, z_{n-1}, \overline{z_{n-1}}, 0) = \varphi_{\lambda, n}(z_1, \overline{z_1}, \ldots, z_{n-1}, \overline{z_{n-1}}, 0)$ for all $\varkappa$, λ and $(z_1, \ldots, z_{n-1}) \in D^{n-1}$;

(iii) for every $\varkappa$, $\varphi_{\varkappa, n}$ is a differentiable function that depends holomorphically on $(z_1, \ldots, z_{n-1}) \in D^{n-1}$, and $\frac{\partial \varphi_{\varkappa,n}}{\partial t}$ vanishes nowhere;

(iv) for every $t \in [0, 1]$, $\frac{\partial \varphi_{\varkappa,n}}{\partial t}$ is a continuous function that depends holomorphically on $(z_1, \ldots, z_{n-1}) \in D^{n-1}$.

As shown in [9], such a local holomorphic family of 1-dimensional polyhedra admits two natural, and hence global, foliations the first of which consists of leaves that are $(n-1)$-dimensional complex manifolds, while the second one consists of 1-dimensional polyhedra of class C^1.

Combining the results of [9] and the proof of Proposition 5.1, we get

Proposition 5.6. *Suppose that X is a complex space and P is a local holomorphic family of 1-dimensional polyhedra. Then the element $t \in \mathfrak{T}^t$ belongs to $\mathfrak{T}^h$, provided the restriction to each leaf of the first foliation is holomorphic and the restriction to each simplex of every leaf in the second foliation is Hölder continuous.*

In order to cover also geometric situations between the two extremes dealt with so far, we make the following definition. The hypersurface S imbedded in the complex manifold V as a closed subset is said to be locally a *k-dimensional complex analytic family of pseudoconvex hypersurfaces at* $x \in S$ if there is a neighborhood U of x and a biholomorphic map χ of U onto D^n such that $\chi(U)$ can be described as

$$z_\lambda = \varphi_\lambda(z_1, \overline{z_1}, \ldots, z_k, \overline{z_k}, t_{2k+1}, \ldots, t_{2n-2k-1}), \quad \lambda = k+1, \ldots, n,$$

where $(z_1, \ldots, z_k) \in D^k, (t_{2k+1}, \ldots, t_{2n-2k+1}) \in I^{2n-2k-1}$, such that

(i) $\begin{pmatrix} \frac{\partial \varphi_\lambda}{\partial z_\varkappa} & \frac{\partial \varphi_\lambda}{\partial \overline{z_\varkappa}} & \frac{\partial \varphi_\lambda}{\partial t_\varrho} \\ \frac{\partial \overline{\varphi_\lambda}}{\partial z_\varkappa} & \frac{\partial \overline{\varphi_\lambda}}{\partial \overline{z_\varkappa}} & \frac{\partial \overline{\varphi_\lambda}}{\partial t_\varrho} \end{pmatrix}$ has maximum rank everywhere;

(ii) $\frac{\partial \varphi_\lambda}{\partial z_\varkappa} = 0$ for all $\varkappa = 1, \ldots, k$ and all $\lambda = k+1, \ldots, n$;

(iii) for every fixed $(z_1^0, \ldots, z_k^0) \in D^k$,

$$z_\lambda = \varphi_\lambda(z_1^0, \bar{z}_1^0, \ldots, z_k^0, \bar{z}_k^0, t_{2k+1}, \ldots, t_{2n-2k-1})$$

is a strongly pseudoconvex hypersurface in $\{(z_1^0, \ldots, z_k^0)\} \times D^{n-k}$ [8].

These conditions give rise (see [*11*]) to two canonical foliations of $U \cap S$, the first of which consists of leaves that are k-dimensional complex manifolds, while the second one consists of leaves that are $(2n - 2k - 1)$-dimensional surfaces which, in suitable holomorphic coordinates, are strongly pseudoconvex hypersurfaces. The first foliation shall be called the c-foliation, while the second one is called the p-foliation. Since both, the c-foliation and the p-foliation, are uniquely determined by intrinsic properties of the tangent space (see [*11*]), these locally defined foliations give rise to global foliations, provided S is locally a k-dimensional complex-analytic family of strongly pseudoconvex hypersurfaces in each point of S. Given a leaf l in the p-foliation and a point x in l, we can speak of the positive (negative side) of l at x in the following sense. In suitable local coordinates, l is a strongly pseudoconvex hypersurface (in the usual sense) in $\{(z_1^0, \ldots, z_k^0)\} \times D^{n-k}$ and possesses as such a positive (negative side). In particular we say that the function t on l is the boundary value of a holomorphic function over the negative side of l if there is a holomorphic function in $\{(z_1^0, \ldots, z_k^0)\} \times D^{n-k}$ that has boundary value t. With these notations we get — without giving an explicit proof — by reexamining the relevant part of [*7*]

Proposition 5.7. *Suppose that X and P satisfy the geometric hypothesis and that for each σ^{2n-1} in P, $\mathring{\sigma}^{2n-1}$ is locally a k-dimensional complex analytic family of strongly pseudoconvex hypersurfaces. Then for every σ^{2n-1} in P, $\mathfrak{T}^h \mid \mathring{\sigma}^{2n-1}$ consists of all those germs of continuous sections in $\mathfrak{L}$ whose restrictions to each complex manifold in the c-foliation is holomorphic and whose restriction to each strongly pseudoconvex hypersurface in the*

[8] According to an oral communication by Professor E. Calabi, the hypersurface S of class C^3 in the complex manifold V is locally a k-dimensional complex analytic family of strongly pseudoconvex hypersurfaces if and only if for every point of S there exists a neighborhood U and a real valued function φ of class C^3 in U such that

(i) $U \cap S = \{\xi : \varphi(\xi) = 0\}$ and $d\varphi$ vanishes nowhere;

(ii) the Levi form $\sum \frac{\partial^2 \varphi}{\partial z^\alpha \partial \bar{z}^\beta} \xi^\alpha \bar{\xi}^\beta$ is positive semidefinite in U;

(iii) the exterior Levi form $\omega = i \sum \frac{\partial^2 \varphi}{\partial z^\alpha \partial \bar{z}^\beta} dz^\alpha \wedge d\bar{z}^\beta$ satisfies $\overset{n-k}{\wedge}\, \omega \neq 0$ everywhere in U and $d\varphi \wedge \overset{n-k}{\wedge}\, \omega \equiv 0$ in U;

(iv) $\left(\frac{\partial^3 \varphi}{\partial z^\alpha \partial \bar{z}^\beta \partial \bar{z}^\gamma} dz^\alpha\right) \wedge \overset{n-k}{\wedge}\, \omega \equiv 0$ in U for all β, γ.

p-foliation is the boundary value of some holomorphic section in $\mathfrak{L}$ over the negative side of the hypersurface.

6. Final remarks

Transmission problems of the type dealt with in this paper have several applications. For instance, the construction of multivalued functions, like abelian integrals or more generally automorphic functions to given factors of automorphy, on a given complex space leads to a transmission problem: one dissects the space X by means of a polyhedron P of topological codimension one in such a way that the function which one wants to construct has a single valued branch over each connected component of $X - |P|$. The transmission function can then be computed easily from the original data.

Another type of application is connected with boundary value problems for holomorphic functions. This case was discussed in [9] for bordered Riemann surfaces. Yet the method and results carry over to higher dimensional bordered complex spaces X as well, provided one can form in some sense the double $2X$ of X. The latter is obviously possible in case X is an analytic polyhedron with border one of the faces of this analytic polyhedron; another case where the double can be formed is a bordered domain in $\boldsymbol{C}^n$ that is the intersection of an open set with the half space $\operatorname{Re} z_1 \leqq 0$.

Theorem 3.1 has obvious applications whenever the base space X has good analytic properties. In particular, the solution space Σ_τ of a linear transmission problem over a compact complex space is of finite dimension.

Finally we should like to remark that transmission problems can also be phrased for categories different from the category of topological spaces resp. complex spaces as stated in this paper. However, in general, one will have to impose additional requirements on the solutions in order to get the results of this paper. As an example, take as category the category of differentiable manifolds. Then it is natural to request that the polyhedron P be of class C^k and that the transmission function t of the transmission problem be also of class C^k, k being the class of the differentiable manifold X. Here one has to impose on a solution s of the transmission problem, in addition to the requirements of this paper, the requisite that for each σ^{2n-1} in P and each $x \in |\sigma^{2n-1}|$ the derivatives up to order k of the solution s in $\hat{x} \in \pi^{-1}(x)$ along the normal at σ^{2n-1} in x are independent of the point $\hat{x} \in \pi^{-1}(x)$. Then the method presented in this paper can be used to investigate differentiable transmission problems and leads to results analogous to the ones for topological resp. holomorphic transmission problems.

References

[1] Bourbaki, N.: Topologie générale. Actualités Sci. Ind. 1142 (Paris) 1951.
[2] Cartan, H.: Espaces fibrés analytiques. Intern. Symp.alg.top., Univ. Nac. Mexico (1958), 97–121.
[3] Giesecke, B.: Simpliziale Zerlegung abzählbarer analytischer Räume. Math. Zschr. **83**, 177–213 (1964).
[4] Grauert, H., und R. Remmert: Komplexe Räume. Math. Ann. **136**, 245–318 (1958).
[5] Hilton, P. J., and S. Wylie: Homology theory. Cambridge Univ. Press 1960.
[6] Kerner, H.: Über die Automorphismengruppe kompakter komplexer Räume. Arch. Math. **11**, 282–288 (1960).
[7] Kneser, H.: Die Randwerte einer analytischen Funktion zweier Veränderlichen. Mh. Math. Phys. **43**, 364–380 (1936).
[8] Lewy, H.: On the local character of the solutions of an atypical linear partial differential equation .... Ann. Math. **64**, 514–522 (1956).
[9] Röhrl, H.: Über das Riemann-Privalovsche Randwertproblem. Math. Ann. **151**, 365–423 (1963).
[10] — Ω-degenerate singular integral equations and holomorphic affine bundles over compact Riemann surfaces, II. (to appear).
[11] Sommer, F.: Komplex-analytische Blätterung reeller Mannigfaltigkeiten im C^n. Math. Ann. **136**, 111–133 (1958).
[12] Whitney, H.: Geometric integration theory. Princeton Univ. Press 1957.

Department of Mathematics
University of Minnesota
Minneapolis, Minn.

Department of Mathematics
University of California at San Diego
La Jolla, Cal.

Attaching Analytic Spaces to an Analytic Space Along a Pseudoconcave Boundary*

By

H. Rossi

1. Introduction

It is sometimes possible to create new mathematics by relaxing by complication the hypotheses of existing theorems so that the proofs still work. That is what has happened in the present case. The essential ideas and techniques used here are to be found among the papers of Bishop [2, 3, 4], Grauert [5, 6] and Andreotti and Grauert [1]. We shall freely refer to details within these papers.

Our analytic spaces are all reduced, and have a countable basis for the topology. For an analytic space X, we shall denote by ${}_X O$ the sheaf of germs of holomorphic functions on X. In this situation $H^0(X, {}_X O)$ is a Fréchet algebra in the topology of uniform convergence on compact

* Received March 19, 1964.

subsets. An important space associated with this algebra is the space S of non-zero continuous complex-valued homomorphisms of $H^0(X, {}_XO)$, to be called the *spectrum* [4, 9]. Evaluation of elements in $H^0(X, {}_XO)$ at a point $x \in X$ gives rise to such a homomorphism φ_x. We shall write $\varphi_x = g(x)$ and the mapping $g: X \to S$ will be called the *Gelfand mapping.* Observe that $g(x) = g(y)$ if and only if $f(x) = f(y)$ for all $f \in H^0(X, {}_XO)$. The best that can happen is that S can be given the structure of a Stein analytic space so that the mapping g is holomorphic, and $\hat{g}: H^0(S, {}_SO) \cong H^0(X, {}_XO)$, where $\hat{g}(f) = f \circ g$. This is sometimes the case [4, 9] and sometimes not [7]. Our first main result is to verify that this is the case when X is a suitable neighborhood of a compact level set of a strongly plurisubharmonic function. We derive this result from certain generalizations by Bishop of his own work ([2], see also [8], Chapter VII). In particular, it is possible to throw away the requirement of relative compactness in the definition of analytic polyhedra and still prove the theorem on approximation by (also noncompact) special analytic polyhedra.

In Section 3 we collect some finiteness-of-cohomology-theorems which are deducible from the arguments of Andreotti and Grauert. We are required to assume that X is everywhere of dimension greater than 2. In dimension 1 the subsequent theorems are trivially false, and in dimension 2 there arise some counterexamples which are of some inherent interest. These are discussed in the final section.

It follows from the main result stated above that, if D is an analytic space of pure dimension $k > 2$ all of whose boundary components are either 1-pseudoconvex or 1-pseudoconcave, then there is only one 1-pseudoconvex boundary component, and the pseudoconcave holes can be filled in so that the result is a finite strongly pseudoconvex Stein space. If there are only 1-pseudoconcave boundary components we can fill these in order to obtain a "compactification" Y of D. If this happens because D is a suitable neighborhood of a positively embedded subvariety A of codimension 1, then the resulting space Y is a projective algebraic subvariety with hyperplane section A.

Throughout this paper we use the notation of [8]. We say that a mapping $f: X \to Y$ of analytic spaces is an injection when f is a biholomorphic mapping of X with a subspace of Y. We wish to thank A. Andreotti for essentially providing the example of Section 6 and for many helpful conversations.

2. Polyhedral domains

Let X be an analytic space of pure dimension k. Let W be an open subset of X. We shall denote by O (or ${}_XO$ when more than one space may be involved) the sheaf of germs of holomorphic functions on X. The ring of sections of O over X will be denoted by $H(W, O)$.

1. Definition. W is a *polyhedral domain* in X if

(i) $B = bW$ is compact,

(ii) there is a neighborhood N of B, and functions $f_1, \ldots, f_n \in H(N, O)$ such that

$$W \cap N = \{x \in N;\ |f_i(x)| < 1,\ 1 \leqq i \leqq n\}.$$

W is a *special polyhedral domain* if $n = k$.

The above notions are slight generalizations of the notions of *analytic polyhedra, special analytic polyhedra*, and the arguments of Bishop [*2, 3*] apply to this generalization. In particular, a polyhedral domain is approximable by special polyhedral domains. Using the ordinary Hartogs' extension theorem in a polydisc, we can obtain the following result.

2. Theorem. *Suppose W is a polyhedral domain in X and $H(W, O)$ has discrete level sets. Then the spectrum S of $H(W, O)$ is a Stein analytic space, dim S = dim W, and the natural (Gelfand) mapping $g: W \to S$ is holomorphic. Further $\hat{g}: H(S, O) \cong H(W, O)$ and $S - g(W)$ is relatively compact.*

Proof. See Theorem VII D 8 of [*8*]. We remark that it is not necessary to assume that $H(W, O)$ has discrete level sets to obtain the above conclusion (except dim S = dim W), as reexamination of the full extent of Bishop's theorems will show. For further applications, we restate Theorem 2 in a slightly more particular case:

3. Theorem. *Suppose W is a polyhedral domain in X, and $H(W, O)$ separates points on W. Then there is a Stein space Y, and an analytic polyhedron S in Y (defined by global functions) and a holomorphic mapping $g: N \to Y$, where N is a neighborhood of $\bar{W}$, such that*

(i) *g is injective on W,*

(ii) *$S - g(W)$ is relatively compact in S,*

(iii) *$\hat{g}: H(S, O) \cong H(W, O)$.*

We can describe this theorem by saying that we can "fill in the holes" of a polyhedral domain in order to make it an analytic polyhedron, as long as the holomorphic functions separate points.

3. The theorems of Andreotti and Grauert

If we consider the unit ball as an alternative to the polydisc, then the analogue of analytic polyhedra are the strongly pseudoconvex spaces. We wish to extend the theorem of Section 2 to a corresponding analogue of polyhedral domains; in order to do so, we need the "théorèmes de finitude" of ANDREOTTI and GRAUERT [1]. For our purposes we need only state a special case of (a trivial generalization of) these theorems.

1. Definition. Let φ be a real-valued C^2 function defined on a domain U in C^n. φ is *strongly q-pseudoconvex* if

$$\frac{\partial^2 \varphi}{\partial z_i \, \partial \bar{z}_j}(p)$$

has $n - q + 1$ positive eigenvalues for all $p \in U$.

For $q = 1$, we say φ is a strictly plurisubharmonic (s. psh.) function. If V is an analytic subvariety of U, and ψ a continuous function on V which has a continuous extension φ to U which is strongly q-pseudoconvex, we shall say that ψ is *strongly q-pseudoconvex on V*. These definitions are invariant under holomorphic coordinate changes, so give well-defined notions on complex manifolds and spaces.

2. Definition. Let X be an analytic space and D an open subset of X. Let $y_0 \in bD$. We say that D is *strongly q-pseudoconvex (strongly q-pseudoconcave)* at y_0 if there exists a neighborhood U of y_0, a real-valued strongly q-pseudoconvex function φ defined in U such that

$$D \cap U = \{x \in U;\ \varphi(x) < \varphi(y_0)\}$$
$$(D \cap U = \{x \in U;\ \varphi(x) > \varphi(y_0)\}).$$

3. Theorem. *Let X be an analytic space of pure dimension k. Let D be a relatively compact domain in X and suppose every boundary point of D is either strongly 1-pseudoconvex or strongly q-pseudoconcave. Let $\mathcal{S}$ be a coherent sheaf on X such that $\mathcal{S}$ is free in a neighborhood of each q-pseudoconcave boundary point of D. Then*

$$\dim H^r(D, \mathcal{S}) < \infty \quad \text{for} \quad 0 < r < k - q.$$

The arguments in [1] culminating in Theorem 11, p. 239 will prove the above theorem (and in fact, much more). Since we shall only consider this strong pseudoconvexity, we shall take the liberty of dropping the adverb "strongly" in front of q-pseudoconvexity. An immediate consequence of Theorem 3 is:

4. Corollary. Let X be an analytic space of pure dimension $k > 2$. Let D be a relatively compact domain in X, and suppose $bD = B_1 \cup B_2$, where B_1 is 1-pseudoconvex, and B_2 is at worst $(k - 2)$-pseudoconcave. If $\mathcal{S}$ is a coherent sheaf on X which is locally free in a neighborhood B_2, then $\dim H^1(D, \mathcal{S}) < \infty$.

Another more subtle consequence of the arguments of ANDREOTTI and GRAUERT which we shall need is the following.

5. Lemma. Let X be an analytic space of pure dimension $k > 2$, and D a relatively compact domain in X such that $bD = B_1 \cup B_2$, where B_1 is 1-pseudoconvex, and B_2 is at worst $(k-2)$-pseudoconcave. Let $x_1, \ldots, x_n$ be finitely many points on B_1, and $\mathcal{S}$ a coherent sheaf on $X - \{x_1, \ldots, x_n\}$ which is locally free in a neighborhood of B_2. Then $\dim H^1(D, \mathcal{S}) < \infty$.

Proof. As in the lemma, p. 237 of [*1*], we can find a cover $\mathfrak{U}$ of $\bar{D}$ such that, if σ is a 0- or 1-simplex in $N(\mathfrak{U})$, $\mathrm{H}^1(|\sigma| \cap D, \mathcal{S}) = 0$. We may assume that there is a neighborhood V_i of x_i, $1 \leqq i \leqq n$, such that $\bar{V}_i$ is contained in one $U \in \mathfrak{U}$ and disjoint from all others. Now, let φ be the s. psh. function defined in a neighborhood N of B_1 such that $B_1 = \{x \in N;\ \varphi(x) = 0\}$, and choose a non-negative function ψ which has support in $\cup V_i$ such that $\varphi + \psi$ is still s. psh. in N. Let $D' = (D - N) \cup \cup \{x \in N;\ \varphi(x) + \psi(x) < 0\}$. Of course, we assume that $\psi(x_i) > 0$, $1 \leqq i \leqq n$, so that no x_i is in D'. Now all of this can be arranged as in [*1*] so that in addition for σ a 0- or 1-simplex of $\mathfrak{U}$ we have $H^1(|\sigma| \cap D', \mathcal{S}) = 0$. Now, by Leray's theorem, $H^1(D, \mathcal{S}) = H^1(N(\mathfrak{U}) \cap D, \mathcal{S})$ and $H^1(D', \mathcal{S}) = H^1(N(\mathfrak{U}) \cap D', \mathcal{S})$. Since $N(\mathfrak{U}) \cap D$ and $N(\mathfrak{U}) \cap D'$ have the same 1-simplices, the restriction mapping $H^1(D, \mathcal{S}) \to H^1(D', \mathcal{S})$ is isomorphic. But $\mathcal{S}$ is coherent in a neighborhood of $\bar{D}'$, so by Corollary 4, $\dim H^1(D', \mathcal{S}) < \infty$, and the lemma is proven.

There are, of course, much more extensive "generalizations" of the theorems of Andreotti and Grauert which we shall not explore, since we now have what is needed in the sequel.

4. Filling in the holes

We are now in a position to apply the above results in order to fill in the holes in domains bounded by compact pseudoconvex hypersurfaces.

1. Definition. Let B be a compact subset of the analytic space X. B is a *1-pseudoconvex hypersurface* if, for every $y \in B$, there is a neighborhood U of y and a s. psh. function φ defined in U such that

$$B \cap U = \{x \in U;\ \varphi(x) = 0\}.$$

2. Lemma. Let B be a 1-pseudoconvex hypersurface in X. There is a neighborhood N of B, and a s. psh. function φ defined in N such that $B = \{x \in N;\ \varphi(x) = 0\}$.

Proof. By definition, we can cover B by finitely many open sets (in X), $U_1, \ldots, U_n$, and find s. psh. functions φ_i in U_i such that $B \cap U_i = \{x \in U_i;\ \varphi_i(x) = 0\}$. We may assume that U_i is identified with a subvariety of a polydisc Δ_i and $\varphi_i = \Phi_i | U_i$ where Φ_i is 1-pseudoconvex in Δ_i. Notice that $\{x \in U_i;\ \varphi_i(x) < 0\}$ is 1-pseudoconvex along B. Since this is just a property of that domain, rather than of φ_i, $\{x \in U_i \cap U_j;\ \varphi_i(x) < 0\} = \{x \in U_i \cap U_j;\ \varphi_j(x) < 0\}$. Let ϱ_i be a non-negative C^2 function of compact support in Δ_i. It is easily verified (see [*8*], Chapter

IX A) that there is a constant $A > 0$ such that $\Psi_i = e^{A\Phi_i} \cdot \varrho_i \cdot \Phi_i$ has a positive semidefinite HESSIAN in some domain Δ_i' such that $\Delta_i' \cap U_i = U_i$, and a positive definite HESSIAN on the support of ϱ_i. We should choose the ϱ_i so that $\bigcup_i \{x \in U_i;\ \varrho_i(x) > 0\}$ covers B. Then $\Psi_i \mid U_i$ defines a C^2 function ψ_i on X. Let $N = \bigcup_i \{x \in U_i;\ \varrho_i(x) > 0\}$ and $\varphi = \sum \psi_i$.

For B such a hypersurface in X, and N such a neighborhood, we shall write $N^- = \{x \in N;\ \varphi(x) < 0\}$, $N^+ = \{x \in N;\ \varphi(x) > 0\}$. We assume (for simplicity) that B is the boundary of N^- in N (the fact that B is the boundary of N^+ in N is automatic).

3. Theorem. *Let X be an analytic space of pure dimension $k > 2$, and let D be a relatively compact subdomain of X such that $bD = B_1 \cup B_2$, where B_1 is 1-pseudoconvex and B_2 is at worst $(k - 2)$-pseudoconcave. Then the spectrum S of $H(D, O)$ is a Stein space and the Gelfand map $g: D \to S$ is holomorphic. Further, if N is a neighborhood of B_1 which is disjoint from B_2, $S - g(N \cap D)$ is relatively compact in S.*

Proof. We shall prove that S can be written as an increasing union of Stein spaces, all of which are $H(S, O)$-convex. It follows that S is STEIN (Theorem VII A 10 of [8]). The rest of the theorem will be seen to follow easily. By the above lemma, let N be a neighborhood of B_1, and φ a s. psh. function defined in N so that $D \cap N = \{x \in N;\ \varphi(x) < 0\}$. Then, for $\varepsilon > 0$ small enough, $D \cup \{x \in N;\ \varphi(x) < \varepsilon\} = D^\varepsilon$ is again a domain of the same type, and by Corollary 3.4, if $\mathscr{S}$ is a locally free sheaf on D^ε, then $H^1(D^\varepsilon, \mathscr{S})$ is finite-dimensional. Now, we can repeat the argument of GRAUERT in [5] to verify the following assertion. For $x_0 \in B_1$, there is an $f_{x_0} \in H(D, O)$ such that $\lim f_{x_0}(x) = \infty$ as $x \in \bar{D} - \{x_0\}$ tends to x_0.

Now, we first verify that $H(D, O)$ has discrete level sets in N^-. If not, let L be a component of a level set in N^- of $H(D, O)$ of positive dimension. Since some function in $H(D, O)$ becomes infinite at any given point in B_1, L is bounded away from B_1, so φ attains a maximum on L. But this contradicts the maximum principle for s. psh. functions ([8], Chapter IX A).

Now, for n large enough, $\{x \in N;\ \varphi(x) = -1/n\} = B^{(n)}$ is a compact subset of N^-. For $x_0 \in B_1$, let $m = \max\{|f_{x_0}(x)|;\ x \in B^{(n)}\}$. There is a neighborhood U_0 of x_0 (in $\bar{D}$) such that $|f_{x_0}(x)| > 2m$ for $x \in U_0$. Then $g = m^{-1} f_{x_0}$ has the property that

$$|g(x)| > 2 \quad \text{for} \quad x \in U_0,$$
$$|g(x)| \leqq 1 \quad \text{for} \quad x \in B^{(n)}.$$

Choose finitely many such neighborhoods U_0 which cover B_1, and let $g_1, \ldots, g_t$ be the associated functions. Then

$$W^{(n)} = (D - N) \cup \{x \in N;\ \varphi(x) \leqq -1/n\} \cup \{x \in N;\ |g_i(x)| < 2,\ 1 \leqq i \leqq t\}$$

is a polyhedral domain. By Theorem 2.2, the spectrum $S^{(n)}$ of $H(W^{(n)}, O)$ is a Stein analytic space. Now, since $W^{(n)}$ is a polyhedral domain defined by functions in $H(D, O)$, $S^{(n)}$ is an analytic polyhedron defined by functions arising from $H(D, O)$, so $H(D, O)$ is dense in $H(W^{(n)}, O)$. Thus if we choose a subsequence $W^{(n)\prime}$ of $W^{(n)}$ such that $W^{(n)\prime} \subset W^{(n+1)\prime}$, it follows that $S^{(n)\prime} \subset S^{(n+1)\prime}$ and $\cup S^{(n)\prime} = S$ is an analytic space such that $H(S, O)$ is dense in $H(S^{(n)\prime}, O)$ for all n.

4. Corollary. Let X be as given in the above theorem, and suppose in addition that $H(D, O)$ separates points. Then there is a Stein space S and an injective mapping $g: D \to S$ such that $S - g(D)$ is relatively compact in S.

Proof. The corollary is easily verified with g the Gelfand mapping.

Notice that as a consequence of the above theorem and [*10*], the set B_1 is always connected. Now it is clear that in the above theorem, $H(D, O)$ need not separate points. For example D may contain some isolated compact subvarieties (blown up points). A worse example is the following: Let $\{x_n\}$ be a sequence of points exterior to the unit ball $B(0, 1)$ in $\boldsymbol{C}^n$ such that $x_n \to (1, 0, \ldots, 0) = x_0$, and if f is holomorphic at x_0 and vanishes on $\{x_n\}$, then $f \equiv 0$. Let D_1 and D_2 be two copies of $B(0, 2) - \bar{B}(0, 1)$, and let D be the analytic space obtained by identifying D_1 with D_2 along the sequence $\{x_n\}$. Then the hypotheses of Theorem 3 apply to D, but the corresponding space S is just $B(0, 2)$ with the Gelfand map identifying every point of D_1 with the corresponding point of D_2. We shall see below that in general there is no such pathology near the 1-pseudoconvex boundary B_1 of D. Consequently, if D is a manifold with no positive dimensional varieties disjoint from B_1, then $H(D, O)$ separates points.

5. Lemma. Let B be a 1-pseudoconvex hypersurface in an analytic space X of pure dimension $k > 2$. There is a neighborhood D of B such that $H(D, O)$ separates points on D.

Proof. Let N be a neighborhood of B and φ a s. psh. function defined in N such that $B \cap N = \{x \varepsilon N;\ \varphi(x) = 0\}$. Choose $\varepsilon > 0$ so that $D^\varepsilon = \{x \varepsilon N,\ -\varepsilon < \varphi(x) < \varepsilon\}$ is relatively compact in N. Then Theorem 3 applies to D^ε; let S be the spectrum of $H(D^\varepsilon, O)$ and $g: D^\varepsilon \to S$ the Gelfand map. Let $W = g(D^\varepsilon)$. The mapping $g: D^\varepsilon \to W$ is light (inverse image of a point is a discrete set) and, as is easily seen, proper. Thus the triple (D^ε, g, W) is an analytic cover ([*8*], Chapter III C), and it follows, as in [*2*] (or [*8*], III C) that there is an integer $\lambda(\varepsilon) > 0$ such that $g^{-1}(w)$ consists of precisely $\lambda(\varepsilon)$ points for almost all $w \in W$. We have to show that $\lambda(\varepsilon) = 1$ for small enough ε. Since λ is a decreasing integer valued function as $\varepsilon \to 0$, we may assume that we are in a range of ε for which λ is constant.

Choose η, $0 < \eta < \varepsilon$, and let $w_0 \in W$ be a regular point for the mapping g such that w_0 is in the image of $\{x \in N, \varphi(x) = \eta\}$. If $\lambda > 1$, there are points $x_0 \neq y_0$ such that $\varphi(x_0) = \varphi(y_0) = \eta$ and $g(x_0) = g(y_0)$. Let $w_n \in g(D^\eta)$ such that $w_n \to w_0$, and choose sequences $\{x_n\}$, $\{y_n\}$ in D^η such that $x_n \neq y_n$ and $x_n \to x_0$, $g(x_n) = g(y_n) = w_n$, $y_n \to y_0$. The sets $V_1 = \{x_n\}$, $V_2 = \{y_n\}$ are closed subvarieties of D^η, and the ideal sheaf $\mathscr{I} = \mathscr{I}(V_1 \cup V_2)$ is a coherent sheaf on $D^\varepsilon - \{x_0, y_0\}$. Then, by Lemma 3.5, $H^1(D^\eta, \mathscr{I})$ is finite dimensional. Let $\mathscr{F}$ be the space of functions on $V_1 \cup V_2$ which vanish on V_2. $\mathscr{F}$ is an infinite dimensional subspace of $H^0(D^\eta, O/\mathscr{I})$ so there is a nonzero function in $\mathscr{F}$ whose image in $H^1(D^\eta, \mathscr{I})$ under the coboundary map is zero. Thus, by exactness of the sequence

$$H^0(D^\eta, O) \to H^0(D^\eta, O/\mathscr{I}) \to H^1(D^\eta, \mathscr{I}),$$

there is an $f \in H^0(D^\eta, O)$ whose restriction to $V_1 \cup V_2$ is non-zero and which vanishes on V_2. Thus there is an n such that $f(x_n) \neq 0 = f(y_n)$. This contradicts the fact that $g(x_n) = g(y_n)$. Thus we must have $\lambda = 1$.

6. Theorem. *Let B be a 1-pseudoconvex hypersurface in an analytic space X of pure dimension $k > 2$. There is a neighborhood D of B, a Stein analytic space Y and an injective holomorphic map $g: D \to Y$ such that $g(B)$ bounds a relatively compact Stein 1-pseudoconvex space, S.*

Proof. For D chosen as in Lemma 5, we need only apply Corollary 4.

If we consider the algebra $\mathscr{A}(B)$ of all continuous functions on B which are approximable by functions holomorphic in a neighborhood of B, then $\mathscr{A}(B)$ is a Banach algebra in the norm $||f|| = \max\{|f(z)|; z \in B\}$. Since Y is the spectrum of $H(D, O)$, it is clear that $\bar{S}$ is the maximal ideal space of $\mathscr{A}(B)$.

Finally, we derive the following consequence of Theorem 6.

7. Corollary. Let X be a 1-pseudoconcave analytic space of pure dimension $k > 2$ (i. e., X is a relatively compact subspace of an analytic space X' with a 1-pseudoconcave boundary). Then there is a compact analytic space Z and a biholomorphic mapping of X into Z.

Proof. Let D be the neighborhood of bX and Y the Stein space determined in Theorem 6. Then $Z = X \cup Y$ with D^+ identified with $g(D^+)$ is a compact analytic space, and X is a subdomain of Z.

Now the restriction $k > 2$ is essential for many of these theorems, as we shall show by example in Section 6. However Theorem 3 may still be true in dimension 2, although the present methods break down completely. A counterexample to Theorem 3 in dimension 2 would have to be an essentially different situation from that presented by Grauert's example [7]. For if S should be covered by an analytic space S', S' would have to be a 1-pseudoconvex space, and thus holomorphically convex.

From that it would follow that S, being also the spectrum of $H(S', O)$ would have the structure of a Stein analytic space.

5. Positive imbeddings of codimension 1

Most of the concepts and techniques of this section are to be found in the work of GRAUERT; only their application is new. We shall refer freely to [5, 6].

1. Definition. Let X be a compact analytic space, and $\pi: L \to X$ a complex analytic line bundle over X. We shall denote by Z the zero-cross section of L. L is *weakly negative (positive)* if there is a relatively compact neighborhood N of Z with a 1-pseudoconvex (1-pseudoconcave) boundary.

Actually, according to GRAUERT, a weakly negative bundle is defined as above, and L is weakly positive if its dual L' is weakly negative. However, it is easily seen that for line bundles the two definitions of positivity are the same (see the lemma, p. 257 [1]). For $L \to X$ a bundle over X, we shall denote by $\mathscr{L}$ the sheaf of germs of holomorphic sections of L. The following theorem is proven by GRAUERT [6].

2. Theorem. *Let A be a compact analytic space and $L \to A$ a weakly positive bundle on A. Then there is an integer $s_0 > 0$ such that for $s \geqq s_0$, $H^1(A, \mathscr{L}^s) = 0$, and the sections of L^s give an imbedding of A into complex projective space.*

Now let A be a compact analytic subspace of a complex space X, and let $\mathscr{I}$ be the ideal sheaf of A in X. Let L_I be the line bundle associated with $\mathscr{I}$ (in the sense of GRAUERT [1; p. 352]). The fiber $L_{I,x}$ is canonically identified with the collection of linear mappings $\lambda: \mathscr{I}_x \to \boldsymbol{C}$ such that, if $g \in \mathscr{I}_x$, $h \in {}_XO_x$, then $\lambda(hg) = h(x)\,\lambda(g)$. The *normal bundle of A in X* is defined to be $N(A, X) = L_I | A$. Thus, if $\mathscr{I}$ is an invertible sheaf, and A is covered by neighborhoods U_α such that $\mathscr{I} \mid U_\alpha = g_\alpha {}_XO | U_\alpha$, then $N(A, X)$ is the line bundle over A with transition functions $g_\beta^\alpha = g_\alpha g_\beta^{-1}$ in $U_\alpha \cap U_\beta \cap A$. In other words, $\mathscr{N}(A, X) = (\mathscr{I}/\mathscr{I}^2)'$. In this case we shall say that A is *of (algebraic) codimension* 1 *in* X.

3. Theorem. *Let A be a compact analytic subspace of an analytic space X of pure dimension $k > 2$. Suppose that A is of algebraic codimension 1 and $L = N(A, X)$ is weakly positive. Then there is a neighborhood Y of A, a projective algebraic variety V of pure dimension k, and an injection g: $Y \to V$ such that $g(A)$ is the hyperplane section of V (i. e., $V - g(A)$ is a subvariety of affine space).*

Proof. We shall first have to make the following assumption:

(A) X has a 1-pseudoconcave boundary.

After applying the theorem in the case that X is a neighborhood of the zero section of L, we shall see that (A) in fact follows from the hypotheses.

Let $\mathscr{I}$ be the ideal sheaf of A in X and $E \to X$ the line bundle associated to $\mathscr{I}$ $(E = L_I)$. Then $L = E|A$, by definition. Let $\mathscr{Q}^s = \mathscr{E}^s \otimes {}_XO/\mathscr{I}^2$ be the sheaf of germs of sections of E^s modulo $\mathscr{I}^2$. We have the exact sequence

$$0 \to \mathscr{E}^s \otimes \mathscr{I}^2 \to \mathscr{E}^s \to \mathscr{Q} \to 0 .$$

Now it is easily verified that $\mathscr{Q} = \mathscr{L}^s \oplus \mathscr{L}^{s-1}$ and $\mathscr{E}^s \otimes \mathscr{I}^2 = \mathscr{E}^{s-2}$. Therefore, for all s we obtain the exact cohomology sequence

$$0 \to H^0(X, \mathscr{E}^{s-2}) \to H^0(X, \mathscr{E}^s) \xrightarrow{r^*} H^0(A, \mathscr{L}^s) \oplus H^0(A, \mathscr{L}^{s-1}) \xrightarrow{\delta}$$
$$\xrightarrow{\delta} H^1(X, \mathscr{E}^{s-2}) \xrightarrow{i^*} H^1(X, \mathscr{E}^s) \to H^1(A, \mathscr{L}^s) \oplus H^1(A, \mathscr{L}^{s-1}) \to \cdots .$$

Let us consider only $s \geqq s_0 + 1$ where s_0 is the integer determined in Theorem 2. We now argue as does GRAUERT in [5]. Since $H^1(A, \mathscr{L}^s) = 0$ for $s \geqq s_0$, the mapping i^* is surjective. Since X is 1-pseudoconcave, the spaces $H^1(X, \mathscr{E}^s)$ are finite dimensional. Thus there is an $s_1 \geqq s_0 + 1$ such that for $s \geqq s_1$, the mapping i^* is an isomorphism. Then by exactness the image of δ is zero, so r^* is surjective for $s \geqq s_1$. Let us fix such an s. Let $d + 1 = \dim H^0(X, \mathscr{E}^s)$, and let $\sigma_0, \ldots, \sigma_d$ be a basis for $H^0(X, \mathscr{E}^s)$. We choose σ_0 as the canonical section of $\mathscr{E}$ (i. e., if $\mathscr{I} | U_\alpha = g_\alpha \cdot {}_XO | U_\alpha$, then σ_0 is given in terms of a local coordinate for $\mathscr{E} | U_\alpha$ by g_α in U_α). $\sigma_0, \ldots, \sigma_d$ can be considered as the homogeneous coordinates of a mapping g of X into $\boldsymbol{P}^d$. Since $H^0(X, \mathscr{E}^s) | A = H^0(X, \mathscr{L}^s)$, it is clear that $g|A$ is injective. Further, since $H^0(X, \mathscr{E}^s)$ maps onto $H^0(X, \mathscr{L}^s) \oplus H^0(X, \mathscr{L}^{s-1})$ modulo $\mathscr{I}^2$, and $s - 1 \geqq s_0$, we can find sections of $H^0(X, \mathscr{E}^s)$ which have non-zero derivatives normal to A at any point $x_0 \in A$. Then g is non-singular along A, so there is a neighborhood Y of A such that $g|Y$ is injective. Now, let $Y_0 = Y - A$, and let $f_j = \sigma_j \sigma_0^{-1}$, $1 \leqq j \leqq d$. Then $f_j \in H^0(Y_0, {}_XO)$ and $F = (f_1, \ldots, f_d)$ defines an injective mapping of Y_0 into $\boldsymbol{C}^d$. For $x_0 \in A$, there is a j such that $\sigma_j(x_0) \neq 0$, thus $f_j(x) \to \infty$ as $x \in Y_0$ tends to x_0. Thus if M is large enough, $D = \{x \in Y_0; |f_j(x)| \leqq M, 1 \leqq j \leqq d\}$ has a compact boundary in Y_0, so $F: Y_0 - D \to \boldsymbol{C}^d - \Delta(0; M)$ properly, and the image is a subvariety of pure dimension k. Now, as in Section 2 (or by Corollary VII D 7 of [8]), there is a closed subvariety V_0 of $\boldsymbol{C}^d$ of pure dimension k such that $V_0 - \Delta(0; M) = F(Y_0 - D)$. Thus $Y - D \cup V_0$ with $Y_0 - D$ identified with $F(Y_0 - D)$ is a closed subvariety of $\boldsymbol{P}^d$, and thus a projective variety. Thus the theorem is proven except for the removal of assumption (A). This is accomplished as follows.

4. Lemma. Under the hypotheses of Theorem 3, (A) follows, i. e., there is a relatively compact neighborhood X_0 of A in X with a 1-pseudoconcave boundary.

Proof. Let Z be the zero cross-section of L. Since $L \to A$ is weakly positive, and $N(Z, L) = L$ (via the isomorphism $Z = A$), the hypotheses hold for Z in L including assumption (A). Thus there is a neighborhood L_0 of Z, a proper injection $F: L_0 - Z \to \boldsymbol{C}^d - \Delta(0, M)$ for some d and M. Let $z_1, \ldots, z_d$ be coordinates for $\boldsymbol{C}^d$, and let $p_n = \sum z_i \bar{z}_i$. Now we can imitate the argument of Satz 8, p. 353 of [*6*] in order to prove that there is a 1-pseudoconcave neighborhood of A in X_0. The only obstacle is the problem of extending $p_n \circ F$ from $L - Z$ to a neighborhood in E, but for this particular psh. function that can be accomplished by Satz 6, p. 350 [*6*].

6. An example

We begin with a theorem of Andreotti concerning complex structures on four dimensional C^∞ manifolds. It should be observed that this theorem is not essential to the subsequent discussion, but it is the motivating idea.

1. Theorem. *Let M be a four (real) dimensional C^∞ manifold. Let φ be a C^∞ complex-valued 2-form on M with the following properties: (i) $d\varphi = 0$, (ii) $\varphi \wedge \varphi = 0$, (iii) $\varphi \wedge \bar{\varphi} > 0$ (in particular M is assumed to be oriented). Then M can be endowed with a 2-dimensional complex analytic structure in a unique way so that φ becomes a holomorphic 2-form.*

Given such a structure we can express the Cauchy-Riemann equations as follows: a function f is holomorphic if and only if $df \wedge \varphi = 0$. Now let M be the C^∞ manifold underlying $\boldsymbol{C}^2 - \{0\}$. For z_1, z_2 coordinates in $\boldsymbol{C}^2$ we define $r = (z_1\bar{z}_1 + z_2\bar{z}_2)^{1/2}$. For any $\varepsilon \in \boldsymbol{C}$, it is easily verified that $\varphi = dz_1 \wedge dz_2 + \varepsilon \partial \partial \log r^2$ has all the properties required by the above theorem. We now assume that M is the corresponding complex manifold. The Cauchy-Riemann equations on M are as follows (with $u = z_2/z_1$, $v = z_1$):

$$f_{\bar{v}} = 0, \; \varepsilon f_v + (1 + u\bar{u})^2 v f_{\bar{u}} = 0\,. \tag{1}$$

Notice that u is a solution, and thus a meromorphic function on M. We can find another solution, namely

$$v' = z_1^2 \left(\frac{1}{2} + \frac{\varepsilon \bar{z}_1}{z_2 r^2}\right). \tag{2}$$

The result is that we obtain the following explicit coordinatization of M. There are two coordinate (actually four) neighborhoods, $U_i = \{z_i \neq 0\}$ with coordinates u_i, v_i, $1 \leqq i \leqq 2$, where

$$\begin{aligned} u_1 &= z_2/z_1\,, \quad v_1 = \frac{z_1^2}{2} - \frac{\varepsilon \bar{u}_1}{1 + u_1\bar{u}_1}\,; \\ u_2 &= z_1/z_2\,, \quad v_2 = \frac{z_2^2}{2} + \frac{\varepsilon \bar{u}_2}{1 + u_2\bar{u}_2}\,. \end{aligned} \tag{3}$$

In fact, it is easily checked that (u_i, v_i) maps U_i into $\boldsymbol{C}^2$ in an exactly 2-1 fashion, the identification being $(z_1, z_2) \sim (-z_1, -z_2)$. Notice that $u_1 = u_2^{-1}$ is meromorphic on M, so that the natural mapping $\boldsymbol{C}^2 - \{0\} \to \boldsymbol{P}^1$ is also holomorphic from $M \to \boldsymbol{P}^1$. Thus M is a fiber space over the Riemann sphere with fiber $\boldsymbol{C} - \{0\}$. v_1, v_2 are holomorphic functions on all of M; so is

$$v_3 = u \cdot v' - \varepsilon = \frac{z_1 z_2}{2} - \frac{\varepsilon u_1 \bar{u}_1}{1 + u_1 \bar{u}_1} \,. \tag{4}$$

We wish to investigate the mapping properties of (v_1, v_2, v_3). Notice that the only identification (v_1, v_2, v_3) makes on M is of a point (z_1, z_2) with its negative, $(-z_1, -z_2)$. This mapping (identifying a point with its negative) maps $\boldsymbol{C}^2 - \{0\}$ onto $K - \{0\}$, where K is the cone $\{x^2 = yz\}$ in $\boldsymbol{C}^3$. In order to proceed, it is best to replace $\boldsymbol{C}^2$ and K by the quadratic transforms Q and Q' of the origin in each case.

Let x_1, x_2 be homogeneous coordinates for $\boldsymbol{P}^1$. Let $U_i = \{x_i \neq 0\}$, $i = 1, 2$. Q is the line bundle over $\boldsymbol{P}^1$ given as follows: U_1, U_2 are coordinate neighborhoods for Q with coordinates y_1, y_2 and in $U_1 \cap U_2$ $y_2 = z_1 y_1$. Q' is the line bundle with coordinate neighborhoods U_1, U_2 and coordinates y_1', y_2' such that $y_2' = x_1^2 y_1'$ in $U_1 \cap U_2$. We can identify $\boldsymbol{C}^2 - \{0\}$ with $Q - \{\text{zero section}\}$ and $K - \{0\}$ with $Q' - \{\text{zero section}\}$. The identification mapping of a point with its negative extends to a holomorphic mapping $\Phi: Q \to Q'$ given by $y_i' = y_i^2$ in U_i, $i = 1, 2$. Now $\Phi: M \to Q'$ is not a holomorphic mapping, but we can change the structure of Q' so that it becomes one. Namely, over U_i choose new coordinates:

$$\begin{aligned} \tilde{y}_1 &= \frac{y_1'}{2} - \frac{\varepsilon \bar{x}_1}{1 + x_1 \bar{x}_1} \text{ over } U_1 \,, \\ \tilde{y}_2 &= \frac{y_2'}{2} + \frac{\varepsilon \bar{x}_2}{1 + x_2 \bar{x}_2} \text{ over } U_2 \,. \end{aligned} \tag{5}$$

In the overlap, $\tilde{y}_2 = x_1^2 \tilde{y}_1 + \varepsilon \tilde{y}_1$. Let $\tilde{Q}$ be Q' with this new structure, then $\tilde{Q}$ is an affine bundle over $\boldsymbol{P}^1$, and $\Phi: M \to \tilde{Q}$ is holomorphic. This explicit representation of the situation will allow us to verify that there is no way to put a compact set in the hole (left by excising the origin in $\boldsymbol{C}^2$) in M so that the result is an analytic space.

2. Theorem. *a) For $\varepsilon \neq 0$, $\tilde{Q}$ is a Stein manifold.*

b) $\tilde{Q}$ is the spectrum of $H(M, O)$; any function holomorphic on M is a holomorphic function of v_1, v_2, v_3.

c) It is impossible to find an analytic space $\tilde{M}$ of dimension 2 and an injection $g: M \to \tilde{M}$ so that if $K \subset M$ is a bounded set, $g(K)$ is relatively compact in $\tilde{M}$.

Proof. Let Z be the zero section of Q'. Since v_1, v_2, v_3 identify the same points as Φ, the v_i determine functions $\tilde{v}_i$ holomorphic on $\tilde{Q} - Z$ which separate points. Comparing (3) and (4) with (5) it can be verified that the $\tilde{v}_i$ have holomorphic extensions to all of $\tilde{Q}$, and $(\tilde{v}_1, \tilde{v}_2, \tilde{v}_3)$ defines a biholomorphic mapping of $\tilde{Q}$ onto a closed submanifold of $\mathbf{C}^3$, namely in coordinates w_1, w_2, w_3 the submanifold $\{w_3(w_3 + \varepsilon) = w_1 w_2\}$. Thus a) is verified.

Note that $\Phi: M \to \tilde{Q} - Z$ is a 2-1 covering map. We can find finitely many functions $f_1, \ldots, f_t \in H(M, O)$ such that for $x, y \in M$, $f(x) = f(y)$ for all $f \in H(M, O)$ if and only if $f_j(x) = f_j(y)$, $1 \leqq j \leqq t$. Now each f_j satisfies a quadratic polynomial equation $P_j(f_j) = 0$ with coefficients holomorphic on $\tilde{Q} - Z$. By Hartogs' theorem [*10*] these coefficients may be assumed to be holomorphic on $\tilde{Q}$. Let $w_1, \ldots, w_t$ be coordinates in $\mathbf{C}^t$. Then $V = \{(x, w) \in \tilde{Q} \times \mathbf{C}^t; P_j(x, w_j) = 0\}$ is a closed analytic subspace, and thus a Stein subspace. It can be shown as in Theorem VII D 8 of [*8*] that V is the spectrum of $H(M, O)$ and $g: M \to V$, $g(z) = (\Phi(z), (f_1(z), \ldots, f_t(z)))$ is the Gelfand mapping. Since Φ is 2-1, g is either 2-1 or an injection. In the first case, $V \cong \tilde{Q}$ under the natural projection so b) is true. It remains to show that the second case is impossible; this will follow from c).

Let us thus assume that an $\tilde{M}$ as described in c) can be found. We may consider M as an open set contained in $\tilde{M}$. Now, by the nature of the explicit coordinatization (3) it is clear that the structure of M tends to the ordinary complex structure as $z \to \infty$. Then, given $\varepsilon > 0$, if R is large enough the boundary of the ball $B_R = \{z \in M; r(z) < R\}$ will be a 1-pseudoconvex hypersurface in M. Then the closure $\tilde{B}_R$ of B_R in $\tilde{M}$ will be a bounded strongly pseudoconvex subdomain of $\tilde{M}$. Outside of some identifications of compact subvarieties of $\tilde{M}$, $\tilde{B}_R$ is thus a Stein space; we may in fact assume that no such identifications are necessary. Now by Hartogs' theorem $H(B_R, O) = H(\tilde{B}_R, O)$ and the mapping $\Phi: B_R \to \tilde{Q}$ extends to $\Phi: \tilde{B}_R \to \tilde{Q}$. Since the singular locus S of the mapping Φ must be a compact subvariety of $\tilde{B}_R$ (it is disjoint from B_R), it is a finite point set.

Let us now recall the meromorphic function $u = \frac{z_2}{z_1} = \frac{v_2}{v_3 + \varepsilon}$. The set $L_c = \{x \in \tilde{B}_R; u(x) = c\}$ for c any point on $\mathbf{P}^1$ is an analytic subvariety of $\tilde{B}_R$, and $\Phi: L_c \cap B_r \to \tilde{Q}$ in 2-1 fashion. Now $\lim_{\substack{z \in L_c \cap B_r \\ z \to 0}} v_i(z)$ is constant, so the functions $v_i(z)$ are constant on $L_c \cap (\tilde{B}_r - B_r)$, therefore this set must be a finite point set (being a compact set on the 1-dimensional space L where a nonconstant function is constant). In fact, since $L_c \cap B_r$ is a punctured plane, $L_c \cap (\tilde{B}_r - B_r)$ must consist of one point, x_c. Since Φ is 2-1 in every neighborhood of x_c, $x_c \in S$. Finally,

$$v_3(x_c) = -\frac{\varepsilon|c|^2}{1+|c|^2}, \quad \text{so if} \quad |c| \neq |c|', \quad x_c \neq x_c .$$

Thus there are infinitely many points $x_c \in S$, contradicting the fact that S must be a finite point set. Thus the theorem is proved.

Now we shall use the above situation to give a counterexample to the theorem of Section 5. Let w_1, w_2, w_3, w_0 be homogeneous coordinates for $\boldsymbol{P}^3$ and V the variety in $\boldsymbol{P}^3$ given by $w_3(w_3 + \varepsilon w_0) = w_1 w_2$. Clearly V is nonsingular. Let y_1, y_2, y_3 be homogeneous coordinates for $\boldsymbol{P}^2$, and let 0 be the point with coordinates (0, 0, 1). Consider the mapping $\Phi: \boldsymbol{P}^2 - 0 \to V$ given by

$$\begin{aligned} w_1 &= y_1^2 - \varepsilon \frac{y_1 \bar{y}_2}{y_1\bar{y}_1 + y_2\bar{y}_2} y_3^2, \\ w_2 &= y_2^2 + \varepsilon \frac{\bar{y}_1 y_2}{y_1\bar{y}_1 + y_2\bar{y}_2} y_3^2, \\ w_3 &= y_1 y_2 + \varepsilon \frac{y_2 \bar{y}_2}{y_1\bar{y}_1 + y_2\bar{y}_2} y_3^2, \\ w_0 &= y_3^2 . \end{aligned}$$

Φ is an everywhere 2-1 mapping. For, it certainly is on $y_3 = 0$, and for $y_3 \neq 0$, using the inhomogeneous coordinates $z_1 = y_1 y_3^{-1}$, $z_2 = y_2 y_3^{-1}$, we are in the situation of the mapping $(v_1, v_2, v_3): \boldsymbol{C}^2 - \{0\} \to \boldsymbol{C}^3$ described above. In particular Φ is a local homeomorphism, so we can give $\boldsymbol{P}^2 - \{0\}$ a unique complex structure which makes Φ locally biholomorphic. Let M be the resulting manifold; as in the above the "hole" in M cannot be filled. In fact, if $A = \{y_3 = 0\}$, no neighborhood of A can be completed to a compact analytic space. However the normal bundle of A in M is positive. For the identity $i: M \to \boldsymbol{P}^2 - \{0\}$ restricted to A induces an isomorphism of ${}_M O/{}_M \mathscr{I}(A)^2 = {}_{\boldsymbol{P}^2} O/\mathscr{I}(A)^2$. For example, in $\{y_1 \neq 0\}$, $u = y_2 y_1^{-1}$, $v = y_3 y_1^{-1}$ are coordinates for $\boldsymbol{P}^2$, and

$$u' = u - \varepsilon \frac{|u|^2}{1+|u|^2} v^2 ,$$

v are coordinates for M. Thus, if f is holomorphic at $x \in A$ in either structure, the first two coefficients of f in a Taylor expansion in powers of v are holomorphic functions of u. Given any n, by considering similar mappings into higher dimensional projective space defined by the monomials in y_1, y_2 of degree $2n$, we can in fact put new structures M on $\boldsymbol{P}^2 - \{0\}$ so that the identity $i: M \to \boldsymbol{P}^2 - \{0\}$ induces an isomorphism

${}_M O/{}_M\mathscr{I}(A)^{2n} = \boldsymbol{P}^2 O/I(A)^{2n}$. The manifold M cannot be completed; in particular, no neighborhood of A in M is isomorphic to any neighborhood of A in $\boldsymbol{P}^n$.

References

[1] Andreotti, A., and H. Grauert: Théorèmes de finitude pour la cohomologie des espaces complexes. Bull. Soc. Math. France **90**, 193—259 (1962).

[2] Bishop, E.: Partially analytic spaces. Amer. J. Math. **83**, 669—693 (1961).

[3] — Mappings of partially analytic spaces. Amer. J. Math. **83**, 209—242 (1961).

[4] — Holomorphic completions, analytic continuation and the interpolation of semi-norms. Ann. of Math. **78**, 468—500 (1963).

[5] Grauert, H.: On Levi's problem and the imbedding of real analytic manifolds. Ann. Math. **68**, 460—472 (1958).

[6] — Über Modifikationen und exzeptionelle analytische Mengen. Math. Ann. **146**, 331—368 (1962).

[7] — Bemerkenswerte pseudokonvexe Mannigfaltigkeiten, Math. Z. **81**, 377—391 (1963).

[8] Gunning, R.C., and H. Rossi: Holomorphic functions of several complex variables. Prentice-Hall 1965.

[9] Rossi, H.: On envelopes of holomorphy. Comm. Pure Appl. Math. **16**, 9—17 (1963).

[10] — Vector fields on analytic spaces. Ann. Math. **78**, 455—467 (1963).

Department of Mathematics
Brandeis University
Waltham, Mass.

Non-compact Complex Lie Groups without Non-constant Holomorphic Functions*,**

By

A. Morimoto

Introduction

In this paper we shall consider, on the one hand, a complex Lie group with sufficiently many holomorphic functions and, on the other hand, a complex Lie group whose holomorphic functions are necessarily constant. The former will be called a Stein group and the latter an (H. C)-group. In the previous paper [3] we considered the complex analytic fibre bundles over Stein manifolds and, among other things, we established a necessary and sufficient condition for a complex Lie group to be a Stein manifold. Using this result, we shall first prove that every connected

* This research was partially supported by the United States Air Force through the Air Force Office of Scientific Research.

** Received May 6, 1964.

complex Lie group G contains the smallest closed complex normal subgroup G^0 such that the factor group G/G^0 is a Stein group. Next we prove that the subgroup G^0 is an (H. C)-group, and so every connected complex Lie group can be obtained by an extension of a Stein group by an (H. C)-group (Theorem 1 in §2). Using this theorem we can characterize a connected complex Lie group to be holomorphically convex by group theoretical conditions. From this characterization we can show that a connected complex Lie group containing no complex torus is a Stein group if and only if it is holomorphically convex.

Now, obviously every compact connected complex Lie group, i.e. complex torus, is an (H. C)-group. In §3 we shall show that the converse of this statement is not true and in fact, we shall construct an (H. C)-group of arbitrary dimension $\geqq 2$, which contains no complex torus of positive dimension. It is known that there exists an example of a non-compact complex manifold without non-constant holomorphic functions (cf. [*1*]); our examples in §3 show that there is a number of such manifolds even in the case of group manifolds.

In §4 we shall consider a non-compact (H. C)-group of dimension 2 and prove that such a group is isomorphic to the factor group $\boldsymbol{C}^2/\Gamma$ of $\boldsymbol{C}^2$ by a discrete subgroup Γ of $\boldsymbol{C}^2$ generated by (1, 0), (0, 1) and (α, β) such that the imaginary part of (α, β) is not zero and that $1, \alpha, \beta$ are linearly independent over the field of rational numbers (Theorem 5). In §5 we shall classify all non-compact (H. C)-groups of dimension 2 and especially we shall show that there exist non-countably many such groups which are not mutually isomorphic. In §6 we classify all connected complex abelian Lie groups of dimension 2 and of rank 3 (cf. §4) which are not (H. C)-groups.

In the last section we sum up tho results obtained in §§4—6 and get the complete classification of all connected complex abelian Lie groups of dimension 2.

§ 1. Preliminaries

For a complex manifold V, we denote by $H(V)$ the ring of all holomorphic functions on V. A complex manifold V is called a Stein manifold if the following three conditions are satisfied.

(S. 1) For $p, q \in V, p \neq q$ there exists $f \in H(V)$ such that $f(p) \neq f(q)$.

(S. 2) For any sequence of points $\{p_\nu\}_{\nu=1}^{\infty}$ of V without accumulation point, there exists $f \in H(V)$ such that $\{f(p_\nu)\}_{\nu=1}^{\infty}$ is unbounded.

(S. 3) For any $p_0 \in V$, there exists a system of functions $f_1, f_2, \ldots, f_n$, $f_i \in H(V)$ such that $\{f_1, f_2, \ldots, f_n\}$ is a system of coordinates in some neighborhood of p_0.

Let G be a connected complex Lie group. Then G is a Stein manifold if the condition (S. 1) is satisfied for $V = G$ (cf. Remarque, [3], p. 147). Such a group will be called a Stein group.

Let N be a connected closed complex normal subgroup of a connected complex Lie group G. Then, if N and G/N are both Stein groups, G is also a Stein group (Théorème 4, [3]).

Let K be a maximal compact subgroup of a connected complex Lie group G, and let $\mathfrak{k}$ and $\mathfrak{g}$ be the Lie algebras of K and G respectively. Naturally, $\mathfrak{k}$ is a real Lie subalgebra of $\mathfrak{g}$. We remark here that the subalgebra $\mathfrak{k} \cap \sqrt{-1}\,\mathfrak{k}$ is contained in the center of $\mathfrak{g}$ (cf. 1. C, [3] p. 139). A Lie subgroup H of G is a complex subgroup of G if and only if the Lie subalgebra corresponding to H is a complex Lie subalgebra of $\mathfrak{g}$.

Throughout this paper, we denote by $\boldsymbol{C}$ the additive group (sometimes, the field) of the complex numbers.

§ 2. Normal subgroup G^0 of a complex Lie group G

Let G be a connected complex Lie group and $H(G)$ be the ring of all holomorphic functions on G. We denote by G^0 the set of all $a \in G$ such that $f(a) = f(e)$ for each $f \in H(G)$, where e denotes the unit element of G. For this set G^0 we have the following lemmas.

Lemma 1. *G^0 is a closed complex Lie subgroup of G. Moreover, G^0 is a characteristic subgroup of G, i.e. every automorphism of G leaves G^0 invariant, especially G^0 is normal in G.*

Proof. For $a, x \in G$ and $f \in H(G)$ we put $f_a(x) = f(ax)$, $f'(x) = f(x^{-1})$. Then obviously $f_a, f' \in H(G)$. Now for $a, b \in G^0$ we see that $f(a \cdot b) = f_a(b) = f_a(e) = f(a) = f(e)$ and $f(a^{-1}) = f'(a) = f'(e) = f(e)$. Hence $a \cdot b$, $a^{-1} \in G^0$, which proves that G^0 is a subgroup of G. Since G^0 is clearly closed in G, G^0 is a Lie subgroup of G. Let $\mathfrak{g}$ and $\mathfrak{g}^0$ be the Lie algebras of G and G^0 respectively. We shall prove that $\mathfrak{g}^0$ is a complex subalgebra of $\mathfrak{g}$. Take an element $X \in \mathfrak{g}^0$ and put $Y = \sqrt{-1}\,X$. Consider the holomorphic mapping ψ of some connected neighborhood V of the origin in $\boldsymbol{C}$ into G defined by

$$\psi(s + \sqrt{-1}\,t) = \exp sX \cdot \exp tY\,,$$

where $\exp tX\,(-\infty < t < \infty)$ denotes the 1-parameter subgroup of G whose tangent vector at e is X. Take $f \in H(G)$. Now we define a function $f_0 \in H(V)$ by $f_0 = f \circ \psi$. Since $\exp sX \in G^0$, we have $f_0(s) = f_0(0)$ for sufficiently small $|s|$. Hence $f_0(z) = f_0(0)$ for $z \in V$, and so $f(\exp tY) = f(e)$ for sufficiently small $|t|$, which implies that $\exp tY \in G^0$ for $-\infty < t < \infty$, and so $Y \in \mathfrak{g}^0$. We have thus proved that G^0 is a complex subgroup of G. Finally for an automorphism σ of G and $f \in H(G)$, we put $f^\sigma(x) = f(\sigma(x))$

for $x \in G$. Then, if $a \in G^0$, $f(\sigma(a)) = f^\sigma(a) = f^\sigma(e) = f(e)$ since $f^\sigma \in H(G)$. This proves that $\sigma(a) \in G^0$ and G^0 is a characteristic subgroup of G.

Lemma 2. *The factor group G/G^0 is a Stein group.*

Proof. By the property of G^0 we can naturally define a homomorphism $\tau: H(G) \to H(G/G^0)$ such that $\tau(f) \cdot \pi = f$ for $f \in H(G)$, where π denotes the natural homomorphism of G onto G/G^0. Now take two distinct elements $\bar{a} = \pi(a)$, $\bar{b} = \pi(b)$ of G/G^0. Then $a^{-1}b \notin G^0$, and so there exists $f \in H(G)$ such that $f(a^{-1}b) \neq f(e)$. Put $\tau(f_{a^{-1}}) = \tilde{f}$. Then we see that $\tilde{f}(\bar{a}) \neq \tilde{f}(\bar{b})$. Hence the manifold $V = G/G^0$ satisfies (S. 1) in §1. Therefore G/G^0 is a Stein group.

Lemma 3. *Let N be a closed complex Lie subgroup of G such that G/N is a Stein group. Then $N \supset G^0$.*

Proof. Take $a \in G$ such that $a \notin N$, then there exists $\tilde{f} \in H(G/N)$ such that $\tilde{f}(\bar{a}) \neq \tilde{f}(\bar{e})$, where $\bar{a} = \pi(a)$, π being the natural homomorphism of G onto G/N. Put $f = \tilde{f} \circ \pi$. Then $f \in H(G)$ and $f(a) \neq f(e)$. Hence $a \notin G^0$. Thus $G^0 \subset N$ is proved.

Lemma 4. *G^0 is connected.*

Proof. Let G_0^0 be the connected component of G containing e. Then G_0^0 is a normal subgroup of G and the factor group G/G_0^0 is a covering group of G/G^0. Hence by a theorem of Stein [4] G/G_0^0 is a Stein group, whence $G_0^0 \supset G^0$ by virtue of Lemma 3. Thus $G_0^0 = G^0$ is proved and so G^0 is connected.

Lemma 5. *G^0 is an* (H. C)*-group.*

Proof. Let $(G^0)^0$ be the set of all element $a \in G^0$ such that $f(a) = f(e)$ for any $f \in H(G^0)$. Since $(G^0)^0$ is a characteristic subgroup of G^0 (Lemma 1), $(G^0)^0$ is a normal subgroup of G. Now, the complex manifolds G/G^0 and $G^0/(G^0)^0$ are Stein groups by Lemma 2. Since $(G/(G^0)^0)/(G^0/(G^0)^0) \cong G/G^0$, $G/(G^0)^0$ is also a Stein Group (cf. §1). Hence by Lemma 3, $(G^0)^0 \supset G^0$, which shows that $(G^0)^0 = G^0$. Thus G^0 is an (H. C)-group.

Lemma 6. *G^0 is contained in the center of G.*

Proof. Let Z be the center of G. Then by the adjoint representation of G, G/Z is a complex linear group and so G/Z is a Stein group (cf. 1. A, [3] p. 193). Hence by Lemma 3, we see $Z \supset G^0$, which proves the Lemma.

Combining Lemmas 1-6, we obtain the following theorem:

Theorem 1. *For every connected complex Lie group G, there exists the smallest closed complex subgroup G^0 of G such that G/G^0 is a Stein group. Moreover G^0 has the following properties:*

(i) *G^0 is a central characteristic connected subgroup of G.*

(ii) *Every holomorphic function on G^0 is a constant.*

Remark. The normal subgroup G^0 should be called the Steinizer of G. Thus every connected complex Lie group can be obtained by an extension of a Stein group by an (H. C)-group. Clearly every compact connected complex Lie group, i.e. complex torus, is an (H. C)-group. But the converse is not true as is shown in the next section. However, every (H. C)-group G is necessarily abelian. In fact, $G = G^0$ is contained in the center of G by Lemma 6. This shows that G is abelian.

Theorem 2. *Let G be a connected complex Lie group and K a maximal compact subgroup of G. Let $\mathfrak{g}$, $\mathfrak{g}^0$, $\mathfrak{k}$ be the Lie algebras of G, G^0, K respectively and K_0 be the Lie subgroup of G corresponding to $\mathfrak{k} \cap \sqrt{-1}\,\mathfrak{k}$. Then $\mathfrak{k} \cap \sqrt{-1}\,\mathfrak{k}$ (and K_0) is independent of the choice of K. Moreover the following conditions are mutually equivalent:*

(1) *G is holomorphically convex, i.e. $V = G$ satisfies* (S. 2) *of* §1.
(2) *G^0 is a complex torus.*
(3) $\mathfrak{g}^0 = \mathfrak{k} \cap \sqrt{-1}\,\mathfrak{k}$.
(4) *K_0 is closed in G.*

Proof. Since $\mathfrak{k} \cap \sqrt{-1}\,\mathfrak{k}$ is contained in the center of $\mathfrak{g}$ (cf. §1) and since all maximal compact subgroups of G are conjugate, it follows that $\mathfrak{k} \cap \sqrt{-1}\,\mathfrak{k}$ is independent of the choice of K.

(1) $\Rightarrow$ (2): Suppose that G^0 is not compact. Then there is a sequence $\{a_\nu\}_{\nu=1}^{\infty}$ of points of G^0 without accumulation point in G^0. Since G^0 is closed in G, $\{a_\nu\}$ has no accumulation point in G. Hence there is a function $f \in H(G)$ such that $\{f(a_\nu)\}_{\nu=1}^{\infty}$ is unbounded. Then f is not a constant on G^0, which is a contradiction. Thus G^0 is compact and so G^0 is a complex torus.

(2) $\Rightarrow$ (1): Take an arbitrary sequence $\{a_\nu\}$ of points of G without accumulation point. Let $\Phi: G \to G/G^0$ be the natural homomorphism. Since G^0 is compact we see that $\{\Phi(a_\nu)\}$ has no accumulation point in G/G^0. Now, by Lemma 2, G/G^0 is a Stein manifold. Hence there is a function $\tilde{f} \in H(G/G^0)$ such that $\{\tilde{f}(\Phi(a_\nu))\}$ is unbounded. Then, for $f = \tilde{f} \circ \Phi \in H(G)$, $\{f(a_\nu)\}$ is unbounded, which proves that G is holomorphically convex.

(2) $\Rightarrow$ (3): Since G^0 is compact there is a maximal compact subgroup K_1 of G such that $G^0 \subset K_1$. Let $\mathfrak{k}_1$ be the Lie algebra of K_1. Then $\mathfrak{g}^0 \subset \mathfrak{k}_1$. Since $\mathfrak{g}^0$ is a complex subalgebra of $\mathfrak{g}$, $\mathfrak{g}^0 \subset \sqrt{-1}\,\mathfrak{k}_1$ and so $\mathfrak{g}^0 \subset \mathfrak{k}_1 \cap \sqrt{-1}\,\mathfrak{k}_1 = \mathfrak{k} \cap \sqrt{-1}\,\mathfrak{k}$. Hence $\mathfrak{g}^0 \subset \mathfrak{k} \cap \sqrt{-1}\,\mathfrak{k}$. Then $\mathfrak{g}^0 = \mathfrak{k} \cap \sqrt{-1}\,\mathfrak{k}$ follows from Lemma 7, which will be proved in the next section.

(3) $\Rightarrow$ (4) is clear.

(4) $\Rightarrow$ (2): Since K_0 is contained in the center of G (cf. §1) we can consider the natural homomorphism Φ of G onto G/K_0. We denote by the same Φ the natural homomorphism of $\mathfrak{g}$ onto $\mathfrak{g}/\mathfrak{k}_0$. First, we prove that

$\Phi(\mathfrak{k}) \cap \sqrt{-1}\Phi(\mathfrak{k}) = 0$. In fact, $\Phi(X) = \sqrt{-1}\,\Phi(Y)(X, Y \in \mathfrak{k})$ implies $X - \sqrt{-1}\;Y \in \mathfrak{k} \cap \sqrt{-1}\;\mathfrak{k}$, whence $X \in \sqrt{-1}\;\mathfrak{k} \cap \mathfrak{k}$ and so $\Phi(X) = 0$. Since $\Phi(K)$ is a maximal compact subgroup of G/K_0 (Lemma 3.15 in [2]), we see that G/K_0 satisfies the condition (P) in [3], and so G/K_0 is a Stein group. Therefore $K_0 \supset G^0$ follows from Lemma 3. Since $K_0 \subset G^0$ by Lemma 7, it follows that $G^0 = K_0$ and G^0 is a complex torus. Thus Theorem 2 is proved.

Corollary. *Let G be a connected complex Lie group containing no complex torus of positive dimension. Suppose that G is holomorphically convex. Then G is a Stein group. (The converse is clearly true.)*

In fact, by Theorem 2, $G^0 = \{e\}$ and so G is a Stein group by Lemma 2.

§ 3. (H. C)-groups containing no complex torus

In this section we shall construct an (H. C)-group of an arbitrary dimension $\geqq 2$ which contains no complex torus of positive dimension. First we begin with some lemmas.

Lemma 7. *Let K be a maximal compact subgroup of a connected complex Lie group G, and let $\mathfrak{g}$, $\mathfrak{g}^0$, $\mathfrak{k}$ be the Lie algebras of G, G^0, K respectively. Then $\mathfrak{k} \cap \sqrt{-1}\;\mathfrak{k} \subset \mathfrak{g}^0$.*

Proof. Put $\mathfrak{k}_0 = \mathfrak{k} \cap \sqrt{-1}\;\mathfrak{k}$, and let K_0 be the Lie subgroup of G corresponding to $\mathfrak{k}_0$. Since $\mathfrak{k}_0$ is contained in the center of $\mathfrak{g}$ (cf. §1) K_0 is an abelian complex Lie group. Let A be the universal covering group of K_0, then A is isomorphic to $\boldsymbol{C}^m$ $(m = \dim_C K)$. Let $\varphi: \boldsymbol{C}^m \to K_0$ be the covering mapping. Now, take a function $f \in H(G)$, and put $f_0 = f \mid K_0$, $f = f_0 \circ \varphi$. Since $K_0 \subset K$ and K is compact f_0 is necessarily bounded. Hence f is also a bounded holomorphic function on $\boldsymbol{C}^m$, and so is a constant, which shows that f_0 is constant, and we see that $f(a) = f(e)$ for $a \in K_0$. This proves that $K_0 \subset G^0$, and hence $\mathfrak{k}_0 \subset \mathfrak{g}^0$.

We can prove easily the following two lemmas.

Lemma 8. *Let N be a connected closed complex normal subgroup of a connected complex Lie group G.*

(i) *If G is an* (H. C)*-group, then G/N is an* (H. C)*-group.*

(ii) *If G/N and N are* (H. C)*-groups, then G is an* (H. C)*-group.*

Lemma 9. *Let G_1 and G_2 be connected complex Lie groups. Then $(G_1 \times G_2)^0 = G_1^0 \times G_2^0$.*

Now consider the elements $v_i (i = 1, 2, \ldots, 5)$ of $\boldsymbol{C}^3$ defined by $v_1 = (1, 0, 0)$, $v_2 = (0, 1, 0)$, $v_3 = (0, 0, 1)$, $v_4 = (\sqrt{-1}\,a_1, \sqrt{-1}\,a_2, \sqrt{-1}\,a_3)$, $v_5 = (0, \sqrt{-1}\,a_4, \sqrt{-1}\,a_5)$, where $a_i (i = 1, 2, \ldots, 5)$ are real and $a_i a_j (i < j)$ are linearly independent over the field $\boldsymbol{Q}$ of rational numbers. We see readily that $v_i (i = 1, 2, \ldots, 5)$ are linearly independent over the field $\boldsymbol{R}$ of real numbers. Let Γ be the discrete subgroup of $\boldsymbol{C}^3$ generated by $v_i (i = 1, 2, \ldots, 5)$. Put $G = \boldsymbol{C}^3/\Gamma$.

Theorem 3. *The complex Lie group G defined above is an* (H. C)-*group. Moreover G contains no complex torus of positive dimension as a subgroup.*

Proof. Let K be the maximal compact subgroup of G. Let $\mathfrak{k}$, $\mathfrak{g}$, $\mathfrak{g}^0$ be the Lie algebras of K, G, G^0 respectively. Put $\mathfrak{k}_0 = \mathfrak{k} \cap \sqrt{-1}\,\mathfrak{k}$. Let K_0 be the subgroup of G corresponding to the Lie subalgebra $\mathfrak{k}_0$. We first prove that K_0 is isomorphic to $\boldsymbol{C}^{*2}$, $\boldsymbol{C}^*$ being the multiplicative group of non-zero complex numbers. In fact, identifying $\mathfrak{g}$ with $\boldsymbol{C}^3$, we see that $K_0 = \Phi(\mathfrak{k}_0)$, where Φ is the natural homomorphism of $\boldsymbol{C}^3$ onto $G = \boldsymbol{C}^3/\Gamma$. Then $K_0 \cong \mathfrak{k}_0/\mathfrak{k}_0 \cap \Gamma$. On the other hand it is easily seen that $\mathfrak{k}_0$ is spanned by u_1 and u_2 over $\boldsymbol{C}$, where $u_1 = (a_1, a_2, a_3)$, $u_2 = (0, a_4, a_5)$. We assert that $\mathfrak{k}_0 \cap \Gamma$ is the discrete subgroup of $\mathfrak{k}_0$ generated by v_4 and v_5, where we have considered $\mathfrak{k}_0$ as a subgroup of $\boldsymbol{C}^3 = \mathfrak{g}$. Take an element $v \in \mathfrak{k}_0 \cap \Gamma$, then it can be written as

$$v = \sum_{i=1}^{5} n_i v_i = z_1 u_1 + z_2 u_2,$$

where $n_i \in \boldsymbol{Z}$ and $z_i \in \boldsymbol{C}$, $\boldsymbol{Z}$ being the additive group of integers. Putting $z_i = x_i + \sqrt{-1}\, y_i$, x_i, y_i being real for $i = 1, 2$, we obtain the following equalities:

$$\begin{cases} a_1 n_2 = a_2 n_1 + a_1 a_4 x_2 \\ a_1 n_3 = a_3 n_1 + a_1 a_5 x_2 \end{cases}$$

Now by the assumption for a_i $(i = 1, 2, \ldots, 5)$ we can see that $n_1 = n_2 = n_3 = 0$. Hence $v = n_4 v_4 + n_5 v_5$, which proves that $\mathfrak{k}_0 \cap \Gamma$ is the discrete subgroup generated by v_4 and v_5. Then, since $\mathfrak{k}_0 = \boldsymbol{C}v_4 + \boldsymbol{C}v_5$ we see that $K_0 \cong \boldsymbol{C}v_4/\boldsymbol{Z}v_4 + \boldsymbol{C}v_5/\boldsymbol{Z}v_5 \cong \boldsymbol{C}^{*2}$.

Next we prove that G contains no complex torus of positive dimension. Suppose G contains a complex torus T, the Lie algebra of T being denoted by $\mathfrak{t}$. Then T is contained in K. Hence $\mathfrak{t} \subset \mathfrak{k}$, which implies $\mathfrak{t} \subset \mathfrak{k} \cap \sqrt{-1}\,\mathfrak{k} = \mathfrak{k}_0$, whence $T \subset K_0$. However, $K_0 \cong \boldsymbol{C}^{*2}$ being a Stein group, K_0 contains no complex torus of positive dimension. Hence $T = \{e\}$, which proves our assertion.

Finally we shall prove that G is an (H. C)-group. We first claim that K_0 is not closed in G. For, if K_0 is closed in G, K_0 must be compact since $K_0 \subset K$. Then K_0 is a complex torus, which contradicts the fact $K_0 \cong \boldsymbol{C}^{*2}$. As G^0 is closed in G, we see $\mathfrak{k}_0 \neq \mathfrak{g}^0$. Now by virtue of Lemma 7, $\mathfrak{k}_0 \subset \mathfrak{g}^0$. Since $\dim_{\boldsymbol{C}} \mathfrak{k}_0 = 2$, we obtain $\dim_{\boldsymbol{C}} \mathfrak{g}^0 = 3$, i. e. $\mathfrak{g}^0 = \mathfrak{g}$, which shows that $G^0 = G$. Thus the proof of Theorem 3 is completed.

Remark. We can also construct a two-dimensional (H. C)-group containing no complex torus as follows. Let $v_i \in \boldsymbol{C}^2$ $i = 1, 2, 3$ be the elements defined by $v_1 = (1, 0)$, $v_2 = (0, 1)$, $v_3 = (\sqrt{-1}, \sqrt{-1}\, a)$, where a is an irrational number and let Γ' be the discrete subgroup of $\boldsymbol{C}^2$ generated by v_1, v_2 and v_3. Put $G' = C^2/\Gamma'$. Then G' has the required properties. Since the proof is similar to the one of Theorem 3, we shall not repeat the proof.

Theorem 4. *For any integer $n \geqq 2$, there exists an* (H. C)*-group of dimension n containing as a subgroup no complex torus of positive dimension.*

Proof. We first remark that if G_1, G_2 are complex Lie groups containing no complex torus then $G_1 \times G_2$ has the same property. For, by the natural projection Φ_1 (Φ_2 resp.) of $G_1 \times G_2$ onto G_1 (G_2 resp.) a complex torus T of $G_1 \times G_2$ is mapped onto a complex torus $\Phi_1(T)$ ($\Phi_2(T)$ resp.) of G_1 (G_2 resp.). Then $\Phi_1(T) = \{e\}$, $\Phi_2(T) = \{e\}$ imply $T = \{e\}$. Now, taking G and G' as in Theorem 3 and its Remark, $G \times G' \times \cdots \times G'$ for odd n and $G' \times \cdots \times G'$ for even n have the required properties as Lemma 9 shows. Thus Theorem 4 is proved.

§ 4. Non-compact (H. C)-groups of dimension 2

In this section we shall characterize two-dimensional non-compact (H. C)-groups. Let G be a connected complex abelian Lie group of dimension n. Then G is (holomorphically) isomorphic to the factor group $\boldsymbol{C}^n/\Gamma$ of $\boldsymbol{C}^n$ by a discrete subgroup Γ of $\boldsymbol{C}^n$. Then it is known that Γ is generated by a finite number of vectors $v_1, v_2, \ldots, v_r$ of $\boldsymbol{C}^n$ which are linearly independent over the field of real numbers $\boldsymbol{R}$. We shall call r the rank of G (or Γ) and denote it $r = r(G)$.

Proposition 1. *Let G be a connected complex abelian Lie group. If $r(G) \leqq 2$, then G is holomorphically convex.*

Proof. If $n = \dim G = 1$, G is compact or a Stein group and so G is holomorphically convex. Suppose $n \geqq 2$. Let $G = \boldsymbol{C}^n/\Gamma$ and let $r(G) = 1$. Then Γ is generated by some element v_1 of $\boldsymbol{C}^n$. Let V be the complex subspace of $\boldsymbol{C}^n$ spanned by v_1 and let W be a complex subspace of $\boldsymbol{C}^n$ such that

$$\boldsymbol{C}^n = V + W \qquad \text{(direct sum)}.$$

Since Γ is contained in V, $G \cong V/\Gamma \times W \cong \boldsymbol{C}^* \times \boldsymbol{C}^{n-1}$, whence G is a Stein group. Let now $r(G) = 2$. Then Γ is generated by two elements v_1 and v_2 which are linearly independent over $\boldsymbol{R}$. First, consider the case v_1 and v_2 linearly independent over $\boldsymbol{C}$. Let V_1 and V_2 be the complex subspace of $\boldsymbol{C}^n$ spanned by v_1 and v_2 respectively, and let W be a complex subspace of $\boldsymbol{C}^n$ such that

$$\boldsymbol{C}^n = V_1 + V_2 + W \qquad \text{(direct sum)}.$$

Let Γ_1 and Γ_2 be the discrete subgroups of Γ generated by v_1 and v_2 respectively. Since $\Gamma = \Gamma_1 \times \Gamma_2$, $G \cong V_1/\Gamma_1 \times V_2/\Gamma_2 \times W \cong C^{*2} \times C^{n-2}$. Hence G is a Stein group. Next consider the case v_1 and v_2 linearly dependent over C. Then the complex subspace V spanned by v_1 and v_2 is of dimension 1. Let again W be a complex subspace of C^n such that

$$C^n = V + W \quad \text{(direct sum)}.$$

Then $G \cong V/\Gamma \times W \cong T \times C^{n-1}$, where T is a complex torus. Hence G is holomorphically convex. Thus the proposition is proved.

Lemma 10. *Let v_1, v_2, v_3 be three elements of C^n which are linearly independent over R. Then there exists a complex linear isomorphism Φ of C^n onto itself such that $\Phi(v_{i_1}) = e_1$, $\Phi(v_{i_2}) = e_2$, where $1 \leqq i_1 < i_2 \leqq 3$ and $e_1 = (1, 0, \ldots, 0)$, $e_2 = (0, 1, 0, \ldots, 0)$.*

Proof. Let V be the complex subspace of C^n spanned by v_1, v_2, v_3. Since v_1, v_2, v_3 are linearly independent over R, $\dim_C V = \frac{1}{2} \dim_R V \geqq \frac{3}{2}$, and so $\dim_C V \geqq 2$. Hence v_{i_1}, v_{i_2} are linearly independent over C for some i_1, i_2 $(1 \leqq i_1 < i_2 \leqq 3)$. Then the existence of the required isomorphism Φ is clear.

Corollary. *Let G be a connected complex abelian Lie group of rank 3 and of dimension n. Then G is isomorphic to C^n/Γ, where Γ is a discrete subgroup of C^n generated by e_1, e_2 and an element v.*

Proof. Let $G = C^n/\Gamma_0$, where Γ_0 is generated by v_1, v_2, v_3 which are linearly independent over R. Then by Lemma 10 we can suppose that there exists an isomorphism Φ of C^n such that $\Phi(v_1) = e_1$, $\Phi(v_2) = e_2$. Let Γ be the discrete subgroup of C^n generated by e_1, e_2 and $v = \Phi(v_3)$. Then clearly Φ induces an isomorphism of $G = C^n/\Gamma_0$ onto C^n/Γ, which proves the Corollary.

Definition 1. *Take an element $v \in C^2 - R^2$, then e_1, e_2 and v are linearly independent over R, where $e_1 = (1, 0)$, $e_2 = (0, 1)$. Let $\Gamma(v)$ be the discrete subgroup of C^2 generated by e_1, e_2 and v. We shall denote by $G(v)$ the factor group $C^2/\Gamma(v)$.*

Proposition 2. *Let $G = G(v)$, and $v = (\alpha, \beta)$, $\alpha = a_1 + \sqrt{-1}\, a_2$, $\beta = b_1 + \sqrt{-1}\, b_2$, a_1, a_2, b_1, b_2 being real numbers. Then G admits a non-constant holomorphic function if and only if v is one of the following types (A_1), (A_2) and (B):*

(A_1) $a_2 \neq 0, \quad b_2 = 0, \quad b_1 \in Q,$
(A_2) $a_2 = 0, \quad b_2 \neq 0, \quad a_1 \in Q,$
(B) $a_2 \cdot b_2 \neq 0, \quad b_2/a_2 \in Q, \quad (a_1 b_2 - a_2 b_1)/a_2 \in Q,$

where Q denotes the field of rational numbers.

Proof. Let $\mathfrak{k}$ be the real subspace of $\boldsymbol{C}^2$ spanned by e_1, e_2 and v. Let $v_0 = (a_2, b_2)$. We shall first show

$$\mathfrak{k} \cap \sqrt{-1}\,\mathfrak{k} = \boldsymbol{C} \cdot v_0 .$$

In fact, $v - a_1 e_1 - a_2 e_2 = \sqrt{-1}\, v_0$, and so $v_0 \in \sqrt{-1}\,\mathfrak{k}$. On the other hand, $v_0 = a_2 e_1 + b_2 e_2 \in \mathfrak{k}$, and hence $v_0 \in \mathfrak{k} \cap \sqrt{-1}\,\mathfrak{k}$. Now, if $v_0 = 0$, we see that e_1, e_2 and v are linearly dependent over $\boldsymbol{R}$. Hence $v_0 \neq 0$ and $\mathfrak{k} \cap \sqrt{-1}\,\mathfrak{k} = \boldsymbol{C} \cdot v_0$, since $\dim_{\boldsymbol{C}}(\mathfrak{k} \cap \sqrt{-1}\,\mathfrak{k}) = 1$. Now we denote by $\Phi : \boldsymbol{C}^2 \to G$ the natural homomorphism.

We shall show that G admits a non-constant holomorphic function if and only if $\Phi(\mathfrak{k}_0)$ is closed in G, where $\mathfrak{k}_0 = \mathfrak{k} \cap \sqrt{-1}\,\mathfrak{k}$. In fact, identifying $\boldsymbol{C}^2$ with the Lie algebra $\mathfrak{g}$ of G, $\Phi(\mathfrak{k}_0)$ is the subgroup K_0 of G corresponding to the subalgebra $\mathfrak{k}_0$ of $\mathfrak{g}$. If $\Phi(\mathfrak{k}_0)$ is closed in G, $\Phi(\mathfrak{k}_0)$ is compact and so G is holomorphically convex by Theorem 2, and hence G admits a non-constant holomorphic function. If $\Phi(\mathfrak{k}_0)$ is not closed, $G^0 = G$ since G^0 is a closed complex subgroup containing $K_0 = \Phi(\mathfrak{k}_0)$ (cf. Lemma 7), where G^0 is the subgroup in Theorem 1. Hence G is an (H. C)-group, which proves our assertion.

Since $\Phi(\mathfrak{k}_0) \cong \mathfrak{k}_0 / \mathfrak{k}_0 \cap \Gamma$, $\Phi(\mathfrak{k}_0)$ is closed (and hence compact) if and only if $\mathfrak{k}_0 \cap \Gamma$ contains two vectors linearly independent over $\boldsymbol{R}$. In order to prove the proposition it is now sufficient to show that $\mathfrak{k}_0 \cap \Gamma$ contains two vectors linearly independent over $\boldsymbol{R}$ if and only if v satisfies one of the three conditions (A_1), (A_2) and (B).

First, suppose $b_2 = 0$, then $a_2 \neq 0$. Let $z \cdot v_0$ be an element of Γ for $z \in \boldsymbol{C}$. Then $z \cdot v_0 = l\, e_1 + m e_2 + n v$ for some $l, m, n \in \boldsymbol{Z}$, $\boldsymbol{Z}$ being the additive group of integers. Hence the following two equalities hold:

$$\begin{cases} l + n(a_1 + \sqrt{-1}\, a_2) = z \cdot a_2 , \\ m + n b_1 \qquad\qquad\quad = 0 . \end{cases}$$

From this, it follows

$$\begin{cases} l + n a_1 = x a_2, \\ n a_2 = y a_2, \\ m + n n_1 = 0 , \end{cases} \tag{1}$$

for $z = x + \sqrt{-1}\, y$, x, y being real. Now the third equality of (1) implies $b_1 \in \boldsymbol{Q}$ since $\mathfrak{k}_0 \cap \Gamma$ contains two vectors which are linearly independent over $\boldsymbol{R}$. Conversely, suppose $b_2 = 0$, $b_1 \in \boldsymbol{Q}$. Then we can find two sets of integers (l_1, m_1, n_1) and (l_2, m_2, n_2) such that $l_1 n_2 - l_2 n_1 \neq 0$, $m_1 + b_1 n_1 = 0$ and $m_2 + b_1 n_2 = 0$. Then, putting $x_i = (l_i + n_i a_1)/a_2$, $y_i = n_i$ and $z_i = x_i + \sqrt{-1}\, y_i$ $(i = 1, 2)$, we see that $z_1 \cdot v_0$, $z_2 \cdot v_0 \in \mathfrak{k}_0 \cap \Gamma$ and that $z_1 \cdot v_0$ and $z_2 \cdot v_0$ are linearly independent over $\boldsymbol{R}$. Hence we have seen that if $b_2 = 0$, $\Phi(\mathfrak{k}_0)$ is closed if and only if $b_1 \in \boldsymbol{Q}$.

In the same way as above, we see that if $a_2 = 0$, $\Phi(\mathfrak{k}_0)$ is closed if and only if $a_1 \in \boldsymbol{Q}$.

Next we consider the case $a_2 \cdot b_2 \neq 0$. Let $z \cdot v_0$ be an element of Γ. Then for some $l, m, n \in \boldsymbol{Z}$ we have:

$$l + n(a_1 + \sqrt{-1}\, a_2) = z \cdot a_2 \quad \text{and} \quad m + n(b_1 + \sqrt{-1}\, b_2) = z \cdot b_2 \,,$$

hence

$$\begin{cases} x = (l + n a_1)/a_2 = (m + n b_1)/b_2 \\ y = n \end{cases} \tag{2}$$

holds for $z = x + \sqrt{-1}\, y$, x, y being real. Suppose $\Phi(\mathfrak{k}_0)$ is closed in G, then $\Gamma \cap \mathfrak{k}_0$ contains two vectors $z_1 \cdot v_0, z_2 \cdot v_0$, linearly independent over $\boldsymbol{R}$. Put $z_i = x_i + \sqrt{-1}\, y_i$ $(i = 1, 2)$, x_i and y_i being real. Then from (2) it follows that

$$\begin{aligned} (l_1 + n_1 a_1) b_2 &= a_2 (m_1 + n_1 b_1) \,, \\ (l_2 + n_2 a_1) b_2 &= a_2 (m_2 + n_2 b_1) \,, \end{aligned} \tag{3}$$

for some integers l_i, m_i and n_i $(i = 1, 2)$ such that $l_1 n_2 - l_2 n_1 \neq 0$. It follows from (3) that $b_2/a_2 \in \boldsymbol{Q}$ and $(a_1 b_2 - a_2 b_1)/a_2 \in \boldsymbol{Q}$.

Conversely, suppose that $a_2 \cdot b_2 \neq 0$, $b_2/a_2 \in \boldsymbol{Q}$ and $(a_1 b_2 - a_2 b_1)/a_2 \in \boldsymbol{Q}$. Put $b_2/a_2 = q/p$, $(a_1 b_2 - a_2 b_1)/a_2 = s/r$, where $p, q, r, s \in \boldsymbol{Z}$. Now let $l_1 = 2p$, $l_2 = p$, $n_1 = n_2 = r$, we define m_1, m_2 by the equalities (3), and we define x_i, y_i $(i = 1, 2)$ by (2) for $(l, m, n) = (l_i, m_i, n_i)$. Then, putting $z_i = x_i + \sqrt{-1}\, y_i$, we see that $z_1 \cdot v_0$ and $z_2 \cdot v_0 \in \Gamma \cap \mathfrak{k}_0$, and these two vectors are linearly independent over $\boldsymbol{R}$. Hence we have shown that if $a_2 \cdot b_2 \neq 0$, $\Phi(\mathfrak{k}_0)$ is closed in G if and only if (B) holds. Thus Proposition 2 is proved.

Corollary. *Let $G = C^2/\Gamma(v)$ and $v = (\alpha, \beta)$. Then G admits a non-constant holomorphic function if and only if $1, \alpha, \beta$ are linearly dependent over $\boldsymbol{Q}$.*

In fact, it is easy to see that $1, \alpha, \beta$ are linearly dependent over $\boldsymbol{Q}$ if and only if v is one of the three types (A_1) (A_2) and (B) in Proposition 2.

Theorem 5. *Let G be a non-compact* (H. C)*-group of complex dimension 2. Then G is (holomorphically) isomorphic to $\boldsymbol{C}^2/\Gamma(v)$ for some $v = (\alpha, \beta) \in \boldsymbol{C}^2$ such that $1, \alpha, \beta$ are linearly independent over $\boldsymbol{Q}$. (The converse is also true.)*

Proof. Since G is a connected abelian Lie group by the Remark following Theorem 1, G is isomorphic to $\boldsymbol{C}^2/\Gamma_1$ for some discrete subgroup Γ_1 of $\boldsymbol{C}^2$. Since G is non-compact $\mathrm{r}(G) \leq 3$. Then by Proposition 1 we have $\mathrm{r}(G) = 3$. Hence by Corollary of Lemma 10, G is isomorphic to $\boldsymbol{C}^2/\Gamma(v)$ for some $v \in \boldsymbol{C}^2$. Now, by the Corollary of Proposition 2, G is an (H. C)-

group if and only if 1, α, β are linearly independent over $\boldsymbol{Q}$. The converse is clearly true by the Corollary of Proposition 2.

§ 5. Classification of non-compact (H. C)-groups of dimension 2

We shall denote by $\mathscr{G} = (\boldsymbol{C}^2 - \boldsymbol{R}^2)'$ the set of all $v = (\alpha, \beta) \in \boldsymbol{C}^2 - \boldsymbol{R}^2$ such that 1, α, β are linearly independent over $\boldsymbol{Q}$. Let $GL(n, \boldsymbol{Z})$ be the group of all matrices of degree n with integer coefficients whose determinants are ± 1. Put $\triangle = GL(3, \boldsymbol{Z})$. For

$$M = \begin{pmatrix} l_1 & l_2 & l_3 \\ m_1 & m_2 & m_3 \\ n_1 & n_2 & n_3 \end{pmatrix} \in \triangle,$$

we define the mapping $\tilde{M} : \mathscr{G} \to \boldsymbol{C}^2$ as follows:

$$\begin{cases} \alpha' = (-l_1\alpha - l_2\beta + l_3)/(n_1\alpha + n_2\beta - n_3) \\ \beta' = (-m_1\alpha - m_2\beta + m_3)/(n_1\alpha + n_2\beta - n_3), \end{cases} \tag{4}$$

where $\tilde{M}(v) = (\alpha', \beta')$ and $v = (\alpha, \beta) \in \mathscr{G}$.

Definition 2. *Let $v, v' \in \boldsymbol{C}^2 - \boldsymbol{R}^2$. We say that v and v' are equivalent and write $v \sim v'$ if $G(v)$ is isomorphic to $G(v')$* (cf. Def. 1).

Lemma 11. *Let $v, v' \in (\boldsymbol{C}^2 - \boldsymbol{R}^2)'$. Then $v \sim v'$ if and only if there exists an element $M \in \triangle$ such that*

$$\tilde{M}(v') = v.$$

Proof. Suppose $v \sim v'$. Then there exists an isomorphism φ_1 of $\boldsymbol{C}^2/\Gamma(v)$ onto $\boldsymbol{C}^2/\Gamma(v')$. Then φ_1 yields an isomorphism φ of $\boldsymbol{C}^2$ onto $\boldsymbol{C}^2$ such that $\pi' \circ \varphi = \varphi_1 \circ \pi$, where π and π' denote the natural homomorphisms of $\boldsymbol{C}^2$ onto $\boldsymbol{C}^2/\Gamma(v)$ and $\boldsymbol{C}^2/\Gamma(v')$ respectively. φ can be considered as an element of $GL(2, \boldsymbol{C})$. Especially $\varphi(\Gamma(v)) = \Gamma(v')$. Since $\varphi(e_1)$, $\varphi(e_2)$ and $\varphi(v)$ generate the group $\Gamma(v')$ there exists a matrix $M = (m_{ij})$ of degree 3 with integer coefficients such that

$$\begin{aligned} e_1 &= \varphi(e_1)m_{11} + \varphi(e_2)m_{21} + \varphi(v)m_{31}, \\ e_2 &= \varphi(e_1)m_{12} + \varphi(e_2)m_{22} + \varphi(v)m_{32}, \\ v' &= \varphi(e_1)m_{13} + \varphi(e_2)m_{23} + \varphi(v)m_{33}. \end{aligned} \tag{5}$$

Since φ is an isomorphism we see that $\det M = \pm 1$. Between the matrices

$$(E\,v') = \begin{pmatrix} 1 & 0 & \alpha' \\ 0 & 1 & \beta' \end{pmatrix} \text{ and } (E\,v) = \begin{pmatrix} 1 & 0 & \alpha \\ 0 & 1 & \beta \end{pmatrix}$$

the following relation holds:

$$(E\,v') = \varphi \cdot (E\,v) \cdot M. \tag{6}$$

Let

$$A = \begin{pmatrix} m_{11} & m_{12} \\ m_{21} & m_{22} \end{pmatrix}, \; B = (m_{31}, m_{32}), \; C = \begin{pmatrix} m_{13} \\ m_{23} \end{pmatrix}$$

and $d = m_{33}$. Then

$$M = \begin{pmatrix} A & C \\ B & d \end{pmatrix},$$

and (6) can be written as follows:

$$(E\, v') = (\varphi(A + v \cdot B)\, \varphi(C + d \cdot v)),$$

and we have $E = \varphi(A + v \cdot B)$ and $v' = \varphi(C + d \cdot v)$, from which we obtain

$$(A + v \cdot B) \cdot v' = C + d \cdot v. \tag{7}$$

Now, (7) is equivalent to the following two equalities

$$\begin{cases} \alpha\,(m_{31}\,\alpha' + m_{32}\,\beta' - m_{33}) = m_{13} - m_{11}\,\alpha' - m_{12}\,\beta', \\ \beta\,(m_{31}\,\alpha' + m_{32}\,\beta' - m_{33}) = m_{23} - m_{21}\,\alpha' - m_{22}\,\beta', \end{cases} \tag{8}$$

which shows $v = \tilde{M}(v')$.

Conversely, suppose $\tilde{M}(v') = v$ for some $M \in \triangle$. Then, since (8) holds, we see that (7) and hence (6), (5) hold for $\varphi = A + v \cdot B$. We shall now see that $\det \varphi \neq 0$. In fact, if $\det (A + v \cdot B) = 0$, we get the following equality:

$$(m_{11} m_{22} - m_{21} m_{12}) + (m_{31} m_{22} - m_{21} m_{32})\alpha + (m_{11} m_{32} - m_{31} m_{12})\beta = 0.$$

Since $1, \alpha, \beta$ are linearly independent over $\boldsymbol{Q}$ we obtain $m_{11} : m_{12} = m_{21} : m_{22} = m_{31} : m_{32}$, which shows that $\det M = 0$. The equations (5) show that φ is an isomorphism of $\boldsymbol{C}^2$ onto itself such that $\varphi(\Gamma(v)) = \Gamma(v')$, and hence φ induces an isomorphism of $\boldsymbol{C}^2/\Gamma(v)$ onto $\boldsymbol{C}^2/\Gamma(v')$ and therefore $v \sim v'$. Thus Lemma 11 is proved.

Corollary. *Let* $v = (\sqrt{-1}, \sqrt{-1}\, a)$; $v' = (\sqrt{-1}, \sqrt{-1}\, a')$; $a, a' \notin \boldsymbol{Q}$. *Then* $v \sim v'$ *if and only if* $a \pm a' \in \boldsymbol{Z}$, $\boldsymbol{Z}$ *being the additive group of integers.*

Proof. Let $M \in \triangle$ and $v' = \tilde{M}(v)$. Then by a simple calculation we get the following equalities:

$$l_1 + l_2\, a = n_3, \tag{9}$$

$$-n_1 - n_2\, a = l_3, \tag{10}$$

$$n_1\, a' + n_2\, a\, a' = -m_3, \tag{11}$$

$$n_3\, a' = m_1 + m_2\, a, \tag{12}$$

where we have put

$$M = \begin{pmatrix} l_1 & l_2 & l_3 \\ m_1 & m_2 & m_3 \\ n_1 & n_2 & n_3 \end{pmatrix}.$$

Since a $\notin \boldsymbol{Q}$, by (9) and (10) we see that $l_2 = 0$, $l_1 = n_3$ and $n_2 = 0$, $-n_1 = l_3$. Then, by (17) we get $n_1 a' = -m_3$ and so $n_1 = m_3 = 0$ since $a' \notin \boldsymbol{Q}$. Then

$$M = \begin{pmatrix} l_1 & 0 & 0 \\ m_1 & m_2 & 0 \\ 0 & 0 & l_1 \end{pmatrix}.$$

Since det $M = \pm 1$ it follows that $l_1 = \pm 1$ and $m_2 = \pm 1$. Therefore by (12) we see that $a \pm a' \in \boldsymbol{Z}$.

Conversely, if $a \pm a' = m_1 \in \boldsymbol{Z}$, put

$$M = \begin{pmatrix} 1 & 0 & 0 \\ \pm m_1 & \mp 1 & 0 \\ 0 & 0 & 1 \end{pmatrix},$$

then we see that $v' = \tilde{M}(v)$ and so $v \sim v'$, which proves the Corollary.

Lemma 12. *Notations being as above we have*

(i) $\tilde{M}(\mathscr{G}) = \mathscr{G}$,

(ii) $\widetilde{M \cdot M'} = \tilde{M} \circ \tilde{M}'$

for all $M, M' \in \triangle$, *i. e.* $\triangle$ *operates on* $\mathscr{G}$ *by* $(M, v) \to \tilde{M}(v)$.

Proof. Take an element $v' \in \mathscr{G}$. Put $\tilde{M}(v') = v$. As in the proof of Lemma 11 we put $M = \begin{pmatrix} A & C \\ B & d \end{pmatrix}$ and $\varphi = A + v \cdot B$. Then det $\varphi \neq 0$, and $\varphi(\Gamma(v)) = \Gamma(v')$, which shows that φ induces an isomorphism of $C^2/\Gamma(v)$ onto $C^2/\Gamma(v')$, and hence $C^2/\Gamma(v)$ is an (H. C)-group and so $v \in \mathscr{G}$, which shows $\tilde{M}(\mathscr{G}) \subset \mathscr{G}$.

(ii) is proved by a direct calculation. Therefore we obtain $\tilde{M}(\mathscr{G}) = \mathscr{G}$, and thus Lemma 12 is proved.

Theorem 6. *Let* $\mathscr{G}_0$ *be the set of all non-compact* (H. C)-*groups of dimension* 2, *two isomorphic ones being of course identified. Then there exists a natural one-to-one correspondence between* $\mathscr{G}_0$ *and the set* $\mathscr{G}/\triangle$ *of equivalence classes of* $\mathscr{G}$ *by the operations of* $\triangle$. *Especially there exists non-countably many non-compact* (H. C)-*groups of dimension* 2 *which are not mutually isomorphic.*

Proof. Take an (H. C)-group G of $\mathscr{G}_0$. Then by Theorem 5, $G = C^2/\Gamma(v)$ for some $v \in \mathscr{G}$. Now, by Lemma 11, $C^2/\Gamma(v)$ is isomorphic to $C^2/\Gamma(v')$ if and only if $v' = \tilde{M}(v)$ for some $M \in \triangle$, which proves the theorem.

§ 6. Connected complex abelian Lie groups of dimension 2 and of rank 3 which are not (H. C)-groups

Let $G = G(v)$ as in Def. 1. Put $v = (\alpha, \beta)$, $\alpha = a_1 + \sqrt{-1}\, a_2$, $\beta = b_1 + \sqrt{-1}\, b_2$, where $a_1, a_2, b_1, b_2 \in \boldsymbol{R}$. Now, by the Corollary of Prop. 2, G is not an (H. C)-group if and only if the vector v is one of the types (A_1), (A_2) and (B). Suppose that v is of type (A_2). Let $\varphi : \boldsymbol{C}^2 \to \boldsymbol{C}^2$ be the automorphism of $\boldsymbol{C}^2$ defined by $\varphi(e_1) = e_2$, $\varphi(e_2) = e_1$, and put $\varphi(v) = v'$. Then it is clear that $G(v)$ is isomorphic to $G(v')$ and that v' is of type (A_1). Next suppose that v is of type (B), then we have the following

Lemma 13. *For any vector v of type (B) there exists a vector v' of type (A_1) such that $G(v) \cong G(v')$.*

Proof. Since α, β, 1 are linearly dependent over $\boldsymbol{Q}$, there exist $n_1, n_2 \in \boldsymbol{Z}$ and $r \in \boldsymbol{Q}$ such that $n_1\alpha + n_2\beta - r = 0$ and that n_1 and n_2 are coprime. Then there exist $p_1, p_2 \in \boldsymbol{Z}$ such that $p_1 n_1 - p_2 n_2 = 1$. Put $\alpha' = p_1\alpha + p_2\beta$, $\beta' = r$, $v' = (\alpha', \beta')$ and $\varphi = \begin{pmatrix} p_1 & p_2 \\ n_1 & n_2 \end{pmatrix} \in GL(2, \boldsymbol{C})$. Then we can easily see that φ is an automorphism of $\boldsymbol{C}^2$ such that $\varphi(\Gamma(v)) = \Gamma(v')$, which proves the lemma.

Lemma 14. *For any vector v of type (A_1) there exists a vector $v' = (\alpha', 0)$ such that $G(v) \cong G(v')$.*

Proof. Let $v = (\alpha, \beta)$, $\beta \in \boldsymbol{Q}$. We can suppose that $\beta = \frac{n}{m} \neq 0$, $m, n \in \boldsymbol{Z}$ and that m, n are coprime. Then we can find $p, q \in \boldsymbol{Z}$ such that $pm - qn = 1$. Put $\alpha' = m\alpha$, $v' = (\alpha', 0)$ and $\varphi = \begin{pmatrix} 1 & mq\alpha \\ 0 & m \end{pmatrix} \in GL(2, \boldsymbol{C})$. Then we see that φ is an automorphism of $\boldsymbol{C}^2$ such that $\varphi(\Gamma(v)) = \Gamma(v')$, which proves our lemma.

Lemma 15. *Let $v = (\alpha, 0)$, $v' = (\alpha', 0)$ such that $\alpha, \alpha' \in \boldsymbol{C} - \boldsymbol{R}$. Then $G(v)$ is isomorphic to $G(v')$ if and only if there exists a matrix*

$$A = \begin{pmatrix} l_1 & l_2 \\ m_1 & m_2 \end{pmatrix} \in GL(2, \boldsymbol{Z})$$

such that

$$\alpha' = \frac{-l_1\alpha + l_2}{m_1\alpha - m_2} \tag{13}$$

holds.

Proof. Suppose that $v \sim v'$. Then by the same argument as in the proof of Lemma 11 there exists a matrix $M = (m_{ij}) \in GL(3, \boldsymbol{Z})$ such that (8) holds, i. e.

$$\begin{cases} \alpha(m_{31}\alpha' - m_{33}) = m_{13} - m_{11}\alpha', \\ \qquad 0 \qquad\quad = m_{23} - m_{21}\alpha'. \end{cases}$$

Then we see that $m_{21}=0$, $m_{23}=0$ and hence $m_{22}=\pm 1$. Now, put

$$A=\begin{pmatrix} l_1 & l_2 \\ m_1 & m_2 \end{pmatrix}=\begin{pmatrix} m_{11} & m_{13} \\ m_{31} & m_{33} \end{pmatrix}^{-1}.$$

Then we see that $A\in GL(2,\boldsymbol{Z})$, and (13) holds. Conversely, if (13) holds, we consider the matrix

$$\varphi=\begin{pmatrix} l_1+l_2\alpha' & 0 \\ & 0 & 1 \end{pmatrix}.$$

Then we can readily see that φ is an automorphism of $\boldsymbol{C}^2$ such that $\varphi(\Gamma(v))=\Gamma(v')$. Thus the lemma is proved.

We can prove easily the following lemma.

Lemma 16. *The group $\triangle'=GL(2,\boldsymbol{Z})$ operates on $\boldsymbol{C}-\boldsymbol{R}$ by the action*

$$\alpha\to\alpha'=\frac{-l_1\alpha+l_2}{m_1\alpha-m_2}\quad for\quad \begin{pmatrix} l_1 & l_2 \\ m_1 & m_2 \end{pmatrix}\in\triangle'.$$

Now, combining Lemmas 13-16 we obtain the following

Theorem 7. *Let $\mathscr{G}_1$ be the set of all connected complex abelian Lie groups of dimension 2 and of rank 3, which are not* (H. C)*-groups (two isomorphic ones being of course identified). Then there exists a natural one-to-one correspondence between $\mathscr{G}_1$ and the set $\boldsymbol{C}-\boldsymbol{R}/\triangle'$ of equivalence classes of $\boldsymbol{C}-\boldsymbol{R}$ by the operations of $\triangle'=GL(2,\boldsymbol{Z})$.*

Summarizing the considerations in §§ 4-6, we obtain the following classification of all connected complex abelian Lie groups of dimension 2.

Theorem 8. *Let $\mathscr{A}$ be the set of all connected complex abelian Lie groups of dimension 2. Then $\mathscr{A}$ is classified as follows:*

$$\mathscr{A}=\mathscr{S}\cup\mathscr{S}'\cup\mathscr{G}_1\cup\mathscr{G}_0\cup\mathscr{T}\quad (disjoint\ sum).$$

$\mathscr{S}$ is the set of all Stein groups $\{\boldsymbol{C}^2, \boldsymbol{C}\times\boldsymbol{C}^*, \boldsymbol{C}^{*2}\}$, $\mathscr{S}'$ is the set of all groups of the form $\boldsymbol{C}\times T$, where T is a one-dimensional complex torus, $\mathscr{G}_1$ is the set of all groups of rank 3 which are not (H. C)*-groups, $\mathscr{G}_0$ is the set of all non-compact* (H. C)*-groups and $\mathscr{T}$ is the set of all complex tori of dimension 2. Moreover, $\mathscr{G}_1$ and $\mathscr{G}_0$ can be classified as follows*

$$\mathscr{G}_1=\boldsymbol{C}-\boldsymbol{R}/\triangle',\quad \mathscr{G}_0=(\boldsymbol{C}^2-\boldsymbol{R}^2)'/\triangle.$$

References

[1] Calabi, E., and B. Eckmann: A class of compact complex manifolds which are not algebraic. Ann. Math. 58, 494—500 (1953).

[2] Iwasawa, K.: On some types of topological groups. Ann. Math. 50, 507—558 (1949).

[3] Matsushima, Y., et A. Morimoto: Sur certains espaces fibrés holomorphes sur une variété de Stein. Bull. Soc. Math. France 88, 137—155 (1960).
[4] Stein, K.: Überlagerungen holomorph-vollständiger komplexer Räume. Arch. Math. 7, 354—361 (1956).

Nagoya University and
University of Minnesota
Minneapolis, Minn.

Uniform Algebras*

By

E. Bishop

Recently concepts from the theory of Banach algebras have been increasingly useful in applications to complex analysis. Here we give a quick survey of some, but by no means all, of these applications. The algebras which occur are algebras of complex valued functions and they are given the uniform norm with respect to some compact set K. Since the usual terms for these algebras (sup. norm algebras or function algebras) are not euphonious, we shall call them *uniform algebras*.

Definition 1. *A uniform algebra $\mathfrak{A}$ is a complex Banach algebra with unit, in which the norm equals the spectral norm.*

This condition means that $\lim\limits_{n\to\infty} \|f^n\|^{1/n} = \|f\|$, for all f in $\mathfrak{A}$. Here $\lim\limits_{n\to\infty} \|f^n\|^{1/n}$ is the spectral norm of f, so called because it equals

$$\sup\{|\varphi(f)| : \varphi \in \text{spectrum } \mathfrak{A}\},$$

where spectrum $\mathfrak{A}$ is defined to be the set of all non-trivial homomorphisms φ of $\mathfrak{A}$ into the complex numbers $\boldsymbol{C}$. Every such homomorphism φ is necessarily continuous. Following Gelfand, we topologize X = spectrum $\mathfrak{A}$ by the weak topology induced by $\mathfrak{A}$, so that for each f in $\mathfrak{A}$ the function $\varphi \to \varphi(f)$ on X is continuous. With this topology X is compact Hausdorff. For convenience we write x (instead of φ) for a generic element of X and write $f(x)$ for $x(f)$, so that we may consider $\mathfrak{A}$ as an algebra of continuous functions on X. It is then a sub-algebra of the uniformly-normed Banach algebra $C(X)$ of all continuous complex-valued functions on X.

The category of all uniform algebras seems to be the natural setting for many of those problems in complex analysis which concern a class of holomorphic functions (or holomorphic functions and their boundary values) known in advance to contain sufficiently many elements. The

* Received June 26, 1964.

presently available methods of the theory do not seem to be suited to problems which require the construction of holomorphic functions from geometrical information, for example the E. E. Levi problem.

The theory begins with the study of the Šilov boundary Γ of $\mathfrak{A}$, which is defined to be the smallest closed subset of X such that

$$\|f\| = \|f\|_\Gamma = \sup\{|f(x)| : x \in \Gamma\}$$

for all f in $\mathfrak{A}$. Šilov proved its existence. In case $\mathfrak{A}$ is separable, there exists a smaller boundary, the minimal boundary M, defined as the smallest subset (not necessarily closed) of X on which all functions in $\mathfrak{A}$ attain their maximum absolute values. It turns out to consist exactly of the peak points of X, i.e. those x in X for which there exists f in $\mathfrak{A}$ for which

$$|f(x)| > |f(y)|$$

for all y in $X - \{x\}$. The closure of M is Γ.

In case $\mathfrak{A} = C(X)$ it is easy to show that $\Gamma = X$ and, when M exists, $M = X$. It is an open problem of whether conversely $M = X$ implies $\mathfrak{A} = C(X)$. In fact there are no general solutions to the problem of finding general conditions which imply that $\mathfrak{A} = C(X)$, although Wermer [27] has recently gotten very pretty results in a special case.

The importance of the Šilov boundary lies in the fact that an element f of $\mathfrak{A}$ is completely determined by its values on Γ (i. e. by its boundary values). In fact we have the possibility of representing the value of a function in $\mathfrak{A}$ at a given point x of X as a boundary integral. More precisely:

Theorem. *To each x in X there exists at least one finite positive Baire measure μ_x on Γ such that*

$$(*) \quad f(x) = \int f\, d\mu_x$$

for all f in $\mathfrak{A}$. In case M exists μ_x can be taken to be a measure on M.

Such a measure μ_x is called a *representing measure* for the point x. Recently Gleason [11], Bungart [8], and others have proved in certain cases the possibility of taking the measure μ_x to be an appropriately well-behaved function of x, thereby obtaining a kernel representation of functions in $\mathfrak{A}$ in terms of their boundary values.

A strengthened form of the last theorem which is sometimes needed is the following.

Theorem. *Let φ be any bounded linear functional on a uniform algebra $\mathfrak{A}$. Then there exists a finite complex-valued Baire measure μ on Γ such that*

$$\|\mu\| = \|\varphi\| \quad \textit{and} \quad \varphi(f) = \int f\, d\mu$$

for all f in $\mathfrak{A}$. In case $\mathfrak{A}$ is separable, μ can be taken to be a measure on M.

Looking at one of the standard examples, the uniform algebra $\mathfrak{A}$ of all continuous functions on the disk

$$D = \{z : |z| \leqq 1\}$$

which are analytic on the interior of D, we see that $X = D$ and

$$M = \Gamma = \{z : |z| = 1\}.$$

In this case the representing measures are unique. The representing measure for the point 0 is $(2\pi)^{-1} d\theta$, where $d\theta$ is the Lebesgue measure on Γ. In this example the representing measures have an important additional property:

$$(\ast\ast) \quad \log |f(x)| \leqq \int \log |f| \, d\mu_x$$

for all x in $\mathfrak{A}$. This is just the classical inequality of Jensen.

With this example as motivation, for a general uniform algebra $\mathfrak{A}$ we call a measure μ_x on Γ such that $(\ast\ast)$ is satisfied for all f in $\mathfrak{A}$ a *Jensen measure* for the point x. Every Jensen measure can be shown to be a representing measure, but simple examples disprove the converse. This raises a natural problem, which is settled by the following theorem.

Theorem. *Every point x in a uniform algebra has a Jensen measure μ_x on the Šilov boundary.*

This was proved in somewhat weakened form by Arens and Singer [*1*], and generally in [*4*]. It is not known whether every point has a Jensen measure on the minimal boundary.

Other important measures on X are the *annihilating measures*, those finite complex Baire measures μ on X for which $\int f \, d\mu = 0$ for all f in $\mathfrak{A}$. The set $\mathfrak{A}^\perp$ of annihilating measures is the annihilator of $\mathfrak{A}$ in the dual of the Banach space $C(X)$. Equally important is the subset of $\mathfrak{A}^\perp$ consisting of those annihilating measures which lie on Γ. Many questions in the theory of uniform algebras can be transferred to the dual space $\mathfrak{A}^\perp$, i.e. reformulated as questions about annihilating measures. Take for example the question of approximation: when can every element of the uniform algebra $\mathfrak{A}$ be uniformly approximated by elements of a given sub-algebra $\mathfrak{B}$ of $\mathfrak{A}$? Elementary considerations show this is true if and only if $\mathfrak{A}^\perp = \mathfrak{B}^\perp$, so that the problem of approximation can be reformulated in terms of annihilating measures. An application of this method of looking at the theory of approximation has recently been given by Glicksberg and Wermer [*14*], who have a simple proof of the well-known result of Mergelyan which states that if K is a compact subset of the complex plane with connected complement then every continuous function on K which is analytic interior to K is a uniform limit on K of entire analytic functions. Mergelyan's theorem has been extended ([*2*] and [*3*]) to the case in which K is replaced by an arbitrary compact subset of a Riemann surface R and the entire functions are replaced by an arbitrary algebra $\mathfrak{B}$ of holomorphic functions on R, but the description of all functions on K which are uniform limits of functions in $\mathfrak{B}$, while complete, is no longer quite so simple.

Another important group of applications of the theory of uniform algebras to one complex variable goes back to a paper of RUDIN [20], who showed that the maximum principle, which is a consequence of analyticity, in certain one-dimensional situations implies the existence of analytic structure.

Theorem. *Let $D = \{z : |z| \leqq 1\}$, and let $\mathfrak{A}$ be a closed subalgebra of $C(D)$ which contains the function z and satisfies the maximum principle in the sense that for every closed subset K of D and every f in $\mathfrak{A}$*

$$\|f\|_K = \|f\|_{\text{bdry} K}.$$

Then every function in $\mathfrak{A}$ is analytic on the interior of D.

WERMER [24] gives the following generalization of this important result.

Theorem. *Let $E = \{z : |z| = 1\}$, and let $\mathfrak{A}$ be a uniformly closed subalgebra of $C(E)$ containing the function z. Then either $\mathfrak{A} = C(E)$ or every f in $\mathfrak{A}$ has an analytic extension to D.*

It is not hard to get RUDIN's result from WERMER's. These theorems are very useful in showing that functions are analytic. For example, WERMER's result has been recently used [13] to give a very short proof of the famous result of RADÓ to the effect that a continuous function on an open set $U \subset \boldsymbol{C}$ which is analytic where it doesn't vanish is everywhere analytic.

These theorems of RUDIN and WERMER have been subjected to heavy generalization. Here is an example.

Theorem. *Let K be a compact Hausdorff space and L a countable closed subset of K such that $K - L$ has the structure of a one-dimensional differentiable manifold. Let $\mathfrak{A}$ be a uniformly closed subalgebra of $C(K)$ which is generated by functions which are differentiable on $K - L$. Let X be the spectrum of $\mathfrak{A}$ and X_0 the image of K in X. Then (essentially) $X - X_0$ can be given the structure of a Riemann surface in such a way that the functions in $\mathfrak{A}$ are analytic on $X - X_0$.*

This theorem, which is essentially due to WERMER but is proved in this generality in [3], can be paraphrased by saying that uniform algebras of differentiable functions on one-dimensional sets are essentially boundary value algebras of analytic functions. It is not known whether differentiability is necessary.

There is another technique which can be used to locate one-dimensional analytic structure in the spectrum X of a uniform algebra $\mathfrak{A}$. To explain it we need the notion of a *part* of $\mathfrak{A}$, due to GLEASON [10].

Definition. *Two points x and y in X are in the same part of $\mathfrak{A}$ if*

$$\varrho(x, y) = \sup\{|f(x) - f(y)| : f \in \mathfrak{A}, \quad \|f\| \leqq 1\} < 2.$$

Since in any case $\varrho(x, y) \leqq 2$, the condition $\varrho(x, y) < 2$ says that the value of f at x is strongly influenced by its value at y. Gleason showed that the relation $\varrho(x, y) < 2$ is an equivalence relation. The equivalence classes are called parts, or Gleason parts. The following useful equivalences are to be found in [*10*] and [*6*].

Theorem. *The following are equivalent conditions for points x and y in the spectrum X of a uniform algebra $\mathfrak{A}$:*

(a) *x and y belong to the same part.*

(b) *There exists a constant θ, $0 < \theta < 1$, such that $|f(x)| \leqq \theta|f(y)| + (1-\theta)\|f\|$, for all f in $\mathfrak{A}$.*

(c) *There exist representing measures μ_x for x and μ_y for y which are mutually absolutely continuous with bounded Radon-Nikodym derivatives.*

(d) *There exist representing measures μ_x for x and μ_y for y which are not mutually singular.*

It was conjectured by Gleason that the parts always have some sort of analytic structure, but this has since been shown wrong. However the feeling persists that this should be true in most reasonable cases. In case $\mathfrak{A}$ is a Dirichlet algebra, which means that the real parts of the functions in $\mathfrak{A}$ are dense in the space of all continuous real-valued functions on the Šilov boundary Γ, Wermer has proved the following

Theorem. *If $\mathfrak{A}$ is a Dirichlet algebra, every part of $\mathfrak{A}$ is either a point or an analytic disk.*

Here an analytic disk is a subset of X which is homomorphic to $\{z : |z| < 1\}$ in such a way that functions in $\mathfrak{A}$ give rise to analytic functions on $\{z : |z| < 1\}$. The proof of this theorem is based on techniques of Helson and Lowdenslager [*16*]. Hoffman [*17*] has extended the theorem to apply to certain more general algebras, which he calls log-modular. This extension is of interest because it applies to the weird and fascinating algebra H_∞ of all bounded analytic functions on $\{z : |z| < 1\}$, which is logmodular but not Dirichlet. The book [*18*] of Hoffman contains information about H_∞. Recently Carleson [*9*] proved the remarkable result that $\{z : |z| < 1\}$ is dense in the maximal ideal space of H_∞. This long-standing problem is known as the corona conjecture, and it is equivalent to the statement that if $f_1, \ldots, f_n \in H_\infty$, $\Sigma|f_i| > r > 0$, then there exists $g_1, \ldots, g_n$ in H_∞ with $\Sigma g_i f_i = 1$. Unfortunately Carleson's methods, which are classical and highly technical, do not give any insight into the corona conjecture for other algebras, for instance the algebra of all bounded analytic functions on the n-fold cartesian product of $\{z:|z| < 1\}$.

It is not known how to give an abstract characterization of those uniform algebras $\mathfrak{A}$ whose parts all have some one-dimensional analytic structure. Presumably some extension of the notion of a Dirichlet algebra is in order. Further information about applications of uniform algebras to one complex variable can be found in the book of WERMER [25].

In a sense the theory of uniform algebras is a branch of the theory of several complex variables. To see how this comes about, call a uniform algebra $\mathfrak{A}$ *finitely generated* if there are elements $f_1, \ldots, f_n$ in $\mathfrak{A}$ such that the smallest closed subalgebra of $\mathfrak{A}$ containing $f_1, \ldots, f_n$ is $\mathfrak{A}$ itself. Associated with generators $f_1, \ldots, f_n$ of $\mathfrak{A}$ is a map $F: X \to C^n$ defined by

$$F(x) = (f_1(x), \ldots, f_n(x)).$$

This map F is a homeomorphism, and the image $K = F(X)$ is *polynomially convex*. This means that the polynomial convexification

$$\tilde{K} = \{z = (z_1, \ldots, z_n) \in \mathbf{C}^n : |f(z)| \leq \sup\{|f(w)| : w \in K\} \text{ for all polynomials } f\}$$

of K is K itself. The algebra $\mathfrak{A}$ is isomorphic to the algebra $\mathscr{P}_K$ obtained by taking the uniform closure in $C(K)$ of the set of all polynomial functions of $z_1, \ldots, z_n$. Conversely, if K is an arbitrary compact subset of $\mathbf{C}^n$ then the uniform algebra $\mathscr{P}_K$ is generated by the coordinate functions $z_1, \ldots, z_n$, and the spectrum of $\mathscr{P}_K$ can be identified with $\tilde{K}$. Thus the study of finitely generated uniform algebras is equivalent to the study of polynomially convex subsets of $\mathbf{C}^n$. In fact however the study of an arbitrary uniform algebra can often be reduced to the finitely generated case by studying appropriate finitely generated subalgebras. ŠILOV first employed this technique to construct analytic functions of finitely many Banach algebra elements. His results were completed by WAELBROECK [23] and ARENS-CALDERÓN, giving the following theorem.

Theorem. *Let $g_1, \ldots, g_k$ be elements of a Banach algebra $\mathfrak{A}$, and let $G(X)$ be the image of the spectrum X of $\mathfrak{A}$ under the map G defined by $G(x) = (g_1(x), \ldots, g_k(x))$ of X into $\mathbf{C}^k$. Let H be a function analytic in some neighborhood of $G(X)$. Then there exists h in B such that $h(x) = H(G(x))$ for all x in X.*

The proof of this theorem makes non-trivial use of the theory of several complex variables, in particular of some integral formula of the type due to BERGMAN or WEIL.

Another result of fundamental importance which makes use of several complex variable theory applied to finitely generated subalgebras is the local maximum modulus principle, proved by Rossi [19] and then by Stolzenberg [22].

Theorem. *Let K be a closed subset of the spectrum X of the uniform algebra $\mathfrak{A}$. Then for every f in $\mathfrak{A}$*

$$\|f\|_K = \|f\|_{\text{bdry}\, K \cup (K \cap \Gamma)}.$$

This result reinforces the belief that the spectrum of a uniform algebra should have some sort of analytic structure on the complement $X - \Gamma$ of the Šilov boundary, but this belief was destroyed by Stolzenberg [21], who gives an example of a polynomially convex subset K in C^2 for which this is not true. In Stolzenberg's example K is the limit of analytic sets, and this suggests the possibility of finding some such limit analytic structure on $X - \Gamma$ in the general case.

One of the most interesting problems in uniform algebras is to find the analytic structure on $X - \Gamma$ when it exists. Gleason [12] gives the following result.

Theorem. *Let x be a point in the spectrum of a uniform algebra $\mathfrak{A}$. Let the maximal ideal I of all functions in $\mathfrak{A}$ vanishing at x be finitely generated, in the algebraic sense. Then there exists a neighbornhood U of x in X which can be given the structure of an analytic space in such a way that the functions in $\mathfrak{A}$ are analytic on U.*

In the cases of most interest it is not possible to verify that I is finitely generated, and other methods are needed. Take the case of an algebra $\mathfrak{A}$ of holomorphic functions on an n-dimensional complex manifold M. The natural topology for $\mathfrak{A}$ is the topology of uniform convergence on compact sets, i. e. the topology determined by the family $\{\| \;\|_K\}$ of norms, where K is a compact subset of M. Let Σ denote the spectrum of $\mathfrak{A}$, consisting of all continuous non-trivial homomorphisms of $\mathfrak{A}$ into C. Then $\Sigma = \cup_K \Sigma_K$, where Σ_K consists of all elements of Σ which are continuous with respect to $\| \;\|_K$. Thus to study the structure of Σ we may look at the structures of the Σ_K. Now Σ_K is the spectrum of the uniform algebra $\mathfrak{A}_K$ obtained by completing $\mathfrak{A}$ in the norm $\| \;\|_K$ (i. e. by taking the closure of $\mathfrak{A}$ in $C(K)$). In this general situation we have the following theorem [4].

Theorem. *Let $\mathfrak{A}$ be an algebra of analytic functions on an n-dimensional complex manifold M. Let K be a compact subset of M. Let $f_1, \ldots, f_n$ be elements of $\mathfrak{A}$. Let the map $F: \Sigma_K \to C^n$ be defined by*

$$F(x) = (f_1(x), \ldots, f_n(x)).$$

Then for almost all z in C^n the set $F^{-1}(z)$ is finite.

Thus Σ_K is not too large. This fact can be applied to give strengthened versions of the result of Oka that the envelope of holomorphy of any domain $M \subset C^n$ is a Stein manifold. The connection with envelopes of holomorphy comes from the fact that the envelope of holomorphy of a subset M of C^n can be identified with the spectrum Σ of the algebra $\mathfrak{A}$ of all holomorphic functions on M. Thus for the case of an arbitrary algebra $\mathfrak{A}$ of holomorphic functions on a complex manifold M it is natural to define the envelope of holomorphy of M relative to $\mathfrak{A}$ to be the spectrum Σ. Much remains to be clarified about the structure of Σ in the general case—only the cases in which M is a domain spread over C^n or in which M is one-dimensional have been thoroughly worked out. GRAUERT [*15*] has shown that the structure of Σ may indeed be very complicated. However there is reason to hope that every point of Σ belongs to a subset of Σ which can be given the structure of a one-dimensional analytic space on which the functions in $\mathfrak{A}$ are analytic.

The problem of analytic continuation falls naturally into this circle of ideas. If $\mathfrak{A}$ is an algebra of analytic functions on a complex manifold M, if K is a compact subset of M, and if $\| \ \|_1$ is some algebra semi-norm on K, one defines interpolated norms $\| \ \|_\theta$, by a process of geometric interpolation between $\| \ \|_K$ and $\| \ \|_1$, for each θ between 0 and 1. For each such θ there is a spectrum $\Sigma(\| \ \|_\theta)$ defined, and it is suggestive to think of the functions in $\mathfrak{A}$ as having been continued analytically to $\Sigma(\| \ \|_\theta)$. In fact a necessary and sufficient condition for one function element to be the analytic continuation of another can be given by means of such interpolated semi-norms.

Some recent applications [*5*] of the theory of uniform algebras involve the notion of capacity, which is the generalization to abstract complex analysis of the notion of logarithmic capacity in C^1.

Definition. *Let Y be a Baire subset of the spectrum X of a uniform algebra $\mathfrak{A}$. For each point x_0 in X we define the capacity of Y with respect to x_0 to be* sup $\{\mu(Y) : \mu$ is a Jensen measure for $x_0\}$. Using this notion we can give a general criterion for the continuation of an analytic set through possible singularities.

Theorem. *Let U be a bounded open set in C^n, B a closed subset of U, A an analytic subset of $U - B$ of pure dimension k such that $B \subset \bar{A}$. Let B be of capacity* 0 *for the algebra $\mathfrak{A}$ of all continuous functions on $\mathfrak{A}$ which are analytic on A, relative to every point x_0 of a dense subset of $\bar{A}$. Let there exist an analytic map π of U onto a connected open subset S of C^k which is proper on B, with $\pi(B) \neq S$. Then $\bar{A} \cap U$ is an analytic subset of U.*

This generalizes now classical results on the removability of singularities due to THIMM, REMMERT-STEIN, ROTHSTEIN, etc.

For many problems in complex analysis the category of uniform algebras seems too large, and it is worthwhile to examine the subcategory of differentiable uniform algebras. Such an algebra $\mathfrak{A}$ is a subalgebra of $C(M)$, where M is a compact differentiable manifold, possibly with boundary, which is generated by finitely many differentiable functions $f_1, \ldots, f_n$. These functions imbed M as a differentiable submanifold of $\boldsymbol{C}^n$. In addition to the topological problem of the nature of the imbedding of M in R^{2n}, there are equally deep and interesting problems concerning the relation of M to the complex structure of $\boldsymbol{C}^n$. Many of these pertain to the structure of the family of all simple closed curves γ in M which form the boundary of some Riemann surface S_γ in $\boldsymbol{C}^n$. For instance, we may ask when the polynomially convex hull of M is the union of the S_γ, or when there are no S_γ, or how the family of all S_γ can be paramaterized, or whether $\cup S_\gamma$ contains an open set. We would also like conditions under which $\mathfrak{A}$ (the closure in $C(M)$ of the polynomials) coincides with the set of all f in $C(M)$ which can be extended analytically to each S_γ. The only deep fact about a general compact differentiable manifold $M \subset \boldsymbol{C}^n$ is due to A. BROWDER [7], who shows that if dim $M \geqq n$ then M is not polynomially convex. This implies, for instance, that the algebra $C(S^2)$ of all continuous complex-valued functions on a two-sphere cannot be generated by two elements. It would be of great interest to find a constructive explanation of Browder's result, for instance to show that under his hypotheses there necessarily exist Riemann surfaces S_γ.

References

[1] ARENS, R., and I. SINGER: *Function values as boundary integrals.* Proc. Amer. Math. Soc. **5**, 735—745 (1954).

[2] BISHOP, E.: *Subalgebras of functions on a Riemann surface.* Pac. J. Math. **8**, 29—50 (1958).

[3] — *Analyticity in certain Banach algebras.* Trans. Amer. Math. Soc. **102**, 507—544 (1962).

[4] — *Holomorphic completions, analytic continuation, and the interpolation of semi-norms.* Ann. Math. **78**, 468—500 (1963).

[5] — *Conditions for the analyticity of certain sets.* Michigan J. Math. **11**, 289—304 (1964).

[6] — *Representing measures for points in a uniform algebra.* Bull. Amer. Math. Soc. **70**, 121—122 (1964).

[7] BROWDER, A.: *Cohomology of maximal ideal spaces.* Bull. Amer. Math. Soc. **76**, 515—516 (1961).

[8] BUNGART, L.: *Holomorphic functions with values in locally convex spaces and applications to integral formulas.* Trans. Amer. Math. Soc. **111**, 317—344 (1964).

[9] CARLESON, L.: *Interpolations by bounded analytic functions and the corona problem.* Ann. Math. **76**, 547—559 (1962).

[10] GLEASON, A.: *Function algebras.* Seminars on Analytic Functions Vol. II, Inst. for Advanced Study, Princeton N.J., 1957.

[11] — *The abstract theorem of Cauchy-Weil.* Pac. J. Math. **12**, 511—525 (1962).

[12] — *Finitely generated ideals in Banach algebras.* J. Math. Mech. **13**, 125—132 (1964).

[13] GLICKSBERG, I.: *Maximal algebras and a theorem of Radó.* (To appear.)

[14] —, and J. WERMER: *Measures orthogonal to a Dirichlet algebra.* Duke Math. J. **30**, 661—666 (1963).

[15] Grauert, H.: to appear.
[16] Helson, H., and D. Lowdenslager: *Prediction theory and Fourier series in several variables.* Acta Math. **99**, 165—202 (1958).
[17] Hoffman, K.: *Analytic functions and logmodular Banach algebras.* Acta Math. **108**, 271—317 (1962).
[18] — *Banach spaces of analytic functions.* Englewood Cliffs: Prentice Hall 1962.
[19] Rossi, H.: *The local maximum modulus principle.* Ann. Math. **72**, 1—11 (1960).
[20] Rudin, W.: *Analyticity and the maximum modulus principle.* Duke Math. J. **20**, 449—458 (1953).
[21] Stolzenberg, G.: *A hull with no analytic structure.* J. Math. Mech. **12**, 103—111 (1963).
[22] — *Polynomially and rationally convex sets.* Acta. Math. **109**, 259—289 (1963).
[23] Waelbroeck, L.: *Le calcul symbolique dans les algèbres commutatives.* J. Math. Pures Appl. **33**, 147—186 (1954).
[24] Wermer, J.: *On algebras of continuous functions.* Proc. Amer. Math. Soc. **4**, 866—869 (1953).
[25] — *Banach algebras of analytic functions.* New York: Academic Press 1961.
[26] — *Function rings and Riemann surfaces.* Ann. Math. **67**, 45—71 (1958).
[27] — *Polynomially convex disks.* (To appear.)

Department of Mathematics
University of California
Berkeley, Calif.

Construction of Kleinian Groups *,**

By

Bernard Maskit

The purpose of this paper is to investigate a special case of the following

Conjecture. *Let D be a plane domain, and let G be a group of conformal autohomeomorphisms of D. Then there exists a schlicht function f, mapping D onto a plane domain D', so that every element of $f \circ G \circ f^{-1}$ is a Möbius transformation.*

If D is simply connected, the above conjecture is an immediate corollary of the classical uniformization theorem. In this paper, we prove the conjecture (Theorem 12) for the case that G is fixed point free and discontinuous in D, and the factor space $D/G = S$ is a *finite* Riemann surface.

In this case, D is a regular covering surface of S, with cover group G, and of course D is a planar surface. One can characterize such coverings

* This research was supported by the National Science Foundation through grants NSF-GP779 and NSF-GP780.
** Received June 26, 1964.

by means of a finite set of simple disjoint loops on S (Theorem 2, this paper; see [9]). In chapter III, we give certain constructions of Kleinian groups, which, in some sense, correspond to the different kinds of loops. These constructions are then put together (Theorem 11) to prove the conjecture in the special case, except that the mapping f is not conformal, but merely topological. We then use the method of "variation of parameters" using quasi-conformal mappings, as developed by AHLFORS and BERS ([2], [3]) (Theorem 3, this paper) to replace the topological mapping by a conformal mapping.

The constructions in chapter III are special cases of Theorems 4 and 6 in chapter II about general discontinuous groups. In chapter II, we define the general notion of discontinuity, and prove the basic theorems. Theorem 4 is a generalization of the classical "combination theorem" of KLEIN [7] (see MACBEATH [8], for another generalization), to free products with amalgamation. There is also a partial converse to this theorem (Theorem 5) and an application to group theory (Corollaries 1 and 2) which gives a characterization of a free product with amalgamation in terms of coset representatives. Theorems 4 and 6 also include the constructions of Fuchsian groups given by NIELSEN-FENCHEL [4].

The author would like to express his gratitude to L. BERS for many informative conversations.

I. Preliminary remarks

1. Let G be a group of Möbius transformations on the Riemann sphere (extended plane) Σ; that is, every element of G is a transformation of the form

$$z \to (az + b)(cz + d)^{-1},$$

where a, b, c, and d are complex numbers with $ad - bc \neq 0$. Each such group G divides Σ into three mutually disjoint sets, defined as follows. A point z is called a *regular point* if there is a neighborhood U of z with $g(U) \cap U = \emptyset$, for all $g \in G$, $g \neq 1$. The set of all regular points is called the *set of regularity* and is denoted by $R_0(G)$.

A point z is called a *limit point* if there is a point $z' \in \Sigma$, and a sequence $\{g_n\}$ of distinct elements of G, with $g_n(z')$ converging to z. The set of all limit points is called the *limit set* and is denoted by $L_0(G)$.

A point z is called an *isolated fixed point* if the subgroup $H_z \subset G$, consisting of those elements of G which leave z fixed, is finite, and if there is a neighborhood U of z, with $g(U) = U$ for $g \in H_z$, and $g(U) \cap U = \emptyset$ for all $g \in G$, $g \notin H_z$. The set of all isolated fixed points is denoted by $F_0(G)$.

It is well known that the three sets $R_0(G)$, $L_0(G)$ and $F_0(G)$ are mutually disjoint, and that every point of Σ belongs to one of these sets.

One easily sees that these three sets are invariant under G, and that $R_0(G)$ is open.

If $R_0(G)$ is not empty, then G is called a *Kleinian group*. In this case, one can, in the usual manner, factor $R_0(G)$ by G. We call the factor space $\bar{S}$, and let $S_1, \ldots, S_n, \ldots$ be the connected components of $\bar{S}$. Each S_i then has a natural conformal structure induced by the projection $p: R_0(G) \to \bar{S}$. With this conformal structure, p is conformal, and each S_i is a Riemann surface.

Let $R_i = p^{-1}(S_i)$. G is called a *simple* Kleinian group if one of the R_i, which we call R, is connected. R is then a connected component of $R_0(G)$, and every element of G maps R onto itself. R is called the *principal region* for G, and $S = p(R)$ is called the *principal surface.*

In what follows we shall be concerned with Kleinian groups which are both *simple* and *finitely generated.* We call such a group an FSK-group, and we denote it by a triple of symbols: (G, R, S), where G is the group, R the principal region, and S the principal surface. (A Kleinian group might have more than one invariant component of the set of regularity, and we want to pick out *one* such component as the principal region.)

If (G, R, S) is an FSK-group, then we denote by L and F the subsets of $L_0(G)$ and $F_0(G)$, respectively, which are boundary points of R. (We remark that $L_0 = L$, but this will not be used in this paper.)

2. The most important notion, for the construction of FSK-groups, is that of a *fundamental domain* (FD). The definition given below is more restrictive (i. e. see property h) than any of the usual definitions. It is not known whether or not there exists an FD for every FSK-group. However it is known that every FSK-group which is a *Fuchsian group of the first kind possesses an* FD (see [6] pp. 285—320), and it is left to the reader to verify that every *elementary group* (i. e. an FSK-group G, for which $L(G)$ consists of 0, 1, or 2 points) possesses an FD.

Definition: A set D is an FD for the FSK-group (G, R, S) if the following properties hold:

(a) The boundary of D consists of a finite number of *open* arcs of circles or straight lines, called *sides* (a side may consist of a complete circle). The endpoints of these arcs are called *vertices*. Every vertex is an endpoint of precisely two sides.

(b) For each side c of the boundary of D, there is a side c', and a $g \in G$, with $g(c) = c'$. Either every point of c is in D, or every point of c' is in D.

(c) The elements of G, which map some side of D onto some other side of D, generate G.

(d) D is connected.

(e) $D \subset R$.

(f) No two points of D are equivalent under G.

(g) Every point of R is equivalent, under G, to some point of D.

(h) If x is a vertex of the boundary of D, and $x \in F$, then the two sides c and c' which meet at x are identified by an element of G (in fact, an element of H_x), and c and c' subtend an angle at x (measured inside D) of $2\pi/\alpha$, where α is the order of H_x. Furthermore, c and c', when extended to full circles, divide Σ into two or four regions; the interior of D lies in precisely one of these regions.

It should be pointed out that the properties listed above are not independent; for example, property (c) follows from the others (see FORD [5] p. 51). It was also shown by FORD ([5] pp. 44—49) that for every FSK-group, there is a set D, satisfying properties (a) through (g).

3. The following is a special case of a theorem of AHLFORS [1].

Theorem 1. *Let (G, R, S) be an FSK-group. Then S is a finite surface (i. e. S is conformally equivalent to a closed surface from which a finite number of points have been removed).*

If (G, R, S) is an FSK-group, then $p: R \to S$ is a regular covering of S, and of course R is a planar surface. It is well known that every regular covering of S can be uniquely described by a normal subgroup of $\pi_1(S)$.

One can also describe a normal subgroup of $\pi_1(S)$ by means of a set of loops on S. That is, if $w_1, w_2, \ldots$ are loops on S, then there is a well-defined smallest normal subgroup N of $\pi_1(S)$ which "contains" all of the loops $w_1, w_2, \ldots$ (Notation: $N = \langle w_1, w_2, \ldots \rangle$.)

The following theorem was proven in [9].

Theorem 2. *Let S be a finite orientable surface and let $w_1, \ldots, w_n$ be a finite set of simple disjoint loops on S. Let $a_1, \ldots, a_n$ be integers. Let $p: \tilde{S} \to S$ be the regular covering of S corresponding to $N = \langle w_1^{\alpha_1}, \ldots, w_n^{\alpha_n} \rangle$. Then $\tilde{S}$ is planar. Furthermore, if $p: \tilde{S} \to S$ is a regular covering of S, where $\tilde{S}$ is planar, then there exists a finite set $w_1, \ldots, w_n$ of simple disjoint loops on S, and there exist integers $\alpha_1, \ldots, \alpha_n$, so that $N = \langle w_1^{\alpha_1}, \ldots, w_n^{\alpha_n} \rangle$ is the normal subgroup of $\pi_1(S)$ corresponding to the given covering.*

Using the above theorems, we see that for every FSK-group (G, R, S), there is a set of simple disjoint loops $w_1, \ldots, w_n$, and a set of integers, $\alpha_1, \ldots, \alpha_n$, so that $p: R \to S$ is the covering of S corresponding to $\langle w_1^{\alpha_1}, \ldots, w_n^{\alpha_n} \rangle$. We then say that (G, R, S) is a *uniformization* of S with *signature* $\langle w_1^{\alpha_1}, \ldots, w_n^{\alpha_n} \rangle$. (Note that the signature is a *normal subgroup* of $\pi_1(S)$.)

If $h: S \to S'$ is a conformal homeomorphism then $h \circ p: R \to S'$ is of course a regular covering of S', corresponding to $\langle v_1^{\alpha_1}, \ldots, v_n^{\alpha_n} \rangle$ where $v_i = h(w_i)$. We then say that (G, R, S) is a *uniformization* of S' with *signature* $\langle v_1^{\alpha_1}, \ldots, v_n^{\alpha_n} \rangle$.

If, in the above, h is merely a homeomorphism (not necessarily conformal), then we say that (G, R, S) is a *topological* uniformization of S' with signature $\langle v_1^{\alpha_1}, \ldots, v_n^{\alpha_n}\rangle$.

4. The following is a simple application of known results in the theory of quasi-conformal mappings (see [2] and [3]).

Theorem 3. *Let the FSK-group (G, R, S) be a topological uniformization of S' with signature $\langle w_1^{\alpha_1}, \ldots, w_n^{\alpha_n}\rangle$. Then there exists an FSK-group (G_1, R_1, S_1) which is a uniformization of S' with signature $\langle w_1^{\alpha_1}, \ldots, w_n^{\alpha_n}\rangle$.*

Proof. By assumption, there is a homeomorphism $h: S \to S'$, where (G, R, S) is a uniformization of S with signature $\langle v_1^{\alpha_1}, \ldots, v_n^{\alpha_n}\rangle$, $v_i = h^{-1}(w_i)$, $i = 1, \ldots, n$. It was shown in [3] that there exists a *quasi-conformal* homeomorphism $f: S \to S'$ which is homotopic to h. Since f is homotopic to h, $h \circ p: R \to S'$, and $f \circ p: R \to S'$ are both described by the same signature. f induces a Beltrami differential $\mu \overline{dz}/dz$ on S. We lift μ to R, to get a function $\mu(z)$, defined in R, where μ is measurable and ess. sup. $|\mu(z)| \leq k < 1$. If $z \notin R$, we set $\mu(z) = 0$. We observe that $\mu(z)$, so defined, is again bounded and measurable, and $\mu(g(z))\overline{g'(z)} = \mu(z) g'(z)$, for all $g \in G$, and all $z \in \Sigma$, where $g' = dg/dz$.

Let $w^\mu(z)$ be the unique solution (see [2]) of the Beltrami equation $w_{\bar z} = \mu w_z$, $w(0) = 0$, $w(1) = 1$, $w(\infty) = \infty$, where $w(z)$ is a homeomorphism of Σ. Set $G^\mu = w^\mu \cdot G \cdot (w^\mu)^{-1}$, $R^\mu = w^\mu(R)$, and $S^\mu = R^\mu/G^\mu$. Then (G^μ, R^μ, S^μ) is again an FSK-group (see Bers [3]). Furthermore, w^μ induces a quasi-conformal homeomorphism $g: S \to S^\mu$. Set $u_i = g(v_i)$, $i = 1, \ldots, n$. Then (G^μ, R^μ, S^μ) is a uniformization of S^μ with signature $\langle u_1^{\alpha_1}, \ldots, u_n^{\alpha_n}\rangle$. Finally observe that $f \circ g^{-1}: S^\mu \to S'$ is a *conformal* homeomorphism and $f \circ g^{-1}(u_i)$ is freely homotopic to w_i, for $i = 1, \ldots, n$ so that (G^μ, R^μ, S^μ) is a uniformization of S' with signature $\langle w_1^{\alpha_1}, \ldots, w_n^{\alpha_n}\rangle$.

II. General discontinuous groups

5. Let X be any topological space, and let $\mathfrak{G}$ be some given group of autohomeomorphisms of X. (For example, $X = \Sigma$, and $\mathfrak{G}$ is the group of all Möbius transformations.)

If G_1 and G_2 are subgroups of $\mathfrak{G}$, then we denote by $\langle G_1, G_2\rangle$ the subgroup of $\mathfrak{G}$ generated by G_1 and G_2. Similarly, if $G \subset \mathfrak{G}$, and $f \in \mathfrak{G}$, $\langle G, f\rangle$ denotes the subgroup of $\mathfrak{G}$ generated by G and f.

A subgroup G of $\mathfrak{G}$ is called *discontinuous* if there is an open non-empty set $U \subset X$, so that $g(U) \cap U = \emptyset$, for all $g \in G$, $g \neq 1$. A point $x \in X$ is called a regular point (notation: $x \in R(G, X)$) if x has a neighborhood U with the above property. One easily sees that $R(G, X)$ is open and invariant under G.

A non-empty set $D \subset R(G, X)$ is called a *partial fundamental set* (PFS) for G if no two points of D are equivalent under G. If, in addition, every point of $R(G, X)$ is equivalent to some point of D, then D is called a *fundamental set* (FS).

6. Theorem 4 (Free product with amalgamation). *Let G_1 and G_2 be discontinuous subgroups of $\mathfrak{G}$, and let H be a common subgroup of G_1 and G_2. Let D_1 and D_2 be* PFS's *for G_1 and G_2, respectively, and let Δ be a* PFS *for H. For $i = 1, 2$, set $E_i = \bigcup_{h \in H} h(D_i)$. Assume that*

$$E_1 \cup E_2 \supset R(G_1, X) \cup R(G_2, X),$$

and $D' = \text{int}(D) \neq \emptyset$, where $D = E_1 \cap E_2 \cap \Delta$. Then $G = \langle G_1, G_2 \rangle$ is discontinuous, no two points of D are equivalent under G, D' is a PFS *for G, and G is the free product of G_1 and G_2 with amalgamated subgroup H.*

Proof. We first remark that since $D \subset \Delta$, if $h \in H$, $h \neq 1$, then

$$h(D) \cap D = \emptyset;$$

i.e. no two points of D are equivalent under H.

If $g \in G$, $g \notin H$, then g can be written, not uniquely, as

$$g = g_n \circ g_{n-1} \circ \cdots \circ g_2 \circ g_1,$$

where $g_{2i+1} \in G_1$, $g_{2i} \in G_2$, and for $i = 2, \ldots, n$, $g_i \notin H$. By interchanging G_1 and G_2, we can also assume that $g_1 \notin H$.

Proposition 1. *Let $x \in E_1$, $g_1 \in G_1$, $g_1 \notin H$. Then $g_1(x) \in E_2 - E_1$.*

Proof. $E_1 \subset R(G_1, X)$, which is invariant under G_1, so

$$g_1(x) \in R(G_1, X) \subset E_1 \cup E_2.$$

If $g_1(x)$ were in E_1, then there would be elements

$$h_1, h_2 \in H,$$

so that $h_1 \circ g_1(x) \in D_1$, and $h_2(x) \in D_1$. Since D_1 is a PFS for G_1, $h_2(x) = h_1 \circ g_1(x)$, or $h_2^{-1} \circ h_1 \circ g_1(x) = x$, and $x \in R(G_1, X)$. Hence $h_2^{-1} \circ h_1 \circ g_1 = 1$ and $g_1 \in H$, which is a contradiction. Therefore

$$g_1(x) \in E_2 - E_1.$$

Similarly, we prove

Proposition 2. *Let $x \in E_2$, $g_2 \in G_2$, $g_2 \notin H$. Then $g_2(x) \in E_1 - E_2$.*

Now let x be any point of D and $g = g_n \circ \cdots \circ g_1 \in G$, as above. Then, using Propositions 1 and 2, $g_1(x) \in E_2 - E_1$, $g_2 \circ g_1(x) \in E_1 - E_2$, and so on, so that $g(x)$ either lies in $E_1 - E_2$ or $E_2 - E_1$. In either case $g(x) \notin D$; i.e. for every $g \in G$, $g \notin H$, $g(D) \cap D = \emptyset$.

Since $D' \subset D$ is open and non-empty, and since $g(D) \cap D = \emptyset$ for all $g \in G$, $g \neq 1$, we know that G is discontinuous, no two points of D are equivalent under G, and D' is a PFS for G. (If we knew that $D \subset R(G, X)$, then we could say that D is a PFS for G, but this need not be true.)

Finally, if G were not the free product of G_1 and G_2 with amalgamated subgroup H, then there would be a word $g = g_n \circ g_{n-1} \circ \cdots \circ g_2 \circ g_1$, as above, with $g = 1$ in G. Then for $x \in D$, we would have $g(x) = x$, and we have shown that this cannot happen.

7. We digress from the main argument to point out that Theorem 4 has a partial converse, and in the next section, to give an application to group theory.

Theorem 5. *Let G be a discontinuous subgroup of $\mathfrak{G}$, and let G be the free product of G_1 and G_2 with amalgamated subgroup H. Then G_1 and G_2 have* PFS's *D_1 and D_2, and H has a* PFS *Δ, so that if we form*

$$E_i = \bigcup_{h \in H} h(D_i), \quad i = 1, 2,$$

then $D' = \operatorname{int}(E_1 \cap E_2 \cap \Delta) \neq \emptyset$, D' is a PFS *for G, and*

$$E_1 \cup E_2 \supset R(G, X).$$

Proof. Let D be an FS for G, where $\operatorname{int}(D) \neq \emptyset$. Let $\Delta = \mathrm{D}$. Let $1, a_\alpha, b_\beta$ be a complete set of right coset representatives for H in G, where each a_α is of the form $a_\alpha = g_2 \circ g_3 \circ \cdots \circ g_n$, $g_i \notin H$, $g_{2i} \in G_2$, $g_{2i+1} \in G_1$, and each b_β is of the form $b_\beta = g_1 \circ g_2 \circ \cdots \circ g_m$, $g_i \notin H$, $g_{2i} \in G_2$, $g_{2+1} \in G_1$.

Let $D_1 = (\cup_\alpha a_\alpha(D)) \cup D$ and $D_2 = (\cup_\beta b_\beta(D)) \cup D$. One easily sees that D_1 and D_2 are PFS's for G_1 and G_2, respectively. We form the corresponding E_1 and E_2, and then $D' = \operatorname{int}(E_1 \cap E_2 \cap \Delta) \subset \operatorname{int}(D) \neq \emptyset$.

If x is any point of $R(G, X)$, then there is a $y \in D$, and a $g \in G$ with $g(y) = x$. If $g \in H$, then since $y \in D \subset D_1 \cap D_2$, $g(y) \in E_1 \cap E_2$. If $g \notin H$, then there is either an a_α or a b_β, and an $h \in H$, so that $ha_\alpha = g$ or $hb_\beta = g$. Assume that we have $ha_\alpha = g$. Then $g(y) = ha_\alpha(y)$. $a_\alpha(y) \in D_1$ and so $ha_\alpha(y) \in E_1$, since $y \in D$. Similarly, if $g = hb_\beta$, then $g(y) \in E_2$. Therefore $R(G, X) \subset E_1 \cup E_2$.

We remark, incidentally, that D_1 is in fact an FS for G_1, *as an action on* $R(G, X)$, and similarly, D_2 is an FS for G_2 as an action on $R(G, X)$.

8. Let G be any group whatsoever. Then we can regard G both as a topological space, with the discrete topology, and as a group of homeomorphisms acting on itself, i. e. g, as a homeomorphism, takes the point $g_1 \in G$ into $g \cdot g_1$. Then G is itself discontinuous, and so is any subgroup. If $G_1 \subset G$, then an *FS* for G_1 is precisely a set of right coset representatives. In this setting Theorems 4 and 5 give the following two corollaries.

Corollary 1. *Let G_1 and G_2 be subgroups of the group G, where G_1 and G_2 together generate G. Let H be a common subgroup of G_1 and G_2. Then G is the free product of G_1 and G_2 with amalgamated subgroup H if and only if* *there exist elements a_α and b_β of G so that*

$$G = G_1 + \Sigma_\alpha G_1 a_\alpha = G_2 + \Sigma_\beta G_2 b_\beta = H + \Sigma_\alpha H a_\alpha + \Sigma_\beta H b_\beta .$$

Corollary 2. *Let G_1 and G_2 be subgroups of the group G, and let G_0 be the subgroup of G generated by G_1 and G_2. Let H be a common subgroup of G_1 and G_2. Then G_0 is the free product of G_1 and G_2 with amalgamated subgroup H if and only if there exist elements a_α and b_β of G, so that*

$$G = G_1 + \Sigma_\alpha G_1 a_\alpha = G_2 + \Sigma_\beta G_2 b_\beta = H \cup \Sigma_\alpha H a_\alpha \cup \Sigma_\beta H b_\beta .$$

In the above, the summation sign stands for disjoint union. The two corollaries are easily proven directly, but they do not appear to be known.

9. Theorem 6 (Closing a handle). *Let G be a discontinuous subgroup of $\mathfrak{G}$, acting on a space X. Let H_1 and H_2 be subgroups of G, and let D be a* PFS *for G. Assume that D can be written as a disjoint union of three sets, D_1, D_2, and D_3, where $\operatorname{int}(D_3) \neq \emptyset$. Assume also that $R(G, X)$ can be written as the disjoint union of three sets, R_1, R_2, and R_3, with $D_i \subset R_i$, $i = 1, 2, 3$, and*

(a) $\bigcup_{h \in H_i} h(D_i) = R_i$, $i = 1, 2$, *and*

(b) *if* $g \in G$, $g \notin H_i$, *then* $g(R_i) \subset R_3$, $i = 1, 2$.

Assume further that there is an $f \in \mathfrak{G}$ with

(c) $f(R_1 \cup R_3) \subset R_1$,

(d) $f^{-1}(R_2 \cup R_3) \subset R_2$, *and*

(e) $f^{-1} \circ H_1 \circ f = H_2$.

Then $G_0 = \langle G, f \rangle$ is again discontinuous, no two points of D_3 are equivalent under G_0, $\operatorname{int}(D_3)$ is a PFS *for G_0, and all relations in G_0 are consequences of the relations in G, together with the relations implicit in hypothesis (e).*

Proof. Let g_0 be some element of G_0. Then g_0 can be written, not uniquely as $g_0 = f^{\alpha_{n+1}} \circ g_n \cdots g_2 \circ f^{\alpha_2} \circ g_1 \circ f^{\alpha_1}$, where $\alpha_i \neq 0, i = 2, \ldots, n$, and $g_i \neq 1$, $i = 1, \ldots, n$. Whenever we have a word of this form, we define the *length* of the word to be $n + \sum_{i=1}^{n+1} |\alpha_i|$. A word is said to have *minimal length* if, whenever $\alpha_i > 0$ and $g_i \in H_1$, then $\alpha_{i+1} \geqq 0$, and if $\alpha_i < 0$ and $g_i \in H_2$, then $\alpha_{i+1} \leqq 0$. It is clear that by using the relations given in (*e*), we can express every $g_0 \in G_0$ by a word of minimal length.

We have to show that if $x \in D_3$, and $g_0 = f^{\alpha_{n+1}} \circ \cdots \circ f^{\alpha_1}$ is a word of minimal length > 0, then $g_0(x) \notin D_3$. The proof of this fact is contained in the following two lemmas.

Lemma 1. *If $\alpha_{n+1} > 0$ (< 0) then $g_0(x) \in R_1 (R_2)$.*

Proof. If $n = 0$, the statement is obvious from conditions (*c*) and (*d*). We assume, for the sake of definiteness, that $\alpha_n > 0$, so that $f^{\alpha_n} \circ \cdots \circ f^{\alpha_1}(x) = y \in R_1$. By condition (*b*) $g_n(y) \in R_3$ unless $g_n \in H_1$, in which case

$g_n(y) \in R_1$. If $g_n(y) \in R_3$, then $f^{\alpha_{n+1}} g_n(y)$ lies either in R_1 or R_2, depending on whether α_{n+1} is positive or negative. If $g_n \in H_1$, then, by the minimality condition as the length of g_0, $\alpha_{n+1} > 0$, and so $f^{\alpha_{n+1}} g_n(y) \in H_1$.

Lemma 2. *If* $\alpha_{n+1} = 0$, *then* $g_0(x) \notin D_3$.

Proof. If $n = 0$, then, since the length of g_0 is positive, $g_1 \neq 1$, and so $g_1(x) = g_0(x) \notin D_3$. If $n > 0$, we again assume that $\alpha_n > 0$. Then by Lemma 1, $f^{\alpha_n} \circ \cdots \circ f^{\alpha_1}(x) = y \in R_1$. If $g_n \in H_1$, then $g_n(y) \in R_1$, and not in D_3. If $g_n \notin H_1$, then $g_n(y) \in R_3$. Since y is equivalent under G to some point of D_1, $g_n(y)$ cannot lie in D_3.

III. Constructions of FSK-groups

10. Theorem 7. *Let S' be a finite Riemann surface, and let $w, u_1, \ldots, u_n, v_1, \ldots, v_m$ be simple, mutually disjoint loops on S', and let $\alpha_1, \ldots, \alpha_n, \beta_1, \ldots, \beta_m$ be positive integers. Assume that w divides S' into two surfaces S_1' and S_2' and that $u_1, \ldots, u_n$ all lie in S_1', and $v_1, \ldots, v_m$ all lie in S_2'. Let S_1^* and S_2^* be finite Riemann surfaces obtained by attaching discs to S_1' and S_2' respectively, along w. Assume that there is an* FSK*-group (G_1, R_1, S_1), which has an* FD D_1, *and which is a topological uniformization of S_1^* with signature $\langle u_1^{\alpha_1}, \ldots, u_n^{\alpha_n} \rangle$, and assume that there is an* FSK-*group (G_2, R_2, S_2), with* FD D_2, *which is a topological uniformization of S_2 with signature $\langle v_1^{\beta_1}, \ldots, v_m^{\beta_m} \rangle$. Then there is an* FSK-*group (G, R, S) which is a topological uniformization of S' with signature $\langle w, u_1^{\alpha_1}, \ldots, u_n^{\alpha_n}, v_1^{\beta_1}, \ldots, v_m^{\beta_m} \rangle$, and which has an* FD D.

Proof. w is a simple, homotopically trivial loop on S_1^*. We lift w to a loop $\bar{w}_1$ in R_1, starting at some point in D_1. $\bar{w}_1$ is then a simple, homotopically trivial loop in R_1. We deform $\bar{w}_1$ to a loop $\bar{w}_1'$ which is a circle lying in the interior of D_1, and normalize G_1 so that w_1 is the unit circle, and so that the set $\{|z| \leqq 1\}$ lies in the interior of D_1. Similarly, we lift w to a loop $\bar{w}_2$ in R_2, deform $\bar{w}_2$ to a circle, and normalize G_2, so that the deformed $\bar{w}_2$ is again the unit circle and so that the set $\{|z| \geqq 1\}$ lies in the interior of D_2.

D_1 and D_2, being FD's, are PFS's in the sense of Section 5. We observe that $\Sigma \subset D_1 \cup D_2$, that $\operatorname{int}(D_1 \cap D_2) \neq \emptyset$, and that there is a Jordan curve, namely the unit circle, which lies in $\operatorname{int}(D_1 \cap D_2)$ and which separates the boundary of D_1 from the boundary of D_2.

The first two statements imply the hypotheses of Theorem 4, where the common subgroup of G_1 and G_2 consists only of the identity, and so we know that $G = \langle G_1, G_2 \rangle$ is a Kleinian group, that $G = G_1 * G_2$, and that $(D_1 \cap D_2)$ is a PFS for G.

With the additional condition stated above, it was shown in [*10*] that $D = D_1 \cap D_2 \subset R_0(G)$, and that, if we set $\Pi = \bigcup_{g \in G} g(D)$, then Π is both open and closed in $R_0(G)$.

From the above remarks, one easily sees that $(G, \Pi, \Pi/G)$ is an FSK-group, with FD D. Namely, the sides of D are the sides of D_1 together with the sides of D_2. The identifications of the sides generate G, and so Π is connected (D is obviously connected). G is finitely generated, since G_1 and G_2 are finitely generated, and so $(G, \Pi, \Pi/G)$ is an FSK-group. To show that D is an FD, we observe that properties (a) through (g) have already been verified. Property (h) follows from the fact that $bd(D_1) \cap bd(D_2) = \emptyset$, and that D_1 and D_2 satisfy property (h).

We still have to show that $(G, \Pi, \Pi/G)$ is a topological uniformization of S' with signature $\langle w, u_1^{\alpha_1}, \ldots, u_n^{\alpha_n}, v_1^{\beta_1}, \ldots, v_m^{\beta_m}\rangle$. From the way G has been constructed, it is clear that Π/G is topologically equivalent to S', where the image of the unit circle is mapped onto w, and so we can regard Π as a regular covering surface of S'. It is also clear that $w, u_1^{\alpha_1}, \ldots, v_m^{\beta_m}$ all lift to *loops* on Π. Let x be any loop on S', which lifts to a loop on Π. We can deform x so that its base point 0 lies on w, and so that $x = y_1 \cdot y_2 \cdots \cdot y_q$ where y_{2i} is a loop lying in S_2', based at 0, and y_{2i+1} is a loop lying in S_1', based at 0. Each y_{2i} then determines an element g_{2i} in G_2, so that $g_q, \ldots, g_2 \cdot g_1 = 1$ in G. Since G is the free product of G_1 and G_2, this implies that each $g_i = 1$, $i = 1, \ldots, q$. $g_1 = 1$ implies that y_i lifts to a loop in Π, and, in fact, y_1 as a loop on S_1, lifts to a loop in R_1. But every loop on S_1^* which lifts to a loop does so as a consequence of the fact that $u_1^{\alpha_1}, \ldots, u_n^{\alpha_n}$ lift to loops, and the fact that every homotopically trivial loop lifts to a loop. Essentially, the only homotopically trivial loop on S_1^*, which is not homotopically trivial on S', is w, and so y_1 lifts to a loop as a consequence of the fact that $w, u_1^{\alpha_1}, \ldots, u_n^{\alpha_n}$ all lift to loops. Similar remarks apply to $y_2, \ldots, y_n$, and so the theorem is proved.

11. Theorem 8. *Let S' be a finite Riemann surface, let $w, u_1, \ldots, u_n$, $v_1, \ldots, v_m$ be simple mutually disjoint loops on S', and let $\gamma, \alpha_1, \ldots, \alpha_n$, $\beta_1, \ldots, \beta_m$ be positive integers with $\gamma > 1$. Assume that w divides S' into two surfaces S_1' and S_2', and that $u_1, \ldots, u_n$ all lie in S_1', and $v_1, \ldots, v_n$ all lie in S_2'. Let S_1^* and S_2^* be finite Riemann surfaces obtained by attaching punctured discs to S_1' and S_2', respectively, along w. Assume that there is an* FSK*-group (G_1, R_1, S_1) which has an* FD *D_1, which is a topological uniformization of S_1^* with signature $\langle w^\gamma, u_1^{\alpha_1}, \ldots, u_n^{\alpha_n}\rangle$, and that there is an* FSK*-group (G_2, R_2, S_2), with* FD *D_2, which is a topological uniformization of S_2^* with signature $\langle \gamma, v_1^{\beta_1}, \ldots, v_m^{\beta_m}\rangle$. Then there is an* FSK*-group (G, R, S) with* FD *D, which is a topological uniformization of S' with signature $\langle w^\gamma, u_1^{\alpha_1}, \ldots, u_n^{\alpha_n}, v_1^{\beta_1}, \ldots, v_m^{\beta_m}\rangle$.*

Proof. It is no loss of generality to assume that γ is the smallest power of w which lies in $\langle w^\gamma, u_1^{\alpha_1}, \ldots, v_m^{\beta_m}\rangle$. Then the lifting of w, as a loop on S_1', to R_1, starting in D_1, is an open arc, and so there is an element $h_1 \in G_1$ corresponding to this lifting. Since w^γ lifts to a closed curve, h_1

is an elliptic Möbius transformation of order γ. Since w bounds a punctured disc on S_1', one easily sees that one of the fixed points x of h_1 is an isolated fixed point on the boundary of D_1. We normalize G_1 so that x is the origin, the other fixed point of h_1 is at ∞, and so that the two sides of D_1, which meet at x, lie on the positive real axis, and the line $\{\arg z = 2\pi/\gamma\}$. We assume the side on the real axis belongs to D_1. We normalize further, and assume that $\{z \mid |z| \leqq 1, 0 < \arg z < 2\pi/\gamma\}$ lies in the interior of D_1.

The same remarks apply to G_2 and D_2; that is, there is an elliptic transformation $h_2 \in G_2$, of order γ, and a fixed point y of h_2 on the boundary of D_2. We normalize G_2 so that y is the point at ∞, the other fixed point of h_2 is at the origin, the sides of D_2 near y lie on the positive real axis and the line $\{\arg z = 2\pi/\gamma\}$, the side lying on the real axis belongs to D_2, and the set $\{z \mid |z| \geqq 1, 0 < \arg z < 2\pi/\gamma\}$ lies in the interior of D_2.

We observe that $h_2 = h_1^{\pm 1}$, so that $H = \langle h \rangle$ is a common subgroup of G_1 and G_2. If we form $E_1 = \bigcup_{h \in H} h(D_1)$ and $E_2 = \bigcup_{h \in H} h(D_2)$, then $E_1 \cup E_2 \supset \Sigma - \{0, \infty\}$. Let $D_3 = \{z \mid 0 \leqq \arg z < 2\pi/\gamma, z \neq 0, z \neq \infty\}$. Then D_3 is an FD for H, $E_1 \cap E_2 \cap D_3 = D_1 \cap D_2 = D$ (note that $D_1 \cap D_3 = D_1$, and $D_2 \cap D_3 = D_2$, by property (h) is the definition of FD), and $\operatorname{int}(D) \neq \emptyset$. Therefore, by Theorem 4, $G = \langle G_1, G_2 \rangle$ is Kleinian, no two points of D are equivalent under G, and G is the free product of G_1 and G_2 with amalgamated subgroup H.

Suppose we knew that (1) $D \subset R_0(G)$, and (2) $\Pi = \bigcup_{g \in G} g(D)$ is both open and closed in $R_0(G)$. Then the arguments given in the proof of Theorem 7 apply equally well to show that $(G, \Pi, \Pi/G)$ is an FSK-group, and, using the fact that the unit circle is a deformation of a lifting of w^ν in both R_1 and R_2, that $(G, \Pi, \Pi/G)$ is a topological uniformization of S' with signature $\langle w^\nu, u_1^{\alpha_1}, \ldots, v_m^{\beta_m} \rangle$.

Lemma 1. $D \subset R_0(G)$.

Proof. If z is an interior point of D, then we already know that $z \in R_0(G)$. If z is an interior point of $E_1 \cap E_2$, then we can find a neighborhood U of z, with $U \subset \operatorname{int}(E_1 \cap E_2)$ so that $h(U) \cap U = \emptyset$ for all $h \in H$, $h \neq 1$, since H is finite. If $g \in G$, $g \notin H$, then $g(\operatorname{int}(E_1 \cap E_2))$ does not intersect $E_1 \cap E_2$ by the argument given in the proof of Theorem 6. Therefore, those points of $\operatorname{bd}(D)$ which lie in $\operatorname{int}(E_1 \cap E_2)$ also lie in $R_0(G)$. The only remaining points are those points lying on the boundary of D_1, or on the boundary of D_2.

Let $z \in D$ be a point on the boundary of D_1. Then $z \in R_1$, and so there is a neighborhood U of z, with $g_1(U) \cap U = \emptyset$ for all $g_1 \in G_1$, $g_1 \neq 1$. If necessary, we make U somewhat smaller, so that U does not intersect the unit circle, and does not intersect any translate of the unit circle under G_1. Now every point of U is equivalent, under G_1, to some point of D_1, which lies inside the unit circle, and hence is a point of $D_1 \cap D_2$. The result now follows since no two points of $D_1 \cap D_2$ are equivalent under G.

Lemma 2. *Π is both open and closed in $R_0(G)$.*

Proof. Π is obviously connected, and if the lemma were false, Π/G would be properly embedded in R_1/G, where R_1 is the connected component of $R_0(G)$ which contains Π. But Π/G is a *finite surface*; that is, the only points on the boundary of D, which are not in $R_0(G)$, are isolated fixed points or limit points of G_1 or G_2, and these points correspond to *removed points, not holes* (this is a purely *local* property).

12. Theorem 9. *Let S' be a finite surface, let $w, u_1, \ldots, u_n$ be simple, mutually disjoint loops on S', and let $\alpha_1, \ldots, \alpha_n$ be positive integers. Assume that w is a non-dividing loop. Let S^* be the finite surface obtained from S' by "cutting" along w, and attaching discs along the two cuts. Assume that there is an* FSK-*group* (G_1, R_1, S_1), *with* FD D_1, *which is a topological uniformization of S^* with signature $\langle u_1^{\alpha_1}, \ldots, u_n^{\alpha_n}\rangle$. Then there is an* FSK-*group* (G, R, S) *with* FD D, *which is a topological uniformization of S' with signature $\langle w, u_1^{\alpha_1}, \ldots, u_n^{\alpha_n}\rangle$.*

Proof. Let w_1 and w_2 be the two loops on S^* obtained from w. Let $\bar{w}_1$ and $\bar{w}_2$ be the liftings of w_1 and w_2, respectively, starting inside D_1. We deform $\bar{w}_1$ and $\bar{w}_2$ to get new loops, which we again call $\bar{w}_1$ and $\bar{w}_2$, which are disjoint circles, both lying in the interior of D_1, and which have the further property that neither circle is separated from the boundary of D_1 by the other circle. Let g be a hyperbolic or loxodromic Möbius transformation which maps $\bar{w}_1$ onto $\bar{w}_2$. Let $G_2 = \langle g\rangle$, and let g be so chosen that the region of Σ bounded by $\bar{w}_1$ and $\bar{w}_2$, together with the circle $\bar{w}_1$ is an FD D_2 for G_2.

Let $G = \langle G_1, G_2\rangle$. The proof that $(G, \Pi, \Pi/G)$ is an FSK-group with FD $D = D_1 \cap D_2$ is the same as that given in theorem 7, where $\Pi = \bigcup_{g \in G} g(D)$.

Π/G is obviously topologically equivalent to S', so that $w, u_1^{\alpha_1}, \ldots, u_n^{\alpha_n}$ all lift to loops in Π. That $(G, \Pi, \Pi/G)$ is a topological uniformization of S' with signature $\langle w, u_1^{\alpha_1}, \ldots, u_n^{\alpha_n}\rangle$ follows from the fact that every loop x on S' can be written in the form $x = y^{\beta_1} z_1 y^{\beta_2} z_2, \ldots$, where y is a loop on S' corresponding to the element $g \in G$, and each z_i is a loop on S^*.

13. Theorem 10. *Let S' be a finite Riemann surface, let $w, u_1, \ldots, u_n$ be simple, mutually disjoint loops on S', where w is non-dividing, and let $\alpha_1, \ldots, \alpha_n, \beta$ be positive integers with $\beta > 1$. Let S^* be a Riemann surface obtained by "cutting" S' along w, to get two boundary curves, w_1 and w_2, and then attaching punctured discs along w_1 and w_2. Assume that there is an*

F S K-*group* (G^*, R^*, S^*) *with* F D D^*, *which is a topological uniformization of* S^* *with signature* $\langle w_1^\beta, w_2^\beta, u_1^{\alpha_1}, \ldots, u_n^{\alpha_n}\rangle$. *Then there is an* F S K-*group* (G, R, S) *with* F D D, *which is a topological uniformization of* S' *with signature* $\langle w^\beta, u_1^{\alpha_1}, \ldots, u_n^{\alpha_n}\rangle$.

Proof. We can assume that β is the smallest positive power of w that lies in $\langle w^\beta, u_1^{\alpha_1}, \ldots, u_n^{\alpha_n}\rangle$. Then there are isolated fixed points x_1 and x_2 on the boundary of D^*, where H_1 and H_2, the isotropy subgroups at x_1 and x_2, respectively, have the same order β. Let h_1 be the identification of the sides A_1 and B_1, near x_1, which generates H_1, $h_1(A_1) = B_1$, and let h_2 be the identification of the sides A_2 and B_2, near x_2, which generates H_2, $h_2(A_2) = B_2$.

Let U_1 be a *closed* circular neighborhood of x_1, with boundary C_1. U_1 is to be chosen so as to be invariant under H_1, and so that if $g \in G^*$, $g \notin H_1$, then $g(U_1) \cap U_1 = \emptyset$. We also require that C_1 intersects both A_1 and B_1. Since C_1 is invariant under H_1, it follows that C_1 is in fact orthogonal to both A_1 and B_1.

Similarly, we can find a circular neighborhood U_2 for x_2, and we choose U_2 to be *open* and disjoint from U_1.

Let $D_1 = U_1 \cap D^*$, $D_2 = U_2 \cap D^*$, $D_3 = D^* - (D_1 \cup D_2)$, $R_1 = \bigcup_{h \in H_1} h(D_1)$, $R_2 = \bigcup_{h \in H_2} h(D_2)$, and $R_3 = R^* - (R_1 \cup R_2)$. We observe that these sets satisfy hypotheses (a) and (b) in Theorem 6. Note that $R_1 = U_1$ and $R_2 = U_2$. Theorem 6 also requires that $\operatorname{int}(D_3) \neq \emptyset$, and one can choose U_1 and U_2 so that this is satisfied. We want to find a Möbius transformation f so that hypotheses (c), (d) and (e) of Theorem 6 are satisfied.

For $i = 1, 2$, let a_i be the point of intersection of C_i with A_i, and let b_i be the point of intersection of C_i with B_i. Then $h_i(a_i) = b_i$, $i = 1, 2$.

If we extend A_1 and B_1 to complete circles, then they intersect at a second point y_1, which is the other fixed point of h_1. Similarly, let y_2 be the second fixed point of h_2. We define the Möbius transformation f by requiring that $f(x_2) = y_1$, $f(y_2) = x_1$, $f(a_2) = a_1$. We observe that A_1 and $f(A_2)$ lie on the same circle. x_2 and y_2 are inverse points with respect to C_2, and C_2 is orthogonal to A_2 at a_2. Hence y_1 and x_1 are inverse points with respect to $f(C_2)$, and $f(C_2)$ is orthogonal to A_1 at a_1; i. e. $f(C_2) = C_1$. Then $f(b_2)$ lies on C_1 and so either $f(b_2) = b_1$, or $f(b_2) = h_1^{-1}(a_1)$. It follows that $f^{-1} \circ h_1 \circ f = h_2^\varepsilon$, $\varepsilon \pm 1$. $f(y_2) = x_1$, y_2 lies outside R_2, and x_1 is an isolated boundary point of R_1, hence $f(R_3 \cup R_1) \subset R_1$. Similarly $f^{-1}(R_3 \cup R_2) \subset R_2$.

The hypotheses of Theorem 6 are satisfied and so we know that $G = \langle G^*, f\rangle$ is Kleinian and that $D = D_3$ is a P F S for G. Set $\Pi = \bigcup_{g \in G} g(D)$. It is left to the reader to verify that $(G, \Pi, \Pi/G)$ is in fact an

FSK-group with FD D, and that $(G, \Pi, \Pi/G)$ is a topological uniformization of S' with signature $\langle w^\beta, u_1^{\alpha_1}, \ldots, u_n^{\alpha_n}\rangle$. The proofs of these facts are essentially the same as those given in the proofs of the preceding three theorems.

IV. Linearization of discontinuous groups

14. Let S' be a finite surface. A signature $\langle u_1^{\alpha_1}, u_2^{\alpha_2}, \ldots, u_n^{\alpha_n}\rangle$ is called *admissible* if the u_i are simple disjoint loops on S', each α_i is a positive integer with the property that α_i is the smallest positive exponent of u_i which lies in the normal subgroup spanned by $u_1^{\alpha_1}, \ldots, u_n^{\alpha_n}$.

If we have an admissible signature $\langle u_1^{\alpha_1}, \ldots, u_n^{\alpha_n}\rangle$, then u_i is called a *boundary loop* if u_i divides S' into two surfaces, and one of these surfaces is a punctured disc.

An admissible signature is said to be *strict* if for every boundary loop u_i, the corresponding α_i is greater than 1.

15. Theorem 11. *Let S' be a finite surface and let $\langle u_1^{\alpha_1}, \ldots, u_n^{\alpha_n}\rangle$ be a strict admissible signature on S'. Then there exists an* FSK-*group* (G, R, S) *with* FD D, *which is a topological uniformization of S' with signature $\langle u_1^{\alpha_1}, \ldots, u_n^{\alpha_n}\rangle$.*

Proof. We first assume that there is a u_i, which we now call u_1, which is non-dividing. We cut S' along u_1, to get a surface S^* with two boundary curves u_{11} and u_{12}. If $\alpha_1 = 1$, we attach discs to S^* along u_{11} and u_{12} to get a new surface S_1' with strict admissible signature $\langle u_2^{\alpha_2}, \ldots, u_n^{\alpha_n}\rangle$. By Theorem 9, it suffices to prove this theorem for S_1' with signature $\langle u_2^{\alpha_2}, \ldots, u_n^{\alpha_n}\rangle$. If $\alpha_1 > 1$, we attach punctured discs to S^* to get a surface S_1' with strict admissible signature $\langle u_{11}^{\alpha_1}, u_{12}^{\alpha_1}, u_2^{\alpha_2}, \ldots, u_n^{\alpha_n}\rangle$, and we can apply Theorem 10.

We can continue in the above manner, until we get a finite surface S_m with strict admissible signature $\langle v_1^{\beta_1}, \ldots, v_q^{\beta_q}\rangle$, where no v_i is non-dividing.

In a similar manner, if say v_1 is not a boundary loop, then we can cut S_m' along v_1 to get two surfaces S_{m+1}' and S_{m+1}'', each with one boundary curve, and attach either a disc or punctured disc to these surfaces along the boundary, depending on whether $\beta_1 = 1$ or $\beta_1 > 1$. We can then apply either Theorem 7 or Theorem 8.

It suffices therefore to prove that if we have a finite surface S' with a *strict admissible signature* $\langle u_1^{\alpha_1}, \ldots, u_n^{\alpha_n}\rangle$ on S', *where each u_i is a boundary loop*, then there is an FSK-group (G, R, S) with FD D, which is a topological uniformization of S' with signature $\langle u_1^{\alpha_1}, \ldots, u_n^{\alpha_n}\rangle$. *This, however, is a classical result.* The desired FSK-groups are precisely the well-known Fuchsian and elementary groups.

16. We observe that in the above theorem, the requirement that the signature be strict is not really essential. Namely, if we have a non-strict

admissible signature $\langle u_1^{\alpha_1}, \ldots, u_n^{\alpha_n}\rangle$ on the finite surface S, then for each boundary loop u_i, with exponent 1, we can "fill in" a point p_i, to get a new surface S^+ on which these loops u_i are contractible. The remaining loops u_j in the signature give us a strict admissible signature for S^+. Then by the above theorem we can find a (G, R, S) uniformizing S^+. We then remove the points p_i, together with the points of R lying over the p_i. This gives

Corollary 3. *Let $\langle u_1^{\alpha_1}, \ldots, u_n^{\alpha_n}\rangle$ be an admissible signature on the finite surface S'. Then there exists a Kleinian group G and an open, connected, invariant subset R' of $R_0(G)$, so that $(G, R', R'/G)$ is a topological uniformization of S' with signature $\langle u_1^{\alpha_1}, \ldots, u_n^{\alpha_n}\rangle$.*

We next recall Theorem 2 which states that every regular, planar covering surface of a finite surface corresponds to an admissible signature. We immediately get

Corollary 4. *Let S be the interior of a compact 2-manifold. Let $p:\tilde{S} \to S$ be a regular covering of S, with cover group G, where S is planar. Then there exists a topological mapping f, of S onto a domain of Σ, so that every element of $f \circ G \circ f^{-1}$ is a Möbius transformation.*

17. Theorem 3 states that if we have a Kleinian group which gives a topological uniformization of a finite Riemann surface with a given signature, then there is another Kleinian group which gives a *conformal* uniformization of this surface with the given signature. We state this as

Corollary 5. *Let S' be a finite Riemann surface and let $\langle u_1^{\alpha_1}, \ldots, u_n^{\alpha_n}\rangle$ be a strict admissible signature on S'. Then there exists an* FSK*-group (G, R, S) which is a (conformal) uniformization of S' with signature $\langle u_1^{\alpha_1}, \ldots, u_n^{\alpha_n}\rangle$.*

We also obviously have

Corollary 6. *Let S' be a finite Riemann surface, and let $\langle u_1^{\alpha_1}, \ldots, u_n^{\alpha_n}\rangle$ be an admissible signature on S'. Then there is a Kleinian group G, and an open, connected, invariant subset R' of $R_0(G)$, so that $(G, R', R'/G)$ is a (conformal) uniformization of S' with signature $\langle u_1^{\alpha_1}, \ldots, u_n^{\alpha_n}\rangle$.*

Using Theorem 2 again, we get the conformal analogue of Corollary 4:

Theorem 12. *Let $p:\tilde{S} \to S$ be a regular covering of the finite Riemann surface S, where p is conformal, let G be the group of cover transformations on $\tilde{S}$, and let $\tilde{S}$ be a planar surface. Then there exists a schlicht function f, mapping $\tilde{S}$ onto a domain in Σ so that $f \circ G \circ f^{-1}$ is a Kleinian group.*

References

[1] Ahlfors, L.: Finitely generated Kleinian groups. (To appear.)

[2] —, and L. Bers: Riemann's mapping theorem for variable metrics. Ann. Math. **72**, 385—404 (1960).

[3] Bers, L. Uniformization by Beltrami equations. Comm. Pure Appl. Math. **14**, 215—228 (1961).

[4] Fenchel, W., and J. Nielsen: Discontinuous groups of non-Euclidean motions. (To appear.)
[5] Ford, L. R.: Automorphic functions. New York: McGraw-Hill 1929.
[6] Fricke, R., und F. Klein: Vorlesungen über die Theorie der automorphen Functionen I. Leipzig: Teubner 1897.
[7] Klein, F.: Neue Beiträge zur Riemannschen Functionentheorie. Math. Ann. **21**, 141—218 (1883).
[8] Macbeath, A. M.: Packings, free products and residually finite groups. Proc. Cambridge Phil. Soc. **59**, 555—558 (1963).
[9] Maskit, B.: A theorem on planar covering surfaces with applications to 3-manifolds. Ann. of Math. (To appear.)
[10] — On Klein's combination theorem. (To appear.)

Institute for Advanced Study
Princeton, New Jersey

The Modular Function and Geometric Properties of Quasiconformal Mappings*

By

L. V. Ahlfors

With 2 Figures

1. We recall the familiar notations

$$e_1 = \wp\left(\frac{\omega_1}{2}\right), \quad e_2 = \wp\left(\frac{\omega_2}{2}\right), \quad e_3 = \wp\left(\frac{\omega_1+\omega_2}{2}\right)$$

associated with Weierstrass' $\wp$-function. The period ratio $\tau = \omega_2/\omega_1$ is always assumed to have positive imaginary part.

The function

$$\lambda(\tau) = \frac{e_3 - e_2}{e_1 - e_2} \tag{1}$$

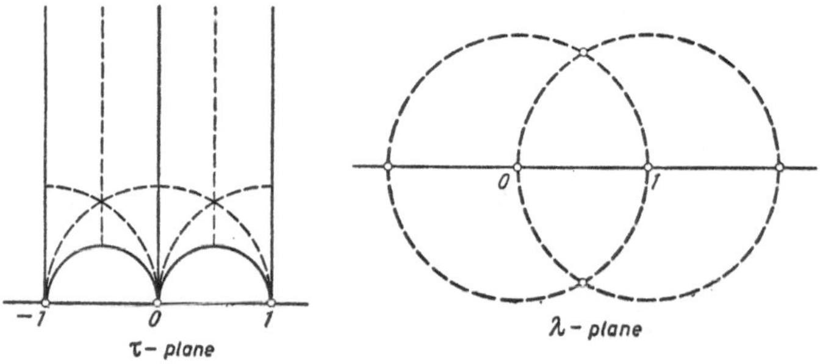

Fig. 1

* Received June 29, 1964.

is analytic and $\neq 0, 1$ in the upper half plane. It is the familiar modular function which maps a noneuclidean triangle with zero angles onto a half plane. The normalization is such that $\lambda(0) = 1, \lambda(1) = \infty, \lambda(\infty) = 0$.

We shall make frequent references to Fig. 1 in which the solid lines in the τ-plane are mapped on the real axis. The dotted lines are axes of symmetry. The symmetries permit us to read off the following values:

$$\begin{aligned} \lambda(i) &= \frac{1}{2}, \\ \lambda\left(\frac{\pm 1 + i\sqrt{3}}{2}\right) &= \frac{1 \pm i\sqrt{3}}{2}, \\ \lambda\left(\frac{\pm 1 + i}{2}\right) &= 2. \end{aligned} \tag{2}$$

Most important, λ is automorphic under modular transformations $\begin{pmatrix} a & b \\ c & d \end{pmatrix} \equiv \begin{pmatrix} 1 & 0 \\ 0 & 1 \end{pmatrix}$ mod 2, and it satisfies

$$\begin{aligned} 1 - \lambda(\tau + 1) &= \frac{1}{1 - \lambda(\tau)}, \\ \lambda\left(-\frac{1}{\tau}\right) &= 1 - \lambda(\tau). \end{aligned} \tag{3}$$

We also recall the explicit representations

$$\begin{aligned} \lambda &= 16q \prod_1^\infty \left(\frac{1 + q^{2n}}{1 + q^{2n-1}}\right)^8, \\ 1 - \lambda &= \prod_1^\infty \left(\frac{1 - q^{2n-1}}{1 + q^{2n-1}}\right)^8, \end{aligned} \tag{4}$$

where $q = e^{\pi i \tau}$.

2. Consider a sense preserving quasiconformal mapping of the finite plane onto itself. If it carries an ordered triple e_1, e_2, e_3 into e_1', e_2', e_3', how large must the maximal dilatation be?

It is no restriction to assume that $e_1 + e_2 + e_3 = 0$, in which case e_1, e_2, e_3 are the e_k of a $\wp$-function whose period ratio is a solution of equation (1). There are infinitely many such τ, congruent under the congruence subgroup mod 2 of the modular group. Similarly, the e_k' determine infinitely many τ'.

Teichmüller has supplied the answer to the question asked above:

Teichmüller's Theorem. *There exists a mapping with maximal dilatation $\leq K$ which carries e_k into e_k' $(k = 1, 2, 3)$ if and only if the noneuclidean distance from τ to the nearest τ' is $\leq \log K$.*

In this connection the noneuclidean metric is normalized as

$$ds = (\operatorname{Im} \tau)^{-1} |d\tau|.$$

We recall the proof very briefly. A mapping of the planes can be lifted to a mapping of the two-sheeted covering surfaces with branch points at e_k and e_k', and from there to a mapping of their universal covering surfaces. The lifted mapping satisfies

$$f(z+\omega_1)=f(z)+\omega_1',$$
$$f(z+\omega_2)=f(z)+\omega_2'.$$

If the dilatation is $\leqq K$ the usual length-area argument shows that

$$\operatorname{Im}\tau' \leqq K \operatorname{Im}\tau. \tag{5}$$

Here τ' is uniquely determined by τ and f. If we replace τ by an equivalent $S\tau$ we must replace τ' by $S\tau'$, and we obtain

$$\operatorname{Im} S\tau' \leqq K \operatorname{Im} S\tau. \tag{6}$$

Let $d(\tau,\tau')$ denote noneuclidean distance. If $S\tau$ and $S\tau'$ are on the same vertical line, $S\tau'$ above $S\tau$, then

$$d(\tau,\tau')=d(S\tau,S\tau')=\log\frac{\operatorname{Im} S\tau'}{\operatorname{Im} S\tau},$$

and (6) would yield

$$d(\tau,\tau')\leqq \log K. \tag{7}$$

Suppose that the noneuclidean line from τ over τ' hits the real axis at τ_0. The images $S\tau_0$ are everywhere dense. We can let $S\tau_0$ tend to ∞, and it is clear that we obtain (7) as a limit relation.

This proves the necessity. The sufficiency is proved by choosing f as an affine mapping.

3. Our aim is to show how Teichmüller's theorem can be applied to derive some known geometric properties of quasiconformal mappings in a very direct manner. With only slight exaggeration it can be said that the results follow from Fig. 1 by geometric inspection.

Let f denote a K-quasiconformal mapping of the finite plane onto itself. We begin with an unpublished result due to Beurling.

Prop. 1. *If $K<\sqrt{3}$ the vertices of any equilateral triangle are mapped onto the vertices of a nondegenerate triangle with the same orientation.*

Colloquially, the triangle does not "flip over".

For the proof, choose e_1, e_2, e_3 as vertices of an equilateral triangle. Then $\lambda=\frac{1}{2}(1\pm i\sqrt{3})$, and hence $\tau=\frac{1}{2}(\pm 1+i\sqrt{3})$. We have to show that λ' cannot be real. But real λ' correspond to points τ' on the solid lines in Fig. 1, and these lines are at noneuclidean distance $\log\sqrt{3}$ from the centers $\frac{1}{2}(\pm 1+i\sqrt{3})$. If λ' were real we should have $d(\tau,\tau')\geqq\log\sqrt{3}$ for all τ', while Teichmüller's theorem guarantees that $d(\tau,\tau')\leqq\log K<\log\sqrt{3}$ for one τ'. We have reached the desired contradiction.

Corollary. *A quasiconformal mapping with $K < \sqrt{3}$ can be approximated by piecewise affine quasiconformal mappings.*

Cover the plane by a net of equilateral triangles and map each triangle in an affine way on the triangle spanned by the images of the vertices. As the net becomes finer the piecewise affine mappings approximate the given mapping.

4. We recall that a quasiconformal mapping of the upper half plane onto itself can be extended to a symmetric quasiconformal mapping of the whole plane. The restriction to the real axis satisfies a condition

$$\varrho(K)^{-1} \leqq \frac{f(x+t)-f(x)}{f(x)-f(x-t)} \leqq \varrho(K). \tag{8}$$

Conversely, if $f(x)$ satisfies a condition of the form (8) it can be extended to a quasiconformal mapping (BEURLING-AHLFORS, [2]).

In this connection we merely point out that the necessity of (8) with a best possible $\varrho(K)$ is an immediate consequence of Teichmüller's theorem. If we choose $e_1 = x - t$, $e_2 = x$, $e_3 = x + t$ we have $\lambda = -1$, $\tau = -1 + i$. The nearest τ' is of the form $-1 + i\eta$ with $K^{-1} < \eta < K$. It follows that (8) holds with

$$\varrho(K)^{-1} = -\lambda(1 + iK) = \frac{\lambda(iK)}{1-\lambda(iK)} = 16\, e^{\pi K} \prod_1^\infty \left(\frac{1+e^{-2n\pi K}}{1-e^{-(2n-1)\pi K}}\right)^8.$$

It is not difficult to show that this is the best possible value[1].

5. We return to the nonsymmetric case and focus our attention on the image L of the real axis. It divides the plane into regions Ω and Ω^* that correspond to the upper half plane H and the lower half plane H^*. The function $s = f[\overline{f^{-1}(\zeta)}]$ defines a quasiconformal reflection across L.

In a recent paper I proved the following geometric theorem:

Prop. 2. *Let $\zeta_1, \zeta_2, \zeta_3$ be points on L such that ζ_3 lies between ζ_1 and ζ_2. A necessary and sufficient condition that L permit a quasiconformal reflection is that*

$$\left|\frac{\zeta_3-\zeta_1}{\zeta_2-\zeta_1}\right| \leqq C \tag{9}$$

for all such triples.

More precisely, if we can find f with dilatation $\leqq K$, then (9) holds with a constant $C(K)$ that depends only on K, and if (9) holds we can find f with a dilatation $\leqq K(C)$.

For the necessity, suppose that $x_2 < x_3 < x_1$ correspond to the three points. Then $\lambda = (x_3 - x_2):(x_1 - x_2)$ lies between 0 and 1 so that τ is on the imaginary axis. According to Teichmüller's theorem there exists a corresponding τ' in the angle $\theta_0 < \arg \tau' < \pi - \theta_0$, $\theta_0 = 2 \arctan K^{-1}$.

---

[1] The result is not new, but I have no reference.

In this angle $\lambda(\tau') \to 1$ as $\tau' \to 0$ and $\lambda(\tau') \to 0$ as $\tau' \to \infty$. It follows that $|\lambda(\tau')|$ lies under a finite bound $C(K)$.

For the opposite conclusion, let Ω and Ω^* be mapped conformally onto H and H^* by g and g^* respectively. We need to show that $g^* \circ g^{-1}$, defined on the real axis, satisfies a condition of the form (8).

The condition can be expressed in terms of extremal length. Consider the subarcs $\alpha_1, \beta_1, \alpha_2, \beta_2$ as shown in Fig. 2. Let Λ_1 be the extremal distance from α_1 to β_1, and Λ_2 the extremal distance from α_2 to β_2 with respect to Ω, and let Λ_1^*, Λ_2^* be the same extremal distances with respect to Ω^*. They satisfy conditions $\Lambda_1 \Lambda_2 = \Lambda_1^* \Lambda_2^* = 1$, and $g(\zeta_1)$, $g(\zeta_2)$, $g(\zeta_3)$ are of the form $x - t, x + t, x$ if $\Lambda_1 = \Lambda_2 = 1$. Under this condition we have to show that Λ_1^* and Λ_2^* are bounded.

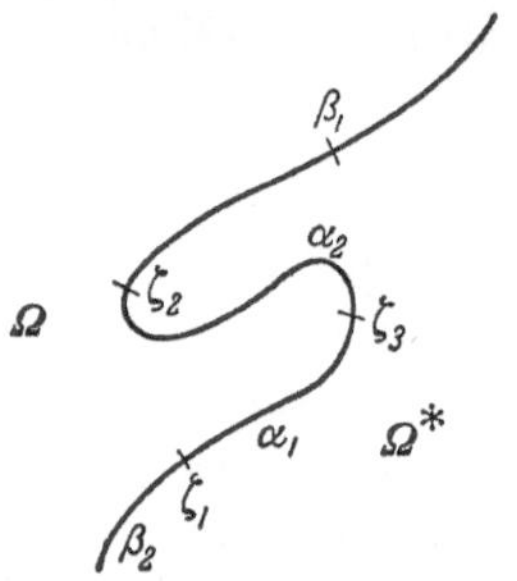

Fig. 2

Let Γ denote the family of simple closed curves that separate ζ_2 and ζ_3 from ζ_1 and ∞, and let $\Lambda(\Gamma)$ denote its extremal length. Because each curve in Γ contains an arc that joins α_1 and β_1 within Ω it is true that $1 = \Lambda_1 \leqq \Lambda(\Gamma)$. But it can be shown, as in the proof of Teichmüller's theorem that

$$\Lambda(\Gamma) = \frac{2}{\operatorname{Im} \tau},$$

where τ is the point *in the fundamental region* that corresponds to $e_k = \zeta_k$ (this is the value with largest $\operatorname{Im} \tau$). It follows that $\operatorname{Im} \tau \leqq 2$, and the same reasoning applied to Λ_2 gives $\operatorname{Im}(-1/\tau) \leqq 2$. In other words, τ is shielded from 0 and ∞.

But condition (9) means that τ is also shielded from ± 1. Therefore, $\operatorname{Im} \tau$ lies above a positive minimum, and we obtain

$$\Lambda_1^* \leqq \frac{2}{\operatorname{Im} \tau} \leqq M(C).$$

The same holds for Λ_2^*.

This proof is considerably simpler than the one given in my earlier papers.

References

[1] AHLFORS, L.: Quasiconformal reflections. Acta Math. **109**, 291—301 (1963).
[2] BEURLING, A., and L. AHLFORS: The boundary correspondence under quasiconformal mappings. Acta Math. **96**, 125—142 (1956).

Department of Mathematics
Harvard University
Cambridge, Mass.

Polynomial Convexity: The Three Spheres Problem*

By

Eva Kallin

A compact subset Y of C^n is called *polynomially convex* if for each point x in $C^n \sim Y$ there is a polynomial p such that

$$|p(x)| > \sup\{|p(y)| : y \in Y\};$$

we say that such a polynomial p *separates* x from Y. A set Y is convex in the ordinary sense just in case each point of the complement can be separated from Y by a polynomial of degree one, so each convex set is polynomially convex. Finite sets are also polynomially convex, so polynomially convex sets need not be connected. Any compact set which lies entirely in the real subspace of points having all real coordinates is polynomially convex because of the Weierstrass approximation theorem. Polynomial convexity, unlike ordinary convexity, is not preserved under real linear transformations or even under general complex linear mappings, though of course it is preserved by complex linear isomorphisms. There is no nice internal description of polynomially convex sets analogous to the condition that a set is convex if and only if it contains the closed line segment joining each pair of its points. Thus although many sets are known to be polynomially convex it can be rather difficult to decide about some fairly simple sets. A single convex set is polynomially convex, as is the union of two disjoint convex sets (this is easy, and it follows from the Separation Lemma below), but what about the union of three disjoint convex sets? What about the very special case of three balls? If they are far enough apart, for instance if they have disjoint projections on some complex line, then the Separation Lemma again provides an easy affirmative answer. Indeed, the answer is always yes, and this is proved in the first part of this paper. The last part consists of an example of three disjoint convex sets in C^3, in fact three "cubical" polycylinders of equal radii, whose union is not polynomially convex.

For any compact set Y in C^n let

$$Y^\wedge = \{x \in C^n : |p(x)| \leq \sup\{|p(y)| : y \in Y\} \quad \text{for all polynomials } p\}.$$

$Y^\wedge$ is called the polynomial hull of Y and is the smallest polynomially convex set containing Y. If A is the Banach algebra of all uniform limits of polynomials on Y, then the maximal ideal space of A is just $Y^\wedge$. Since any finitely generated semi-simple commutative Banach algebra with the spectral norm is essentially just such an algebra A, some knowledge of how Y "spans" $Y^\wedge$ is necessary for the study of such algebras, variously called function algebras or uniform algebras.

* Received July 16, 1964.

A related problem involving polynomial convexity is the determination of Runge domains in C^n, i. e. open sets in which every holomorphic function can be uniformly approximated on compact subsets by polynomials. It was in this connection that H. GRAUERT asked about the three spheres some time ago; the question was repeated to me by several people. It follows from known facts that the union of three disjoint open balls is always a Runge domain just in case the polynomial hull of three disjoint spheres is just the three closed balls they span individually. Each sphere spans its interior because each point of the interior lies on some complex line and inside a circle whose circumference is part of the sphere.

A compact set X in C^1 is polynomially convex if and only if $C^1 \sim X$ is connected. If $X = X^\wedge$, then $C^1 \sim X$ must consist of just the unbounded component, since by the maximum modulus theorem any bounded component of $C^1 \sim X$ must be contained in $X^\wedge$. The converse is a consequence of RUNGE's Theorem ([*1*], page 177) since if $C^1 \sim X$ is connected and $y \in C^1 \sim X$ a function which is 1 in a small neighborhood of y and 0 in a neighborhood of X can be uniformly approximated on $X \cup \{y\}$ by polynomials in z. Thus we see that in C^1 we obtain the polynomial hull $X^\wedge$ just by "filling in the holes" in X. An attractive conjecture to the effect that in C^n any hull could be obtained by some similar "filling in of holes" on all possible analytic surfaces was disproved by G. STOLZENBERG [*3*], who constructed a set in C^3 which is not polynomially convex and whose polynomial hull contains no pieces of analytic surface at all; in fact its projections have measure zero in all directions.

From the description of polynomially convex sets in C^1 we see that any finite disjoint union of such sets is again polynomially convex. However in C^2 the union of two disjoint polynomially convex sets may fail to be polynomially convex: e. g., the two rims of the annulus

$$\{z \in C^2 : z_1 z_2 = 1, \quad 1 \leqq |z_1| \leqq 2\}$$

are each separately polynomially convex but together they span the annulus. Since the inner rim is contained in the closed unit ball and the outer rim is disjoint from it, we also see that in general the disjoint union of a polynomially convex set and one closed ball is not polynomially convex. A criterion for the polynomial convexity of a union of two disjoint polynomially convex sets is found in the following lemma. It may be recognized as a kind of simple converse of the much deeper SHILOV Idempotent Theorem [*1*].

Separation Lemma. *If X_1 and X_2 are compact sets in C^n, f a polynomial such that $(f[X_1])^\wedge \cap (f[X_2])^\wedge = \emptyset$, then $(X_1 \cup X_2)^\wedge = X_1^\wedge \cup X_2^\wedge$.*

Note that the condition on f is simply that neither of the disjoint compact plane sets $f[X_1]$, $f[X_2]$ surrounds any part of the other. We will say that such an f *separates* X_1 from X_2. Because of Runge's theorem this usage is consistent with our earlier definition of separation of a point from a set.

Proof. Suppose $y \notin X_1^\wedge \cup X_2^\wedge$. If $f(y) \notin f[X_1]^\wedge \cup f[X_2]^\wedge$, then by Runge's theorem there is a polynomial p in one variable so that $p(f(y))$ is near 1 and $p \circ f$ is near 0 on $X_1 \cup X_2$, so $y \notin (X_1 \cup X_2)^\wedge$. If $f(y) \in f[X_1]^\wedge \cup f[X_2]^\wedge$, say $f(y) \in f[X_1]^\wedge$, then since $y \notin X_1^\wedge$ there is a polynomial q with $q(y) = 1$, $|q| < \frac{1}{2}$ on X_1. Let $M = \sup\{|q(x)| : x \in X_2\}$. Again use Runge's theorem to find a polynomial r so that $|r - 1| < 1/3$ on $f[X_1]$ and $|r| < 1/2M$ on $f[X_2]$. Then $h = (r \circ f) \cdot q$ is a polynomial satisfying $|h(y) - 1| < 1/3$, $|h| < 2/3$ on $X_1 \cup X_2$, so again $y \notin (X_1 \cup X_2)^\wedge$. Thus we have shown $(X_1 \cup X_2)^\wedge \subset X_1^\wedge \cup X_2^\wedge$, and the reverse inclusion is always true.

Corollary. *If X_1 and X_2 are disjoint convex compact sets, then $X_1 \cup X_2$ is polynomially convex.*

Proof. A polynomial of degree one will separate X_1 from X_2.

Theorem. *The polynomial hull of three disjoint spheres S_i in C^n is just $\cup\{S_i^\wedge : i = 1, 2, 3\}$ where $S_i^\wedge$ is the closed ball whose surface is S_i.*

Proof. We may as well assume the $S_i^\wedge$ are disjoint also. The theorem will follow from the separation lemma if we can find a polynomial which separates one sphere from the other two, since we know that two spheres together span nothing but their interiors. We may assume that the radius of S_1 is 1, that the other radii t and r of S_2 and S_3 respectively are at most 1, and that the center of S_1 is at the origin. In the subspace of dimension at most two spanned by the centers of S_2 and S_3 choose coordinates so that the center of S_2 is on the real z_1-axis. Then rotate separately in z_1 and z_2 so that the coordinates (α, β) of the center of S_3 are both real; this leaves the center of S_1 at $(0, 0)$ and that of S_2 at some point $(\gamma, 0)$. Now the polynomial $f(z) = z_1^2 + z_2^2$ will separate S_1 from $S_2 \cup S_3$. Clearly $|f(z)| \leqq 1$ on S_1, the unit sphere. In the plane of values $w = f(z)$ the compact set $f[S_3]$ lies to the right of the vertical line $\operatorname{Re} w = 1$, and the set $f[S_2]$ lies on the opposite side of another line tangent to the unit disk, so together $f[S_2]$ and $f[S_3]$ cannot surround $f[S_1]$. Explicitly we have the following

Lemma. $\operatorname{Re}(z_1^2 + z_2^2) > 1$ *on the set* $\{z \in C^2 : |z_1 - \alpha|^2 + |z_2 - \beta|^2 \leqq r^2\}$ *if* $\alpha^2 + \beta^2 > (1 + r)^2$, *$\alpha$ and β are real, and* $r \leqq 1$. *Thus also* $\operatorname{Re}(\theta(z_1^2 + z_2^2)) > 1$ *on* $\{z \in C^2 : |z_1 - \gamma|^2 + |z_2|^2 \leqq t^2\}$ if $|\gamma| > 1 + t$, $t \leqq 1$, *and* $\theta = |\gamma|^2/\gamma^2$.

Proof. Write $z_j = x_j + i y_j$, $j = 1, 2$. Then if

$$(x_1 - \alpha)^2 + y_1^2 + (x_2 - \beta)^2 + y_2^2 \leqq r^2 \quad \text{and} \quad \eta = (\alpha^2 + \beta^2)^{1/2} - 1 - r$$

and
$$\varepsilon = r - ((x_1 - \alpha)^2 + (x_2 - \beta)^2)^{1/2}$$
we have $\operatorname{Re}(z_1^2 + z_2^2)$
$$= x_1^2 + x_2^2 - (y_1^2 + y_2^2) \geqq x_1^2 + x_2^2 + (x_1 - \alpha)^2 + (x_2 - \beta)^2 - r^2 \geqq$$
$$(1 + \eta + \varepsilon)^2 + (r - \varepsilon)^2 - r^2 = 1 + 2(1 - r)\varepsilon + 2\eta + (\eta + \varepsilon)^2 + \varepsilon^2 > 1$$
since $\eta > 0$ and $\varepsilon > 0$. The second inequality follows from the first by replacing z_1, z_2, α, β, and r by $(|\gamma|/\gamma)z_1$, $(|\gamma|/\gamma)z_2$, $|\gamma|$, 0 and t respectively. The inequalities $\alpha^2 + \beta^2 > (1 + r)^2$ and $|\gamma| > 1 + t$ express the fact that the balls $\hat{S_3}$ and $\hat{S_2}$ do not meet $\hat{S_1}$.

An example of three disjoint convex sets whose union is not polynomially convex:

Let R be the surface in C^3 defined by $z_1 z_2 = 1$, $z_3(1 - z_1) = 1$. Let K_1, K_2, K_3 be the curves lying on R defined by $|z_1| = 1/M$, $|z_1 - 1| = 1/M$, $|z_1| = M$ respectively, where M is a fixed real number greater than 2, so the projections on the z_1-plane of the K_i are three disjoint circles, one large one and two small disjoint ones inside it. Because linear fractional transformations take circles into circles, the projections of the K_i on the z_2- and z_3-planes are of the same sort, the only difference being that while K_3 is the big outside circle in the z_1-view, K_1 is outside in the z_2-projection and K_2 is outside in the z_3-projection. Their convex hulls are thus certainly disjoint, since for example the convex hull of K_1 must lie in the cylinder $|z_1| \leqq 1/M$ while that of K_2 lies in $|z_1 - 1| \leqq 1/M$, and the other coordinate projections serve to separate the other pairs. However that part of the surface R which is bounded by the three curves is clearly in their polynomial hull and hence in the polynomial hull of the union of their convex hulls, or indeed in the polynomial hull of any set containing $\cup K_i$. Now it is possible to arrange three disjoint polycylinders P_i of radius M in each direction so that each P_i contains the corresponding K_i. To accomplish this just take the center of P_1 at $(-M + 1/M, 0, M + M/(M + 1))$, the center of P_2 at $(M + 1 - 1/M, M + M/(M+1), 0)$, and the center of P_3 at $(0, -M + 1/M, -M + 1/(M + 1))$. Thus $\cup P_i$ is not polynomially convex. By taking slightly larger open polycylinders containing the P_i which are still disjoint one obtains a union of three disjoint cubical open polycylinders which is not a Runge domain.

References

[1] SAKS, S., and A. ZYGMUND: Analytic functions. Warsaw 1952.
[2] SHILOV, G. E.: On the decomposition of a commutative normed ring into a direct sum of ideals. Mat. Sb. N.S. **32**, 353—364 (1953) (in Russian).
[3] STOLZENBERG, G.: A hull with no analytic structure. J. Math. Mech. **12**, 103—111 (1963).

Department of Mathematics
Mass. Inst. of Techn.
Cambridge, Mass.

Appendix: Problems Submitted

1. Let X and Y be two-dimensional complex spaces and $f: Y \to X$ a proper branched covering map. Are there proper modifications $X' \to X$ and $Y' \to Y$, together with a covering map $Y' \to X'$ such that

1) $$\begin{array}{ccc} Y' & \to & Y \\ \downarrow & & \downarrow \\ X' & \to & X \end{array}$$ is commutative,
2) Y' is nonsingular,
3) X' is Jungian?

(Abhyankar)

2. Represent some automorphic functions in terms of the kernel function (generalized kernel function) and its derivatives.
(Bergman)

3. a) Give more exact bounds for Euclidean measures in pseudo-conformal transformations than those obtained by using domains of comparison. b) Classify the boundary points in more general cases than those considered in Bergman, Mem. Sci. Math. **108**, 12 and 25. See also Bergman, Mem. Sci. Math. **106**, 53; Bergman, Mem. Sci. Math. **108**, 48–63; Bergman, Annali Mat. LVII 295–309 (1962); Maschler, Pacific J. Math. **6**, 501 (1956); Stark, Pacific J. Math. **6**, 565, (1956); Hahn, Pacific J. Math., to appear; Hörmander, Existence theorems for the $\bar{\partial}$ operator by L^2 methods, to appear in Acta Math.
(Bergman)

4. Study the properties of the functions $v^{(10)}(z, t)$ and $v^{(01)}(z, t)$ (see Mem. Sci. Math. **108**, 39, Eq. (21)), which map the given domain B onto the representative domain.
(Bergman)

5. A region (open set) in a complex linear topological space X is called pseudo-convex if and only if every intersection with any finite dimensional translated linear subspace of X is pseudo-convex. A function is called holomorphic in a region D if and only if its restriction to every intersection of D with a finite dimensional translated linear subspace of X is holomorphic.

One defines as usual "a holomorphic function is singular at a boundary point" and "region of holomorphy". If D is a region of holomorphy D is pseudo-convex. Problem: Is the converse true? (Levi problem for infinite dimension.)
(Bremermann)

6. Let X be a complex space, S an analytic subset of X, and t a tangent vector to S at a regular point of S. Is t a limit of tangent vectors to X at regular points?
(Bungart)

7. Let X be a complex space. Then $x \in X$ is called a k-simple point ($0 \leqq k \leqq \dim X$) if x has a neighborhood U that is a holomorphic family of k-dimensional analytic subsets with parameter space a polycylinder of dimension $\dim X - k$, but fails to be such a family for $k - 1$ instead of k. How large is the set of k-simple points? Is it the complement of a variety of co-dimension greater or equal to k?

Note: A regular point is a 0-simple point.

(Bungart)

8. Let X be a compact, complex manifold with a differentiable boundary B_0 and consider the diffeotopy class $\mathscr{B}$ of B_0 in X. If B_1 is a hypersurface that minimizes among all $B \in \mathscr{B}$ the oscillation of the Levi invariant from point to point, can one estimate the range of the Levi invariant for B_1 from the geometry of B_0? An analogous question might be formulated for complex spaces.

(Calabi)

9. Let X be an n-dimensional finite complex manifold with at least two boundary components B_1, B_2. If the Levi invariant in B_1 has everywhere at least p positive roots ($1 \leqq p < n$), does there necessarily exist a $p_2 \in B_2$ where the Levi invariant has at least p negative roots?

(Calabi-Kohn-Rossi)

10. Let $(X, \mathscr{O}_X)$ be a compact non-reduced complex space and $\mathscr{G}$ the Lie algebra of derivations of $\mathscr{O}_X$. Is the connected component of the automorphism group of $(X, \mathscr{O}_X)$ a complex Lie group with Lie algebra $\mathscr{G}$? (This is true in the algebraic case as well as in the reduced complex case; for the latter see: Kerner: Arch. Math. XI, 282—288 (1960); Kaup: Schriftenreihe Math. Inst. Münster 24 (1963).)

(Matsumura)

11. Let X be a projective algebraic variety and M_1 and M_2 coherent sheaves on X. Assume that for every complex line bundle L on X, $\dim_C H^0(X, M_1 \otimes L) = \dim_C H^0(X, M_2 \otimes L)$ holds. Is M_1 isomorphic to M_2?

(Röhrl)

12. Let X be a projective algebraic variety and G a connected complex Lie group. Is the canonical map $H^1(X, G_\omega) \to H^1(X, G_c)$ surjective? (This is true for X a compact Riemann surface: Topology 2, 247—252 (1963).)

(Röhrl)

13. Let X_0 be a Stein manifold with $H_q(X, \boldsymbol{Z}) = 0$ for $q > 1$. Let $p: X \to X_0$ be a finite, proper holomorphic map. Suppose that every local ring on X is a Macaulay ring. Is then the ring $\mathscr{H}(X)$ of holomorphic functions on X isomorphic to $\mathscr{H}(X_0)/(P(t))$ where $P(t)$ is a polynomial

in one variable over $\mathcal{H}(X_0)$ with leading coefficient 1? (This is true if either $p: X \to X_0$ is unramified or else has only two sheets.)
(RÖHRL)

14. Does there exist a connected complex manifold without a dense countable subset?
(STOLL)

15. Does there exist a Jensen-Poisson formula for open, connected, relatively compact subsets with C^∞ boundary of Stein manifolds?
(STOLL)

16. Let A be a pure dimensional analytic subset of a complex Stein manifold X. When can a bounded holomorphic function on A be extended to a bounded holomorphic function on X?
(STOLL)

17. Let ν_λ, $\lambda \in \Lambda$, be a normal family of non-negative principal divisors on a Stein manifold X. Is there a normal family of holomorphic functions h_λ, $\lambda \in \Lambda$, on X such that h_λ has divisor ν_λ, $\lambda \in \Lambda$, and no convergent subsequence of the family h_λ, $\lambda \in \Lambda$, has limit 0 resp. ∞?
(STOLL)

18. Let N be the set of non-negative principal divisors on the Stein manifold X and $\mathcal{H}(X)$ the Fréchet space of holomorphic functions on X. Is there a continuous map $\gamma: N \to \mathcal{H}(X) - \{0\}$ such that for every $\nu \in N$, ν is the divisor of $\gamma(\nu)$?
(STOLL)

19. Let A be a pure p-dimensional analytic set in $\mathbf{C}^n$ and $V(r)$ the $2p$-dimensional measure of $A \cap \{z: |z| \leqq r\}$. Then $n_A(r) = p! \pi^{-p} r^{-2p} V(r)$ is monotonically increasing. If $n_A(r)$ is of finite order λ, is there an entire function f of finite order $\leqq \lambda$ that vanishes on A but does not vanish identically? (The answer is affirmative for $p = 0$ (trivial) and $p = n - 1$: STOLL: Math. Zeitschr. 57, 211—237 (1953).)
(STOLL)

20. Let U be a connected neighborhood of $\mathcal{O} \in \mathbf{C}^n$, whose intersection A with the hyperplane $z_n = 0$ is connected. Let $G = U - A$ and suppose that $\varphi = (\varphi_1, \ldots, \varphi_n)$ is a holomorphic mapping of U into $\mathbf{C}^n$ that is biholomorphic on G and maps A into $\mathcal{O} \in \mathbf{C}^n$. Write $\varphi_\nu(z) = z_n^{p_\nu} \psi_\nu(z)$, $\nu = 1, \ldots, n$, such that ψ_ν does not vanish identically on A. Is there an index ν for which $\psi_\nu(\mathcal{O}) \neq 0$? (This is true for $n = 2$: HOPF, Comm. Math. Helv. 29, 132—156 (1955).)
(STOLL)

21. Let D be the unit disk and $f_1, \ldots, f_n$ be holomorphic functions in $D - \{0\}$. Let A be the subset $\{(z, f_1(z), \ldots, f_n(z)): z \in D - \{0\}\}$ of $D \times \mathbf{C}^n$. Describe the set $\bar{A} \cap (\{0\} \times \mathbf{C}^n)$.
(STOLL)

22. Let $\exp_n$ denote the exponential function on the space $\boldsymbol{C}^{n^2}$ of $n \times n$ complex matrices. Is there an entire map $f: \boldsymbol{C}^{n^2} \to \boldsymbol{C}^{n^2}$ such that $\exp_n = f \circ f$ (For $n = 1$ the answer is negative: BAKER: Math. Ann. **129**, 174—182 (1955).)
(STOLL)

23. Let A be a thin analytic subset of the complex manifold X. Suppose that φ is a holomorphic map of $X - A$ into some complex manifold such that for every one-dimensional complex submanifold L of X satisfying $L \cap A = \bar{L} \cap A = \{p\}$ the limit $\lim_{L \cap A \ni z \to p} \varphi(z)$ exists. Is then φ meromorphic in the sense of REMMERT?
(STOLL)

24. Let P^2 be the complex projective plane. Let Q be a complete quadrilateral. Is the universal covering space of $P^2 - Q$ isomorphic to some open subset of $\boldsymbol{C}^2$?

25. Let X be a finite, strongly pseudo-convex complex Stein space. Suppose that X is nonsingular in a neighborhood of the boundary bX and that bX is homeomorphic to S^{2n-1}. Is X a complex manifold?

26. Let X be an n-dimensional complex space and M a coherent sheaf on X. Is then $H^p(X, M) = 0$ for $p > n$? If, in addition, X is Stein, is then $H^p(X, \boldsymbol{C}) = 0$ for $p > n$? (This is true for complex manifolds: MALGRANGE: Bull. Soc. Math. France **83**, 231—237 (1955); SERRE: Coll. fonctions plusieurs var. Bruxelles 57—68 (1953).)